T0280974

Classical and Quantum Mechanics with Lie Algebras

Classical and Quantum Mechanics with Lie Algebras

Yair Shapira

Technion - Israel Institute of Technology, Israel

NEW JERSEY · LONDON · SINGAPORE · BEIJING · SHANGHAI · HONG KONG · TAIPEI · CHENNAI · TOKYO

Published by

World Scientific Publishing Co. Pte. Ltd.

5 Toh Tuck Link, Singapore 596224

USA office: 27 Warren Street, Suite 401-402, Hackensack, NJ 07601

UK office: 57 Shelton Street, Covent Garden, London WC2H 9HE

Library of Congress Control Number: 2021028960

British Library Cataloguing-in-Publication Data
A catalogue record for this book is available from the British Library.

Cover image: NASA's Hubble Celebrates 21st Anniversary with "Rose" of Galaxies. Original from NASA. Digitally enhanced by rawpixel. [In Public Domain]

CLASSICAL AND QUANTUM MECHANICS WITH LIE ALGEBRAS

ISBN 978-981-124-005-8 (hardcover)
ISBN 978-981-124-145-1 (paperback)
ISBN 978-981-124-006-5 (ebook for institutions)
ISBN 978-981-124-007-2 (ebook for individuals)

For any available supplementary material, please visit
https://www.worldscientific.com/worldscibooks/10.1142/12364#t=suppl

Printed in Singapore

Contents

Part III The Binomial Formula and Quantum Statistical Mechanics

Part V Introduction to Quantum Physics and Chemistry

Part VII   Appendix: Background in Calculus

Preface

To master physics, you need a little bit of math. After all, this is the language of nature. Still, keep math to a minimum: algebra rather than calculus. With a good preparation work in algebra and geometry, physics flows as smoothly as ever. Math will translate into a vivid physical intuition.

How to see physics in its full glory? This book offers a new approach: start from math, in its simple and elegant tools: discrete math, geometry, and algebra, avoiding heavy analysis that might obscure the true picture. This will get you ready to see how physics develops gradually, from Newtonian to quantum mechanics and relativity.

Indeed, Part I uses elementary calculus to introduce Newtonian mechanics. Part II uses simple discrete math to introduce a key concept in classical mechanics: stability. This marks the start of modern physics: initial data are not quite enough to predict the future. This will be emphasized even more in quantum mechanics and relativity.

Part III introduces the (extended) binomial formula, useful in quantum statistical mechanics. Part IV introduces special (and general) relativity from an algebraic-geometrical point of view.

Part V uses elementary linear algebra to introduce quantum physics and chemistry, including energy levels and spin. Finally, Part VI uses just mathematical induction to introduce Lie algebras, and decomposes them in terms of simple ideals, useful in quantum and Hamiltonian mechanics.

The book could be used as a textbook in four undergraduate courses, three at a physics department, and the fourth at a math department:

- Introduction to Newtonian mechanics and special relativity (with a focus on Chapters 1–3, 10–11, and 23).
- Introduction to Hamiltonian mechanics and stability (Chapters 4–7 and 22).
- Introduction to quantum physics and chemistry (Chapters 8–9, 12–13, and 23).
- Introduction to Lie algebras with applications in physics (Chapters 14–23).

Each chapter ends with a lot of relevant exercises, with solutions or at least guidelines. The exercises are not boring at all: they help learn new material, step by step, exercise by exercise, in an active pedagogical way.

The only prerequisites are linear algebra and calculus. To make the book self-contained, this background is available in the appendix.

How to pronounce the title of the book? Write "physics," but say "phyZics." This is a phonetic law: use the easiest way to pronounce, no matter how the word is written. Likewise, write "tensor" and "isomorphism," but say "tenZor" and "iZomorphiZm," and so on.

Yair Shapira
Haifa, Israel

Part I

Introduction
to Newtonian Physics

Introduction to Newtonian Physics

In this part, we discuss Newtonian mechanics. To do this, we use just elementary math: geometry and elementary calculus (available in the appendix). This gives us a good starting point. Later on, we'll see how modern physics comes on top, and gives us a yet better understanding.

Thanks to geometry and calculus, we can go ahead and discuss motion and dynamics. For this, we need an important physical quantity: speed (or velocity).

Speed is not so innocent as it looks. As a matter of fact, it is quite tricky. After all, it is often nonconstant, and changes in time. To see how, we must differentiate it, and obtain its derivative: acceleration. This will lead to Newton's law: force is proportional not only to mass but also to acceleration.

Moreover, we can also integrate force along a path, and obtain a new fundamental quantity: work. Work has a yet more fundamental face: energy, decomposed into two parts: kinetic and potential.

Thanks to energy, we can tell a lot about the physical process, even before solving for its explicit dynamics. This is quite useful in gravity and electromagnetics.

To describe motion, we have three fundamental concepts: energy, momentum, and angular momentum. Energy tells us the capacity to do work. Momentum, on the other hand, tells us how difficult it is to stop the motion. Angular momentum, on the other hand, tells us how difficult it is to stop a rotation. All three are conserved: in a closed system, they never change or get lost. This may tell us a lot about the dynamics, even before the explicit motion is known.

Later on, we'll put this in a deeper context: Hamiltonian mechanics. This will give physics not only geometrical but also algebraic face: Lie algebras.

1

Introduction
to Newtonian Mechanics:
Energy and Work

To understand motion, we must "solve" Zeno's paradox, and introduce a new fundamental concept: speed (or velocity). Likewise, we can go ahead and define acceleration and force. These are the basic ingredients in the laws of nature.

Thanks to integration, we can then go ahead and define yet another fundamental quantity: energy. This is indeed Newtonian mechanics: a comprehensive mathematical theory to explain motion and dynamics [14, 33].

This theory has its own limits: in astronomic macroscale, it is less accurate than Einstein's relativity. In subatomic microscale, on the other hand, it is less accurate than quantum mechanics. Fortunately, it is still sufficiently reliable and accurate in intermediate scales, used often in practical engineering problems. Moreover, it offers new conceptual tools, and open horizons towards a better understanding of many other physical processes that govern the entire nature around us.

1.1 Motion: Zeno's Paradox

1.1.1 Speed – Velocity

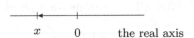

x 0   the real axis

Fig. 1.1. The infinite real axis. The arrow leading from 0 to the negative number $x < 0$ makes an angle of 180° (or π) with the positive part of the real axis.

To understand motion, we need a new fundamental concept: speed (velocity). For this purpose, assume that a particle moves along the real axis (Figure 1.1). At time $t \geq 0$, the particle is at $x \equiv x(t)$. Thus, x depends on t, whereas t depends on nothing. This way, t is the independent variable, and $x \equiv x(t)$ is a function of it.

At the initial time $t = 0$, assume that the particle lies at the origin:

$$x(0) = 0.$$

(After all, in theory, we could place the origin wherever we like.) Then, the particle starts to move. At time $t > 0$, it reaches the point $x(t)$ on the real axis.

The velocity is a ratio: the distance that the particle makes per second. In general, The velocity is nonconstant: it may change in time. In other words, the velocity is a function of t as well.

For example, at the beginning, the particle may be at rest: zero velocity. Then, it may accelerate, and gain more and more speed. Still, there is a paradox here: since it had zero velocity, how could it start moving in the first place? This is Zeno's paradox.

1.1.2 Average vs. Momentary Velocity

We must therefore be careful. First, define the *average* velocity. For this purpose, pick some parameter $h > 0$. Then, define the average velocity from time t to time $t + h$:

$$\frac{x(t + h) - x(t)}{h}.$$

Still, this is not good enough: we'd like to have not only the average velocity in the entire time interval $[t, t + h]$ but also the *momentary* velocity at the precise time t. Fortunately, for this purpose, we could use a limit process.

How to define momentary velocity? At first glance, this might seem impossible. After all, at the precise time t, there is no motion at all, so how could velocity be relevant? This is indeed Zeno's paradox.

Later on, we'll "solve" Zeno's paradox in the context of quantum mechanics (Chapter 7, Section 7.12.6). Furthermore, in the context of general relativity, we'll see how it is related to Riemann normal coordinates. Here, on the other hand, we look at it both geometrically and analytically.

1.1.3 Zeno's Paradox in Geometrical Terms

Zeno's paradox also has a geometrical version: how could the real axis be covered by dimensionless points? After all, each individual point has no size at all, so even infinitely many are not enough!

The ancient Greeks were puzzled with Zeno's paradox. Why? Because they avoided the infinity. Indeed, in Euclidean geometry, a line is not *made* of points. On the contrary: it is a fundamental object on its own right, defined implicitly by some plausible axioms.

1.1.4 Line and Points in Euclidean Geometry

The line may contain many points: every two distinct points on it must have a third point in between. Still, these points never *make* the original line. They have a different nature, and remain isolated and discrete. As such, they could never produce the continuous line in the first place. This is why Zeno's paradox is so relevant: there is no way to "jump" from point to point to model a physical motion. With no analytic terms at hand, the original language of Euclidean geometry seems too poor.

Moreover, in Euclidean geometry, there is no notion of length or dimension at all ([15] and Chapter 6 in [43]). An interval is not "one-dimensional" in any way, and has no length. In fact, intervals are never measured by any length unit. Length is just a relative property, useful to compare intervals: one interval could be longer than another, or shorter.

1.1.5 Line and Points in Analytical Geometry

In analytical geometry, on the other hand, we can already use infinite sets. The line is no longer interpreted as a pure abstract object, but as a set of those points that satisfy some linear equation. Thus, the line is a high-level object, made of lower-level objects: points. This is how the line indeed takes its one-dimensional nature used to model a one-dimensional motion. Later on, in Chapter 7, the concept of dimension will be defined in some more detail.

1.1.6 Line and Points in Set Theory

How many points are needed to make a line? Thanks to set theory, we already know the answer: $\aleph$. This kind of infinity is bigger than the discrete infinity, known as $\aleph_0$:

$$\aleph > \aleph_0.$$

$\aleph$ is big enough: it helps introduce a new dimension, with a substantial size or length. This is how points of zero measure could join to make a new continuous interval of a positive measure.

Why isn't $\aleph_0$ big enough? Because an infinite sequence of points, placed one by one in a row, could never make a continuous line. Fortunately, we can still use "more" points: a "greater" infinity: $\aleph$.

More precisely, why is $\aleph_0$ too small? Because the natural numbers make an ordinal: every subset must start somewhere. $\aleph$, on the other hand, does not. Indeed, the real axis contains many subsets that start nowhere. For example, let t be fixed, and consider an open semiaxis, containing those points larger than t:

$$(t, \infty) \equiv \left\{ \tilde{t} \mid t < \tilde{t} < \infty \right\}.$$

Because t itself is excluded, this semiaxis is open, and has no start at all. For this reason, we can now use a one-sided limit process, and approach t from the right.

1.1.7 Limit Process

How to do this? Easy: let $h > 0$ be a positive parameter. This way, $t + h$ already belongs to our semiaxis, as required. Now, make h smaller and smaller. This way, $t + h$ is still in our semiaxis, and gets ever so close to t. This way, $t + h$ indeed approaches t from the right. This is denoted by

$$t + h \to_{h \to 0+} t^+.$$

Likewise, t could also be approached from the left. This will be useful to define the velocity at t.

1.1.8 Concept and Word

The ancient Greeks were not yet ready to use infinite sets or analytical geometry. Why? Because, in their philosophy, a concept or idea has a life of its own: it mustn't depend on any other object of any other kind. For example, the line mustn't depend on a point.

Still, the ancient Greeks managed to "solve" Zeno's paradox in their own way: the line has a "soul" or a nature of its own. This is what makes it more than just a collection of points. It has a new meaning, planted into it by the "god" of linearity. This is what makes it not just the sum of its own ingredients, but a little more.

How would a dog view a line? Well, a dog can understand only a concrete line: a long straight path to run freely on. We humans, on the other hand, can speak. For this purpose, we must use words, which describe not only concrete but also abstract ideas. After all, the word "line" doesn't necessarily stand for a straight path: it could stand for any other straight thing.

1.1.9 Plato's Philosophy

In this context, the word "line" is an atom word. Where does it get its unique meaning from? In Plato's theory, every term must have its own private spiritual origin (or source, or spirit, or god) to define it uniquely, and give it its general characteristics and features.

How to comprehend the idea of "being linear?" In Plato's theory, this must be represented by the god of linearity, who gives each particular line its own linear nature. This is how the line can indeed model a one-dimensional physical motion. Zeno's paradox is solved!

1.1.10 Aristotle's Philosophy

In Aristotle's theory, on the other hand, there is no such god. Every term has its own private nature, built in it. For example, the nature of the line is to be straight and long, to allow one-dimensional motion along it.

We can now see the mathematical roots behind Aristotle's theory. Indeed, analytical geometry is more free and relative: everyone is free to draw his/her own coordinate system. No need to wait for anyone to tell us where the origin is, or what a line is.

Plato and Aristotle answered Zeno's paradox philosophically. Newton, on the other hand, answered it analytically.

1.1.11 The Language of Analysis

To answer Zeno's paradox in mathematical terms, Newton used a powerful analytic tool: the limit process. With this tool at hand, Newton was ready to enrich the language of mathematics yet more, and introduce the new concept of derivative (Chapter 24, Section 24.2.5).

1.2 Velocity and Acceleration

1.2.1 Average Velocity and Its Limit

Thanks to the derivative, the momentary velocity can now be defined as the limit of the average velocity in shorter and shorter intervals:

$$v(t) \equiv \lim_{h \to 0} \frac{x(t+h) - x(t)}{h} = x'(t) = \frac{dx}{dt}(t).$$

In summary, velocity is the derivative of position. This answers Zeno's paradox once and for all.

In reality, however, the particle may move not only in one but also in three dimensions. This way, $x(t)$ could be not just a scalar but actually a three-dimensional vector, with three components:

$$x(t) \equiv \begin{pmatrix} x_1(t) \\ x_2(t) \\ x_3(t) \end{pmatrix}.$$

In this case, the particle may move in every spatial direction. This is where the term velocity is better than speed. Indeed, speed is just a scalar, whereas velocity could also be a vector:

$$v(t) \equiv \begin{pmatrix} v_1(t) \\ v_2(t) \\ v_3(t) \end{pmatrix} \equiv \begin{pmatrix} x_1'(t) \\ x_2'(t) \\ x_3'(t) \end{pmatrix}.$$

For simplicity, however, we stick here to our original particle, confined to one dimension only. (After all, in theory, we could always pick our axis system to align with the motion.) This way, velocity is the same as speed: a scalar.

1.2.2 Laws of Nature

Fig. 1.2. Newton's first law: if there is no force, then the velocity v remains constant.

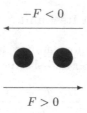

Fig. 1.3. Newton's third law: the left bullet applies force F upon the right bullet. As a reaction, the right bullet applies force $-F$ back upon the left bullet.

To govern motion, we have three fundamental laws. The first law says that, so long as no force is applied to it, the particle moves undisturbed, at a constant velocity:

$$v(t) \equiv v(0), \quad t > 0$$

(Figure 1.2). Why is this law first? Because it is most fundamental, and talks about most elementary circumstances.

The second law says that a force that is applied to the particle affects it in one way only: changing its momentum (to be discussed later). For a particle of a constant mass, this means changing the velocity. This is indeed acceleration.

Finally, the third law talks about action and reaction: if a force is applied to our particle by some other particle that hits it, then our particle reacts symmetrically: it applies the same amount of force back upon the other particle, in the opposite direction (Figure 1.3). This way, the entire system remains in balance: the forces cancel out, and sum to zero. After all, no force could ever come from nothing.

1.2.3 Acceleration

What is acceleration? It tells us how the velocity changes per second. In other words, it is the derivative of the velocity with respect to time:

$$a \equiv a(t) \equiv v'(t) \equiv \frac{dv}{dt}(t).$$

This way, the acceleration is momentary as well: it is nonconstant, and may change in time. In the beginning, for example, you might want to put your foot on the gas, and accelerate a lot. Once you gained a good velocity, you might want to relax, and stop accelerating, keeping the same velocity for a while. Finally, you need to break, have a negative acceleration, and make a complete stop.

This is why the acceleration $a \equiv a(t)$ is a function of time in its own right. Now, recall that velocity is the derivative of position. Thus, acceleration is actually the second derivative of position (with respect to time):

$$a(t) \equiv v'(t) \equiv x''(t) \equiv \frac{d^2x}{dt^2}(t).$$

1.2.4 Towards a Three-Dimensional Motion

In the real three-dimensional world, in which both x and v have three components, a has three components as well:

$$a(t) \equiv \begin{pmatrix} a_1(t) \\ a_2(t) \\ a_3(t) \end{pmatrix} \equiv \begin{pmatrix} v_1'(t) \\ v_2'(t) \\ v_3'(t) \end{pmatrix} \equiv \begin{pmatrix} x_1''(t) \\ x_2''(t) \\ x_3''(t) \end{pmatrix}.$$

This means that the three-dimensional velocity vector may change not only magnitude but also direction, leading the particle along a complicated curve (or trajectory) in the three-dimensional space.

Here, however, we stick to our original one-dimensional model, in which the velocity (and the acceleration) is just a scalar, not a vector. This way, the velocity may change in two ways only: either increase (positive acceleration), or decrease (negative acceleration).

1.3 Force: Derivative of Momentum

1.3.1 Mass

Here, we assume that our particle has a constant mass:

$$m \equiv \text{const.}$$

Still, this is not a must. In general, the mass may depend on time: $m \equiv m(t)$. For example, think of a spaceship, traveling in space. During its journey, it burns fuel, and loses mass. This way, its mass has a negative time derivative: $m' < 0$.

1.3.2 Inertial Mass

The mass m is also called inertial mass: the amount by which the particle resists any change to its original velocity. Thanks to this resistance, a force is needed: with no force, the velocity would never change.

Still, the full resistance comes not from mass but from a more fundamental property: momentum. This really tells us how the particle resists any change. This is why, with no external force, momentum could never change. This is indeed a conservation law. Later on, we'll see this from a geometrical point of view.

1.3.3 Momentum

The momentum tells us how difficult it is to change the course of the system. In our case, it tells us how difficult it is to make the particle change its original motion.

To do this, the momentum should better be proportional to both mass and velocity. In other words, it should be defined as their product: mv. Again, in general, this is a function of time: $m(t)v(t)$.

In this form, the momentum is nonconstant, and may change in time. How fast? This is determined by the force.

1.3.4 Force: Derivative of Momentum

The force tells us how fast the momentum changes. In other words, the force is the derivative of the momentum. In fact, at each individual time t, the force is

$$F = (mv)' = m'v + mv' = m'v + ma.$$

1.3.5 The Second Law of Nature

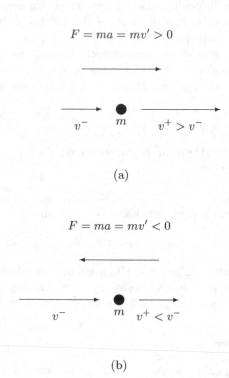

$$F = ma = mv' > 0$$

(a)

$$F = ma = mv' < 0$$

(b)

Fig. 1.4. Newton's second law: the force $F(t)$ is linearly proportional to both the mass m and the acceleration $a(t) \equiv v'(t) \equiv x''(t)$. This could be either positive, increasing the velocity from v^- to v^+ (a), or negative, decreasing the velocity from v^- to v^+ (b).

Could this be simplified yet more? Recall that our particle has a constant mass:

$$m \equiv \text{const. or } m' \equiv 0.$$

Thus, the force takes the simpler form of

$$F = m'v + ma = ma.$$

This is Newton's second law of nature (Figure 1.4): at each individual time t,

$$F = ma = mv' = mx'',$$

where m is the constant mass of the particle, and F is the force applied to it.

What is the physical meaning of this? Well, the force is linearly proportional to both mass and acceleration. For example, assume that the force is applied from the left ($F > 0$). (Say, our particle is hit by some other particle from the left.) In this case, the velocity of our particle increases, and the acceleration is positive ($a > 0$). If, on the other hand, the force is applied from the right ($F < 0$), then the velocity decreases, and the acceleration is negative ($a < 0$).

Assume now that there is no force at all. In this case, momentum never changes: it is conserved. Let's see this geometrically.

1.4 Momentum and Its Conservation

1.4.1 Elastic Particles

Consider a system of M particles, with velocities $v_i(t)$ and masses m_i ($1 \leq i \leq M$). The particles may hit each other, thus applying forces upon each other. Still, they are completely elastic: a particle that is hit by some other particle may change shape only temporarily, and soon get back to its original shape. In the process, it applies the same amount of force (in the opposite direction) back upon the particle that hits it.

This way, the forces never make any permanent (plastic) change in any particle. On the contrary: the forces reflect (or return) fully, as in Newton's third law (Figure 1.3). No friction or heat is generated. This could model balls on a billiards (or pool, or snooker) table.

1.4.2 Closed System

The system is closed, or isolated from the outer world: no external force is applied from the outside. Thus, the total force in the entire system must be zero:

$$0 = F = \left(\sum_{i=1}^{M} m_i v_i \right)' = \sum_{i=1}^{M} m_i v_i' = \sum_{i=1}^{M} m_i a_i.$$

Now, let's look at two discrete times: the beginning, and the end. Let the ith particle have velocity v_i^- at the initial time, and v_i^+ at the end. Thanks to the above equation, the total momentum remains unchanged:

$$\sum_{i=1}^{M} m_i v_i^- = \sum_{i=1}^{M} m_i v_i^+.$$

1.4.3 Conservation of Momentum

This is indeed conservation of momentum. In Figure 1.5, this is illustrated for two particles ($M = 2$):

$$m_1 v_1^- + m_2 v_2^- = m_1 v_1^+ + m_2 v_2^+.$$

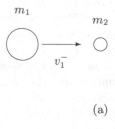

m_1

m_2

v_1^-

(a)

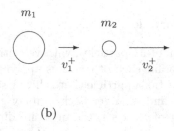

m_1

m_2

v_1^+ v_2^+

(b)

Fig. 1.5. Conservation of momentum: before the collision, the little ball on the right is at rest, so the momentum $m_1 v_1^-$ is concentrated in the big ball on the left (a). After the collision, on the other hand, the original momentum splits into two parts: $m_1 v_1^+$ in the big ball, and $m_2 v_2^+$ in the little ball (b). This way, the velocity of the big ball decreases from v_1^- to v_1^+. Still, momentum is never lost, but only transferred to the little ball, to give it the new velocity v_2^+.

Furthermore, some particles may collide too hard, and even break and split, so the total number of particles may increase from M to $N > M$:

$$\sum_{i=1}^{M} m_i v_i^- = \sum_{i=1}^{N} m_i v_i^+$$

(Figure 1.6).

1.4.4 Equilibrium: Zero Total Momentum

Consider now a static football that lies at rest. Assume that the temperature is above the absolute zero $(-273°C)$. This means that the molecules in the football are never at a complete rest: they constantly move, and have nonzero momenta. Furthermore, they often hit each other, and change direction. How come the football stands still? Is there no momentum?

Well, on average, there is no momentum at all. After all, the molecules move randomly, towards no specific direction. Later on, we'll call this Brownian motion

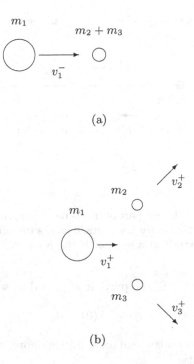

Fig. 1.6. At the collision, the little ball on the right may break into two yet smaller balls. This way, the original momentum $m_1 v_1^-$ in (a) splits into three parts: $m_1 v_1^+$ is the momentum left in the big ball, whereas $m_2 v_2^+$ and $m_3 v_3^+$ are the new momenta introduced in the small balls (b). The horizontal momentum remains positive, and the vertical momentum remains zero.

(Chapter 8, Sections 8.4.1–8.4.2). Therefore, their momenta cancel each other, and sum to zero in each and every direction.

Even when molecules hit each other, their momenta are never lost, but only transferred to one another, as in Figure 1.5 or 1.6. For this reason, the momentum of each individual molecule remains random, with no favorite direction, and the total momentum remains zero. This is why the football remains at rest, at least until someone comes and kicks it. This opens the system up: it introduces a new external force, and gives all molecules a new (uniform, nonrandom) velocity and momentum, in the same spatial direction.

1.5 Harmonic Oscillator

1.5.1 A Spring

Consider now a more specific problem. Assume that the particle is connected to the origin by a spring (or an elastic wire, made of rubber). This makes a well-known model: harmonic oscillator.

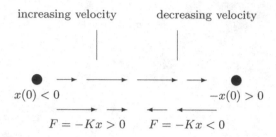

Fig. 1.7. Hooke's law: in the left part of the x-axis, the spring pulls the particle rightwards at a force of $F = -Kx > 0$. Once the particle passes the origin on its way rightwards, the spring pulls it back leftwards at a force of $F = -Kx < 0$.

At the initial time $t = 0$, the particle is placed at some negative point

$$x(0) < 0,$$

to the left of the origin on the real axis. Furthermore, the particle is held firmly, so it has no initial velocity:

$$v(0) = 0.$$

Suddenly, we let the particle go free (Figure 1.7). This way, the spring applies a positive force to it, to make it move rightwards.

For simplicity, we also assume that there is no friction, so no heat is generated. This way, the force contributes only to increasing the velocity. This is a fair assumption: after all, in many realistic cases, the friction is indeed negligible.

At the initial time $t = 0$, the spring is well drawn, so the force is strong. As time goes by, though, the spring gets loose, so the force gets weaker and weaker. At some unknown time, the particle reaches the origin. At this point, the spring gets so loose that it applies no force at all any more. Still, the velocity accumulated so far makes the particle keep going rightwards, pass the origin, into the right part of the real axis.

There, the spring applies a negative force, in an attempt to pull the particle back to the origin. This also introduces a negative acceleration that gradually reduces the velocity. Thanks to symmetry, the rightmost point that the particle can reach is $-x(0) > 0$.

At this point, the velocity has decreased to zero. Still, the negative force doesn't stop: it keeps reducing the velocity yet more, and make it even negative, pulling the particle leftwards, back to the origin.

Once the particle passes the origin on its way leftwards, the force becomes positive again, and the velocity increases again. By the time the particle finally returns to the initial point $x(0) < 0$, the velocity has increased back to its initial value: zero.

This completes one cycle or period. The process may then restart, and repeat time and again forever.

Once we understand the physical process, we're ready to write it mathematically.

1.5.2 Hooke's Law

Fortunately, the formal language of math is much more concise and precise than our own human language, and much more suited to model the physical process. Indeed, the above long story is summarized in Hooke's law:

$$F(t) = -Kx(t),$$

where K is a positive constant, dependent only on the material the spring is made of. At the initial time $t = 0$, F is positive, so it forces the particle to start moving rightwards:

$$F(0) = -Kx(0) > 0$$

(Figure 1.7). As time goes by, $x(t)$ decreases in absolute value, so $F(t)$ decreases as well. Once the particle reaches the origin, both x and F vanish. Still, the velocity accumulated so far moves the particle ahead, into the right part of the real axis, where $x(t) > 0$. There, F becomes negative: it now pulls the particle back leftwards.

However, it takes some time before the velocity decreases to zero. This happens at the rightmost point $-x(0) > 0$. At this point, the velocity becomes negative. The negative force wins: it reverses the motion, forcing the particle to move leftwards rather than rightwards.

Once the particle passes the origin on its way leftwards, $x(t)$ becomes negative again, so $F(t)$ becomes positive again. This way, the force produces a positive acceleration again, making the particle slow down gradually on its way leftwards, until arriving safely back home: the initial position $x(0) < 0$. The entire period is complete, and is ready to start again.

All this story is now summarized in just one formula: Hooke's law. Let's use the language of math yet more, to find out the explicit solution $x(t)$ at each and every time t. How to do this? Combine Hooke's law with Newton's second law. This will form the differential equation that describes the complete physical process. Before getting into this, we must first make sure that the physical units are consistent.

1.5.3 Physical Units

In math, we often use numbers: pure abstract objects. In physics, on the other hand, numbers are not good enough. To measure a physical quantity, they must be accompanied by a suitable physical unit.

One popular system of units is cgs (centimeters, grams, seconds). This system uses rather small units: length is measured in centimeters, weight in grams, and time in seconds. Velocity (change in distance per time unit) is measured in centimeters per second, or cm/s. Acceleration (change in velocity per time unit), is measured in (cm/s)/second, or cm/s^2.

In a physical equation, both sides must have the same physical unit. In Newton's second law, for example, force must be measured in terms of weight times acceleration, or $g \cdot cm/s^2$.

This unit must also be used in Hooke's law. For this purpose, K must have a new unit: g/s^2. This way, both sides have the same unit: $g \cdot cm/s^2$.

The cgs system is often used in science. A more practical system, on the other hand, is MKS: meter, kilogram, second.

1.5.4 The Differential Equation

We are now ready to combine Hooke's law with Newton's second law:

$$-Kx(t) = F(t) = mx''(t) = m\frac{d^2x}{dt^2}(t),$$

or

$$x''(t) = -\frac{K}{m}x(t).$$

This is a differential equation. The unknown function is $x(t)$: the position of the particle at time t.

To make sure that there is no mistake, let's compare the physical units. On the left, $x(t)$ is differentiated twice with respect to time, so the units are cm/s^2. Fortunately, on the right, the units are cm/s^2 as well: after all, K has units g/s^2 (Section 1.5.3).

1.5.5 Frequency

Since both K and m are positive, we can now introduce the frequency:

$$\omega \equiv \sqrt{\frac{K}{m}}.$$

Why is it called frequency? We'll soon see why. Thanks to the frequency, We can now simplify our differential equation to read

$$x''(t) = -\omega^2 x(t).$$

Let's go ahead and solve it for $x(t)$.

1.5.6 How to Solve The Differential Equation?

We are now ready to solve the differential equation. Fortunately, we already know how the solution should look like:

$$x(t) = a\sin(\omega t) + b\cos(\omega t),$$

where a and b are some constant parameters, to be specified later. Indeed, the first derivative is

$$v(t) = x'(t) = \frac{dx}{dt}(t) = \omega\left(a\cos(\omega t) - b\sin(\omega t)\right).$$

Furthermore, the second derivative is

$$a(t) = v'(t) = \frac{dv}{dt}(t) = \omega^2\left(-a\sin(\omega t) - b\cos(\omega t)\right) = -\omega^2 x(t),$$

as required.

Actually, this is a family of solutions. Indeed, for any a and b, this is a legitimate solution. How to specify a and b uniquely? For this purpose, we must use more physical information.

1.5.7 Initial-Value Problem

To specify both a and b uniquely, we must use the initial state. Assume that the particle is initially placed at $-L$, for some $L > 0$:

$$-L = x(0) = a\sin(0) + b\cos(0) = b.$$

This uncovers b. To uncover a as well, recall that the initial velocity vanishes:

$$0 = v(0) = \omega\left(a\cos(0) - b\sin(0)\right) = \omega a.$$

In summary, we have

$$a = 0 \quad \text{and} \quad b = -L,$$

so the unique solution is

$$x(t) = -L\cos\left(\omega t\right).$$

In summary, once supplemented with initial conditions as above, the original differential equation makes a new initial-value problem, with the unique solution $x(t)$. Let's go ahead and use it.

1.5.8 The Period Time

Thanks to the above solution, we can now calculate the period time:

$$T \equiv \frac{2\pi}{\omega}.$$

Indeed, after time T, the particle is back at

$$x(T) = -L\cos\left(\omega T\right) = -L = x(0).$$

Furthermore, after time T, the velocity is again

$$v(T) = x'(T) = \omega L\sin\left(\omega T\right) = 0 = v(0).$$

So, the particle is ready to start the next cycle, and so on.

1.5.9 Muzzle Velocity

After a quarter of a cycle, at time

$$\frac{T}{4} = \frac{\pi}{2\omega},$$

the particle arrives at the origin:

$$x\left(\frac{T}{4}\right) = -L\cos\left(\omega\frac{T}{4}\right) = -L\cos\left(\frac{\pi}{2}\right) = 0.$$

Its velocity is then at its maximum:

$$v\left(\frac{T}{4}\right) = x'\left(\frac{T}{4}\right) = L\omega\sin\left(\omega\frac{T}{4}\right) = L\omega.$$

Assume now that, at time $T/4$, the spring is suddenly removed. This may model a slingshot.

In this case, the above solution is valid until time $T/4$ only. After this time, there is no force any more. Thus, as in Newton's first law (Figure 1.2), the particle may continue to fly freely rightwards, at the same sped: the muzzle speed.

As a matter of fact, to calculate the muzzle speed, there is no need to solve the differential equation at all! To see this, we must first introduce a new fundamental concept: energy.

1.6 Work and Energy

1.6.1 Work

How to obtain the period time? In Section 1.5.8, we did this in the hard way, using the original differential equation, and its explicit solution. Still, is there a better way? After all, in many physical phenomena, the differential equation is too difficult to solve explicitly!

Fortunately, even before being solved, the original mathematical model may hint and direct us in the right way. To read between the lines, we need two new fundamental concepts: work and energy.

To define work, note that F depends on t only indirectly, through the dependent variable $x(t)$:

$$F \equiv F(x) \equiv F(x(t)).$$

Now, consider a path (or interval) along the real axis. How much work does the force do along this path? This is the work: the force times the length of the path. This indeed makes sense: the work is now linearly proportional to both the force and the length of the path.

Still, the force is not necessarily constant: it may change along the path. Therefore, the original path must break into many subpaths of length $\triangle x$ each. This way, the total work is actually the sum of terms of the form $F(x)\triangle x$.

More precisely, $\triangle x$ should actually be infinitesimal: dx. This way, the work is not the sum, but the integral of the force, over the entire path (Chapter 24, Section 24.13.1). For example, the work done by the spring in Sections 1.5.1–1.5.2 to get the particle from $x(0) = -L$ to the origin is

$$\int_{-L}^{0} F(x)dx = -K \int_{-L}^{0} xdx = -K\frac{x^2}{2}\Big|_{-L}^{0} = -\frac{K}{2}(0^2 - L^2) = \frac{KL^2}{2}.$$

1.6.2 Energy

Energy mirrors work. In fact, energy is the ability to do work. At the initial time $t = 0$, the particle has the ability to travel to the origin. (Here, the origin is used as a reference point.) Thus, the particle must have the potential energy $KL^2/2$, required to do this work. Once it arrives at the origin, its original potential energy has been transformed into kinetic energy, incorporated in its new velocity (and momentum).

1.6.3 Energy in a Closed System

In a closed (isolated, insulated) system, energy is never lost: it is only transformed from one form to another. This is the first law of thermodynamics: conservation of energy. Later on, we'll illustrate this geometrically as well.

1.6.4 From Potential to Kinetic Energy

How is potential energy transformed into kinetic energy? To see this, differentiate the composite function v^2 (with respect to t):

$$\frac{d\left(v^2\right)}{dt} = 2v(t)v'(t)$$

(Chapter 24, Section 24.7.2). Next, divide this by 2:

$$\left(\frac{v^2(t)}{2}\right)' = v(t)v'(t).$$

In other words, $v^2/2$ is the primitive function (or the indefinite integral) of the function $v'v$ (with respect to the independent variable t). Let's use this to get rid of t, and calculate the work (and indeed the energy) in terms of v only. After all, it is v that tells us the physical state, not t. Indeed, v tells us what really happens. t, on the other hand, tells us when. But who cares?

Under a general force F, the particle travels from its initial position $x(0)$ to some new position $x(s)$ (at time $s > 0$). Thanks to Newton's second law (Figure 1.4), we can now calculate the work along this path:

$$
\begin{aligned}
\int_{x(0)}^{x(s)} F(x)dx &= \int_0^s F(x(t))dx(t) \\
&= \int_0^s F(x(t))\frac{dx(t)}{dt}dt \\
&= \int_0^s F(x(t))v(t)dt \\
&= m\int_0^s a(t)v(t)dt \\
&= m\int_0^s v'(t)v(t)dt \\
&= \frac{m}{2}v^2(t)\Big|_0^s \\
&= \frac{m}{2}\left(v^2(s) - v^2(0)\right) \\
&= \frac{m}{2}v^2(s)
\end{aligned}
$$

(assuming that the initial velocity was zero).

What does this mean physically? On the left-hand side, we have the work that the force has invested in the particle. This work has now been transformed into the new kinetic energy on the right-hand side.

1.6.5 Kinetic Energy at The Muzzle

The above is true for any kind of force invested in the particle. Consider, for example, the case of a slingshot. Let's specify s to be the time by which the particle arrives at the origin (the muzzle). Actually, we already know what this time is:

$$s = \frac{T}{4}.$$

Still, let's pretend we don't. After all, we want to get rid of time altogether, and focus on energy.

Thanks to Section 1.6.1,

$$\frac{KL^2}{2} = \frac{mv^2(s)}{2},$$

or

$$v^2(s) = \frac{KL^2}{m} = \omega^2 L^2.$$

This gives us the same muzzle velocity as in Section 1.5.9:

$$v(s) = L\omega.$$

This calculation uses energy considerations only. There is no need to solve the differential equation any more.

1.7 Hamiltonian and The Phase Plane

1.7.1 The Phase Plane

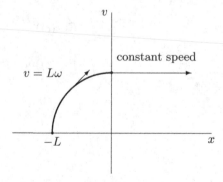

Fig. 1.8. The (x, v)-phase plane. The horizontal axis tells us the location of the particle, and the vertical axis tells us its velocity. At the initial state $(-L, 0)$, the particle is at $-L$, with no velocity at all. Then, The force pushes the particle rightwards to 0, increasing its velocity gradually to $L\omega$. This makes a curve, leading from $(-L, 0)$ to $(0, L\omega)$. (This is a nonphysical or virtual curve: physically, the particle remains on the horizontal x-axis.) Then, the force stops, and the particle proceeds rightwards at a constant speed.

When talking about state of phase, we only ask "what," not "when." We only ask what happened, but don't care when. This is why, in the above, we don't care what s is any more. We are only interested in the energy, transformed from potential to kinetic.

Thus, we better drop t altogether, and focus on the physical state (or phase) of the particle: its location x, and its velocity v. Both may depend on time, but we don't care how: This could remain. hidden and implicit.

Thus, t is gone, and x and v take its place as the new independent variables. Together, they span a new plane: the (x, v) phase plane (Figure 1.8).

In the phase plane, each state is illustrated as a point with two coordinates: the horizontal x-coordinate tells us the location of the particle on the real axis, and the vertical v-coordinate tells us its velocity at the same time. Thus, in the phase plane, the particle "travels" along the curve (or trajectory) leading from the initial state

$$(x(0), x'(0)) = (-L, 0)$$

to the muzzle state

$$(x(s), x'(s)) = (0, L\omega).$$

This is just a virtual (nonphysical) curve. Where is the particle really? To find out, just project vertically downwards, onto the x-axis. After all, the motion is one-dimensional: the particle never leaves the x-axis.

In the phase plane, the virtual curve tells us more: not only position but also velocity. Indeed, to have the velocity, just project horizontally rightwards, onto the vertical v-axis. (Compare this with Figures 7.14–7.15.)

1.7.2 Virtual Curve

From the initial state to the muzzle state, the virtual curve passes through infinitely many intermediate states, without telling us when. As a matter of fact, to draw the virtual curve, there is no need to solve the differential equation any more: energy considerations are good enough.

Indeed, consider an intermediate time $0 \le t \le s$. By this time, the force has invested some work in the particle. This work has already been converted into kinetic energy. To calculate this, go back to Section 1.6.4, and work the other way around, with t playing the role of s:

$$m\frac{v^2(t)}{2} = m\left(\frac{v^2(t)}{2} - \frac{v^2(0)}{2}\right)$$

$$= \int_{-L}^{x(t)} F(x)dx$$

$$= -K \int_{-L}^{x(t)} x dx$$

$$= -K\frac{x^2}{2}\Big|_{-L}^{x(t)}$$

$$= -\frac{K}{2}\left(x^2(t) - L^2\right)$$

$$= \frac{K}{2}\left(L^2 - x^2(t)\right).$$

How does this look like geometrically? To see this more clearly, better drop the irrelevant time variable t, and use x and v as new independent variables:

$$\frac{m}{2}v^2 + \frac{K}{2}x^2 = \frac{K}{2}L^2.$$

What is this geometrically? In the (x, v)-plane, this is the upper-left part of an ellipse (Figure 1.8). We see once again how easy it is to draw the virtual curve, using energy considerations only, and without solving for $x(t)$ any more.

1.7.3 Conservation of Energy

On the other hand, what is this physically? Conservation of energy! Indeed, at each state on the virtual curve, the potential and kinetic energies sum to the same constant number: the total energy in the entire system. Energy just changes a little: from potential (at the beginning) to kinetic (at the muzzle).

What happens after the particle passed the muzzle at time s? Nothing! After all, there is no force any more, so the particle keeps flying freely at the same speed forever, as in Newton's first law. In the phase plane, this makes a straight horizontal line from the muzzle state and rightwards, on which v remains constant.

1.7.4 Hamiltonian

So, we could forget about t altogether! After all, in Figure 1.8, t remains implicit: it is just a parameter that tells us how to advance along the curve. This may be relevant for making a movie to show the (virtual) progress along the curve.

To see the picture more clearly, let's define a new function – the Hamiltonian:

$$H(x, v) \equiv \frac{m}{2}\left(v^2 + \omega^2 x^2\right).$$

This way, H is defined on the phase plane only: it is completely independent of t. Furthermore, H could also be written in terms of the momentum mv rather than the velocity v:

$$H(x, mv) = \frac{1}{2m}(mv)^2 + \frac{m}{2}\omega^2 x^2 = \frac{m\omega^2}{2}\left(\frac{1}{m^2\omega^2}(mv)^2 + x^2\right).$$

Later on, in quantum mechanics, this style will prove better. Why? Because momentum is more fundamental than velocity. In either style, the elliptic curve in Figure 1.8 makes a level set, on which H is constant:

$$H = \text{const.}$$

This is indeed conservation of energy in its geometrical form: on the elliptic curve, the total energy H (kinetic and potential alike) never changes. This principle governs not only motion but also many other physical phenomena.

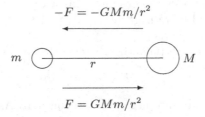

Fig. 1.9. Gravity force: the balls attract each other symmetrically, at a force of $F = GMm/r^2$. This is as in Newton's third law.

1.8 Gravity

1.8.1 Gravity

Newton had yet another important contribution: a new mathematical model of gravity, to help explain Kepler's laws that govern the motion of planets in the solar system. In this model, every two objects in the outer space attract each other at a force that is linearly proportional to their masses M and m, and inversely proportional to r^2, where r is the distance between their mass centers:

$$F = \frac{GMm}{r^2},$$

where G is a positive universal constant, with suitable physical units (Section 1.5.3).

In Figure 1.9, for example, the big ball (of mass M) attracts the small ball (of mass $m < M$) at a positive force of $F = GMm/r^2 > 0$, where r is the distance between their centers. As a reaction, the small ball attracts the big ball at a negative force of $-F = -GMm/r^2 < 0$. This is as in Newton's third law. Let's approximate this formula, and obtain the gravity on the face of the Earth.

1.8.2 Gravity on The Earth

Let's use this formula to see how objects fall to the ground. For this purpose, let M be the mass of the entire Earth, and R the radius of the Earth. Consider an object of mass $m \ll M$ that lies on the face of the Earth. Because the object is far smaller, its distance from the center of the Earth is approximately R. Thus, thanks to the above formula, the Earth pulls it at a force of

$$F = -\frac{GMm}{R^2}.$$

Here, the minus sign tells us that the force is downwards, towards the ground. From Newton's second law, the object falls to the ground at a constant acceleration, denoted by $-g$:

$$-g = -\frac{GM}{R^2}.$$

Again, the minus sign tells us that the acceleration is downwards, (This way, g is a positive constant.) towards the ground. This way, as it approaches the ground, the object gains more and more negative velocity. As it hits the ground, it reaches its maximal velocity downwards.

In summary, the Earth pulls the object at force

$$F = ma = -mg.$$

(This way, both F and a are negative, as required.) As a reaction, the object also pulls the entire Earth upwards, at force

$$-F = mg.$$

This is as in Newton's third law (Figure 1.9).

1.8.3 An Airplane

The constant acceleration $-g$ is relevant not only on the face of the Earth but also slightly higher. For example, an airplane that flies at the moderate height of $H \ll R$ feels roughly the same gravity force: $F = -mg$. (See exercises below.)

This is good enough for an airplane, but not for a rocket that flies much higher, far above the atmosphere. Indeed, it feels a nonconstant force that decreases quadratically as the rocket gets higher and higher, as in the original formula.

1.8.4 Rocket and Its Energy

Consider now a rocket of a constant mass m, lunged from the Earth upwards, towards the sky. Assume that the rocket is far away from any other star, so it feels gravity from the Earth only. What is the minimal velocity it must have to escape from the Earth, and never fall back to the ground?

To answer this, there is no need to solve any differential equation: energy considerations are good enough. Indeed, the initial velocity $v(0)$ should give the rocket just enough kinetic energy to overcome the Earth's gravity, and fly to infinity. During this journey, the initial kinetic energy will be invested in the work required to fly against the Earth's gravity. In the end, the entire kinetic energy will be transformed into potential energy.

Thus, as in Section 1.6.4, no energy will be lost or wasted. It will only be used to do work:

$$\frac{mv^2(0)}{2} = -\frac{m}{2}\left(v^2(\infty) - v^2(0)\right)$$

$$= -\int_R^\infty F(r)dr$$

$$= \int_R^\infty \frac{GMm}{r^2}dr$$

$$= GMm\int_R^\infty \frac{1}{r^2}dr$$

$$= -GMm\frac{1}{r}\Big|_R^\infty$$

$$= -GMm \left(0 - \frac{1}{R} \right)$$

$$= \frac{GMm}{R}.$$

Thus, the minimal initial velocity must be

$$v(0) = \sqrt{\frac{2GM}{R}}.$$

1.8.5 Laws of Thermodynamics

The process is reversible symmetrically: if we let the rocket fall from infinity all the way back to the Earth, then the above potential energy would transform back into kinetic energy, and the rocket would eventually hit the ground at speed $-v(0)$. This energy would then split into three new kinds of energy: heat, sound, and some momentum to the entire Earth downwards. After all, energy is never lost: just changes form. This is the first law of thermodynamics.

Still, both heat and sound have a higher entropy than before. Thus, although the total energy remains the same, it gets less useful: it has now a higher entropy. This is indeed the second law of thermodynamics. We'll come back to this later.

1.9 Exercises: Galileo's Odd Numbers

1.9.1 Speed and Its Derivative

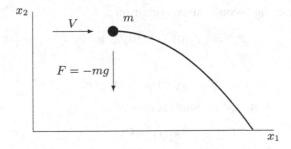

Fig. 1.10. The entire path of the bomb in the (x_1, x_2)-plane, from drop time to hit time.

1. Let R be the radius of the Earth.
2. For simplicity, assume that the Earth has a flat face.
3. On the face of the Earth, draw the x_1-axis: a straight horizontal line.

4. Above the x_1-axis, imagine an aircraft, flying ahead at a constant height $H \ll R$, at a constant speed V.
5. Once the aircraft is above the origin $x_1 = 0$, it suddenly drops a bomb of mass m.
6. At time $t \geq 0$, where is the bomb? Hint: in the (x_1, x_2)-plane, the bomb is at

$$x(t) \equiv (x_1(t), x_2(t)).$$

In other words, the bomb is at height $x_2(t)$ above $x_1(t)$ (Figure 1.10).
7. What force does the bomb feel? Hint: thanks to Newton's second law (Figure 1.4), the gravity force is

$$F \equiv (F_1, F_2) \equiv (0, -mg)$$

(Section 1.8.2).
8. What acceleration does this force produce? Hint:

$$a \equiv (a_1, a_2) \equiv (0, -g).$$

9. Write the differential equation for the horizontal component. Hint: thanks to Newton's second law,
$$x_1''(t) = a_1 = 0.$$
10. What is the general solution of this differential equation? Hint:

$$x_1(t) = b + ct,$$

for some unspecified parameters b and c (Chapter 24, Sections 24.5.2–24.5.3).).
11. Use the initial condition
$$x_1(0) = 0$$

to uncover b:
$$0 = x_1(0) = b + c \cdot 0 = b.$$

12. Furthermore, use the second initial condition

$$v_1(0) = x_1'(0) = V$$

to uncover c:
$$V = v_1(0) = x_1'(0) = c.$$
13. Conclude that, at time t, the bomb is above

$$x_1(t) = Vt.$$

14. Conclude that the horizontal component of the velocity is constant:

$$v_1(t) = x_1'(t) \equiv V.$$

15. Is this as expected from Newton's first law? Hint: yes! See Figure 1.2.
16. Obtain the horizontal velocity $v_1(t)$ in yet another way, using its own differential equation. Hint: thanks to Newton's second law,

$$v_1'(t) = a_1 = 0.$$

17. What is the general solution of this differential equation? Hint:

$$v_1(t) \equiv c,$$

for some unspecified parameter c (Figure 24.9).

18. To specify c, use the initial condition

$$V = v_1(0) = c.$$

19. Does this give the same horizontal velocity as before?

20. Show that the kinetic energy thanks to the horizontal velocity is constant as well:

$$\frac{mv_1^2(t)}{2} \equiv \frac{mV^2}{2},$$

and is never transformed to any other kind of energy.

21. Did you know this even before solving for $v_1(t)$ explicitly? Hint: since there is no horizontal force, no work is ever done in the horizontal direction:

$$\int F_1 dx_1 = \int 0 \cdot dx_1 = 0.$$

22. Write the differential equation for the vertical component as well. Hint: thanks to Newton's second law,

$$x_2''(t) = a_2(t) = -g.$$

23. Write the general solution of this differential equation. Hint:

$$x_2(t) = b + ct - g\frac{t^2}{2},$$

for some unspecified parameters b and c (Chapter 24, Sections 24.5.1–24.5.3).

24. Use the initial condition

$$x_2(0) = H$$

to uncover b:

$$H = x_2(0) = b.$$

25. Furthermore, use the second initial condition

$$x_2'(0) = v_2(0) = 0$$

to uncover c:

$$0 = x_2'(0) = c.$$

26. Conclude that, at time t, the bomb is at height

$$x_2(t) = H - g\frac{t^2}{2}.$$

27. In particular, look at discrete seconds:

$$t = l = 1, 2, 3, 4, \ldots.$$

28. During the lth second, how much closer did the bomb get to the ground? Hint:

$$x_2(l-1) - x_2(l) = \frac{g}{2}l^2 - \frac{g}{2}(l-1)^2 = \frac{g}{2}(2l-1).$$

29. Note that this is proportional to the odd number

$$2l - 1 = 1, 3, 5, 7, \ldots.$$

30. This is Galileo's law of odd numbers.
31. Write the two-dimensional position of the bomb at time t:

$$x(t) \equiv (x_1(t), x_2(t)) = \left(Vt, H - g\frac{t^2}{2}\right).$$

32. Draw the parabola that the positions $(x_1(t), x_2(t))$ make in the (x_1, x_2)-plane. Hint: see Figure 1.10.
33. Write the vertical component of the velocity:

$$v_2(t) = x_2'(t) = -gt.$$

34. Obtain $v_2(t)$ in yet another way, using its own differential equation. Hint: thanks to Newton's second law,

$$v_2'(t) = a_2(t) = -g.$$

35. What is the general solution of this differential equation? Hint:

$$v_2(t) = c - gt,$$

for some unspecified parameter c.
36. To specify c, use the initial condition

$$0 = v_2(0) = c.$$

37. Could you now obtain $x_2(t)$ in yet another way? Hint: it solves the new differential equation

$$x_2'(t) = v_2(t) = -gt.$$

38. What is the general solution of this differential equation? Hint:

$$x_2(t) = b - g\frac{t^2}{2},$$

for some unspecified parameter b.
39. To specify b, use the initial condition

$$H = x_2(0) = b.$$

40. Does this agree with the above?
41. When does the bomb hit the ground? Hint: at time

$$s \equiv \sqrt{\frac{2H}{g}}.$$

Indeed, s satisfies

$$x_2(s) = 0.$$

42. What is the hit velocity? Hint: at time s, the vertical velocity is

$$v_2(s) = -gs = -g\sqrt{\frac{2H}{g}} = -\sqrt{2Hg}.$$

43. What is the hit kinetic energy? Hint: the kinetic energy thanks to the vertical velocity is

$$m\frac{v_2^2(s)}{2} = m\frac{2Hg}{2} = mHg.$$

44. What is the total work made by the Earth's gravity on the bomb?

$$\int_H^0 F_2 dy = \int_H^0 (-mg)dy = mg\int_0^H dy = mgy|_0^H = mg(H - 0) = mgH.$$

45. What is the physical meaning of this? Hint: the potential energy of the bomb thanks to its initial height H was converted into new (vertical) kinetic energy at hit time.

46. Did you know this even before solving for $v_2(t)$ explicitly?

47. At hit time, what will happen to this (vertical) kinetic energy? Hint: it will split into three new kinds of energy: heat, sound, and some momentum to push the entire Earth downwards a little.

48. Is the same true not only at hit time but also at any intermediate time $0 \le t \le s$? Hint: yes! The work made by the Earth's gravity so far is subtracted from the original potential energy, and converted into the new (vertical) kinetic energy gained so far.

49. Prove this mathematically. Hint: as in Section 1.6.4,

$$
\begin{aligned}
m\frac{v_2^2(t)}{2} &= \frac{m}{2}\left(v_2^2(t) - v_2^2(0)\right) \\
&= \int_H^{x_2(t)} F_2 dy \\
&= \int_H^{x_2(t)} (-mg)dy \\
&= -mg\int_H^{x_2(t)} dy \\
&= -mgy\,|_H^{x_2(t)} \\
&= mgy\,|_{x_2(t)}^H \\
&= mg\left(H - x_2(t)\right).
\end{aligned}
$$

50. To simplify this, get rid of t, and use x_2 and v_2 as new independent variables:

$$m\frac{v_2^2}{2} + mgx_2 = mgH.$$

51. Did you know this already?

52. What is the physical meaning of this? Hint: conservation of energy: at any position that the bomb may take in the (x_1, x_2)-plane, the (vertical) kinetic and potential energies sum to the initial potential energy.

53. In the above, was there any need to have $x_2(t)$ or $v_2(t)$ in their explicit form?

54. Conclude that there was actually no need to solve for $x(t)$ or $v(t)$ explicitly: energy considerations are good enough! After all, energy is more fundamental.

55. Still, to confirm the above equation, substitute the known solution $x_2(t)$ and $v_2(t)$, and make sure that the equation indeed holds. Hint: the loss in potential energy is indeed the same as the gain in (vertical) kinetic energy:

$$mg\left(H - x_2(t)\right) = mg\left(H - \left(H - g\frac{t^2}{2}\right)\right)$$
$$= m\frac{g^2 t^2}{2}$$
$$= m\frac{v_2^2(t)}{2}.$$

56. In Figure 1.8, where is the time variable? Why is it missing? Hint: this is a phase plane. In it, time is implicit: just a parameter that "pushes" along the virtual curve.

57. To see this more vividly, make a movie that shows the advance along the virtual curve.

2

Angular Momentum and Its Conservation

In this chapter, we introduce two key terms in geometrical mechanics: angular velocity and angular momentum. This way, we can also see the geometrical relation between them.

Furthermore, we also introduce new coordinates: the rotating coordinates. This is a good preparation work for relativity. Indeed, the rotating coordinates are related to the self-coordinates in special relativity, and also to Riemann's normal coordinates in general relativity.

Why are the new coordinates so useful? Because, in the new coordinate system, we have a few new forces: Euler force, Coriolis force, and more. We show how to use them in practice.

The rotating coordinate system is also related to the principal axes and the inertia matrix, discussed in the exercises (and their solutions). Thanks to them, we can see quite vividly how angular momentum is indeed conserved.

Later on, we'll put this in a deeper context: Hamiltonian mechanics. This will put the geometrical face in a proper algebraic framework: Lie algebras.

2.1 Inner Product and Orthogonality

2.1.1 Inner (Scalar) Product

Let's start with some background in linear algebra. Let u and v be column vectors in the three-dimensional Cartesian space:

$$u \equiv \begin{pmatrix} u_1 \\ u_2 \\ u_3 \end{pmatrix} \in \mathbb{R}^3$$

$$v \equiv \begin{pmatrix} v_1 \\ v_2 \\ v_3 \end{pmatrix} \in \mathbb{R}^3.$$

The transpose vector, on the other hand, is a row vector:

$$u^t = (u_1, u_2, u_3)$$
$$v^t = (v_1, v_2, v_3).$$

The inner product of u and v is a new number, defined by

$$(u, v) = u^t v = u_1 v_1 + u_2 v_2 + u_3 v_3.$$

2.1.2 Vector Norm

For example, look at the inner product of u with itself:

$$(u, u) \equiv u^t u = u_1^2 + u_2^2 + u_3^2 \geq 0.$$

This could vanish only if u is the zero vector (the origin). We can now define the norm (or length, or magnitude) of u as

$$\|u\| \equiv \sqrt{(u, u)} \geq 0.$$

In particular, every vector of norm 1 is called a unit vector.

2.1.3 Orthogonal Vectors and Matrix

If their inner product vanishes:

$$(u, v) = 0,$$

then our vectors are orthogonal to each other. If they are also unit vectors, then they are also orthonormal. A matrix with orthonormal columns is called orthogonal.

2.1.4 Orthogonal Projection

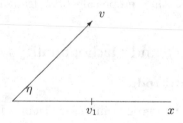

Fig. 2.1. The vector v makes angle η with the positive part of the x-axis: $\cos(\eta) = v_1/\|v\|$.

Let η be the angle between u and v. The inner product takes now a new geometrical meaning:

$$(u, v) = \|u\| \cdot \|v\| \cos(\eta).$$

In particular, orthogonal vectors that have zero inner product must also be perpendicular: have angle 90° in between.

To prove the above formula, let's look first at a very simple case, in which u is a special vector: a standard unit vector that points in the x-direction:

$$u \equiv \begin{pmatrix} 1 \\ 0 \\ 0 \end{pmatrix}.$$

In this case, the inner product just filters out the x-coordinate of v:

$$(u, v) = v_1 = \|v\| \cos(\eta)$$

(Figure 2.1). In other words, the inner product with u is just the orthogonal projection on u.

By now, we've proved our formula in the special case in which u points in the x-direction. But, is this case so special? Not at all! After all, we could always apply the same orthogonal matrix to both u and v, and rotate them together, until u matches the x-direction. (See exercises below.) This would change neither their inner product nor the angle between them. (See Chapter 26, Section 26.3.2.)

2.2 Vector Product and the Right-Hand Rule

2.2.1 Vector (Cross) Product

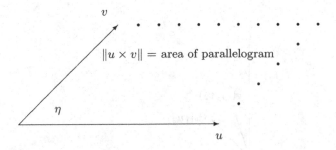

Fig. 2.2. $\|u \times v\|$ is the area of the parallelogram made by u and v.

Inner product tells us to what extent two vectors are nearly the same. Indeed, if they point in (nearly) the same direction, then their inner product must be big. If, on the other hand, they point in completely different directions, then they could be (nearly) perpendicular, and their inner product could (almost) vanish.

Vector product, on the other hand, works the other way around: it tells us how much it takes to turn from one vector to another. If they are nearly the same, then this doesn't take much, so their vector product should be small. If, on the other hand, they are nearly perpendicular, then this takes a lot, so their vector product should be big. This is why vector product is so useful to measure angular momentum. For this purpose, however, it can't be just a number: it must be a new vector, with a direction of its own.

Inner product is just a scalar (number). Vector product, on the other hand, is more than that: a complete three-dimensional vector, with a direction of its own. For our three-dimensional vectors u and v, for example, their vector product is a new vector, denoted by $u \times v$. What is its direction? Well, it is perpendicular (or orthogonal) to both u and v. In other words, it has zero inner product with either of them:

$$(u \times v, u) = (u \times v, v) = 0.$$

Thus, $u \times v$ is perpendicular not only to u and v but also to every linear combination of them, and indeed to the entire plane that they span. Thus, it could serve as an axis of rotation to turn from u to v. Later on, we'll see that this must be done with your right hand. (Otherwise, pick a minus sign.)

By now, we already have an idea about the direction of $u \times v$. Still, what is its length (or norm)? This is the same as the area of the parallelogram that u and v draw (Figure 2.2):

$$\|u \times v\| = \|u\| \cdot \|v\| \sin(\eta)$$

(where η is the angle in between). For this reason, if u and v coincide, then we get

$$u \times u = \mathbf{0}$$

(the zero vector). This is called alternativity.

2.2.2 The Right-Hand Rule

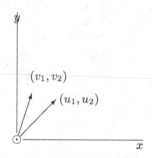

Fig. 2.3. The right-hand rule: take your right hand, and match your thumb to (u_1, u_2), and your index finger to (v_1, v_2). Then, your middle finger will point upwards, towards your own eyes, as indicated by the '$\odot$' at the origin.

By now, the vector product is specified up to sign. After all, it could be multiplied by -1, and still satisfy the above conditions. What should the correct sign be? To find out, use the right-hand rule: take your right hand, and match your thumb to u, and your index finger to v. This way, your middle finger will point in the correct direction of $u \times v$ (Figure 2.3).

To turn from u to v, turn your thumb towards your index finger. While doing this, point your middle finger towards your own eyes. The turn will then be counterclockwise: a positive mathematical turn.

Thanks to the right-hand rule, we also have antisymmetry:

$$u \times v = -v \times u.$$

2.2.3 Linearity: The Distributive Law

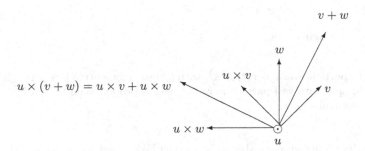

Fig. 2.4. The distributive law. u points upwards, towards your eyes, as indicated by the '⊙.' Thus, taking the vector product with u is linear: it just rotates by 90°, and multiplies by $\|u\|$.

As defined above, is the vector product linear? To check on this, let α be a real number. Then, α could be pulled out:

$$(\alpha u) \times v = u \times (\alpha v) = \alpha(u \times v).$$

Indeed, the parallelogram in Figure 2.2 is just multiplied by factor α.

Next, let w be yet another three-dimensional vector. Then, the distributive law holds – one may open parentheses:

$$u \times (v + w) = u \times v + u \times v.$$

To see this, project v, w, and $v + w$ onto the plane perpendicular to u. This has no effect on their vector product with u: the parallelogram in Figure 2.2 (viewed from the side) just transforms into a rectangle of the same area. Now, the distributive law follows from Figure 2.4.

Thanks to linearity, vector product is an attractive algebraic operation: it is ready to serve as Lie brackets, in a Lie algebra. This will be discussed later in the book. Before doing this, let's use it in geometrical physics.

2.3 Linear and Angular Momentum

2.3.1 Linear Momentum

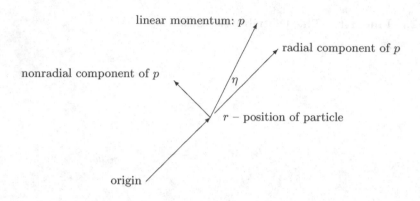

Fig. 2.5. At time t, the particle is at $r \equiv r(t) \in \mathbb{R}^3$, with a linear momentum $p \equiv p(t) \in \mathbb{R}^3$. This momentum could split into two parts: the radial part is proportional to r, whereas the other part is perpendicular to r.

The vector product introduced above is particularly useful in geometrical physics. To see this, consider a particle of constant mass m, traveling in the three-dimensional Cartesian space. At time t, it is at position

$$r \equiv r(t) \equiv \begin{pmatrix} x(t) \\ y(t) \\ z(t) \end{pmatrix} \in \mathbb{R}^3.$$

Later on, in quantum mechanics, we'll see that this is not so simple. Still, for the time being, let's accept this.

To obtain the velocity of the particle, differentiate with respect to time:

$$r' \equiv r'(t) \equiv \begin{pmatrix} x'(t) \\ y'(t) \\ z'(t) \end{pmatrix} \in \mathbb{R}^3.$$

To obtain the linear momentum, multiply by m (the mass of the particle):

$$p \equiv p(t) \equiv mr'(t) \in \mathbb{R}^3.$$

Later on, in special relativity, we'll redefine the linear momentum more carefully. Still, for the time being, let's accept this.

Finally, differentiate p, to obtain the force: the new vector $p'(t)$. In summary, at each particular time t, the linear momentum p describes the full motion of the particle, telling us how it moves in each and every direction. Still, we'll soon see that only two directions matter: the third one vanishes.

Indeed, p will soon split into two orthogonal components. The first will tell us how fast the particle gets farther and farther away from the origin, or how fast $\|r\|$ grows. This component must be radial: proportional to r, pointing in the same direction. In fact, it is just the orthogonal projection of p onto the unit vector $r/\|r\|$:

$$\frac{(r,p)}{\|r\|^2} r = \left(\frac{r}{\|r\|}, p\right) \frac{r}{\|r\|} = \cos(\eta)\|p\|\frac{r}{\|r\|},$$

where η is the angle between p and r (Figures 2.1 and 2.5). This radial component indeed tells us how fast the particle gets away from the origin, or how fast $\|r\|$ grows in time, as required.

2.3.2 Angular Momentum

Still, this is not the whole story. After all, as time goes by, r may not only change magnitude but also change direction. To understand this part of the motion, let's look at the other component of p, which will tell us how fast the particle rotates around the origin.

Where does this rotation take place? Well, infinitesimally, it takes place in the r-p plane: the plane spanned by r and p. In other words, this is the plane orthogonal to the vector product $r \times p$.

So, the second component of p must be nonradial: orthogonal (or perpendicular) to r. This way, it will indeed tell us how fast the particle rotates around a new vector: the angular momentum $r \times p$.

What is the norm of $r \times p$? We already know what it is:

$$\|r \times p\| = \|r\| \cdot \|p\| \sin(\eta)$$

(Section 2.2.1). To rotate around $r \times p$, the particle must make a small (infinitesimal) arc in the r-p plane. For this purpose, the second component of p must be tangent to this arc, or orthogonal not only to $r \times p$ but also to r (Figure 2.5). In summary, it must be proportional to

$$(r \times p) \times r.$$

Fortunately, we already know the norm of this vector:

$$\|(r \times p) \times r\| = \|r \times p\| \cdot \|r\| \sin\left(\frac{\pi}{2}\right) = \|r \times p\| \cdot \|r\| = \|p\| \sin(\eta)\|r\|^2.$$

The second (nonradial) component of p should be proportional to this vector, but could have a different norm. Indeed, to have the correct norm, just divide by $\|r\|^2$:

$$\frac{r \times p}{\|r\|^2} \times r = \frac{r \times p}{\|r\|} \times \frac{r}{\|r\|}.$$

Indeed, the norm of this vector is

$$\left\|\frac{r \times p}{\|r\|^2} \times r\right\| = \sin(\eta)\|p\|,$$

as required in Pythagoras' theorem. In summary, we now have the complete orthogonal decomposition of the original linear momentum:

$$p = \frac{(r,p)}{\|r\|^2}r + \frac{r \times p}{\|r\|^2} \times r = \frac{(r,p)}{\|r\|} \cdot \frac{r}{\|r\|} + \frac{r \times p}{\|r\|} \times \frac{r}{\|r\|}.$$

What is so nice about this decomposition? Well, it is uniform: both terms are written in the same style. The only difference is that the former uses inner product, whereas the latter uses vector product.

Furthermore, both are orthogonal (perpendicular) to one another. After all, this is how they were designed in the first place: the former is proportional to r, whereas the latter is perpendicular to r. This is why they also satisfy Pythagoras' theorem: the sum of squares of their norms is

$$\left\| \frac{(r,p)}{\|r\|} \cdot \frac{r}{\|r\|} \right\|^2 + \left\| \frac{r \times p}{\|r\|} \times \frac{r}{\|r\|} \right\|^2 = \left(\frac{(r,p)}{\|r\|} \right)^2 + \left\| \frac{r \times p}{\|r\|} \right\|^2$$

$$= \cos^2(\eta)\|p\|^2 + \sin^2(\eta)\|p\|^2$$

$$= \left(\cos^2(\eta) + \sin^2(\eta) \right)\|p\|^2$$

$$= \|p\|^2,$$

as required.

2.4 Angular Velocity

2.4.1 Angular Velocity

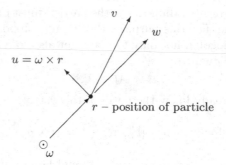

Fig. 2.6. For the sake of better visualization, we assume that the angular velocity ω is perpendicular to r. This way, it points from the page towards your eyes, as indicated by the '$\odot$' at the origin. (Don't confuse ω with the other vector w!)

What is momentum? It is mass times velocity. This is true not only in one but also in three dimensions. So, to obtain the velocity vector, just divide by the mass of the particle:

$$v = \frac{p}{m} = \frac{r \times p}{m\|r\|^2} \times r + \frac{(r,p)}{m\|r\|^2}r.$$

This is just a special case of a more general (not necessarily orthogonal) decomposition:

$$v = u + w$$

(Figure 2.6). Here,

$$u \equiv \omega \times r,$$

and

$$\omega \equiv \omega(t) \equiv \begin{pmatrix} \omega_1(t) \\ \omega_2(t) \\ \omega_3(t) \end{pmatrix} \in \mathbb{R}^3$$

is a new vector: the angular velocity. (Don't confuse it with the other vector w.) This way, the particle rotates around ω. The norm $\|\omega\|$ tells us how fast: by what angle per second. By definition, u must be perpendicular to both ω and r.

The angular velocity could be time-dependent: ω may change in time, not only in magnitude but also in direction. For simplicity, however, we often look at some fixed time. Thus, we often drop the argument '(t)' for short.

2.4.2 The Rotating Axis System

In general, ω may point in just any direction, not necessarily perpendicular to r or v. (See exercises below.) For simplicity, however, we often assume that ω is perpendicular to r:

$$(\omega, r) = 0.$$

Otherwise, just redefine the origin, and shift it along the ω-axis, until obtaining new orthogonal ω and r. This way, ω-r-u make a new right-hand system, rotating around the ω-axis. In terms of these new coordinates, the particle may "feel" a few new forces.

2.4.3 Velocity and Its Decomposition

As we've seen so far, at time t, the particle rotates (infinitesimally) around the ω-axis (at a rate of angle $\|\omega\|$ per second), making an infinitesimal arc. In our velocity

$$v = u + w,$$

the former term

$$u \equiv \omega \times r$$

is tangent to this arc. The remainder

$$w \equiv v - u,$$

on the other hand, may be nontangential, and even perpendicular to the arc.

For the sake of better visualization, we often assume that w is perpendicular not only to r but also to w (Figure 2.6). Still, this is not a must: in our final example below, w is nonradial, and contains a component parallel to ω. In most of our discussion, on the other hand, w is orthogonal to w (and v) too. In this case, it makes sense to define

$$\omega \equiv \frac{r \times p}{m\|r\|^2},$$

in agreement with the formula at the beginning of Section 2.4.1. This way, w is radial, so ω-w-u make the same right-hand system as ω-r-u: the rotating coordinate system (Section 2.4.2).

2.5 Real and Fictitious Forces

2.5.1 The Centrifugal Force

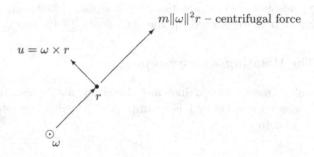

Fig. 2.7. The fictitious centrifugal force: $-m\omega \times (\omega \times r)$. If ω and r are orthogonal to each other, then it is also radial: $m\|\omega\|^2 r$.

What is the centrifugal force? This is the force that the particle "feels" in its own ideal "world:" the rotating coordinate system.

In general (even if ω is not orthogonal to r), the centrifugal force is

$$-m\omega \times (\omega \times r).$$

To illustrate, it is convenient to assume that ω and r are orthogonal to each other:

$$(\omega, r) = 0.$$

In this case,

$$\|\omega \times r\| = \|\omega\| \cdot \|r\|,$$

and the centrifugal force is radial:

$$-m\omega \times (\omega \times r) = m\|\omega\|^2 r$$

(Figure 2.7). Still, this is "felt" in the rotating system only. In reality, on the other hand, there is no centrifugal force at all. Indeed, in the static system used in Figures 2.5–2.7, the real force is just $p' = mv'$. So long as $v' = \mathbf{0}$, there is no force at all: Newton's first law holds, and the linear momentum is conserved. As a result, v

can never change physically: what could change is just its writing style in rotating coordinates. This nonphysical "change" is due to the fictitious centrifugal "force."

Still, the rotating coordinates are legitimate too, and we might want to work in them. In fact, if the particle stayed at the same rotating coordinates $(0, \|r\|, 0)$ all the time, then it would rotate physically round and round forever. For this purpose, however, a new counter-force must be applied, to cancel the centrifugal force out.

2.5.2 The Centripetal Force

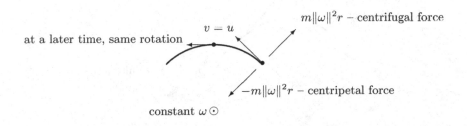

Fig. 2.8. Here, the particle is connected to the origin by a wire. This supplies the centripetal force required to cancel the original centrifugal force, and keep the particle at the constant distance $\|r\|$ from the origin, rotating at the constant angular velocity ω.

How to balance (or cancel) the centrifugal force? For this purpose, let's go ahead and "connect" the particle to the origin by a wire. In rotating coordinates, this reacts to the centrifugal force. Indeed, thanks to Newton's third law, this supplies the required counter-force – the centripetal force:

$$m\omega \times (\omega \times r).$$

If ω and r are orthogonal to each other, then this force is radial as well:

$$m\omega \times (\omega \times r) = -m\|\omega\|^2 r$$

(Figure 2.8). This is indeed how the wire must pull the particle back towards the origin. After all, Newton's third law must work in just any coordinate system, rotating or not.

In the rotating system, the centripetal force cancels the original centrifugal force, leaving the particle at the same (rotating) coordinates $(0, \|r\|, 0)$ forever. In the static coordinates, on the other hand, the centripetal force has a more physical job: to make u turn.

In fact, the centripetal force pulls the particle towards the origin in just the correct amount: it keeps the particle at a constant distance $\|r\|$ from the origin, rotating at a constant angular velocity ω. This way, the particle makes not only infinitesimal but also global arc around the ω-axis (Figure 2.8). In fact, this arc is as big as a complete circle of radius $\|r\|$. As a result, the original velocity vector is always tangent to this circle: there is no nontangential component any more:

$$v = u \quad \text{and} \quad w = \mathbf{0}.$$

2.5.3 Euler Force

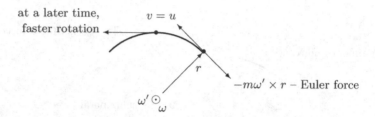

at a later time,
faster rotation

$v = u$

r

$-m\omega' \times r$ – Euler force

$\omega' \odot$
ω

Fig. 2.9. Here, the particle rotates counterclockwise faster and faster, so ω' points in the same direction as ω. In this case, Euler force pulls the particle clockwise, opposing the original rotation, and slowing it down.

In the above example, the particle rotates at the constant angular velocity ω. But what if ω changed in time? In this case, ω might have a nonzero time derivative:

$$\omega' \equiv \omega'(t) \equiv \begin{pmatrix} \omega_1'(t) \\ \omega_2'(t) \\ \omega_3'(t) \end{pmatrix} \neq \mathbf{0}.$$

In the rotating coordinate system, this introduces a new force – Euler force:

$$-m\omega' \times r.$$

What is the direction of this new force? In general, we can't tell. After all, ω' could point in just any direction. Indeed, as time goes by, the angular velocity may change not only magnitude but also direction, making the particle rotate in all sorts of new r-u planes.

Still, for simplicity, assume that ω' keeps pointing in the same direction as the original ω (Figure 2.9). This way, ω keeps pointing in the same direction all the time: it only gets bigger and bigger in magnitude:

$$\|\omega\|' \equiv \frac{d\|\omega\|}{dt} > 0.$$

What happens physically? The particle rotates counterclockwise faster and faster. What could possibly supply the energy required for this? Well, assume that there is some angular accelerator that keeps increasing the frequency (the angle that the particle makes per second). This way, the particle keeps rotating counterclockwise in the same r-u plane, at a bigger and bigger angle $\|\omega(t)\|$ per second.

In the rotating coordinates, the Euler force pulls the particle back clockwise, in an attempt to oppose this motion and slow it down. So, the accelerator must balance Euler force, and supply a force in the amount of

$$m\|\omega' \times r\| = m\|\omega'\| \cdot \|r\|$$

counterclockwise.

In the original static coordinate system, this accelerates the particle angularly, as required. In the rotating axis system, on the other hand, the particle remains at rest. It always keeps the same (rotating) coordinates: $(0, \|r\|, 0)$. This way, in its own subjective (rotating) "world," it keeps floating effortless, allowing the rotating axis system to carry it round and round, faster and faster.

2.5.4 The Earth and Its Rotation

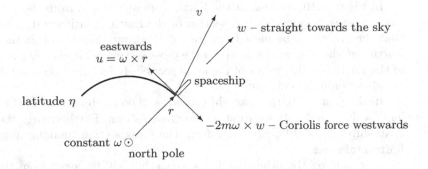

Fig. 2.10. The Earth – a view from above: the north pole, surrounded by a horizontal cross section of the northern hemisphere at latitude $0 \le \eta < \pi/2$. Here, we place the origin not at the center of the Earth but above it: at the center of the cross section. As a result, r is horizontal as well: it lies in the cross section. But w (the direction of the spaceship) is not: it makes angle η with the horizontal r-axis. The Coriolis force pulls the entire spaceship westwards.

The rotating coordinates are not just theoretical. They may also be quite real, and easy to work with. Let's go ahead and use them in practice.

In Figures 2.8–2.9, the particle makes a closed circle around the w-axis. In fact, the velocity is tangent to this circle, with no nontangential component at all:

$$v = u \quad \text{and} \quad w = \mathbf{0}.$$

Let us now consider a more complicated case, in which v does contain a nontangential component as well:

$$v = u + w, \quad \text{where} \quad w \neq \mathbf{0}.$$

As a matter of fact, we've already seen such a case. In Figure 2.6, however, w might seem radial. Here, on the other hand, w makes angle η with the horizontal r-axis, and angle $\pi/2 - \eta$ with the vertical ω-axis:

$$(\omega, w) \geq 0$$

(Figure 2.10), Consider, for example, a spaceship standing somewhere in the northern hemisphere of the Earth, at latitude $0 \leq \eta < \pi/2$. Its nose points upwards: straight towards the sky.

The entire Earth rotates eastwards: this is why we see the sun rising from the east. For this reason, in Figure 2.10, the angular velocity ω points northwards: from the page upwards, straight towards your eyes.

In reality, ω is not quite constant: it changes direction, although very slowly. In fact, as the Earth rotates, ω rotates too. Why? Because the Earth is not a perfect sphere. For this reason, the north pole is not fixed: it loops clockwise. This is very slow: the loop takes $27,000$ year to close. This is called precession.

Besides, the north pole makes yet another (small) loop clockwise. This loop is quicker: it takes 1.3 year to close. Still, it is very small: just ten meters in radius. So, both loops could be ignored.

In Figure 2.10, we look at the Earth from above: say, from the North star. This way, we see a horizontal cross section of the Earth at latitude $0 \leq \eta < \pi/2$. Recall that the origin can be picked arbitrarily. Here, we place it not at the center of the Earth but above it: at the center of the cross section. This way, $\|r\|$ is not the radius of the Earth but the radius of the cross section. Indeed, r is horizontal: it lies in the cross section in its entirety.

Initially, at $t = 0$, the spaceship stands still on the boundary of the cross section. This way, r marks its location in the cross section. Furthermore, the body of the spaceship points obliquely away from the cross section, making angle η with the horizontal r-axis.

At $t > 0$, on the other hand, the spaceship will fly away from the Earth. Fortunately, the cross section is part of an infinite horizontal plane. In this plane, r will denote the orthogonal projection (or "shadow") that the spaceship will make on the plane. This way, r will always be horizontal: it will remain in the plane at all times.

If $\eta = 0$, then the situation is simple. The cross section is surrounded by the equator, and its center coincides with the center of the Earth. Since r lies in the cross section, it coincides with the radius of the Earth. This way, r is perpendicular to the face of the Earth: it points straight into the sky. Fortunately, the centrifugal force in this direction is well-balanced by gravity, which supplies the required centripetal force.

By the way, this answers an interesting question: why isn't the Earth a perfect sphere? Well, if it were, then some gravity would be wasted at the equator, to balance the centrifugal force there. So, at the equator, gravity would be a little weaker. This is probably how the Earth evolved in the first place: at the equator, due to weaker gravity, it got a bit wide and "fat:" an oblate spheroid.

If, on the other hand, $\eta > 0$, then the situation is more complicated. The cross section passes not through the center of the Earth but above it. Since r is still horizontal, it is now shorter than before:

$$\|r\| = \cos(\eta) \text{ times the radius of the Earth.}$$

Furthermore, r is no longer perpendicular to the face of the Earth: it also has a new component that points southwards.

What is the norm of this new component? Clearly, it is $\sin(\eta)\|r\|$. This produces a new centrifugal force in the amount of $m\|\omega\|^2\sin(\eta)\|r\|$ southwards, which is not balanced by gravity any more. After all, gravity pulls downwards, towards the ground, not northwards.

Recall that here we work in the rotating axis system: we stand on the face of the Earth, unaware of any rotation. Therefore, to us, the above force is real, and is truly felt. Likewise, the spaceship feels it as well, and its route could be affected, including the shadow r it makes on the (extended) cross section.

Fortunately, the above force can never affect ω, which produced it in the first place. Can it affect the "shadow" r? Not much: it needs time to act. Therefore, for a small t, its effect on r is as small as t^2. (See exercises at the end of Chapter 1.) For this reason, it hardly affects the original motion, illustrated in Figure 2.10 at the initial time of $t = 0$. Still, after a while, the effect may accumulate and grow, and should be taken into account. To balance this, the nose of the spaceship should better point a little obliquely northwards, from the start.

2.5.5 Coriolis Force

Together with the entire Earth, the spaceship rotates eastwards, at a rate of angle $\|\omega\|$ per second. This produces one component of its velocity – the tangential part:

$$u \equiv \omega \times r.$$

Here, though, we work in the rotating axes, spanned by ω, r, and u. In rotating coordinates, there is no tangential motion at all. This is why we on the Earth are never aware of the Earth's rotation.

Now, at time $t = 0$, the spaceship suddenly gets an initial velocity w: upwards, straight towards the sky. This is liftoff: the spaceship really feels it. Still, at the same time, it also feels a new force, which mustn't be ignored:

$$-2m\omega \times w.$$

This is the Coriolis force (Figure 2.10).

What is its direction? Well, it must be perpendicular to the entire ω-w plane. Thanks to the right-hand rule, this must be westwards.

What is the norm of the Coriolis force? Well, this depends on the angle $\pi/2 - \eta$ between ω and w:

$$2m\|\omega \times w\| = 2m\|\omega\|\cdot\|w\|\sin\left(\frac{\pi}{2} - \eta\right) = 2m\|\omega\|\cdot\|w\|\cos(\eta).$$

Fortunately, this force doesn't have much time to act: for a small t, its effect on w is as small as t. (See exercises at the end of Chapter 1.) Thus, it hardly affects the motion, illustrated in Figure 2.10 at the initial time of $t = 0$. Still, as time goes by, it may accumulate, and mustn't be ignored. To balance it, the spaceship should better point a little obliquely eastwards, from the start.

2.6 Exercises: Inertia and Principal Axes

2.6.1 Determinant of a 2 × 2 Matrix

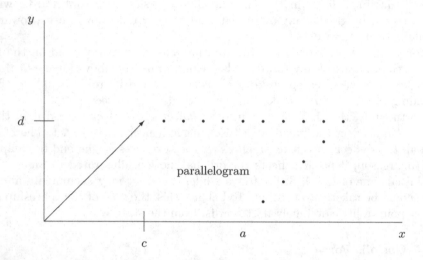

Fig. 2.11. For simplicity, the first vector$(a, b)^t$ is horizontal: $b = 0$. The second vector $(c, d)^t$, on the other hand, is oblique. Together, they make a parallelogram of area $ad - bc$. This is also the determinant of the matrix that they make.

1. What is a determinant? Hint: later on, we'll define the determinant in general terms (Chapter 13). For the time being, we only need to define it for a 2×2 matrix, containing two column vectors:

$$\det\left(\begin{pmatrix} a & c \\ b & d \end{pmatrix}\right) \equiv \det\left(\begin{pmatrix} a \\ b \end{pmatrix} \begin{pmatrix} c \\ d \end{pmatrix}\right) \equiv ad - bc.$$

2. What is this geometrically? Hint: for simplicity, assume that the first column vector $(a, b)^t$ is horizontal: $b = 0$, as in Figure 2.11. This way, it is easy to see that the determinant $ad - bc$ has a geometrical meaning: the area of the parallelogram made by the above column vectors.
3. Use this to sum the areas of two parallelograms of the same basis. Hint: in Figure 2.11, draw a new oblique vector: $(\tilde{c}, \tilde{d})^t$. This makes a new parallelogram of the same basis. How to sum the areas of these two parallelograms? There are two equivalent methods:
   - calculate each area separately, and then sum up.
   - sum the oblique vectors: $(c + \tilde{c}, d + \tilde{d})^t$. This makes a new parallelogram of the same basis. Calculate its area, using its determinant.
4. Why are these methods equivalent to each other? Hint: the distributive law.

5. How to extend the above geometrical interpretation to a more general column vector $(a, b)^t$, not necessarily horizontal? Hint: pick another first column vector $(\tilde{a}, \tilde{b})^t$, vertical rather than horizontal: $\tilde{a} = 0$ and $\tilde{b} < 0$. Check that the above geometrical interpretation still holds: the determinant $\tilde{a}d - \tilde{b}c$ is the same as the area of the new parallelogram. Finally, extend it linearly, as in the previous exercise. (This time, $(c, d)^t$ serves as a basis.)

6. What is the determinant of a product of two matrices? Hint: the same as the product of their individual determinants.

7. How to define the determinant of a bigger 3×3 matrix, in such a way that the above property still holds? Hint: see Chapter 13.

8. How to define the determinant of a yet bigger 4×4 matrix, in such a way that the above property still holds? Hint: see Chapter 13.

2.6.2 Rotation and Euler Angles

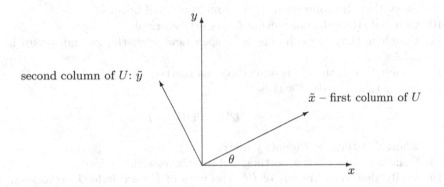

second column of U: $\tilde{y}$

$\tilde{x}$ – first column of U

Fig. 2.12. The orthogonal matrix U rotates the entire x-y plane by angle θ counterclockwise, and maps it to the new $\tilde{x}$-$\tilde{y}$ plane.

1. Let Q be an $n \times n$ orthogonal matrix. Show that it preserves inner product: for every two n-dimensional vectors u and v,

$$(Qu, Qv) = (u, Q^t Qv) = (u, Iv) = (u, v)$$

(where I is the $n \times n$ identity matrix).

2. Conclude that Q also preserves norm:

$$\|Qv\|^2 = (Qv, Qv) = (v, v) = \|v\|^2.$$

3. Let O and Q be two orthogonal matrices of the same order. Show that their product OQ is an orthogonal matrix as well. Hint: thanks to associativity,

$$(OQ)^t(OQ) = (Q^t O^t)(OQ) = Q^t (O^t O) Q = Q^t I Q = Q^t Q = I.$$

4. Let $0 \le \theta < 2\pi$ be some (fixed) angle . Let U be the following 2×2 matrix:

$$U \equiv U(\theta) \equiv \begin{pmatrix} \cos(\theta) & -\sin(\theta) \\ \sin(\theta) & \cos(\theta) \end{pmatrix}.$$

5. Show that U rotates the x-axis by angle θ counterclockwise. Hint: apply U to the standard unit vector that points rightwards:

$$U \begin{pmatrix} 1 \\ 0 \end{pmatrix} = \begin{pmatrix} \cos(\theta) \\ \sin(\theta) \end{pmatrix}.$$

   The result is the $\tilde{x}$-unit vector in Figure 2.12.

6. Show that U rotates the y-axis by angle θ counterclockwise as well. Hint: apply U to the standard unit vector $(0, 1)^t$.

7. Conclude that U rotates the entire x-y plane by angle θ counterclockwise. Hint: extend the above linearly. For this purpose, write a general two-dimensional vector as a linear combination of the standard unit vectors $(1, 0)^t$ and $(0, 1)^t$.

8. Show that the columns of U are orthogonal to each other.

9. Show that the columns of U are unit vectors of norm 1.

10. Conclude that the columns of U are orthonormal.

11. Conclude that the columns of U span new axes: the $\tilde{x}$- and $\tilde{y}$-axes in Figure 2.12.

12. Conclude also that U is an orthogonal matrix.

13. Verify that U indeed satisfies

$$U^t U = U U^t = I,$$

   where I is the 2×2 identity matrix.

14. Conclude that U^t is an orthogonal matrix as well.

15. Verify that the columns of U^t (the rows of U) are indeed orthogonal to each other as well.

16. Verify that the columns of U^t (the rows of U) are indeed unit vectors of norm 1.

17. Recall that U has a geometrical meaning: once applied to a two-dimensional vector, it rotates it by angle θ counterclockwise. Hint: check this for the standard unit vectors $(1, 0)^t$ and $(0, 1)^t$. Then, extend this linearly.

18. Interpret U^t geometrically as the inverse rotation: once applied to a vector, it rotates it by angle θ clockwise.

19. Interpret the equation $U^t U = I$ geometrically. Hint: the composition of U^t on top of U is the identity mapping that changes nothing. Indeed, rotating clockwise cancels rotating counterclockwise.

20. Work the other way around: interpret the equation $U U^t = I$ geometrically as well.

21. Show that U has determinant 1:

$$\det(U) = \det\left(U^t\right) = 1.$$

22. Introduce a third spatial dimension: the z-axis. This makes the new x-y-z axis system in $\mathbb{R}^3$.

23. Consider yet another (right-hand) axis system: the $\tilde{x}$-$\tilde{y}$-$\tilde{z}$ axis system. (Assume that it shares the same origin.) How to map the original x-y-z system to the new $\tilde{x}$-$\tilde{y}$-$\tilde{z}$ system?

24. To do this, use three stages:
   - Rotate the entire $\tilde{x}$-$\tilde{y}$ plane by a suitable angle ψ clockwise, until the $\tilde{x}$-axis hits the x-y plane.
   - Then, rotate the entire x-y plane by a suitable angle ϕ counterclockwise, until the x-axis matches the up-to-date $\tilde{x}$-axis.
   - By now, the up-to-date x- and $\tilde{x}$-axes align with each other. So, all that is left to do is to rotate the up-to-date y-z plane by a suitable angle θ counterclockwise, until the y- and z-axes match the up-to-date $\tilde{y}$- and $\tilde{z}$-axes, respectively.

25. Conclude that, to map the original x-y-z axis system to the original $\tilde{x}$-$\tilde{y}$-$\tilde{z}$ axis system, one could use three stages:
   - Rotate the entire x-y plane by angle ϕ counterclockwise.
   - Then, rotate the up-to-date y-z plane by angle θ counterclockwise.
   - Finally, rotate the up-to-date x-y plane by angle ψ counterclockwise.

26. The angles ϕ, θ, and ψ are called Euler angles.

27. Show that this is a triple product of three orthogonal matrices.

28. Conclude that this triple product is orthogonal too.

29. What is its determinant? Hint: 1.

30. What matrix represents the first stage? Hint:

$$\begin{pmatrix} U(\phi) & \\ & 1 \end{pmatrix}.$$

31. What matrix represents the second stage? Hint: note that this stage rotates the *up-to-date* y-z plane. Therefore, its matrix is

$$\begin{pmatrix} U(\phi) & \\ & 1 \end{pmatrix} \begin{pmatrix} 1 & \\ & U(\theta) \end{pmatrix} \begin{pmatrix} U(\phi) & \\ & 1 \end{pmatrix}^{-1}.$$

32. What is the product of these two matrices? Hint:

$$\begin{pmatrix} U(\phi) & \\ & 1 \end{pmatrix} \begin{pmatrix} 1 & \\ & U(\theta) \end{pmatrix}.$$

33. What matrix represents the third stage? Hint: note that this stage rotates the *up-to-date* x-y plane. Therefore, its matrix is

$$\begin{pmatrix} U(\phi) & \\ & 1 \end{pmatrix} \begin{pmatrix} 1 & \\ & U(\theta) \end{pmatrix} \begin{pmatrix} U(\psi) & \\ & 1 \end{pmatrix} \left(\begin{pmatrix} U(\phi) & \\ & 1 \end{pmatrix} \begin{pmatrix} 1 & \\ & U(\theta) \end{pmatrix} \right)^{-1}.$$

34. What matrix represents the entire mapping? Hint: multiply the matrices of the three stages:

$$\begin{pmatrix} U(\phi) & \\ & 1 \end{pmatrix} \begin{pmatrix} 1 & \\ & U(\theta) \end{pmatrix} \begin{pmatrix} U(\psi) & \\ & 1 \end{pmatrix}.$$

35. What is its transpose? Hint:

$$\begin{pmatrix} U^t(\psi) & \\ & 1 \end{pmatrix} \begin{pmatrix} 1 & \\ & U^t(\theta) \end{pmatrix} \begin{pmatrix} U^t(\phi) & \\ & 1 \end{pmatrix}.$$

36. How does this matrix map the original x-y-z system onto the new $\tilde{x}$-$\tilde{y}$-$\tilde{z}$ system? Hint: symbolically:

$$\begin{pmatrix} \tilde{x}\text{-axis} \\ \tilde{y}\text{-axis} \\ \tilde{z}\text{-axis} \end{pmatrix} = \begin{pmatrix} U^t(\psi) & \\ & 1 \end{pmatrix} \begin{pmatrix} 1 & \\ & U^t(\theta) \end{pmatrix} \begin{pmatrix} U^t(\phi) & \\ & 1 \end{pmatrix} \begin{pmatrix} x\text{-axis} \\ y\text{-axis} \\ z\text{-axis} \end{pmatrix}.$$

37. Indeed, take the transpose:

$$(x\text{-axis},\, y\text{-axis},\, z\text{-axis}) \begin{pmatrix} U(\phi) & \\ & 1 \end{pmatrix} \begin{pmatrix} 1 & \\ & U(\theta) \end{pmatrix} \begin{pmatrix} U(\psi) & \\ & 1 \end{pmatrix}$$
$$= (\tilde{x}\text{-axis},\, \tilde{y}\text{-axis},\, \tilde{z}\text{-axis}).$$

38. Multiply this formula by $(1,0,0)^t$ from the right:

$$(x\text{-axis},\, y\text{-axis},\, z\text{-axis}) \begin{pmatrix} U(\phi) & \\ & 1 \end{pmatrix} \begin{pmatrix} 1 & \\ & U(\theta) \end{pmatrix} \begin{pmatrix} U(\psi) & \\ & 1 \end{pmatrix} \begin{pmatrix} 1 \\ 0 \\ 0 \end{pmatrix}$$
$$= (\tilde{x}\text{-axis},\, \tilde{y}\text{-axis},\, \tilde{z}\text{-axis}) \begin{pmatrix} 1 \\ 0 \\ 0 \end{pmatrix} = \tilde{x}\text{-axis}.$$

39. How does the new $\tilde{x}$-axis look like? Hint: thanks to the above formula, it is spanned by the first column of the above matrix:

$$\begin{pmatrix} U(\phi) & \\ & 1 \end{pmatrix} \begin{pmatrix} 1 & \\ & U(\theta) \end{pmatrix} \begin{pmatrix} U(\psi) & \\ & 1 \end{pmatrix} \begin{pmatrix} 1 \\ 0 \\ 0 \end{pmatrix}.$$

40. How does the new $\tilde{y}$-axis look like? Hint: likewise, it is spanned by the second column:

$$\begin{pmatrix} U(\phi) & \\ & 1 \end{pmatrix} \begin{pmatrix} 1 & \\ & U(\theta) \end{pmatrix} \begin{pmatrix} U(\psi) & \\ & 1 \end{pmatrix} \begin{pmatrix} 0 \\ 1 \\ 0 \end{pmatrix}.$$

41. How does the new $\tilde{z}$-axis look like? Hint: likewise, it is spanned by the third column:

$$\begin{pmatrix} U(\phi) & \\ & 1 \end{pmatrix} \begin{pmatrix} 1 & \\ & U(\theta) \end{pmatrix} \begin{pmatrix} U(\psi) & \\ & 1 \end{pmatrix} \begin{pmatrix} 0 \\ 0 \\ 1 \end{pmatrix}.$$

42. Does the above matrix have determinant 1? Hint: this matrix is the product of three rotation matrices of determinant 1. Now, the determinant of a product is the product of determinants (Chapter 13, Section 13.6.7).

43. In terms of group theory, what does the above discussion tell us? Hint: the group of three-dimensional rotations is *generated* by two-dimensional rotations of individual planes.

2.6.3 Principal Axes

1. Recall that

$$r \equiv \begin{pmatrix} x \\ y \\ z \end{pmatrix}$$

   is the position of the particle in the Cartesian space. Assume that r is fixed. Show that rr^t is a 3×3 matrix.

2. Write it explicitly. Hint:

$$rr^t \equiv \begin{pmatrix} x \\ y \\ z \end{pmatrix} (x, y, z) = \begin{pmatrix} x^2 & xy & xz \\ yx & y^2 & yz \\ zx & zy & z^2 \end{pmatrix}.$$

3. Use this to show that rr^t is symmetric.

4. Show this in yet another way. Hint:

$$\left(rr^t \right)^t = \left(r^t \right)^t r^t = rr^t.$$

5. What are the eigenvalues and eigenvectors of rr^t? Hint: see below.

6. Show that r is an eigenvector of rr^t, with the eigenvalue $\|r\|^2$. Hint: thanks to associativity,

$$\left(rr^t \right) r = r \left(r^t r \right) = (r, r)r = \|r\|^2 r.$$

7. Let q be a vector that is orthogonal to r. Show that q is an eigenvector of rr^t, with the zero eigenvalue. Hint: thanks to associativity,

$$\left(rr^t \right) q = r \left(r^t q \right) = (r, q)r = 0r = \mathbf{0}.$$

8. Design two different q's, both orthogonal to r, and also orthogonal to each other. Hint: pick two linearly independent q's, both orthogonal to r. Then, apply the Gram-Schmidt process to them.

9. Show that this could be done in many different ways. Still, pick arbitrarily one particular pair of orthogonal q's.

10. Conclude that rr^t is positive semidefinite: its eigenvalues are greater than or equal to zero.

11. Conclude that rr^t has the following diagonal form:

$$rr^t = O \begin{pmatrix} \|r\|^2 & & \\ & 0 & \\ & & 0 \end{pmatrix} O^t,$$

   where O is an orthogonal matrix, with columns that are proportional to r and the above q's. These columns are called principal axes. They span the entire Cartesian space, using new principal coordinates.

12. In terms of principal coordinates, where is the particle? Hint: its principal coordinates are: $(\|r\|, 0, 0)$.

13. Show that there are two principal axes that could be defined in many different ways. Still, pick arbitrarily one particular choice.

2.6.4 The Inertia Matrix

1. Let I be the 3×3 identity matrix. Define the new 3×3 matrix

$$A \equiv A(r) \equiv \|r\|^2 I - rr^t.$$

2. Once A is multiplied by the mass, we obtain mA: the inertia matrix of the particle. It will be useful below.
3. Show that A is symmetric.
4. Show that r is an eigenvector of A, with the zero eigenvalue.
5. Show that the above q's are eigenvectors of A, with the eigenvalue $\|r\|^2$.
6. Conclude that A is positive semidefinite: its eigenvalues are greater than or equal to zero.
7. These eigenvalues (times m) are called moments of inertia. They will be useful below.
8. Show that A has the diagonal form

$$A = O \begin{pmatrix} 0 & & \\ & \|r\|^2 & \\ & & \|r\|^2 \end{pmatrix} O^t,$$

   where O is as above.
9. Recall that

$$\omega \equiv \begin{pmatrix} \omega_1 \\ \omega_2 \\ \omega_3 \end{pmatrix} \in \mathbb{R}^3$$

   is the angular velocity. To help visualize things better, we assumed so far that ω was perpendicular to r. This, however, is not a must: from now on, let's drop this assumption. Show that, even if ω is no longer perpendicular to r, $A\omega$ still is:

$$(A\omega, r) = 0.$$

   Hint: in $A\omega$, there is no radial component any more.
10. Show that

$$\|\omega \times r\|^2 = \|\omega\|^2 \|r\|^2 - (\omega, r)^2.$$

   Hint: see Sections 2.1.1 and 2.2.1.
11. Conclude that

$$\|\omega \times r\|^2 = \omega^t A \omega.$$

   Hint: thanks to associativity,

$$\begin{aligned} \|\omega \times r\|^2 &= \|\omega\|^2 \|r\|^2 - (\omega, r)^2 \\ &= \|r\|^2 \omega^t \omega - (\omega^t r)(r^t \omega) \\ &= \|r\|^2 \omega^t I \omega - \omega^t (rr^t) \omega \\ &= \omega^t A \omega. \end{aligned}$$

12. Show that the entire principal axis system could rotate around the ω-axis, carrying the "passive" particle at angle $\|\omega\|$ per second.

13. What is the necessary condition for this? Hint: to have such a rotation, we must have no nontangential velocity: $w = \mathbf{0}$.
14. Give an example where this condition holds, and another example where it doesn't. Hint: in Figure 2.10, for example, the entire Earth rotates around ω. The spaceship, on the other hand, spirals towards the outer space, until escaping from the Earth's gravity. Indeed, it has two kinds of motion. On one hand, it rotates around ω, with the entire atmosphere. On the other hand, it also has a nontangential velocity: w, towards the sky.

2.6.5 Triple Vector Product

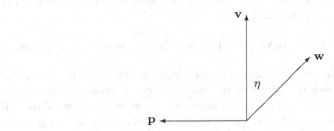

Fig. 2.13. In the v-w plane, p is orthogonal to v, but not to w.

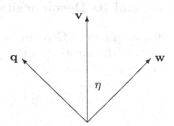

Fig. 2.14. In the v-w plane, q is orthogonal to w, but not to v.

1. Let $\mathbf{v}$ and $\mathbf{w}$ be linearly independent vectors in $\mathbb{R}^3$. This means that $\mathbf{v}$ is not a scalar multiple of $\mathbf{w}$, but points in a different direction (Figure 2.13). Let η be the angle between $\mathbf{v}$ and $\mathbf{w}$ ($0 < \eta < \pi$). Let $\mathbf{p}$ be yet another vector, perpendicular to $\mathbf{v}$ in the $\mathbf{v}$-$\mathbf{w}$ plane. Show that

$$\mathbf{p} \times (\mathbf{v} \times \mathbf{w}) = (\mathbf{p}, \mathbf{w})\mathbf{v}.$$

   Hint: use the fact that
$$\cos\left(\frac{\pi}{2} + \eta\right) = -\sin(\eta).$$

2. Conclude that
$$\mathbf{p} \times (\mathbf{w} \times \mathbf{v}) = -(\mathbf{p}, \mathbf{w})\mathbf{v}.$$

3. In the latter formula, interchange the roles of $\mathbf{v}$ and $\mathbf{w}$.
4. Conclude that
$$\mathbf{q} \times (\mathbf{v} \times \mathbf{w}) = -(\mathbf{q}, \mathbf{v})\mathbf{w},$$

   where $\mathbf{q}$ is a new vector, perpendicular to $\mathbf{w}$ in the $\mathbf{v}$-$\mathbf{w}$ plane (Figure 2.14).
5. Add these formulas to each other:

$$(\mathbf{p} + \mathbf{q}) \times (\mathbf{v} \times \mathbf{w}) = (\mathbf{p}, \mathbf{w})\mathbf{v} - (\mathbf{q}, \mathbf{v})\mathbf{w} = (\mathbf{p} + \mathbf{q}, \mathbf{w})\mathbf{v} - (\mathbf{p} + \mathbf{q}, \mathbf{v})\mathbf{w},$$

   where $\mathbf{p}$ is perpendicular to $\mathbf{v}$ and $\mathbf{q}$ is perpendicular to $\mathbf{w}$ in the $\mathbf{v}$-$\mathbf{w}$ plane.
6. Show that every vector $\mathbf{u}$ in the $\mathbf{v}$-$\mathbf{w}$ plane could be written as $\mathbf{u} = \mathbf{p} + \mathbf{q}$, where $\mathbf{p}$ is perpendicular to $\mathbf{v}$ and $\mathbf{q}$ is perpendicular to $\mathbf{w}$ in the $\mathbf{v}$-$\mathbf{w}$ plane. Hint: because $\mathbf{v}$ and $\mathbf{w}$ are linearly independent of each other, so are also $\mathbf{p}$ and $\mathbf{q}$.
7. What does this mean geometrically? Hint: this is just the parallelogram rule.
8. Conclude that
$$\mathbf{u} \times (\mathbf{v} \times \mathbf{w}) = (\mathbf{u}, \mathbf{w})\mathbf{v} - (\mathbf{u}, \mathbf{v})\mathbf{w},$$

   where $\mathbf{u}$ is just any vector in the $\mathbf{v}$-$\mathbf{w}$ plane.
9. Extend the above formula to a yet more general $\mathbf{u}$ that may lie outside the $\mathbf{v}$-$\mathbf{w}$ plane as well. Hint: the new component of $\mathbf{u}$ that is perpendicular to the $\mathbf{v}$-$\mathbf{w}$ plane contributes nothing to either side of the above formula.
10. Let's use this in practice.

2.6.6 Linear Momentum and Its Decomposition

1. How to use this formula in practice? Give an example.
2. For instance, set $\mathbf{u} \equiv \mathbf{v} \equiv r$ (the position of the particle) and $\mathbf{w} \equiv p$ (the linear momentum). What do you get? Hint:

$$r \times (r \times p) = (r, p)r - (r, r)p,$$

   or
$$(r, r)p = -r \times (r \times p) + (r, p)r = (r \times p) \times r + (r, p)r.$$

3. Divide both sides by $\|r\|^2$.
4. Is this familiar? Hint: this is the orthogonal decomposition of the linear momentum p. (See the end of Section 2.3.2.)

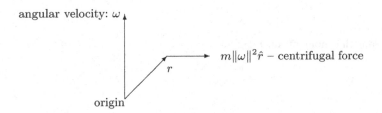

Fig. 2.15. A more general case, in which r is no longer perpendicular to w. (This is a view from the side.) Let $\hat{r}$ be the part of r that is perpendicular to w. Then, the centrifugal force is $m\|w\|^2\hat{r}$ rightwards.

2.6.7 Centrifugal and Centripetal Forces

1. For our triple-vector-product formula, give yet another practical example.
2. Use this formula to write the centrifugal force (Figure 2.7) in a more general case, in which w and r are not necessarily perpendicular to each other, so the rotating axes not necessarily align with them any more. Hint: set $\mathbf{u} = \mathbf{v} \equiv w$ and $\mathbf{w} \equiv r$:

$$w \times (w \times r) = (w, r)w - (w, w)r.$$

This way, the centrifugal force (divided by m) takes the form

$$-w \times (w \times r) = -(w, r)w + (w, w)r$$
$$= \|w\|^2 \left(r - \frac{(w, r)}{\|w\|^2} w \right)$$
$$= \|w\|^2 \hat{r},$$

where $\hat{r}$ is the part of r that is perpendicular to w (Figure 2.15).

3. Prove this in a more geometrical way. Hint: note that $w \times r = w \times \hat{r}$. Therefore,

$$-w \times (w \times r) = -w \times (w \times \hat{r}) = \|w\|^2 \hat{r}.$$

4. Does the centrifugal force really exist? Hint: only in the rotating axis system! In the static system, on the other hand, it has no business to exist. Indeed, in Figures 2.5–2.6, the velocity is often constant, meaning equilibrium: no force at all.
5. Does the centripetal force exist? Hint: only if supplied by some source, such as gravity.
6. What is the role of the centripetal force? Hint: in the rotating axis system, it balances the centrifugal force, and cancels it out. This way, there is no force at all, so the particle always has the same (rotating) coordinates, and is carried effortless by the rotating axis system round and round forever. In the static

system, on the other hand, the centripetal force has a more "active" role: to make u turn. This indeed makes the particle go round and round, as required.

2.6.8 Inertia Matrix times Angular Velocity

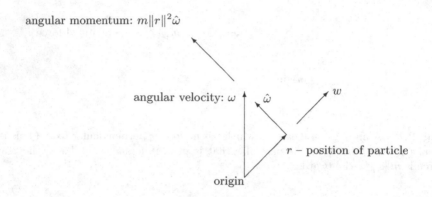

angular momentum: $m\|r\|^2\hat{\omega}$

angular velocity: ω

$\hat{\omega}$

w

r – position of particle

origin

Fig. 2.16. Let $\hat{\omega}$ be the part of ω that is perpendicular to r. If w is radial, then the angular momentum is $m\|r\|^2\hat{\omega}$. This is nonconstant: it must change direction to point not only upwards but also obliquely inwards. Why isn't the angular momentum conserved? Because the system is not closed or isolated: a horizontal centripetal force must be supplied from the outside.

1. At the end of Section 2.4.3, there is a suggestion for defining the angular velocity to align with the angular momentum:
$$\omega \equiv \frac{r \times p}{m\|r\|^2}.$$
This makes ω orthogonal to r. Here, on the other hand, we don't need this kind of orthogonality, and don't assume it any more.

2. Work the other way around: assume that the angular velocity is given, and use it to uncover the angular momentum (assuming that w in Figure 2.6 is radial):
$$r \times p = m \cdot r \times v$$
$$= m \cdot r \times (u + w)$$
$$= m \cdot r \times u$$
$$= m \cdot r \times (\omega \times r)$$
$$= m \left((r,r)\omega - (r,\omega)r \right)$$
$$= m\|r\|^2 \left(\omega - \frac{(r,\omega)}{\|r\|^2} r \right)$$
$$= m\|r\|^2\hat{\omega},$$
where $\hat{\omega}$ is the part of ω that is perpendicular to r (Figure 2.16). Hint: in our triple-vector-product formula, set $\mathbf{u} \equiv \mathbf{w} \equiv r$ and $\mathbf{v} \equiv \omega$.

3. Prove this in a more geometrical way. Hint: note that $w \times r = \hat{w} \times r$. Therefore,

$$r \times (w \times r) = r \times (\hat{w} \times r) = \|r\|^2 \hat{w}.$$

4. Conclude that, if w in Figure 2.6 is radial, then the angular momentum has a yet simpler form:

$$r \times p = mA\omega,$$

where $mA \equiv mA(r)$ is the inertia matrix of the particle.

5. What is the physical meaning of the inertia matrix? Hint: what is linear momentum? It is mass times velocity. In other words, what is mass? It is what connects velocity to momentum. Likewise, the inertia matrix is a kind of "mass:" it is the missing link that connects angular velocity to angular momentum (if w is radial).

6. In Figure 2.10, how to make w radial? Hint: shift the origin downwards, until it meets the center of the Earth. This way, the new r is no longer horizontal, but oblique: it leads from the center of the Earth to the spaceship, at the same direction as w. As a result, w is now radial, as required. Thus, the angular momentum is not vertical but oblique: it is proportional to $\hat{w}$, as in Figure 2.16.

7. Look at things the other way around. Assume now that the angular momentum is already available. How to define the angular velocity? Hint: design w in two stages: first, let it have some radial component. Then, define its other component by

$$\hat{w} \equiv \frac{r \times p}{m\|r\|^2}.$$

8. This way, in Figure 2.6, is w radial? Hint: thanks to the above definition of $\hat{w}$,

$$
\begin{aligned}
r \times p &= m \cdot r \times v \\
&= m \cdot r \times (u + w) \\
&= m \cdot r \times u + m \cdot r \times w \\
&= m \cdot r \times (w \times r) + m \cdot r \times w \\
&= m\|r\|^2 \hat{w} + m \cdot r \times w \\
&= r \times p + m \cdot r \times w.
\end{aligned}
$$

Thus,

$$0 = m \cdot r \times w,$$

so

$$r \times w = 0.$$

This means that w must indeed be radial.

2.6.9 Angular Momentum and Its Conservation

1. In Figure 2.16, the angular momentum is not conserved. On the contrary: it changes direction all the time. Indeed, it is proportional to $\hat{w}$, which changes direction to keep pointing obliquely (upwards and inwards). Why? Isn't the angular momentum supposed to be constant? Hint: only in a closed (isolated) system is the angular momentum conserved. The particle, however, is not isolated: there is an external force acting upon it – a horizontal centripetal force that makes it rotate around the vertical w-axis.

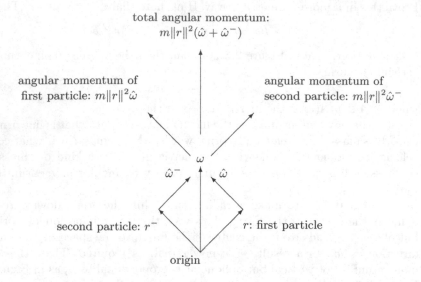

Fig. 2.17. The particles at r and r^- attract each other just enough to supply the horizontal centripetal force required to make them rotate together around the vertical ω-axis. This is a closed isolated system: no external force acts upon it. This is why the total angular momentum is now conserved: $\hat{\omega} + \hat{\omega}^-$ keeps pointing straight upwards all the time.

2. To supply this centripetal force, introduce a second particle at position

$$r^- \equiv \begin{pmatrix} -x \\ -y \\ z \end{pmatrix},$$

on the other side of the vertical ω-axis (Figure 2.17).

3. What is its inertia matrix? Hint: $mA(r^-)$.

4. What are the eigenvectors of $A(r^-)$? Hint: one eigenvector is r^- itself. The two other eigenvectors are orthogonal to r^- and to each other.

5. Could the eigenvectors of $A(r^-)$ be the same as those of $A(r)$? Hint: only if r^- is proportional or perpendicular to r.

6. What are the principal axes of the second particle? Are they the same as those of the first particle?

7. In terms of its own principal coordinates, where is the second particle? Hint: it is always at the same principal coordinates: $(\|r\|, 0, 0)$.

8. Let $\hat{\omega}^-$ be the part of ω perpendicular to r^-. Show that the angular momentum of the second particle is

$$m\|r^-\|^2\hat{\omega}^- = mA(r^-)\omega.$$

9. Together, these particles make a closed system: for a suitable $\|r\|$, they attract each other just enough to supply the horizontal centripetal force required to make them rotate together around the vertical ω-axis.

10. This is called the two-body system.

11. Around what point does it rotate? Hint: around its center of mass: the midpoint in between the particles.
12. What is the total angular momentum of the two-body system?
13. What is the direction of this vector?
14. Must it be vertical?
15. Show that the total angular momentum is now conserved. Hint: the sum $\hat{\omega} + \hat{\omega}^-$ always points straight upwards, with no inward component any more.
16. Define the inertia matrix of the two-body system:

$$B \equiv B(r) \equiv mA(r) + mA(r^-).$$

17. Write the total angular momentum as

$$m\|r\|^2 \left(\hat{\omega} + \hat{\omega}^- \right) = m \left(A(r)\omega + A(r^-)\omega \right) = B\omega.$$

18. Without calculating B explicitly, show that it is symmetric.
19. Conclude that B has three orthogonal eigenvectors. Hint: see Chapter 26, Section 26.4.5.
20. Could these new eigenvectors be the same as those of $A(r)$? Hint: only if r^- is proportional or perpendicular to r.
21. Show that B is still positive semidefinite: its eigenvalues are greater than or equal to zero. Hint: every three-dimensional vector q could be decomposed as a linear combination of eigenvectors of $A(r)$ (or $A(r^-)$). Therefore,

$$(q, Bq) = (q, mA(r)q) + (q, mA(r^-)q) = m(q, A(r)q) + m(q, A(r^-)q) \geq 0.$$

22. Show that, if $r^- \neq -r$, then either $(q, A(r)q) > 0$ or $(q, A(r^-)q) > 0$.
23. Conclude that, in this case, $(q, Bq) > 0$ as well.
24. Conclude that, in this case, B is positive definite: its eigenvalues are strictly positive.
25. There are the moments of inertia of the two-body system.
26. Design an eigenvector of B that is perpendicular to both r and r^-. Hint: if $r^- \neq -r$, then take their vector product: $r \times r^-$.
27. What is its eigenvalue? Hint: $2m\|r\|^2$.
28. Still, this eigenvector depends on r. Design yet another eigenvector that doesn't depend on r, and remains the same all the time. Hint: take the vertical vector $(0, 0, 1)^t$.
29. What is its eigenvalue? Hint: $2m(\|r\|^2 - z^2) = 2m(x^2 + y^2)$.
30. Could this be negative or zero? Hint: no! If it were zero, then $r^- = r$, which is impossible.
31. Conclude that our vertical ω remains an eigenvector of B all the time.
32. Conclude that the total angular momentum $B\omega$ remains vertical all the time.
33. Conclude once again that the total angular momentum is indeed conserved.

2.6.10 Rigid Body

1. The above is just a special case of a more general setting: a closed system, containing many particles, rotating around a constant eigenvector ω of the inertia matrix of the system. This way, the total angular momentum remains constant, as required.

2. How to use this in practice? Hint: the same discussion is valid not only for two but also for (infinitely) many particles that make a new rigid body. In this case, use an integral rather than a sum:

$$B \equiv \int \int \int A(r)\rho(r)dxdydz.$$

This defines the new B: the total inertia matrix. Indeed, at $r = (x, y, z)^t$, ρ is the local density, so the local matrix $A(r)$ is multiplied by the local mass $\rho dxdydz$.

3. Is B still symmetric?

4. Conclude that B still has three orthogonal eigenvectors: the new principal axes.

5. Is B still positive semidefinite? Hint: the proof is as in the two-body system.

6. If our rigid body rotates around a constant ω (an eigenvector of B that doesn't depend on B), what happens to the angular momentum? Hint: $B\omega$ is constant as well: the conserved angular momentum.

7. In the rigid body, is the system closed? Hint: yes, no force acts on the rigid body from the outside.

8. Is the angular velocity constant? Hint: yes, we assume here that ω is an eigenvector of B that doesn't depend on B. Therefore, ω may remain constant, even if B changes during the rotation.

9. Is the angular momentum conserved? Hint: yes, $B\omega$ remains constant as well.

10. What are the principal axes algebraically-geometrically? Hint: each principal axis is spanned by an eigenvector of B.

11. What are the principal axes physically? Hint: look at a principal axis spanned by an eigenvector of B that is independent of B. This principal axis marks the direction of constant angular velocity and momentum, around which the entire rigid body could rotate round and round forever.

12. Could B be singular (have a zero eigenvalue)? Hint: this could happen only in a very simple rigid body: a straight wire, in which all atoms lie on the same line. This way, B has a zero eigenvalue. The corresponding eigenvector points in the same direction as the wire itself. Around it, no rotation could take place, because the wire is too thin: it has no width at all.

13. In the wire, where should the origin be placed? Hint: best place it in the middle.

14. Does the wire have a more genuine principal axis? Hint: issue a vertical from the origin, in just any direction that is perpendicular to the wire. This could serve as a better principal axis, around which the entire wire could rotate round and round forever.

15. Consider yet another example of a rigid body: a thin disk. Where should the origin be placed? Hint: at its center.

16. What principal axis must the disk have? Hint: the vertical at its center. Around it, the disk could indeed rotate round and round forever.

17. What other principal axes could the disk have? Hint: every diameter could serve as a new principal axis, around which the entire disk could rotate round and round forever.

18. In Figure 2.7, even if ω is not perpendicular to r, show that the centrifugal force could be written simply as

$$-m\omega \times (\omega \times r) = mA(\omega)r.$$

2.6.11 The Percussion Point

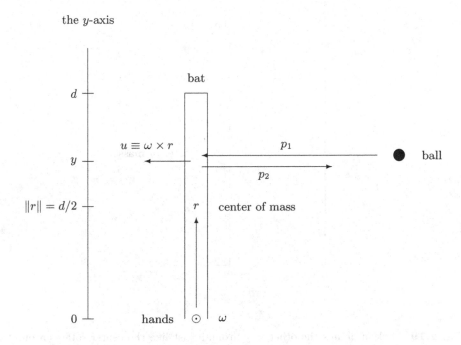

Fig. 2.18. A vertical baseball bat of height d. The ball meets it at height y. This way, the ball loses a linear momentum of $p_1 - p_2$. This is transferred fully to the center of mass of the bat (at height $\|r\| = d/2$), producing a new velocity of $u \equiv \omega \times r$.

1. Hold your baseball bat vertically in your hands, ready to hit a ball.
2. Assume that the bat has mass m and length d.
3. For simplicity, assume that the bat is uniform. This way, its center of mass is at the midpoint, at height $d/2$ (Figure 2.18). Furthermore, its density is m/d kilogram per meter. (It is easy to extend this to a more standard, nonuniform bat.)
4. Let r be the vertical vector leading from the bottom to the center of mass at the midpoint.
5. What is the norm of r? Hint: $\|r\| = d/2$.
6. What happens if you hit the ball at the midpoint r? Hint: this is no good: the hit would push the entire bat horizontally into your hands, and hurt you.
7. On the other hand, what happens if you hit the ball at the tip at $2r$? Hint: this is no good either: the hit would rotate the bat around its center of mass, pulling it away from your hands.

the r-axis

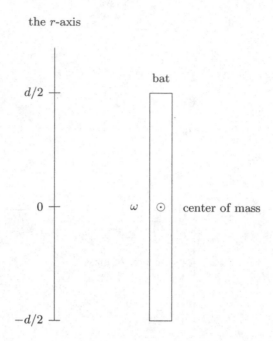

Fig. 2.19. Look at things the other way around: just like the center rotates around the bottom, the bottom also rotates around the center (at the same angular velocity w, issuing now from the center, not the bottom). From the perspective of the center, the bat rotates around it, and has a new moment of inertia.

8. Where is the best place to hit the ball? Hint: somewhere in between, where no angular momentum is wasted on rotating the bat needlessly around its center of mass. This is called the percussion point.
9. Where exactly is the percussion point? Hint: see below.
10. Let a ball come from the right, at a (horizontal) linear momentum p_1. Upon meeting the bat at height y, the ball returns rightwards, at a smaller linear momentum p_2. How much linear momentum did the ball lose? Hint: $p_1 - p_2$.
11. What happened to this linear momentum? Hint: thanks to conservation of linear momentum, it goes fully to the center of mass of the bat, to give it an initial linear momentum of

$$p_1 - p_2,$$

and an initial velocity of

$$\frac{p_1 - p_2}{m} = u = w \times r$$

leftwards (Figure 2.18).

12. What is w? Hint: this is the angular velocity at which the bat rotates counter-clockwise around its bottom. This way, w points from the page towards your eyes, as usual.

13. Conclude that w-r-u is a right-hand system.

14. Use this in the above formula, and take the norm of both sides. Hint:

$$\frac{1}{m}\|p_1 - p_2\| = \|u\| = \|w\| \cdot \|r\| = \|w\|\frac{d}{2}.$$

15. Multiply both sides by m:

$$m\|w\|\frac{d}{2} = \|p_1 - p_2\|.$$

This will be useful later.

16. Now, look at things the other way around. Just as the midpoint rotates around the bottom, the bottom also rotates counterclockwise around the midpoint, at the same angular velocity w (issuing now from the midpoint, as in Figure 2.19).

17. What is the moment of inertia around the midpoint? Hint:

$$\frac{m}{d}\int_{-d/2}^{d/2} \tilde{y}^2 d\tilde{y} = 2\frac{m}{d}\int_0^{d/2} \tilde{y}^2 d\tilde{y}$$

$$= 2\frac{m}{d}\cdot\frac{1}{3}\left(\frac{d}{2}\right)^3$$

$$= 2\frac{m}{d}\cdot\frac{d^3}{24}$$

$$= \frac{md^2}{12}.$$

18. Again, look at things from the perspective of the midpoint. How to write conservation of angular momentum? Hint: at hit time, the original angular momentum of the ball around the midpoint is used fully to produce a new angular momentum in the bat: the above moment of inertia times w:

$$\left(y - \frac{d}{2}\right)\frac{r}{\|r\|} \times (p_1 - p_2) = \frac{md^2}{12}w.$$

19. Take the norm of both sides:

$$\left(y - \frac{d}{2}\right)\|p_1 - p_2\| = \frac{md^2}{12}\|w\|.$$

20. Recall that

$$\|p_1 - p_2\| = m\|w\|\frac{d}{2}.$$

21. Plug this in:

$$\left(y - \frac{d}{2}\right)m\|w\|\frac{d}{2} = \frac{md^2}{12}\|w\|.$$

22. Divide both sides by $m\|w\|d/2$:

$$y - \frac{d}{2} = \frac{d}{6}.$$

23. Conclude that

$$y = \frac{d}{2} + \frac{d}{6} = \frac{2}{3}d.$$

24. This is the percussion point: the best point to hit the ball.
25. What is so good about this point? Hint: at this point, thanks to conservation of angular momentum, there is no extra angular momentum around the midpoint, but only the exact amount required to rotate the bat around its bottom. This way, your hands will feel no strain at all. After all, rotating around the bottom is fine: only an extra rotation around the midpoint is bad, because it could pull the bat away from your hands. Fortunately, at the above y, there is none.

2.6.12 Bohr's Atom and Its Stability

1. Look at two physical phenomena: gravity and electrostatics. What do they have in common? Hint: both make a force of $\|r\|^{-2}$, and a potential of $\|r\|^{-1}$ (times a constant, which we disregard).
2. Look at the solar system. The sun is at the origin: $(0,0,0)$. Around it, there is a planet at r, orbiting the sun.
3. What is the centrifugal force? Hint: it is proportional to $\|\omega\|^2\|r\|$. (Forget about constants.)
4. What is the direction of the centrifugal force? Hint: radial, outwards.
5. What keeps the planet in orbit? In other words, what prevents the planet from escaping to the outer space? Hint: the centripetal force, which balances the centrifugal force.
6. What is the direction of the centripetal force? Hint: radial, inwards.
7. What is its magnitude? Hint: the same as that of the centrifugal force.
8. What supplies the centripetal force? Hint: gravity.
9. What is the angular velocity? Hint: proportional to $\|r\|^{-3/2}$. This way, the centripetal force agrees with the gravitational force:

$$\|\omega\|^2\|r\| \sim \left(\|r\|^{-3/2}\right)^2 \|r\| = \|r\|^{-2}.$$

10. What is the angular momentum? Hint:

$$\|r \times p\| \sim \|A(r)\omega\| \sim \|r\|^2\|\omega\| \sim \|r\|^2\|r\|^{-3/2} = \|r\|^{1/2}.$$

11. Is it conserved? Hint: yes $-$ $\|r\|$ is constant in the orbit. Furthermore, $r \times p$ keeps pointing in the same direction as ω: perpendicular to the plane containing the orbit (Figure 2.8).
12. What is the kinetic energy? Hint: kinetic energy is half momentum times velocity. In the context of circular motion, on the other hand, kinetic energy is half angular momentum times angular velocity:

$$\frac{1}{2}\|r \times p\| \cdot \|\omega\| \sim \|r\|^{1/2}\|r\|^{-3/2} = \|r\|^{-1}.$$

13. Is it conserved? Hint: yes $-$ $\|r\|$ is constant in the orbit.
14. Where did this energy come from? Hint: from the gravitational potential, which is proportional to $\|r\|^{-1}$ as well. (See below.)

15. Why doesn't the Earth spiral or fall into the sun? Hint: because it has just enough kinetic energy to remain in orbit all the time.

16. Why is a Neptune-year much longer than an Earth-year? Hint: because the angular velocity decreases sharply with $\|r\|$:

$$\|\omega\| \sim \|r\|^{-3/2}.$$

17. Next, look at the atom. In Bohr's model, the nucleus is at the origin, and the electrons orbit it.

18. To have an idea about the atom, use the same model as in the solar system. Hint: forget about constants, as above.

19. Later on, in quantum mechanics, we'll see that angular momentum comes in discrete quanta: say, n quanta, where n is an integer number.

20. Assume that the electron is at r, orbiting the nucleus. What could the radius $\|r\|$ be? Hint: not every radius is allowed. Indeed, the angular momentum must be n (times a constant):

$$\|r \times p\| \sim \|r\|^{1/2} \sim n.$$

Thus, the radius must be n^2 (times a constant):

$$\|r\| \sim n^2.$$

21. What energy level is allowed? Hint: the energy level must be n^{-2} (times a constant):

$$-\|r\|^{-1} \sim -n^{-2}.$$

22. Where did the minus sign come from? Hint: the electron has a negative charge. This way, it is attracted to the positive charge of the proton in the nucleus. How to model this mathematically? Better insert a minus sign. This makes the potential increase monotonically away from the nucleus. This way, the minimal potential is near the nucleus. To have a minimal potential, the electron must therefore travel inwards. This is indeed attraction.

23. Where did the electron get its kinetic energy from? Hint: assume that the electron was initially at infinity, where the potential is zero. Then, it started to spiral and fall towards the nucleus, gaining negative potential and positive kinetic energy, until reaching its current position: r. During its journey, it lost potential energy of

$$0 - \left(-\|r\|^{-1}\right) = \|r\|^{-1}.$$

Still, this wasn't lost for nothing: it was converted completely from potential to kinetic energy.

24. According to Maxwell's theory, while orbiting the nucleus, the electron should actually radiate an electromagnetic wave (light ray), and lose its entire energy very quickly. Why doesn't this happen? Hint: in terms of quantum mechanics, its position in its orbit is only nondeterministic: a periodic wave function, with an integer wave number. This is a standing wave that radiates nothing. This is how Debroglie proved stability in Bohr's model.

25. By now, the electron is at r, at radius $\|r\|$ from the nucleus. Could it get yet closer to the nucleus? In other words, could it "jump" inwards, from its allowed radius to an inner radius? Hint: only if the inner radius is allowed as well.

26. In the inner radius, how much potential energy would it have? Hint: less than before. Indeed, $-\|r\|^{-1}$ increases monotonically with $\|r\|$.

27. What happened to the potential energy that was lost? Hint: it transforms into light: the electron emits a photon.

28. On the other hand, could the electron get away from the nucleus, and "jump" outwards, to an outer radius? Hint: only if the outer radius is allowed as well.

29. In the outer radius, how much potential energy would it have? Hint: more than before. Indeed, $-\|r\|^{-1}$ increases monotonically with $\|r\|$.

30. What could possibly supply the extra energy required for this? Hint: a photon (light ray) comes from the sun, and hits the electron hard enough.

31. Could the photon hit the electron ever so hard, and knock it off the atom, sending it all the way to infinity? Hint: only if the light ray is of high frequency, with energy as high as n^{-2} (the old energy level). Only this could give the electron sufficient energy to escape to $\|r\| = \infty$, and have zero potential.

32. This is the photo-electric effect, discovered by Einstein. We'll come back to this later.

3

Stability
in Geometrical Optics:
The Infinity Point

To model a physical phenomenon, we often use an infinite geometrical structure: the real axis, the two-dimensional plane, or the three-dimensional space. Still, it makes sense to add one more "point:" the infinity point. This closes the original real axis, tying its "endpoints" with one another, and making a new compact "circle." Likewise, this also closes the original two-dimensional plane into a new sphere, and the original three-dimensional spaces into a compact hypersphere. In this chapter, we use this in geometrical optics.

3.1 The Infinity Point

3.1.1 The Infinity Paradox

The adjective "infinite" contains two words: not finite. Thus, it is actually a negative adjective: it assumes that we already know what finite is, and tells us that it is not.

This is why the property of being infinite can't be verified by any practical experiment. For example, assume that we're given a set of items. How to find out whether it is finite or not? Let's try a naive algorithm: count the items, one by one. If you finish, then the set is finite. Otherwise, it is not.

Is this any good? Well, for a finite set, this indeed works: you will finish, and find out that the set is indeed finite. But what about an infinite set? In this case, you never finish, and never know the answer. This is the infinity paradox.

Thus, to comprehend the property of being infinite, we must use our own imagination or faith. This is why one of the elementary axioms in set theory is the existence of an infinite set.

3.1.2 Mathematical Induction

To help comprehend infinity, one could also accept yet another useful axiom: mathematical induction. This could help define countless natural numbers: first, define the natural number 1. Then, assume that some natural number has already been defined, and define its successor as well.

Could this process ever stop? No! Indeed, if we ever got tired and stopped, then we could rest for a while, and then continue to define the next successor as well.

Being unstoppable is again a negative property: we know what a stoppable process is, so we can also believe or accept that some process is not. We know what an end is, so we can trust that some process is endless. This is why the above proof is by contradiction: if the process had ended, then a contradiction would emerge.

3.1.3 Natural and Real Numbers

The set of natural numbers, although infinite, is still discrete. This is the smallest kind of infinity: $\aleph_0$. The real axis $\mathbb{R}$, on the other hand, is "more" infinite (Figure 1.1). This is denoted by $\aleph > \aleph_0$. In fact, even the unit interval is of type $\aleph$: it contains a continuum of real numbers.

3.1.4 The Infinity Point: An Abstract Object

Since the real axis has no left or right end, it makes sense to introduce the extra object ∞ to stand for the imaginary infinity point. The infinity point is not a real number: it is just an abstract object, assumed to be as large as any real number.

Clearly, the infinity point mustn't lie on the real axis, or there would be yet larger real numbers on its right. Thus, it must "lie" outside of the real axis. In fact, it is an auxiliary point that "exists" in our minds only.

3.1.5 The Extended Real Axis

Let's extend the real axis by one auxiliary object: the infinity point. This is no point or number in the usual sense, but just an artificial abstract object, denoted by '∞'.

In set theory, the real axis (the set of real numbers) is denoted by

$$\mathbb{R} \equiv \{x \mid x \text{ is a real number}\}.$$

Now, define a small set, containing one item only:

$$\{\infty\}.$$

Indeed, this set contains one element only: the infinity point ∞. This is denoted by

$$\infty \in \{\infty\}$$

(meaning that ∞ belongs in $\{\infty\}$). We can now extend the original real axis by just one point: the infinity point. The result is

$$\mathbb{R} \cup \{\infty\}.$$

(Here, '$\cup$' stands for the union of two sets.) In some sense, this is the same as a circle.

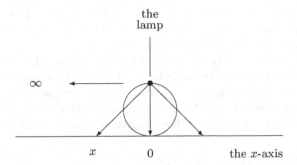

Fig. 3.1. The lamp mapping: each point on the circle projects to its shadow on the x-axis. The lamp itself maps to ∞.

3.2 The Lamp Mapping

3.2.1 The Lamp Mapping

By now, we've extended the real axis by one more point: the infinity point. What is the result? In a sense, it is the same as a circle.

Indeed, look at the Cartesian x-y plane (Figure 3.1). In it, consider the unit circle of radius 1, centered at the origin $(0,0)$. Now, shift the circle upwards, so its center is now at $(0,1)$ rather than $(0,0)$. This way, the circle sits now on of the x-axis, with its bottom at $(0,0)$, and its top at $(0,2)$.

Now, at $(0,2)$, place a tiny lamp. Each light ray issuing from the lamp passes through one point from the circle, and maps it to its shadow on the x-axis. Together, all light rays map the entire circle onto its shadow: the entire x-axis, or

$$\{(x,0) \mid -\infty < x < \infty\}.$$

This is the lamp mapping: each point on the circle projects to its shadow on the x-axis below.

There is just one point that maps to itself: the origin $(0,0)$. After all, it coincides with its own shadow.

3.2.2 The Lamp Maps to Infinity

Still, one point has no shadow at all: the lamp itself, at $(0,2)$. Where could it map? Well, it must map to ∞. This is done by the horizontal ray, tangent to the circle at $(0,2)$.

Why does this definition make sense? Well, in Figure 3.1, look at the lower-left oblique ray that makes angle $45°$ with the x-axis. This ray hits the x-axis at some $x < 0$. This is where the shadow is.

Now, let x move leftwards, slowly and smoothly:

$$x \to -\infty.$$

In this limit process, the ray changes as well: the angle it makes with the x-axis gets more and more acute, decreasing gradually, and approaching $0°$. At this limit, the ray becomes horizontal: it maps the lamp to ∞.

This completes the definition of the lamp mapping. The entire circle has been mapped onto the extended real axis by a one-to-one mapping: each individual point on the circle has been mapped to a distinct point on the extended real axis.

Furthermore, the extended real axis is completely covered: the mapping is not only *into* but also *onto* it. Thus, the original mapping is also invertible.

3.2.3 Topological Homeomorphism

In the above limit process, x travels leftwards:

$$x \to -\infty$$

(Figure 3.1). Instead, it could also travel rightwards:

$$x \to \infty.$$

What is the difference? Well, this time, the ray points down-right rather than down-left. Still, in the limit, it becomes horizontal again, tangent to the circle at $(0,2)$, pointing rightwards rather than leftwards.

As before, one can still assume that this horizontal ray points to the infinity point ∞. After all, in the extended real axis, $+\infty$ is the same as $-\infty$: the unique infinity point ∞.

Thus, in the extended real axis, $+\infty$ and $-\infty$ coincide with each other, to make the same infinity point:

$$+\infty = -\infty = \infty.$$

What is the advantage of this approach? Well, it makes the lamp mapping smooth and continuous. To see this, consider a particle that lies on the circle in Figure 3.1. The lamp maps it to its own shadow. Where is this? At x, on the x-axis below.

Initially, the particle lies at the origin. Now, let it move slowly clockwise on the circle. At the same time, its shadow x moves leftwards on the x-axis.

Once the particle approaches the lamp from the left, its shadow approaches $-\infty$ on the x-axis, from the right. Furthermore, once the particle hits the lamp, its shadow hits the infinity point ∞ in the extended real axis.

Moreover, once the particle passes the lamp on its way clockwise, its shadow proceeds smoothly, "through" the infinity point, to the far right part of the x-axis. Finally, once the particle travels along the right part of the circle back to the origin, its shadow travels leftwards on the x-axis, back to the origin.

Thus, the lamp mapping makes a complete topological homeomorphism between the circle and the extended real axis. Indeed, every continuous move of the particle on the circle is mirrored by a continuous move of its shadow on the extended real axis, and vice versa. Topologically, the circle and the extended real axis are just the same. They are homeomorphic, or indistinguishable from each other: both have the same continuity properties. This will be useful in geometrical optics.

3.2.4 Reaching The Infinity

Reaching the infinity? Sounds fantastic. After all, it is too far away... Still, we just did! Well, not directly, but only indirectly.

Indeed, the lamp itself mirrors the infinity, and projects onto it. This makes the infinity much more friendly and close to mind.

Reaching the infinity requires infinite time, which we don't have. Still, in our imagination, we can make the impossible. Instead of stepping into infinity, we use a trick: just step into the lamp, and pass it smoothly and continuously. At the same time, your shadow goes through the infinity point, as required.

Unfortunately, this is too geometrical: the precise location of the shadow is not easy to calculate. Is there a more algebraic version? Fortunately, there is: to mirror the infinity, use not a lamp, but the origin itself.

3.3 The Hyperbolic Mapping

3.3.1 The Hyperbolic Mapping on an Interval

On the extended real axis, we can now define another useful mapping: the hyperbolic mapping. For this purpose, consider the closed interval

$$[-1, 1] \equiv \{x \mid |x| \leq 1\}$$

(including the endpoints). On this interval, define the hyperbolic mapping:

$$x \to \frac{1}{x}$$

(Figure 3.2).

Let's start from $x = 1$. At this point, $1/x = 1$ as well. Now, let x decrease, gradually and continuously. At the same time, $1/x$ increases continuously. Once x approaches 0 from the right, $1/x$ approaches $+\infty$:

$$\frac{1}{x} \to_{x \to 0^+} +\infty.$$

Fortunately, on the extended real axis, the infinity point is unique:

$$-\infty = +\infty = \infty.$$

Therefore, once x hits 0, $1/x$ hits the infinity point ∞ on the extended real axis.

3.3.2 Topological Homeomorphism

Now, let x decrease yet more, pass 0, into the negative part of the original interval. At the same time, $1/x$ passes "through" the infinity point, and continues smoothly into the far left part of the real axis.

Finally, let x decrease yet more, and reach -1 from the right. At the same time, $1/x$ increases smoothly, and reaches -1 from the left.

Thus, the original interval $[-1, 1]$ is homeomorphic to the complementary set

$$\{y \mid |y| \geq 1\} \cup \{\infty\}.$$

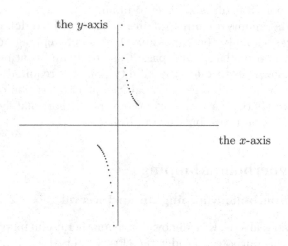

Fig. 3.2. The hyperbolic mapping maps the interval $[-1,1]$ onto the complementary set $\{y \mid |y| \geq 1\} \cup \{\infty\}$.

3.3.3 The Inverse Mapping

Indeed, the above mapping is invertible: the inverse mapping is just the same:

$$x \to \frac{1}{x}.$$

In fact, the inverse mapping maps the set

$$\{y \mid |y| \geq 1\} \cup \{\infty\}$$

back onto the original interval $[-1,1]$ (Figure 3.3).

3.3.4 The Extended Hyperbolic Mapping

In summary, the hyperbolic mapping $x \to 1/x$ could actually be defined not only in the original interval $[-1,1]$ but also in the extended real axis. This way, it maps the extended real axis continuously onto itself (Figure 3.4). This is useful in geometrical optics.

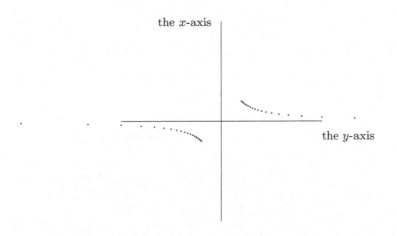

the x-axis

the y-axis

Fig. 3.3. The inverse hyperbolic mapping maps the set $\{y \mid |y| \geq 1\} \cup \{\infty\}$ back onto the interval $[-1, 1]$.

3.4 Optical Focus

3.4.1 Plano Convex Lens

The hyperbolic mapping is particularly useful in geometrical optics [13, 21]. Consider, for instance, the plano convex lens in Figure 3.5. Why is it called so? Because its right face is flat, straight, and vertical. Its left face, on the other hand, is round and convex. This way, the left face could be viewed as part of a circle of radius $r > 0$, centered at the bullet in the middle of Figure 3.5.

Assume now that two parallel horizontal light rays approach the lens from the left. Upon entering from the left, they break, and change direction towards the focus (marked by $f > 0$ in Figure 3.5).

This is just a geometrical model. Still, it models the true physical phenomenon quite well. To test this, place the lens with its convex face against the sun. At f, on the other side, place a piece of paper. This way, the sun rays will break, focus at f, and eventually burn the paper completely!

The focus is also called the focal point. The positive number f is also called the focal length: the distance between the lens and the focus. In a lens of glass, for example, f satisfies

$$\frac{1}{f} = \frac{1}{2r},$$

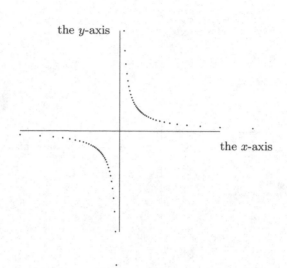

the y-axis

the x-axis

Fig. 3.4. The extended hyperbolic mapping maps the extended real axis onto itself.

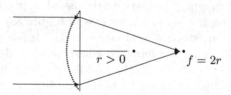

$r > 0$ $f = 2r$

Fig. 3.5. A plano convex lens, made of glass. Upon entering from the left, the parallel horizontal rays break, and meet each other at the focus $f = 2r$.

or

$$f = 2r.$$

Let's use this formula in a special case.

3.4.2 Flat Window of Glass

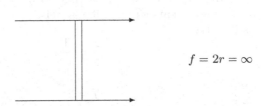

$$f = 2r = \infty$$

Fig. 3.6. A transparent window of glass. The parallel horizontal rays that enter it from the left never break: they remain horizontal, and "meet" each other at $f = \infty$.

Now, let the radius r increase, gradually and continuously. At the same time, the focus travels rightwards smoothly as well, approaching the infinity point ∞:

$$f = 2r \to_{r \to \infty} \infty.$$

What happens when r reaches the infinity point ∞? Well, at the same time, f reaches ∞ as well. This means that the focus is at infinity.

What happens physically here? Both faces are now parallel: flat, straight, and vertical. Thus, the original horizontal rays never break any more. On the contrary: they keep traveling horizontally rightwards: from $-\infty$, through the lens, to the focus at ∞ (Figure 3.6). This is just a transparent window of glass.

3.4.3 Virtual Focus

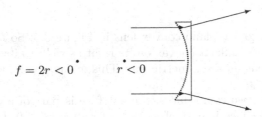

$$f = 2r < 0 \qquad r < 0$$

Fig. 3.7. A plano concave lens. Upon entering from the left, the parallel horizontal rays break away from each other, and never meet any more. To a viewer who looks from the right, they seem to come from the virtual focus at $f = 2r < 0$.

Let the above process proceed yet more: r (and $f = 2r$) passes the infinity point, smoothly into the far left part of the real axis. What does this mean physically?

Well, the left face gets now slightly concave: it is now part of a circle of radius $-r$, centered at the bullet marked by $r < 0$, in the middle of Figure 3.7. Why is r negative? Because the circle is now on the left, not on the right.

In this case, what happens to the horizontal rays that approach from the left? They break away from each other, and never meet any more. Still, to a viewer who looks from the right, they seem to come from one point: the virtual focus at $f = 2r < 0$ (Figure 3.7, on the left). For this reason, every object placed on the left may seem to the viewer smaller than it really is.

Thus, f still satisfies

$$\frac{1}{f} = \frac{1}{2r},$$

or

$$f = 2r,$$

only this time both f and r are negative, not positive. After all, they are now to the left of the lens, not to the right!

In summary, the focus completed its journey: from $f > 0$, rightwards, through the infinity point, emerging at $f < 0$ on the far left. On the extended real axis, this is indeed a smooth journey. This is why optics changes smoothly as well.

3.4.4 Convex Lens

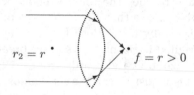

Fig. 3.8. The convex lens. Upon entering from the left, the parallel horizontal rays break. Upon exiting from the right, they break once again, to meet each other at the focus $f = r = r_2 > 0$.

Let us now go back to the plano convex lens in Figure 3.5. So far, it is convex on its left only. Now, let's make it convex on its right as well: to its right, stick yet another plano convex lens, in a symmetric way. This makes a complete convex lens (Figure 3.8).

The convex lens has two round faces: the left face is part of a circle of radius $r > 0$, whereas the right face is part of a circle of radius $r_2 > 0$. Unlike r and f, however, r_2 is measured the other way around: from the lens and leftwards, not rightwards. This is why r_2 is positive as well.

What happens to parallel horizontal rays that approach the lens from the left? They break twice. Upon entering from the left, they break as in Figure 3.5. Upon exiting from the right, they break yet more, to meet each other at a closer focus. Indeed, f satisfies now

$$\frac{1}{f} = \frac{1}{2}\left(\frac{1}{r} + \frac{1}{r_2}\right).$$

As a matter of fact, this formula applies to a plano convex lens as well: just pick $r_2 = \infty$. Here, however, we pick $r_2 < \infty$, so $1/f$ gets larger, so f gets smaller. In Figure 3.8, for example, we pick $r_2 = r$. This makes a symmetric convex lens, with $f = r_2 = r$.

3.4.5 Nonflat Window of Glass

Thus, the convex lens could be obtained from a plano convex lens by reducing r_2 from ∞ to $r_2 < \infty$. Now, let's go the other way around: let r_2 increase again, gradually and continuously. To see what's going on, we have now an algebraic tool: the above formula.

What happens physically? Well, the right face gets less and less convex. From the above formula, this makes f increase as well.

Once r_2 reaches the infinity point ∞, the right face gets completely flat and vertical. From the above formula, f gets as large as $2r$. This is no surprise: we've seen this already (Figure 3.5).

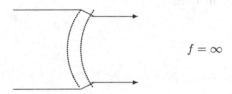

$f = \infty$

Fig. 3.9. A nonflat window of glass. The parallel rays remain horizontal, so they "meet" each other at the focus at the infinity point ∞.

But this is not the end of it. After all, in the extended real axis,

$$-\infty = +\infty = \infty.$$

Thus, r_2 may proceed smoothly: through the infinity point, to the far left (negative) part of the real axis.

What happens physically? The right face gets now slightly concave, not convex. Thanks to the above formula, this makes f increase yet more.

Next, let r_2 increase yet more, traveling from $-\infty$ rightwards. At the same time, f increases yet more. Once r_2 reaches $-r$ from the left, we have

$$\frac{1}{f} = \frac{1}{2}\left(\frac{1}{r} + \frac{1}{r_2}\right) = \frac{1}{2}\left(\frac{1}{r} - \frac{1}{r}\right) = 0,$$

so

$$f = \pm\infty.$$

This is indeed a transparent (but not flat) window of glass. Indeed, horizontal rays remain horizontal, and "meet" each other at infinity (Figure 3.9).

In this case, the focus lies at the infinity point ∞, which can be interpreted as either $+\infty$ (real focus) or $-\infty$ (virtual focus). Indeed, to a viewer who looks from the right, the rays seem to come from $-\infty$.

3.4.6 Virtual Focus: Beyond The Infinity

Finally, let r_2 increase a little more, so $-r < r_2 < 0$. Thanks to the above formula, f must now be negative: it passed through the infinity point, and emerged from the far left part of the real axis.

What happens physically? The right face gets so concave that it wins: it is stronger than the left face, and overrides its effect. This way, upon entering from the left, the parallel horizontal rays still break towards each other. Still, upon exiting from the right, they break away even more sharply. The total effect is, thus, a divergence: to a viewer looking from the right, the rays seem to come from a virtual focus, to the left of the lens.

In summary, like r_2, f completed its journey. Indeed, it traveled rightwards smoothly: from r and $2r$, through the infinity point, to emerge from the far left part of the real axis.

3.4.7 Symmetric Lens

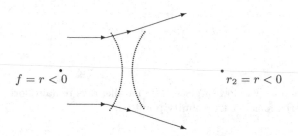

Fig. 3.10. A symmetric concave lens. Upon entering from the left, the parallel horizontal rays break away from each other. Upon exiting from the right, they break away even more, and never meet one another any more. To a viewer who looks from the right, they seem to come from the virtual focus at $f = r < 0$.

The above process could also be carried out from the other side. Instead of increasing r_2, increase r. This has the same effect on f.

What happens physically? Instead of the right face, the left face gets now less and less convex. This has the same effect as before: the focus travels rightwards again, pass the infinity point, to become virtual at the far left part of the real axis, as before.

Now, why not carry out both processes together, simultaneously at the same time? This makes a new symmetric process, using a symmetric lens all the time:

start from a symmetric convex lens, with $r_2 = r$ (Figure 3.8). Then, increase both $r_2 = r$ at the same time. This way, $f = r_2 = r$ increases as well.

3.4.8 From Convex to Flat to Concave

What happens physically? The symmetric lens gets less and less convex. At the same time, its focus travels rightwards as well. At the infinity point $f = r_2 = r = \infty$, the lens gets completely flat, like a window of glass (Figure 3.6).

This is not the end of it. Carry on: make the symmetric lens concave, with two concave faces (Figure 3.10). This way, it has now a new virtual focus, at the left part of the real axis.

3.4.9 Convex and Concave: New Definition

In summary, to tell whether a (symmetric or nonsymmetric) lens is convex or concave, just look at the sign of f. If $f > 0$, then the lens is convex. (This actually defines a convex lens.) If, on the other hand, $f < 0$, then the lens is concave. (This actually defines a concave lens.) Finally, if $f = \infty$, then the lens is transparent: a window of glass.

3.5 Magnification

3.5.1 A Convex Lens

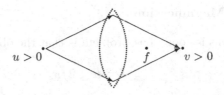

Fig. 3.11. No magnification: the object is at $u = 2f$, so its image at $v = 2f = u$ has the same size. Indeed, thanks to symmetry, the image is the same as the original object.

Let's use our convex lens in practice. For this purpose, place an object on the left, at distance u from the lens. Here, u is positive: it is measured from the lens leftwards, not rightwards (Figure 3.11).

Consider now a few light rays, issuing from the object, and passing through the lens, to produce a new image on the right, at distance v. Unlike u, v is measured from the lens rightwards. This way, $v > 0$ as well.

Fortunately, v could be obtained algebraically, from the following formula:

$$\frac{1}{u} + \frac{1}{v} = \frac{1}{f}.$$

This will be useful below. Moreover, the magnification is

$$\frac{|v|}{u}.$$

This means that a viewer whose eye is at v (looking leftwards towards the lens) may think that the object is $|v|/u$ times as big as it really is.

3.5.2 No Magnification

Let's "play" with u. After all, this is our degree of freedom, which we can control and change. To calculate the magnification, we can now use the above formula.
    Assume first that

$$u = 2f$$

(Figure 3.11). From the above formula, we then have

$$\frac{1}{v} = \frac{1}{f} - \frac{1}{u} = \frac{2}{u} - \frac{1}{u} = \frac{1}{u},$$

so we have complete symmetry:

$$v = u.$$

It is therefore no surprise that the magnification is

$$\frac{|v|}{u} = 1.$$

So, there is no gain: the image has the same size as the original object.

3.5.3 Moderate Magnification

Don't worry: the lens is good. The problem is that the object is too far. Indeed, let u decreases a little, so

$$f < u < 2f.$$

This way, the object gets a little closer to the lens. Could this help? Well, since $1/u$ increases,

$$\frac{1}{v} = \frac{1}{f} - \frac{1}{u}$$

decreases, so v must increase. As a result, the magnification is now larger than before:

$$\frac{|v|}{u} > 1.$$

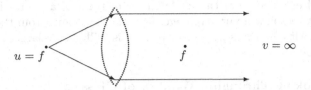

Fig. 3.12. Maximal magnification: the object is at $u = f$, so its image at $v = \infty$ has an "infinite" size.

3.5.4 Maximal Magnification

Now, let u decrease yet more, approaching f^+ from above. At the same time, v will increase, improving magnification yet more. Finally, once u hits f, we have

$$\frac{1}{v} = \frac{1}{f} - \frac{1}{u} = 0,$$

so

$$v = \infty.$$

This is maximal magnification:

$$\frac{|v|}{u} = \infty.$$

Indeed, the infinitesimal object at $u = f$ "grows" into a sizable image at $v = \infty$ (Figure 3.12).

Of course, this is just a theoretical model. A real object could never be so small, and fit in a single point. Still, the above geometrical model teaches us two plausible things:

- The magnification is pretty good, as evident from experiment.
- Physics is stable: the magnification remains good even "beyond" infinity.

Let's see stability geometrically. Later on, we'll also discuss it algebraically, in the context of chaos theory.

3.6 Stability

3.6.1 Virtual Image

To prove stability, let the object get even closer to the lens:

$$\frac{f}{2} < u < f.$$

This way,

$$\frac{1}{v} = \frac{1}{f} - \frac{1}{u} < 0.$$

What happened to v? It passed through the infinity point, smoothly and continuously, and emerged from the far left part of the real axis. Looking from the right, the viewer sees now a virtual image: as if the rays came from the far left part of the real axis. Although v is now negative, $|v|$ is still pretty large, so the magnification is still good.

3.6.2 Looking Through a Window of Glass

So far, we've discussed a convex lens. What about a transparent window of glass (Figure 3.6 or 3.9)? In this case,

$$f = \infty.$$

After all, this is what transparency means: horizontal rays remain horizontal, and "meet" each other at infinity. Thus,

$$\frac{1}{v} = \frac{1}{f} - \frac{1}{u} = -\frac{1}{u},$$

so

$$v = -u.$$

This is good: the viewer views the rays as coming from the object itself, as they really do. There is no magnification at all, as expected.

3.6.3 Looking Through a Concave Lens

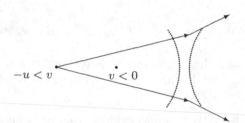

Fig. 3.13. A concave lens. To a viewer looking from the right, the virtual image at $v < 0$ seems smaller than the original object.

Finally, consider a concave lens, with a virtual focus: $f < 0$ (Figure 3.13). (After all, in Section 3.4.9, this is a legitimate definition.) In this case,

$$\frac{1}{v} = \frac{1}{f} - \frac{1}{u} < -\frac{1}{u},$$

so v is negative, but not as negative as $-u$:

$$-u < v < 0.$$

For this reason,

$$\frac{|v|}{u} < 1.$$

Thus, a viewer looking from the right will see a virtual image smaller than the original object (Figure 3.13). This is indeed as evident from experiment.

3.6.4 How to Reach The Infinity?

Why is the hyperbolic mapping so useful in geometrical optics? Because it lets us "reach" infinity, and "tie" the real axis at infinity.

Is this possible? How to "reach" infinity, or "step" through it? Isn't it too far? Fortunately, math lets us "do" this, using the power of imagination.

Indeed, in the hyperbolic mapping, infinity is mirrored at zero. For example, upon approaching zero from above, we have

$$\frac{1}{x} \to_{x \to 0^+} +\infty.$$

Likewise, upon approaching zero from below, we have

$$\frac{1}{x} \to_{x \to 0^-} -\infty.$$

To preserve continuity (and stability), both limits must be the same. This is why, in the extended real axis, we have just one infinity point:

$$-\infty = \lim_{x \to 0^-} \frac{1}{x} = \lim_{x \to 0^+} \frac{1}{x} = +\infty.$$

So, to "tie" the real axis at infinity, there is no need to actually reach infinity. Instead, just let zero mirror it, using the hyperbolic mapping.

As a matter of fact, we've already used this trick: in the lamp mapping, the lamp mirrors the infinity (Section 3.2.4 and Figure 3.1). To "tie" the real axis at infinity, just tie the circle well at the lamp, and then project the lamp to infinity, either rightwards or leftwards:

$$-\infty = +\infty.$$

3.7 The Cotangent Function

3.7.1 From an Interval to a Circle

In the lamp mapping, the lamp is placed at the top of the circle. This maps the entire circle: each point projects to its own shadow. Still, this is not the only way. Why not place the lamp at the center, and project the lower semicircle only? This makes a familiar function: the cotangent function. Still, before doing this, we must first "tie" the semicircle, and produce a complete circle.

For this purpose, consider the closed interval

$$[0, 2\pi] \equiv \{x \mid 0 \le x \le 2\pi\}$$

(including the endpoints). Now, identify both endpoints with one another, and consider them as one and the same:

$$0 = 2\pi.$$

In other words, map this interval to a circle:

$$x \to \eta,$$

where η is an angle in the circle. Since $\eta = 0$ is the same angle as $\eta = 2\pi$, we can also identify $x = 0$ with $x = 2\pi$, and treat them as one and the same.

3.7.2 The Composite Cotangent Function

Next, map this interval onto the shorter interval

$$[0, \pi] \equiv \{x \mid 0 \leq x \leq \pi\}.$$

This could be done linearly:

$$x \to \frac{x}{2}.$$

On top of this, apply the cotangent function:

$$\cotan(x/2) \equiv \frac{\cos(x/2)}{\sin(x/2)}.$$

This composite function is applied to the original interval $[0, 2\pi]$ (or the circle), and maps it onto the extended real axis. In particular, $0 = 2\pi$ is mapped to $-\infty = \infty$, as required.

3.7.3 Project Onto The Cotangent Line

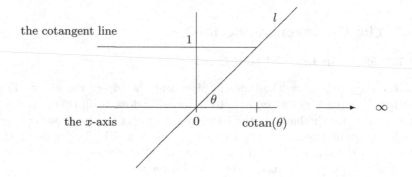

Fig. 3.14. To define the cotangent function geometrically, just project onto the horizontal cotangent line. For this purpose, draw the oblique line l that makes angle θ with the positive part of the x-axis. Then, look at the point where l meets the cotangent line. Look at its x-coordinate: this is $\cotan(\theta)$. For $\theta = 0$, on the other hand, l "meets" the cotangent line at ∞.

What is the geometrical meaning of this? To see this, look at the two-dimensional x-y plane. In it, draw a horizontal line, parallel to the x-axis:

$$\{(x, 1) \mid \quad -\infty < x < \infty\}.$$

This is the cotangent line.

How to define the cotangent function geometrically? To do this, pick some angle $0 < \theta < \pi$. Now, in the x-y plane, draw the oblique line l that passes through the origin $(0, 0)$, and makes angle θ with the positive part of the x-axis (Figure 3.14). This way, l could be written in terms of a new parameter α:

$$l \equiv \{\alpha(\cos(\theta), \sin(\theta)) \mid \quad -\infty < \alpha < \infty\}.$$

We can now redefine the cotangent function geometrically: just project on the cotangent line. In other words, find the unique point where l meets the cotangent line. For this purpose, pick α cleverly:

$$\alpha \equiv \frac{1}{\sin(\theta)}.$$

This way, the y-coordinate is 1, so we are indeed on the cotangent line:

$$\alpha(\cos(\theta), \sin(\theta)) = \frac{1}{\sin(\theta)}(\cos(\theta), \sin(\theta)) = (\cotan(\theta), 1).$$

At this particular point, l indeed meets the cotangent line. After all, this point belongs to both l and the cotangent line. Now, look at its x-coordinate. This is what we wanted: $\cotan(\theta)$. Finally, to have our composite function, substitute $x/2$ for θ:

$$\cotan(x/2).$$

3.7.4 Project to Infinity

All this is very nice, but there is still one problem: what about

$$x = 0 = 2\pi,$$

or

$$\theta = 0 \text{ or } \pi?$$

In this case, l coincides with the x-axis:

$$l \equiv \{\alpha(1, 0) \mid \quad -\infty < \alpha < \infty\} = \{\alpha(-1, 0) \mid \quad -\infty < \alpha < \infty\}.$$

In this case, how to define the cotangent? Easy: let l approach the x axis. This could be done in two ways: θ could approach either 0 or π. Fortunately, this is just the same:

$$\cotan(0) \equiv \lim_{\theta \to 0^+} \cotan(\theta) = \infty = -\infty = \lim_{\theta \to \pi^-} \cotan(\theta) = \cotan(\pi).$$

After all, in the extended real axis, there is only one infinity point:

$$-\infty = \infty.$$

(This is useful in projective geometry: see Chapter 6 in [45].) In summary, the special angle $x = 0 = 2\pi$ indeed mirrors the infinity well, as required.

3.8 The Extended Plane

3.8.1 The Extended Cartesian Plane

So far, we worked in the one-dimensional real axis. To it, we added one more point: the infinity point. Next, let's move on to the two-dimensional Cartesian plane:

$$\mathbb{R}^2 \equiv \{(x, y) \mid \quad -\infty < x, y < \infty\}.$$

To it, add one more point: the infinity point. Together, we have the extended Cartesian plane:

$$\mathbb{R}^2 \cup \{\infty\}.$$

The infinity point "ties" the plane, and produces a complete sphere. To see this, we need to extend the original lamp mapping.

3.8.2 The Lamp Mapping in The Sphere

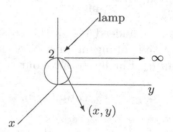

Fig. 3.15. The lamp mapping in the sphere: place the lamp at the top of the sphere, at $(0, 0, 2)$. Each point on the sphere projects to its own shadow on the x-y plane. The lamp itself maps to ∞.

Consider the three-dimensional x-y-z Cartesian space. In it, consider the unit sphere of radius 1, centered at the origin $(0, 0, 0)$. Now, shift it upwards, so that its center is now at $(0, 0, 1)$ rather than $(0, 0, 0)$. This way, it sits on the x-y plane, with its bottom at $(0, 0, 0)$, and its top at $(0, 0, 2)$ (Figure 3.15).

Now, at the top of the sphere, at $(0, 0, 2)$, place a tiny lamp. Each light ray issuing from the lamp passes through one point from the sphere, and maps it to its shadow on the x-y plane. Together, all light rays map the entire sphere onto its shadow: the entire x-y plane, or

$$\{(x, y, 0) \mid \quad -\infty < x, y < \infty\}.$$

For example, the origin $(0, 0, 0)$ maps to itself. After all, it coincides with its own shadow. Still, what about the lamp itself? Where should it map?

To find out, consider a particle, placed at the origin $(0, 0, 0)$. Let it travel along some longitude, slowly and continuously, towards the lamp at $(0, 0, 2)$. At the same time, its shadow travels as well: from the origin, straight towards infinity. To preserve continuity, the lamp itself must therefore map to the unique infinity point: ∞.

This way, the entire sphere indeed maps onto the extended Cartesian plane, as required. Furthermore, this is a one-to-one mapping: two distinct points on the sphere map to two distinct points in the extended Cartesian plane.

Moreover, the extended Cartesian plane is completely covered, so the mapping is not only *into* but also *onto* it. As a result, the original mapping is also invertible. As such, it indeed makes a topological homeomorphism. This is useful in the complex plane.

3.9 The Complex Plane

3.9.1 The Imaginary Axis

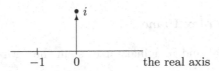

Fig. 3.16. The imaginary number i. The arrow leading from the origin to i makes a right angle of 90° with the real axis. In $i^2 = -1$, on the other hand, this angle doubles, to make a flat angle of 180° with the positive part of the real axis.

The negative number -1 has no square root: there is no real number whose square is -1. Fortunately, we already know how to introduce a new auxiliary (not real) number. Let's use this trick once again, and introduce a new imaginary number (denoted by i), satisfying

$$i^2 = -1,$$

or

$$i \equiv \sqrt{-1}.$$

Because it lies outside the real axis, i may now span a new vertical axis, perpendicular to the original real axis (Figure 3.16).

For this purpose, place i at $(0, 1)$, above the origin. This way, i spans the entire imaginary axis:

$$\{bi = (0, b) \mid -\infty < b < \infty\}.$$

Here, b could be any real number. In the above, the algebraic multiple bi also takes the geometrical form $(0, b)$: a new point on the vertical imaginary axis. In particular, if $b = 1$, then we obtain the original imaginary number one again: $i = (0, 1)$.

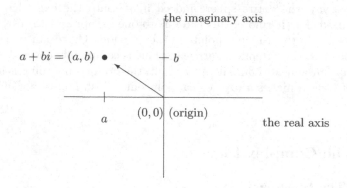

Fig. 3.17. The complex plane $\mathbb{C}$. The imaginary number $i \equiv \sqrt{-1}$ is at $(0,1)$. A complex number $a + bi$ is at (a, b).

3.9.2 The Complex Plane

Together, the real and imaginary axes span the entire complex plane, denoted by $\mathbb{C}$:

$$\mathbb{C} \equiv \{a + bi \equiv (a, b) \mid \quad -\infty < a, b < \infty\}.$$

Here, a and b are any real numbers. The new complex number $a + bi$ occupies the geometrical point (a, b) in $\mathbb{C}$ (Figure 3.17).

3.9.3 Arithmetic Operations

So far, the complex plane was defined geometrically only. What is its algebraic meaning? Do complex numbers enjoy arithmetic operations between each other? Fortunately, they do. Indeed, addition is easy enough: it could be carried out geometrically, as in the parallelogram rule. What about multiplication? Well, we already know how to multiply i by itself:

$$i^2 = -1.$$

After all, this is how i was defined in the first plane.

3.9.4 Multiplication in The Complex Plane

This equation is our algebraic key. What is its geometrical face? Well, i points vertically upwards (Figure 3.16). This way, i makes a right angle of 90° with the positive part of the real axis. Now, taking the square means doubling the angle. For this reason, i^2 makes a flat angle of 180° with the positive part of the real axis. This way, i^2 must point leftwards on the real axis. The result is $i^2 = -1$, as required.

So, we already have one algebraic equation. Let's go ahead and extend it linearly to the entire complex plane. For this purpose, let a, b, c, and d be some real numbers. Let's use the distributive and commutative laws to multiply $a + bi$ times $c + di$:

$$(a + bi)(c + di) = a(c + di) + bi(c + di)$$
$$= ac + adi + bci + bdi^2$$
$$= ac - bd + (ad + bc)i.$$

3.9.5 The Complex Conjugate

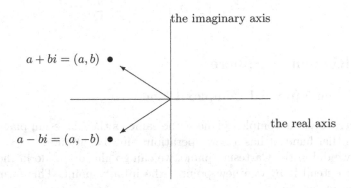

Fig. 3.18. The original complex number $a + bi$ is mirrored by its complex conjugate $a - bi$.

How to make the latter term vanish? Set

$$c = a \quad \text{and} \quad d = -b.$$

This way,

$$c + di = a - bi$$

is nearly the same as the original complex number. There is only one change: the imaginary part picks a minus sign. This mirrors the original complex number: the real axis acts like a mirror (Figure 3.18).

This is the complex conjugate of $a + bi$. It is denoted by $\overline{a + bi}$, with a small bar on top:

$$a - bi = \overline{(a + bi)}.$$

In this case, the above product is real, and has no imaginary part at all:

$$(a + bi)(a - bi) = a^2 + b^2.$$

3.9.6 Absolute Value

This has a familiar geometrical interpretation: the squared length (or magnitude, or norm) of the two-dimensional vector (a, b). In the complex plane, (a, b) is just the same as the complex number $a + bi$, so its length should be the same as the absolute value $|a + bi|$. Therefore, it makes sense to define

$$|a + bi|^2 \equiv (a + bi)(a - bi) = a^2 + b^2.$$

3.9.7 Reciprocal or Inverse

If this is nonzero, then it could be placed in the denominator:

$$(a+bi)\frac{a-bi}{a^2+b^2} = 1.$$

So, we now also have the reciprocal (or inverse) of $a + bi$:

$$(a+bi)^{-1} = \frac{a-bi}{a^2+b^2}.$$

3.10 Riemann's Sphere

3.10.1 The Extended Complex Plane

Geometrically, the complex plane is the same as the Cartesian plane. Algebraically, on the other hand, it has a new operation: multiplication.

As we did in the Cartesian plane, we can go ahead and do in the complex plane as well: extend it by one new point – the infinity point. This forms the extended complex plane:

$$\mathbb{C} \cup \{\infty\}.$$

Thanks to the lamp mapping, this is topologically homeomorphic to the sphere. This is why it is also called Riemann's sphere.

3.11 Towards Differential Geometry

3.11.1 Tangent Plane: Chart

Why is the lamp mapping so important? To see this in yet another way, consider the face of the Earth. At the north pole, place a tiny lamp. This maps the face of the Earth (excluding the north pole) onto the Cartesian plane, tangent to the Earth at the south pole. Thanks to the lamp, each latitude has a circular shadow, and each longitude has a radial shadow. Together, they make new polar coordinates on the tangent plane, which is now ready to serve as a new chart to help describe the Earth geographically.

3.11.2 Atlas of Charts

But what about the north pole? It is still excluded! To cover it as well, work also the other way around: mirror the above from the south. This time, the south pole plays the role of the north pole. This makes a new chart, tangent at the north rather than the south.

Together, both charts make a new atlas to describe the Earth fully. Indeed, together, they cover the entire face of the Earth, including both poles, as required. Clearly, they have a big overlap. Fortunately, they are compatible there: smoothly transferable from one to another.

3.11.3 Differentiable Manifold

This is a rather small atlas: it contains two charts only. Later on, we'll design yet another atlas. Thanks to set theory, we'll also prove that there is a maximal atlas. Thanks to this, the sphere makes a new differentiable manifold. This is the starting point in differential geometry. This will be useful in general relativity later on.

3.12 Exercises

1. Show that the lamp mapping (Figure 3.1) is a one-to-one mapping: two distinct points map to two distinct points.
2. Show that it is not only *into* but also *onto* the extended real axis: it covers it completely.
3. Conclude that it is invertible.
4. Show that it is a topological homeomorphism.
5. Show that the inverse mapping is one-to-one as well.
6. Show that it is not only *into* but also *onto* the original circle.
7. Show that the lamp mapping in the sphere (Figure 3.15) is one-to-one: two distinct points map to two distinct points.
8. Show that it is not only *into* but also *onto* the extended Cartesian plane.
9. Conclude that it is invertible.
10. Show that it is a topological homeomorphism.
11. Show that the inverse mapping is one-to-one as well.
12. Show that it is not only *into* but also *onto* the original sphere.

Part II

Towards Stability
in Classical Mechanics

Towards Stability in Classical Mechanics

In this part, we introduce some discrete math, and use it in physics. In discrete math, there is no continuum: the mathematical objects are discrete, and well-separated from each other. In a mathematical graph, for example, two nodes may be connected to each other by an edge. Still, this doesn't tell us how "close" they are to each other. In fact, there is no notion of distance at all.

In a general graph, the inner structure is not obvious. A tree, on the other hand, is easier to understand: it develops hierarchically, level by level. In fact, it starts from a unique root: the head, at the very top. From the root, a few branches may issue downwards. At the end of each branch, there is a new node. Together, these nodes form the next lower level. This process may then continue recursively. This way, the tree may grow and develop, level by level, until the bottom level, which contains the leaves. This is indeed a multilevel hierarchy.

A binary tree, for example, grows exponentially: each level contains twice as many nodes as the previous level. This kind of growth also appears in stability analysis in classical mechanics.

Furthermore, a binary tree is closely related to Cantor set. Together, they help form fractals, and introduce chaos theory from a geometrical point of view.

Chaos theory studies instability in dynamical processes. It teaches us that the initial data are not always good enough to predict the future. Indeed, however accurate they are, they may still lead to a substantial inaccuracy in the long run. This teaches us to be more humble, and not always trust our own eyes. Better have a little bit of doubt, and never take things as an absolute truth. This will be emphasized even more in quantum mechanics and relativity.

4

Poincare Stability
in Classical Mechanics

For ages, people used the pigeonhole principle without noticing. Here, we give it the honor it deserves.

In its original version, the pigeonhole principle talks about a finite set. Even so, it can still help study stability in classical mechanics. Furthermore, it also has an infinite version, useful to prove theorems in calculus. What's so good about this proof? It will help design the Cantor set, and study stability and instability in classical mechanics.

Thus, the pigeonhole principle lies on the interface between finite and infinite. Is infinity physical? No! Indeed, later on, we'll use Cantor set to show that the universe cannot possibly be infinite.

4.1 Geometrical Preliminaries

4.1.1 The Unit Interval

$$0 \qquad\qquad\qquad\qquad 1$$

Fig. 4.1. The closed unit interval contains both endpoints: 0 and 1. For this reason, it is denoted by $[0, 1]$.

$$0 \qquad\qquad\qquad\qquad\qquad\qquad\qquad 1$$

Fig. 4.2. The open unit interval, on the other hand, doesn't contain its endpoints. For this reason, it is denoted by $(0, 1)$.

We start with some geometrical background. In Chapter 3, Section 3.7.1, we've already looked at a closed interval, and "tied" its endpoints to one another, to produce a complete circle. Here, we'll extend this to higher dimensions as well.

The closed unit interval contains all real numbers between 0 and 1, including the endpoints (Figure 4.1). This is denoted by square brackets:

$$[0, 1] \equiv \{x \text{ is a real number } \mid 0 \le x \le 1\}.$$

Why is this a closed interval? Because it contains both endpoints: 0 and 1. Indeed, in the above formula, one could legitimately pick $x = 0$ or $x = 1$. For this reason, both 0 and 1 are included:

$$0 \in [0, 1] \quad \text{and} \quad 1 \in [0, 1].$$

In fact, we could form a small set, containing two numbers only: 0 and 1 alone. This set is denoted by braces: $\{0, 1\}$. It is then *included* in the closed unit interval:

$$\{0, 1\} \subset [0, 1].$$

However, it is not included in the *open* unit interval. Indeed, the open unit interval doesn't contain its endpoints (Figure 4.2). This is why it is denoted by round parentheses: $(0, 1)$. In fact, it could be obtained from the closed unit interval by dropping (or subtracting) both endpoints:

$$(0, 1) \equiv \{x \text{ is a real number } \mid 0 < x < 1\} = [0, 1] \setminus \{0, 1\}.$$

Don't get confused! In the Cartesian plane, $(0, 1)$ stands for just one point: the x-coordinate is 0, and the y-coordinate is 1. In the context of the real axis, on the other hand, there is no y-coordinate at all. This is why $(0, 1)$ stands for the open unit interval, as discussed above.

Finally, the closed unit interval could be obtained from the open unit interval by adding the endpoints:

$$[0, 1] = (0, 1) \cup \{0, 1\}.$$

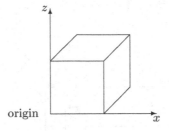

Fig. 4.3. The unit cube in the three-dimensional Cartesian space.

4.1.2 The Unit Cube

Consider now the three-dimensional Cartesian space, spanned by the standard x-, y-, and z-axes. This way, each individual point could be written as (x, y, z), where x, y, and z are its spatial coordinates in our coordinate system.

There is nothing special about these axes: you could pick them as you like, provided that they are perpendicular to each other. In particular, pick one point to serve as the origin: $(0, 0, 0)$. What's so special about the origin? It is the unique point where the axes meet (and cross) each other.

Next to the origin, we have the unit cube of length 1, width 1, and height 1 (Figure 4.3). It contains those points with coordinates between 0 and 1: $0 \leq x, y, z \leq 1$. For this reason, it is often written in terms of the closed unit interval:

$$[0, 1]^3 \equiv [0, 1] \times [0, 1] \times [0, 1] \equiv \{(x, y, z) \mid 0 \leq x, y, z \leq 1\}.$$

Let's convert it into a more complicated domain: a three-dimensional torus.

4.1.3 Three-Dimensional Torus

Now, let's "stretch" the above cube, and "bend" it: the left side loops clockwise, until it meets the right side and sticks to it (Figure 4.4).

For example, look at some fixed y and z. On the left side of the cube, where $x = 0$, we have the point $(0, y, z)$. Across from it, on the right side, where $x = 1$, we have the point $(1, y, z)$. Now, let $(0, y, z)$ "travel," and make a complete loop clockwise in the x-y plane, until meeting $(1, y, z)$ from the right (Figure 4.4).

Once these twin points are stuck together, these are just two different names for the same geometrical point:

$$(0, y, z) = (1, y, z), \quad 0 \leq y, z \leq 1.$$

Likewise, do the same in the z-coordinate as well – identify the top with the bottom:

$$(x, y, 0) = (x, y, 1), \quad 0 \leq x, y \leq 1.$$

Finally, do the same in the y-coordinate as well:

$$(x, 0, z) = (x, 1, z), \quad 0 \leq x, z \leq 1.$$

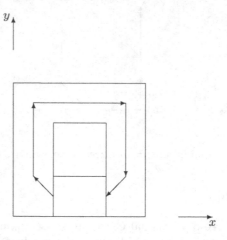

Fig. 4.4. Stretching and bending the unit cube (a view from above): the left side makes a complete loop clockwise, to meet the right side from the right and stick to it. The result is a thick square torus.

This is a bit tricky to visualize. Indeed, to make the latter identification, three dimensions are not enough. A new leap of the imagination is needed: go outside the physical world, and loop through a fourth dimension. Alternatively, keep working in our world: stick to the original cube in Figure 4.3, but keep the above identifications in mind.

So, what have we got? A three-dimensional "torus." It has no boundary any more: each point in it is an inner point, surrounded by other points from it. To imagine this, better stick to Figure 4.3, but keep in mind that sides that lie across from each other are now considered as one and the same.

4.1.4 Vector and Its Direction

A point (x, y, z) can also be viewed as a vector: an arrow issuing from the origin, with its head pointing at the point (x, y, z). From Pythagoras theorem, its length (or magnitude, or norm) is

$$\|(x, y, z)\| = \sqrt{x^2 + y^2 + z^2}.$$

Still, this is not so important. What is more important is its direction. Fortunately, the direction is rather stable: it doesn't change if all coordinates are multiplied by the same nonzero scalar:

$$\alpha(x, y, z) = (\alpha x, \alpha y, \alpha z), \quad \alpha \neq 0.$$

After all, to specify the direction uniquely, the important things are the ratios (proportions) between the coordinates, which remain the same for all $\alpha \neq 0$.

4.2 Integrable System and Its Stability

4.2.1 Rational Proportion

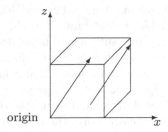

Fig. 4.5. In the three-dimensional torus, the ceiling coincides with the floor. For this reason, the light ray issuing from the origin in direction $(1, 1, 2)$ will make two complete loops. Indeed, upon hitting the ceiling at the middle, it will reappear at the bottom, and reenter from below. Then, upon reaching the ceiling for the second time, it will hit the corner $(1, 1, 1)$, which coincides with the origin.

Consider now a light ray issuing from the origin at some direction. In our three-dimensional torus, the light ray will loop. For example, upon hitting the ceiling of the original cube, it will reappear at the bottom and reenter, and so on (Figure 4.5). This way, it may go on making a lot of loops, never exiting the three-dimensional torus.

Will the light ray hit the origin ever again? Only if the direction vector could be written with integer coordinates:

$$(x, y, z) = \alpha(l, m, n),$$

where l, m, and n are some integer numbers, at least one of which is nonzero. We then say that x, y, and z have a common length unit: α. In this case, the ratios between the coordinates could be written as rational numbers. For example, if $n \neq 0$, then

$$\frac{x}{z} = \frac{l}{n} \quad \text{and} \quad \frac{y}{z} = \frac{m}{n}.$$

For this reason, once the light ray has coordinate $z = n$, it must also have coordinates $x = l$ and $y = m$, so it must hit the point (l, m, n), which coincides with the origin.

4.2.2 Irrational Proportion

But what if the ratio between two coordinates was irrational, and could never be written in such a way? In this case, the coordinates of the direction vector have no common length unit. As a result, the light ray will never hit the origin any more, however many times it may loop. Fortunately, it will still get arbitrarily close to the origin. This is indeed stability.

4.2.3 Poincare Recurrence Theorem

To see this, pick an arbitrarily small number $\varepsilon > 0$. Consider a small $\varepsilon \times \varepsilon \times \varepsilon$ cube: a small cell of length ε, width ε, and height ε. Let's use many such cells to cover the entire unit cube. For this purpose, we may need as many as ε^{-3} cells. For simplicity, assume that ε^{-3} is an integer number. Otherwise, just pick a slightly smaller ε.

This number, however big, is still finite. For this reason, the light ray can't go on visiting new cells only. After sufficiently many loops, it will have to step once again in an old cell, where it has already been before. Since ε is very small, the ray will eventually get ever so close to a point where it has already been before.

Now, there is nothing special about this point: it could be easily shifted to the origin. After all, in the three-dimensional torus, the origin is just an ordinary inner point. Thus, after sufficiently many loops, the ray will get ever so close to the origin. In other words, since ε is arbitrarily small, the ray will eventually get arbitrarily close to the origin. This is Poincare recurrence theorem.

4.2.4 Approaching The Infinity

In the above proof, infinity was never mentioned! After all, however small, ε is still nonzero. Likewise, however big, ε^{-3} is still finite.

Only at the end of the above proof was the infinity used, implicitly and indirectly. Indeed, by saying "arbitrarily close," we actually used the implicit limit process

$$\varepsilon \to 0.$$

Since ε must remain positive, this is often written as

$$\varepsilon \to 0^+.$$

In this process, as ε gets smaller and smaller, the total number of cells gets bigger and bigger, with no bound:

$$\varepsilon^{-3} \to_{\varepsilon \to 0^+} \infty.$$

Still, for each fixed $\varepsilon > 0$, ε^{-3} is fixed as well. However big, this is still finite, and can be used in the above proof. It approaches infinity, but never reaches it!

4.3 The Pigeonhole Principle

4.3.1 The Finite Version

So, the above proof is purely discrete and algebraic: it uses finite sets only, and never mentions infinity. As a matter of fact, it is based on a very simple principle in discrete math and set theory: the pigeonhole principle.

Assume that ten pigeons need to occupy nine holes. There is not enough room for all! As a result, at least one hole must host two pigeons or more. This seems trivial, yet it is a fundamental principle. Who said math wasn't simple?

4.3.2 The Pigeonhole Principle in Stability

How was this principle used in the above proof? Well, instead of nine holes, we had ε^{-3} cells. Instead of ten pigeons, we had ε^{-3} loops (or even more). For this reason, once the light ray loops ε^{-3} times, it must revisit an old cell, where it has already been before.

4.3.3 The Infinite Version

So far, we've stated the pigeonhole principle in its finite version. Now, let's see an infinite version as well. For this purpose, assume that there are infinitely many pigeons, but only a finite number of holes. In this case, at least one hole must host infinitely many pigeons. Let's use this extended version as well.

4.4 Application: The Bolzano–Weierstrass Theorem

4.4.1 Infinite Sequence

Let's use the infinite version in calculus and chaos theory. For this purpose, consider an infinite sequence:
$$x_1, x_2, x_3, x_4, \ldots, x_n, \ldots.$$

Assume that the entire sequence is contained in the closed unit interval. This means that, for all n, $0 \le x_n \le 1$.

4.4.2 Splitting The Interval

$$\tilde{x} = 0 \qquad\qquad \tfrac{1}{2} \qquad\qquad 1$$

Fig. 4.6. The first step: split the closed unit interval into two nonoverlapping subintervals. Pick one subinterval (say, the left one). Initialize $\tilde{x}$ to be its left endpoint (say, $\tilde{x} \equiv 0$).

Now, let's split the unit interval into two equal subintervals (Figure 4.6). In other words, let's write it as the union

$$[0,1] = \left[0, \frac{1}{2}\right] \cup \left[\frac{1}{2}, 1\right].$$

Here, '∪' stands for the union of two sets.

Now, where are the x_n's located? Well, they must lie somewhere: either in the left subinterval, or in the right one. Thanks to the pigeonhole principle (in its infinite version), at least one subinterval must contain infinitely many x_n's. Let's pick it, and drop the other one.

More precisely, if the left subinterval contains infinitely many x_n's, then let's pick it, and drop all the other x_n's that are not in it. In this case, let's mark its left endpoint by

$$\tilde{x} \equiv 0.0 \equiv 0 \cdot 1 + 0 \cdot \frac{1}{2}.$$

This is a binary representation: it tells us that $\tilde{x}$ indeed lies in the left subinterval. This is just an initialization: later on, we may add more binary digits on the right.

If, on the other hand, the left subinterval contains just a finite number of x_n's, then drop them, and pick the right subinterval. In this case, $\tilde{x}$ is initialized as its left endpoint:

$$\tilde{x} \equiv 0 \cdot 1 + 1 \cdot \frac{1}{2} \equiv 0.1.$$

Here, the binary digit 1 behind the point stands for 1 times 1/2, as required. Later on, we may add more binary digits behind it, to update $\tilde{x}$ further.

4.4.3 The Induction Step

Fig. 4.7. The second step: split one subinterval (say, the left one) into two nonoverlapping subsubintervals. Pick one subsubinterval. Update $\tilde{x}$ to be its left endpoint.

Next, let's look at our subinterval, and do the same to it: split it into two subsubintervals, and pick one of them (Figure 4.7). If the left subsubinterval is picked, then add 0 to the binary representation of $\tilde{x}$, on the far right. If, on the other hand, the right subsubinterval is picked, then add 1 instead. In either case, our subsubinterval contains infinitely many x_n's, and the up-to-date $\tilde{x}$ indeed marks its left endpoint, as required.

4.4.4 Accumulation Point

This is actually the induction step. By mathematical induction, it can be repeated on and on forever, producing a hierarchy of nested intervals, contained in one another. The final (possibly irrational) $\tilde{x}$ will be an accumulation point of our original sequence: every open interval that contains $\tilde{x}$, however short, must also contain infinitely many x_n's.

This is indeed Bolzano–Weierstrass theorem: every infinite bounded sequence must have an accumulation point.

4.5 Compact Domain in 3-D

4.5.1 Extension to 3-D

Let's extend the above to three dimensions as well. For this purpose, let D be a bounded three-dimensional domain. This way, D can be confined in a box:

$$D \subset [x_{min}, x_{max}] \times [y_{min}, y_{max}] \times [z_{min}, z_{max}].$$

Here, x_{min} stands for the minimal possible x-coordinate in any point in D, and so on.

Consider now an infinite sequence of points in D:

$$(x_1, y_1, z_1), \ (x_2, y_2, z_2), \ \dots, (x_n, y_n, z_n), \ \dots.$$

Does it have an accumulation point?

4.5.2 Splitting The Interval

To answer this, let's apply one step as in Figure 4.6: split $[x_{min}, x_{max}]$ into two subintervals, pick one of them, and initialize $\tilde{x}$ to mark its left endpoint. While doing this, drop some (x_n, y_n, z_n)'s from the original sequence, yet leave infinitely many.

4.5.3 The Induction Step

Then, do the same in the y-direction: split $[y_{min}, y_{max}]$ into two subintervals, pick one of them, and initialize $\tilde{y}$ to mark its left endpoint. While doing this, drop some more (x_n, y_n, z_n)'s from the up-to-date sequence, yet leave infinitely many.

Finally, do the same in the z-coordinate as well: split $[z_{min}, z_{max}]$ into two subintervals, pick one of them, and initialize $\tilde{z}$ to mark its left endpoint. While doing this, drop some more (x_n, y_n, z_n)'s, yet leave infinitely many.

4.5.4 Accumulation Point

This completes the induction step. What have we done in it? We've actually split the original box into eight subboxes, and picked one of them, which still contains infinitely many points from the original sequence. All other points that are not in this subbox have been dropped, never to be regarded any more.

The above step can now repeat time and again in a mathematical induction, updating $\tilde{x}$, $\tilde{y}$, and $\tilde{z}$ by adding a suitable binary digit to each of them on the far right. This way, these binary representations are in $[0, 1]$. In the end, they may need to shift, to make the required accumulation point:

$$(x_{min} + (x_{max} - x_{min})\tilde{x}, \ y_{min} + (y_{max} - y_{min})\tilde{y}, \ z_{min} + (z_{max} - z_{min})\tilde{z}).$$

Indeed, every open cell that contains this point, however small, must also contain infinitely many points from the original sequence. This extends the original theorem to three dimensions as well.

4.5.5 Compact Domain

Assume now that D is not only bounded but also closed: it contains all its accumulation points. In this case, we say that D is compact. So, thanks to the pigeonhole principle (in its infinite version), we've just proved that, in a compact domain, every infinite sequence must have an accumulation point.

4.6 Nonintegrable System: Poincare Stability

4.6.1 Domain and Its Mapping

So far, we've considered a three-dimensional torus. This is a linear model. In classical mechanics, it is called an integrable system. In it, a light ray issuing from some point could shift, and produce a new (parallel) light ray, issuing from a new point. This is why Poincare recurrence theorem is so easy to prove: the pigeonhole principle (in its finite version) is good enough (Section 4.2.3).

But what about a nonintegrable system? In this case, instead of the three-dimensional torus, we have now a more complicated domain. Furthermore, instead of a straight light ray, we have now a curved trajectory, describing the flow of a particle from the initial point v_0. After one second, it arrives at the new point $T(v_0)$. After two seconds, on the other hand, it arrives at $T^2(v_0)$, and so on.

Thus, T actually maps the entire domain. It describes the entire dynamics in terms of discrete math and set theory. The system is now ready to benefit from the pigeonhole principle.

4.6.2 Mapping and Its Origin

In a flow as above, is T one-to-one? In other words, could two particles occupy the same point? If not, then T is invertible: one could time-travel, from the position after one second, back to the initial position. In this case, the inverse mapping is denoted by T^{-1}.

Here, however, we drop this assumption. Indeed, we never assume that T is invertible. On the contrary: two different trajectories may meet each other. Thus, T^{-1} is *not* a mapping any more. What is T^{-1}? Well, if A is a set in the domain D, then $T^{-1}(A)$ contains all those points that are mapped into A:

$$T^{-1}(A) \equiv \{v \in D \mid T(v) \in A\}.$$

We then say that $T^{-1}(A)$ is the origin of A under T.

4.6.3 Intersection and Its Origin

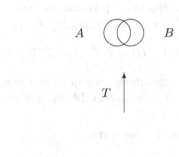

$T^{-1}(A)$ 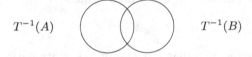 $T^{-1}(B)$

Fig. 4.8. The lower-left circle is mapped to the upper-left circle. The lower-right circle is mapped to the upper-right circle. As a result, the lower overlap is mapped to the upper overlap.

Consider now two subsets of D:

$$A, B \subset D.$$

Their intersection (or overlap) contains those common points that are in both A and B:

$$A \cap B \equiv \{v \in D \mid v \in A \text{ and } v \in B\}.$$

In other words, v is in the intersection if (and only if) v is in both A and B:

$$v \in A \cap B \Leftrightarrow v \in A \text{ and } v \in B.$$

In other words, $A \cap B$ is the maximal set included in both A and B.

For example, if A and B are disjoint from each other, then they have no overlap at all:

$$A \cap B = \emptyset$$

(the empty set). Still, this is a rather rare case. Usually, A and B may overlap, and have common points.

What is the origin of $A \cap B$ under T? Well, as in Figure 4.8,

$$
\begin{aligned}
T^{-1}(A \cap B) &\equiv \{v \in D \mid T(v) \in A \cap B\} \\
&= \{v \in D \mid T(v) \in A \ \text{ and } \ T(v) \in B\} \\
&= \{v \in D \mid T(v) \in A\} \cap \{v \in D \mid T(v) \in B\} \\
&= T^{-1}(A) \cap T^{-1}(B).
\end{aligned}
$$

Thus, we have a sort of "distributive" law, teaching us how to open parentheses. It tells us that order doesn't matter: instead of intersecting and then applying T^{-1}, we can first apply T^{-1}, and then intersect.

In summary, the origin of the overlap is the overlap of the origins. This works not only for T but also for the composite mapping T^n $(n \geq 0)$:

$$T^{-n}(A \cap B) = T^{-n}(A) \cap T^{-n}(B).$$

We're now ready to use the pigeonhole principle (in a new finite version) to prove Poincare recurrence theorem in a nonintegrable system as well.

4.6.4 Stationarity: Volume Preservation

Assume now that our domain has a finite volume:

$$V(D) < \infty.$$

Furthermore, assume also that our mapping is stationary: it preserves volume. In other words, the origin has the same volume:

$$V\left(T^{-1}(A)\right) = V(A)$$

for any subset $A \subset D$.

In theory, A could be as thin as a sheet of paper. In this case, it would have zero volume:

$$V(A) = 0.$$

Here, on the other hand, we assume that A is thick and substantial, so it has a positive volume:

$$V(A) > 0.$$

Now, use A to form an infinite sequence of sets: the origin of A, the origin of the origin, and so on:

$$A, \ T^{-1}(A), \ T^{-2}(A), \ \ldots, \ T^{-n}(A), \ \ldots.$$

Thanks to stationarity, all these sets have the same volume: $V(A) > 0$. Thanks to the pigeonhole principle, they can't be completely disjoint from each other: there is just not enough room for all. Thus, there must be two sets that overlap substantially with each other. In other words, there must be two indices $j > i \geq 0$ for which

$$V\left(T^{-i}(A) \cap T^{-j}(A)\right) > 0.$$

Thanks to our "distributive" law in Section 4.6.3, we can now factor T^{-i} out:

$$T^{-i}(A) \cap T^{-j}(A) = T^{-i}\left(A \cap T^{-(j-i)}(A)\right).$$

In summary, thanks to volume preservation, we now have

$$V\left(A \cap T^{-(j-i)}(A)\right) = V\left(T^{-i}\left(A \cap T^{-(j-i)}(A)\right)\right)$$
$$= V\left(T^{-i}(A) \cap T^{-j}(A)\right)$$
$$> 0.$$

Is this familiar? The same process has already been used on the light ray in Section 4.2.3, including the shift in the end. Here, however, T is not necessarily one-to-one: this is why there is a need to use the origins.

Finally, define

$$N \equiv j - i \geq 1.$$

In summary, we've used the pigeonhole principle to prove that, for every A of positive volume, there exists a natural number $N \geq 1$ for which $T^{-N}(A)$ overlaps with A substantially:

$$V\left(A \cap T^{-N}(A)\right) > 0.$$

Note that each A may have its own N, associated with it. In other words, two different A's may have two different N's. Thus, N depends on A:

$$N \equiv N(A).$$

Still, so long as it is clear what A we're talking about, we often drop the '(A),' and write N on its own for short.

Let's use the above result to show that a nonintegrable system is in principle the same as our original integrable system: it enjoys stability as well.

4.6.5 Poincare Stability

So far, infinity was never mentioned. Furthermore, we've talked about sets only, not individual points. This is a good sign: it indicates that we're on the right track. After all, sets are more fundamental than points. Now, it is time to talk about individual points as well.

In an integrable system (our three-dimensional torus), we've already established stability: for every cell, however small, every light ray issuing from it will eventually return to it (Section 4.2.3). Let's see that this is also the case for almost every point in a nonintegrable system. For this purpose, we need the infinity.

Let $A \subset D$ be a substantial set of positive volume: $V(A) > 0$. (Still, this volume could be as small as you like.) Let $B \subset A$ contain those points that never return to A:

$$B \equiv \{v \in A \mid T^n(v) \notin A, \ n = 1, 2, 3, \ldots\}.$$

The rest of the points in A, on the other hand, do return to A eventually. Now, what is stability? Well, stability means that almost all points in A do return to A eventually. In other words, B is unsubstantial: as thin as a sheet of paper.

Let's prove stability by contradiction. Assume momentarily that B was substantial:

$$V(B) > 0.$$

In this case, thanks to Section 4.6.4, there must be a natural number

$$N \equiv N(B) \geq 1$$

for which

$$V\left(B \cap T^{-N}(B)\right) > 0.$$

But these sets are disjoint:

$$B \cap T^{-N}(B) = \emptyset.$$

Indeed, consider a particular point $v \in T^{-N}(B)$. After N applications of T, v returns to $B \subset A$. Therefore, $v \notin B$.

Thus, our momentary assumption must have been false: B could never have been substantial. On the contrary:

$$V(B) = 0.$$

This is indeed Poincare stability: almost all $v \in A$ must eventually return to A.

Where was the infinity used? Only implicitly, in the very definition of B. Indeed, to define B, one must have a mechanism to check whether a given point $v \in A$ ever returns to A. If it does ($v \notin B$), then this is easy enough to verify: just apply T time and again, until getting back into A. This way, you get your answer in a finite time. If, on the other hand, $v \in B$, then you may apply T time and again forever, and got no answer in any finite time. This is the infinity paradox. To believe in infinity, you must trust your imagination. This is indeed an axiom in set theory: there exists an infinite set. (See Chapter 3, Section 3.1.1.)

4.7 Exercises

4.7.1 Three-Dimensional Torus in $\mathbb{R}^4$

1. Define the real axis:
$$\mathbb{R} \equiv \{x \mid -\infty < x < \infty\}.$$

2. Define the three-dimensional Cartesian space:

$$\mathbb{R}^3 \equiv \{(x, y, z) \mid -\infty < x, y, z < \infty\}.$$

3. To the three-dimensional Cartesian space, add also a fourth dimension: the w-coordinate:

$$\mathbb{R}^4 \equiv \{(x, y, z, w) \mid -\infty < x, y, z, w < \infty\}.$$

4. State the pigeonhole principle in its finite version.
5. Prove it.
6. Use it to prove Poincare recurrence theorem in the three-dimensional torus. Hint: see Section 4.2.3.

7. State the pigeonhole principle in its infinite version.
8. Prove it.
9. Use it to prove the Bolzano–Weierstrass theorem. Hint: see Sections 4.4.1–4.4.4.
10. Extend it to a three-dimensional compact domain. Hint: see Sections 4.5.1–4.5.5.
11. Extend it to a four-dimensional compact domain as well.
12. Consider a three-dimensional compact domain $D \subset \mathbb{R}^3$. Does it have a finite volume? Hint: since D is bounded, its volume could be defined as

$$V(D) \equiv \int \int \int_D dxdydz < \infty.$$

13. Establish Poincare stability in a nonintegrable system as well: a compact domain $D \subset \mathbb{R}^3$. Hint: see Sections 4.6.4–4.6.5.
14. Extend this to a four-dimensional compact domain $D \subset \mathbb{R}^4$ as well.
15. Conclude once again that the original three-dimensional torus indeed enjoys Poincare stability. Hint: it is indeed a compact domain in $\mathbb{R}^4$.
16. What is its volume? Hint: it is still 1, just like the unit cube. Indeed, the integration could still take place in 3-D: the sides are only two-dimensional, so sticking them together (as in Figure 4.4) has no effect.
17. To define B in Section 4.6.5, how to "solve" the infinity paradox? Hint: see Section 4.6.5.

5

Cantor Set
and Its Applications

In its infinite version, the pigeonhole principle helped design an accumulation point. Let's use the same process to design a fundamental set: the Cantor set.

Later on, we'll use the Cantor set in chaos theory. Before that, let's use it to approximate a real number by a rational number. For this purpose, we design the Cantor set not only in one but also in two dimensions.

Is nature infinite? No! Indeed, later on, we'll use Cantor set to show that the universe cannot possibly be infinite. As a matter of fact, nature is finite in both the macroscale and microscale, as we'll see in quantum mechanics and relativity later on.

5.1 Probability

5.1.1 Throwing a Dice

Let's play a game: throw a dice. If you get 1 or 2 or 3 or 4 or 5, then you win. If, on the other hand, you get 6, then I win.

How likely are you to win? Well, in five of six cases, you win. So, the probability that you win is as high as 5/6. The probability that I win, on the other hand, is as low as 1/6.

Now, let's play the game twice. How likely are you to win twice? Clearly, the probability for this is lower:

$$\frac{5}{6} \cdot \frac{5}{6} = \frac{25}{36}.$$

Now, let's play n times, for some large n. How likely are you to win n times? Unfortunately for you, the probability for this is already very low. In fact, as n increases, it gets closer and closer to zero. In other words, it approaches zero:

$$\left(\frac{5}{6}\right)^n \rightarrow_{n \rightarrow \infty} 0.$$

In summary, if we play sufficiently many times, then I'm highly likely to win at least once.

Where is this useful? Pick an arbitrarily long decimal number at random. How likely is it that the digit 7 never appears in it? The probability for this is as low as zero! We'll discuss this later from a geometrical pint of view.

5.1.2 Throwing Two Dices

Now, let's move on to a yet more interesting game: throw two dices together at the same time. If you get $(6, 6)$, then I win. Otherwise, you win.

The probability that I win is now as low as $1/36$. The probability that you win, on the other hand, is now as high as $35/36$. Still, how likely are you to win n times in a row? Well, as n increases, this approaches zero again:

$$\left(\frac{35}{36}\right)^n \to_{n \to \infty} 0.$$

So, if we play many times, then I'm highly likely to win at least once.

Later on, we'll use this to model two decimal numbers, and their individual digits. For this purpose, let's look at things from a geometrical point of view.

5.2 Geometrical Setting

5.2.1 Splitting The Interval

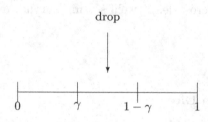

Fig. 5.1. In the first step, drop the middle subinterval: from γ to $1 - \gamma$.

In set theory, we get to see a few kinds of infinity. For instance, look at a rather strange kind of infinity: the Cantor set. In what way is it strange? Well, it seems to contain nothing, yet contains infinitely many points!

To design the Cantor set, start from the closed unit interval $[0, 1]$: the real numbers between 0 and 1 (including the endpoints). (See Chapter 4, Section 4.1.1.) Let γ be a fixed number (parameter) between 0 and $1/2$:

$$0 < \gamma < \frac{1}{2}.$$

(For example, $\gamma = 1/3$.) This way, $2\gamma < 1$, so $1 - 2\gamma > 0$, and

$$1 - \gamma > \gamma.$$

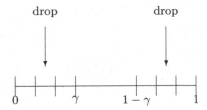

Fig. 5.2. In the second step, apply the above procedure twice: look at the left and right subintervals. From each of them, drop its own middle subsubinterval.

This is good: we can now start our geometrical process (Figure 5.1). Split the original interval into three disjoint subintervals: the first one is from 0 to γ, the second from γ to $1 - \gamma$, and the third from $1 - \gamma$ to 1:

$$[0, 1] = [0, \gamma] \cup (\gamma, 1 - \gamma) \cup [1 - \gamma, 1].$$

Here, the left and right subintervals are closed, whereas the middle one is open: it doesn't contain its own endpoints.

This is indeed a symmetric splitting: the left and right subintervals have the same length: γ. The middle subinterval, on the other hand, may have a different length: $1 - 2\gamma$. Let's go ahead and drop it!

We are now left with two subintervals only: the right one, and the left one. To each of them, apply the same procedure recursively. This way, we are now left with four subsubintervals, of length γ^2 each (Figure 5.2).

This is the induction step. The same procedure of splitting and dropping may now go on and on (recursively) time and again, infinitely many times.

What is left in the end? Well, the points that are never dropped make a new set: the Cantor set. How does it look like? Well, to illustrate it, better look at a new version, in a more familiar setting: decimal. After all, who says we must drop the middle subinterval? Why not drop a new subinterval, of length 1/10?

5.2.2 Decimal Base

A real number $0 \leq x \leq 1$ can be written as an infinite decimal fraction:

$$x = 0.a_1 a_2 a_3 \ldots,$$

where each a_i is some digit between 0 and 9. For example, 1/10 could be written in two different styles: either a finite style:

$$\frac{1}{10} \equiv 0.1 \equiv 0.1000000000000000000 \ldots,$$

or an infinite style:

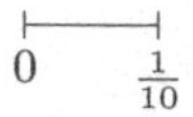

Fig. 5.3. In the first step, one subinterval drops from the unit interval.

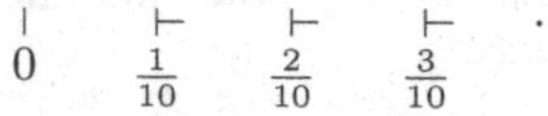

Fig. 5.4. The second step: from each subinterval, drop one subsubinterval. In total, nine subsubintervals are dropped.

$$\frac{1}{10} \equiv 0.0999999999999999\ldots.$$

In this case, pick the former style. Only for the number 1 must we pick the infinite style:

$$1 \equiv 0.999999999999999\ldots.$$

In set theory, one could ask an interesting question: pick a random $0 \le x \le 1$. How likely is x to have no zero digit at all? In other words: how many x's have nonzero digits only?

The surprising answer is: next to none! Indeed, let's consider the half-open unit interval:

$$[0, 1) \equiv \{x \text{ is a real number} \mid 0 \le x < 1\}.$$

Why is it called half-open? Because it contains only its left endpoint, but not its right one (Chapter 4, Section 4.1.1).

What is the size (or length) of this interval? Just take the right endpoint, and subtract the left one:

$$1 - 0 = 1.$$

As a matter of fact, the closed (or open) unit interval has the same length as well. After all, in terms of length, excluding an endpoint has no effect whatsoever. Still, a half-open interval is easier to work with.

Now, drop those x's with some zero digit. This could be done step by step. In the first step, drop the first (half-open) subinterval:

$$\left[0, \frac{1}{10}\right) = \left\{x \text{ is a real number} \mid 0 \le x < \frac{1}{10}\right\}$$

(Figure 5.3). This drops those x's with $a_1 = 0$. Clearly, the length (or size) of this subinterval is:

$$\frac{1}{10} - 0 = \frac{1}{10}.$$

So, we are left with 90% only: the remaining (half-open) subinterval

$$\left[\frac{1}{10}, 1\right) = \left\{x \text{ is a real number} \mid \frac{1}{10} \le x < 1\right\},$$

whose length is

$$1 - \frac{1}{10} = \frac{9}{10}.$$

Now, split this into nine disjoint (half-open) subintervals, of size $1/10$ each:

$$\left[\frac{1}{10}, 1\right) = \left[\frac{1}{10}, \frac{2}{10}\right) \cup \left[\frac{2}{10}, \frac{3}{10}\right) \cup \cdots \cup \left[\frac{9}{10}, 1\right).$$

These subintervals are now forwarded to the next step. In this step, from each subinterval of length $1/10$, drop the first subsubinterval of length $1/100$ (Figure 5.4). This drops those x's with $a_1 > 0$ and $a_2 = 0$. Again, we are left with 90% only, or 81% of the original interval.

This is the induction step. The process may now continue step by step in the same way. In each step, the total remaining size gets 0.9 times as small.

Thanks to mathematical induction, we can now go ahead and carry out infinitely many steps. Indeed, assuming that n steps have already been made, go ahead and do the next step as well. In the "end," we are left with a size as small as

$$\lim_{n \to \infty} 0.9^n = 0.$$

This is Cantor's set (in base 10): a sparse "sieve" of those x's that never use 0 in their decimal representation. Although it is infinite, this set is as thin as air: zero total length, or zero measure.

Thus, real numbers with no zero digit are very rare. If you pick a random $0 \le x < 1$, then it is highly likely to have at least one zero behind the decimal point. Let's extend this result to a higher dimension as well.

5.2.3 Two Dimensions

Let us extend the above to two dimensions as well. For this purpose, consider the x-y Cartesian plane. In it, consider the (half-open) unit square, which contains points with coordinates between 0 and 1:

$$[0, 1)^2 \equiv \{(x, y) \mid 0 \le x, y < 1\}.$$

Clearly, the area of the unit square is 1. Now, let (x, y) be some point in it. This way, both x and y have their own decimal representations:

$$x = 0.a_1 a_2 a_3 \ldots$$
$$y = 0.b_1 b_2 b_3 \ldots,$$

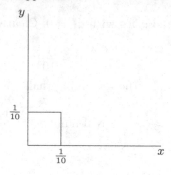

Fig. 5.5. In the first step, the lower-left subsquare drops from the unit square.

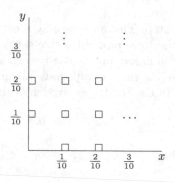

Fig. 5.6. The second step: from each subsquare, drop its own lower-left subsubsquare. In total, 99 subsubsquares are dropped.

where the a_i's and the b_i's are some digits between 0 and 9. As we've seen in Section 5.2.2, x is most likely to have some index $k(x)$ for which $a_{k(x)} = 0$. Likewise, y is most likely to have some other index $k(y)$ for which $b_{k(y)} = 0$.

Still, $k(x)$ and $k(y)$ may be different from each other. Is there yet another index $k \equiv k(x, y)$, for which both $a_k = b_k = 0$?

Most likely, there is! To see this, drop one (half-open) subsquare: the lower-left subsquare, of size $(1/10) \times (1/10)$:

$$\left[0, \frac{1}{10}\right)^2 = \left\{ (x, y) \mid 0 \le x, y < \frac{1}{10} \right\}$$

(Figure 5.5). This drops those (x, y)'s with $a_1 = b_1 = 0$. The area of this subsquare is as small as

$$\frac{1}{10} \cdot \frac{1}{10} = \frac{1}{100}.$$

So, we are left with 99% only. This completes the first step.

The remaining 99 subsquares are now forwarded to the second step: from each of them, drop the lower-left subsubsquare (Figure 5.6). This drops those (x, y)'s with $a_2 = b_2 = 0$ (but $a_1 > 0$ or $b_1 > 0$). Again, we are left with 99% only, or 98.01% of the original square.

This is the induction step. The process may now continue step by step in the same way. After each step, the total remaining area gets 0.99 times as small.

Thanks to mathematical induction, we can now go ahead and carry out infinitely many steps. Indeed, assuming that n steps have already been made, go ahead and do the next step as well. Eventually, we are left with area as small as

$$\lim_{n \to \infty} 0.99^n = 0,$$

or with a set as thin as air: it has zero area or size or measure.

Thus, if you pick a random point (x, y) from the unit square, then both x and y are highly likely to have at least one joint zero digit $a_k = b_k = 0$ (for some $k \geq 1$). Let's use this observation to approximate both x and y by rational numbers.

5.3 Application: How to Approximate a Real Number?

5.3.1 Rational Approximation

Let's pick a random point (x, y) from the unit square. Let $k \geq 1$ be the first index for which both

$$a_k = b_k = 0.$$

As discussed above, at probability 1, k indeed exists. Furthermore, in our dropping process, in the kth step, (x, y) must drop from the unit square. This way, both x and y have a decimal approximation of length $k - 1$:

$$|x - 0.a_1 a_2 a_3 \ldots a_{k-1}| \leq 10^{-k}$$
$$|y - 0.b_1 b_2 b_3 \ldots b_{k-1}| \leq 10^{-k}.$$

In other words, both x and y have rational approximations that share the same denominator $m \equiv 10^{k-1}$, but may have different numerators $l \equiv a_1 a_2 a_3 \ldots a_{k-1}$ and $n \equiv b_1 b_2 b_3 \ldots b_{k-1}$:

$$\left| x - \frac{l}{m} \right| \leq \frac{1}{10m}$$
$$\left| y - \frac{n}{m} \right| \leq \frac{1}{10m}.$$

For instance, if $k = 1$, then $l = n = 0$.

As a matter of fact, one could even improve on this: pick a yet bigger k, and construct many more rational approximations, with m as large as you like. Alternatively, one could pick a bigger base.

5.3.2 Nondecimal Base

So far, we've used decimal digits only. After all, we humans have ten fingers to count with, so base 10 suits us best. The computer, on the other hand, has just two "fingers:" 0 and 1. This is why it uses base 2 rather than 10: not decimal but binary digits.

Base 2 is the smallest base possible. In our application, however, better use a much larger base. To design such a base, let $\varepsilon > 0$ be an arbitrarily small parameter. Without loss of generality, assume that ε^{-1} is an integer number. (Otherwise, just make ε a little smaller.)

Now, in the discussion in Sections 5.2.2–5.3.1, replace base 10 by base ε^{-1}. This way, the now integer numbers a_i and b_i are different from before: they are no longer between 0 and 9, but between 0 and $\varepsilon^{-1} - 1$.

With these changes, the new k, l, n, and m are also different from before. This leads to a new (and better) rational approximation:

$$\left| x - \frac{l}{m} \right| \leq \frac{\varepsilon}{m}$$

$$\left| y - \frac{n}{m} \right| \leq \frac{\varepsilon}{m}.$$

This result will be used in the next chapter.

5.4 Exercises: Cantor Null Set

5.4.1 Cantor Set – Big or Small?

1. After the first step in Figure 5.1, how many subintervals remain? Hint: 2.
2. What is their total length? Hint: 2γ.
3. After the second step in Figure 5.2, how many subsubintervals remain? Hint: 4.
4. What is their total length? Hint: $4\gamma^2$.
5. Show that, after each step, the total remaining length gets 2γ times as short.
6. After the nth step, how many intervals remain? Hint: 2^n.
7. What is their total length? Hint: $(2\gamma)^n$.
8. Use the assumption that $\gamma < 1/2$ to conclude that Cantor's set has zero length (or size, or measure). Hint: since $2\gamma < 1$,

$$(2\gamma)^n \to_{n\to\infty} 0.$$

9. For instance, set $\gamma \equiv 1/3$. In this case, what is the measure of Cantor's set? Hint:

$$\left(\frac{2}{3} \right)^n \to_{n\to\infty} 0.$$

10. Pick some $0 \leq x \leq 1$. In base 3, write x as a new trinary fraction:

$$x = 0.a_1 a_2 a_3 \ldots,$$

where the new a_i's are now different from before: each new a_i is now a trinary digit: either 0 or 1 or 2.

11. For example, to write 1/3, use the infinite style:

$$\frac{1}{3} \equiv 0.02222222222222222222222222222\ldots.$$

12. To write 2/3, on the other hand, use the finite style:

$$\frac{2}{3} \equiv 0.2 \equiv 0.2000000000000000000000000000\ldots.$$

13. Note that, in either of these examples, the digit '1' is never used!

14. Assume that x belongs to Cantor's set, constructed with $\gamma \equiv 1/3$. How does the trinary representation of x look slike? Hint: for each i, it may use $a_i \equiv 0$ or $a_i \equiv 2$, but never $a_i \equiv 1$.

15. Conversely, show that every trinary fraction that looks like this indeed belongs to this Cantor set.

16. Map this Cantor set as follows: in the above trinary fraction, change each '2' into '1'. Then, interpret the result as a binary fraction, not trinary.

17. Conclude that the above mapping covers the entire unit interval.

18. Conclude that Cantor's set is a rather strange kind of infinity: it has zero measure, yet contains as many numbers as the entire unit interval!

19. Explain this paradox. Hint: an infinite set may still be sparse, and have zero size.

20. Extend the construction in Section 5.2.3 to three dimensions as well.

21. What is the volume of the resulting Cantor set? Hint: zero.

22. Could this Cantor set serve as the set B in Chapter 4, Section 4.6.5? Hint: yes! Indeed, B should have zero volume. As proved above, this Cantor set indeed has zero volume.

6

Is The Universe Infinite?

In ancient philosophy, it was often assumed that the universe was infinite and unbounded. Nowadays, on the other hand, we know better: the universe contains only a finite number of galaxies, each containing only a finite number of stars. Could this be proved with elementary tools only, with no background in physics at all?

In modern physics, the universe is clearly bounded. As a matter of fact, one can even tell the "size" of the observable universe. This follows from Einstein's principle: the speed of light is maximal, and can never be exceeded. Furthermore, the speed of light is the same in every coordinate system, moving or not. In your own (private) system, you'd also measure the same speed of light: 300,000 kilometer per second (in terms of your own seconds and kilometers). We'll discuss this widely later, in the chapters about relativity.

Thanks to Einstein's principle, the observable universe is actually a three-dimensional "hypersphere." 13.7 billion year ago, at the big bang, it started as a singular point. Since then, it has expanded as fast as light, until today.

What is the "radius" of this hypersphere? It is the long line made by a light ray issuing from this singularity and traveling since the big bang until today. During this time, the ray made quite a long journey: 13.7 billion light year: 13.7 billion year, times $365 \cdot 24 \cdot 3600$ second per year, times 300,000 kilometer per second.

However big, this radius is still finite. Still, can this also be proved using simple math only, assuming no background in physics at all? In this chapter, we prove this from scratch, using elementary math only.

For this purpose, we need just one simple observation. This is indeed the power of experimental physics. This is also the difference between physics and math. Math is easy: it requires no experiment at all. Physics, on the other hand, is more advanced: it must be supported by some experiment, and contradicted by no other experiment.

The universe is finite not only in the macroscale but also in the microscale: matter is discrete, not continuous. (Later on, we'll discuss this in the chapters about quantum mechanics.) In fact, each piece of matter contains only a finite (albeit huge) number of molecules. Each molecule contains only a few atoms. Each atom contains only a few subatomic particles: protons, neutrons, and electrons. Still, in this chapter, we focus on the macroscale only: the universe is bounded, not infinite.

6.1 Flux of Light

6.1.1 The Geometrical Model

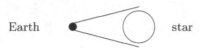

Fig. 6.1. The light coming from a star to the Earth makes a complete cone, full of light.

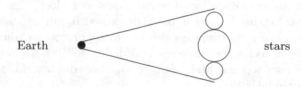

Fig. 6.2. The star is now twice as far. From it, the Earth looks now four times as small, and receives four times less light. To compensate for this, there are now more stars nearby, to make a complete "layer" of stars. Together, they shed on the Earth the same light flux as before.

How much light do we receive here on the Earth? To help estimate this, let's draw a geometrical model. For this purpose, consider a star shining at the Earth (Figure 6.1). Assume that the star is (nearly) as big as the sun: much bigger than the Earth. Assume also that the star is far away from the Earth: nearly as far as the sun. This way, the light rays make a complete cone (three-dimensional sector), full of light.

The total amount of light that the Earth receives from the star is called the light flux. How does it change with distance? To see this, let's imagine a thought experiment.

Suddenly, the star jumps away, and gets twice as far. For an observer who sits on the star, the Earth looks now four times as small: it occupies now four times less area in the sky. Since light shines uniformly in all directions, the Earth receives now four times less light.

To compensate for this, place more stars nearby (Figure 6.2). This makes a vertical "layer" of stars, filling the cone with as much light as before. In fact, the vertical cross-section area of the entire layer is about four times as before. This

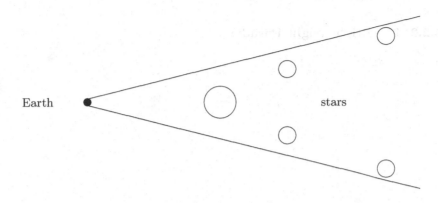

Fig. 6.3. The star still sheds four times less light onto the Earth. To compensate for this, there are now more stars behind it, to fill the entire cone with light, and give the Earth the same light flux as before.

indeed compensates: we have the same cone as before, full of light. So, the light flux remains unchanged: people here on the Earth would see nothing unusual.

Better yet, the layer could get nonuniform: some stars may go ahead and "jump" farther, provided that more stars are added nearby or behind, to make sure that there are no dark "holes" in the sky (Figure 6.3). This way, each individual star that jumped away and gives now less light to the Earth also triggered an immediate compensation: new stars nearby or behind, to make sure that no flux is lost: people on the Earth still see the same cone, full of light.

6.2 Olbers' Paradox

6.2.1 A Wrong Model

In ancient days, it was often assumed that the universe was infinite and unbounded, with infinitely many stars in it. Let's prove this wrong.

The proof is by contradiction. Assume momentarily that the universe contained infinitely many stars. They must have been distributed rather uniformly throughout the outer space. After all, there is no reason why one place should be denser than another, and contain more stars per unit volume.

For example, consider a three-dimensional infinite grid. Each point of the form (l, n, m) (where l, n, and m are some integer numbers) has a small star centered at it. Only at the origin $(0, 0, 0)$ is there no star: the viewer sits there, staring at the sky.

Each individual star is rather small compared to the distance between stars. Still, a star is not infinitesimal: it must still confine a little cube of size $(2\varepsilon) \times (2\varepsilon) \times (2\varepsilon)$, for some small $\varepsilon > 0$.

But this model is wrong! Indeed, if it were true, then we'd have a cone full of light, as in Figure 6.3, even at night!

6.2.2 The Bright-Night Paradox

Fig. 6.4. Olbers' bright-night paradox: if the universe were infinite, then every ray issuing from the Earth would eventually hit some star.

Indeed, in the above infinite grid, consider a viewer who sits on the Earth at $(0,0,0)$, staring at the sky in some direction (Figure 6.4). Let's prove that the ray issuing from his/her eye must eventually hit some star, and get some light from it, even at night!

Indeed, let (x, y, z) be some point on this ray. For simplicity, assume that all three coordinates are nonnegative. Furthermore, assume also that z is positive and maximal:

$$z \geq \max(x, y).$$

(Otherwise, just interchange the role of z with that of x or y.) Consider now the point

$$\left(\frac{x}{z}, \frac{y}{z}, 1\right)$$

on the ray. (This trick is often used in projective geometry.) From Chapter 5, Section 5.3.2, the pair $(x/z, y/z)$ is highly likely to have a rational approximation of the form

$$\left|\frac{x}{z} - \frac{l}{m}\right| \leq \frac{\varepsilon}{m}$$
$$\left|\frac{y}{z} - \frac{n}{m}\right| \leq \frac{\varepsilon}{m},$$

or

$$\left|m\frac{x}{z} - l\right| \leq \varepsilon$$
$$\left|m\frac{y}{z} - n\right| \leq \varepsilon.$$

Thus, the point

$$m\left(\frac{x}{z}, \frac{y}{z}, 1\right) = \left(m\frac{x}{z}, m\frac{y}{z}, m\right)$$

must lie in the star centered at (l, n, m) in the grid. As a matter of fact, the existence of such a star also follows from the Poincare recurrence theorem in a three-dimensional torus (Chapter 4, Section 4.2.3).

Once we've established this, we can now work the other way around: the light from this star could now travel all the way back, along the same line, but in the opposite direction, to meet the viewer's eye. Thus, people here on the Earth would see light everywhere. As in Figure 6.3, there would be no dark hole, even at night!

6.2.3 The Observable Universe

Evidently, night is dark, not bright. Why? Because the universe could never be modeled as an infinite grid as above.

Here one may ask: why not? Perhaps there are infinitely many stars, only they are as far as 13.7 billion light year away from us, so we can never see them! After all, no light could make such a long journey in time to reach our eyes here today!

The answer is simple: such stars, if existed, could never be considered as part of our own observable universe. After all, they are completely detached from us, and have no effect on us whatsoever. As such, they actually belong to a hidden part of the universe, completely irrelevant to us. This part also came from the big bang, so it must be finite as well. Still, this is no longer a scientific question: we'll never know anything about this part, because we'll never get any information from it.

6.3 Energy and Entropy: Laws of Thermodynamics

6.3.1 Stars and Galaxies

By now, we've seen that the universe could never be both infinite and uniform at the same time. Still, could it be infinite and nonuniform? For example, could it be denser near the origin, and sparser elsewhere? Fortunately not. After all, in our infinite grid, the origin is picked arbitrarily. Physically, all places have the same opportunity to attract matter. Why should one place be denser than another?

In a small scale, there could be some nonuniformity. In a large scale, on the other hand, this could be averaged off, and wiped out. For example, in the above model, instead of individual stars, work with galaxies. This way, thanks to the same paradox, the universe could contain only a finite number of galaxies. Furthermore, each galaxy could contain only a finite number of stars. Thus, even in a hierarchical setting, the universe must still be finite and bounded.

6.3.2 Big Bang and Inflation

In fact, the universe must be not only finite but also nonuniform. We've already seen two levels: in the global level, the universe contains galaxies. Within each galaxy, there is a smaller level: stars. Why do we have this hierarchy and nonuniformity?

One theory is like this. Right after the big bang, the universe expanded exponentially fast, in what is known as inflation. If it were completely uniform, then matter could never form. Fortunately, thanks to the uncertainty principle, there were some subtle quantum fluctuations in energy level, at which elementary particles formed.

Fortunately, these fluctuations were so tiny that inflation didn't last for too long. (There could be parallel universes with bigger fluctuations and longer inflation, but no stars or galaxies could possibly form or survive in them for too long.) Later on, the universe cooled down, and gravity helped form clouds of particles and gas, galaxies, and stars.

6.3.3 Symmetry Had Broken

Ideally, the universe should actually have been completely uniform. Right after the big bang, no matter should actually have formed at all. After all, why should a particle form here, not there? Why is one place better than another? Everywhere should have been just the same, in a complete symmetry and uniformity. This could be quite boring.

Fortunately, thanks to the principle of uncertainty, some quantum fluctuations occurred. This way, right after the big bang, some nonsymmetry has also developed. In the inflation that followed, uniformity had broken down. Around small islands of different energy levels, particles formed.

6.3.4 The First Law of Thermodynamics

Thanks to gravity, when the universe cooled down, clouds of particles and gas formed as well, and then galaxies and stars. Thanks to the first law of thermodynamics, energy is never lost: all the original energy that was at the big band remains with us until today. Still, energy may change form: from a well-ordered (useful) energy in the beginning, to a poorly-ordered (used) energy later on.

6.3.5 The Second Law of Thermodynamics

Indeed, the second law of thermodynamics tells us that energy may change form, but only in one direction: on average, good useful energy (with low entropy) could get more and more random and useless (high entropy). For example, the original pure energy, concentrated at the singularity, was used to produce elementary particles and atoms, containing a lot of potential nuclear energy. This energy is still of high quality: it still has low entropy. Still, in the process, a lot of heat was also generated. This is random energy, with high entropy. Thus, the overall entropy always increase: on average, energy can only get less and less useful. This is the second law of thermodynamics.

The first atoms were rather light: they contained only a few protons, neutrons, and electrons. Billions of years later, in a process of nuclear fusion, they merged into heavier atoms, with yet more potential nuclear energy of yet higher quality (lower entropy). Still, this generated a lot of heat, increasing entropy on average. This way, in the entire system, the total entropy increased, as in the second law of thermodynamics. Later on, we'll use the binomial coefficients to discuss entropy from an algebraic-geometrical point of view.

6.3.6 Nature Is Finite and Discrete

Nature is finite in two ways. In terms of astronomic macroscale, the universe contains only a finite number of galaxies, each containing only a finite number of stars. To prove this, we've used mathematical tools only: geometry and set theory. No background in physics was used.

Still, nature is finite not only in the macroscale but also in the microscale and nanoscale: each piece of matter contains only a finite (albeit large) number of molecules, each containing only a few atoms, each containing only a few subatomic particles.

Thus, matter is discrete rather than continuous. As a matter of fact, the distance between two atoms is far larger than the size of an individual atom. In other words, most of matter is just vacuum or empty space. In between the atoms, there is nothing.

Still, to help model nature, the continuum is vital. It gives us the mathematical tools needed to model mechanics, electromagnetics, and relativity.

6.3.7 Continuous Models

Although matter is discrete, it is better described in a continuous geometrical model, with infinitely many points: the real axis, the two-dimensional plane, or the three-dimensional space. Each point could be interpreted as a vector: this makes a linear vector space.

To describe the elementary laws of nature, we better use infinitely many numbers: the real numbers, with their attractive (algebraic and analytic) properties. In this context, differentiation and integration are often used to describe motion, flow, and dynamics in general.

6.4 Exercises: What Is Time?

6.4.1 Entropy and The Time Arrow

1. Is the universe infinite? Why?
2. Use Olbers' bright-night paradox to prove that it is not.
3. In this proof, use the rational approximation in Chapter 5, Section 5.3.2. Hint: see Section 6.2.2.
4. Alternatively, use Poincare recurrence theorem. Hint: see Chapter 4, Section 4.2.3.
5. What is time? Hint: some agreed process that, in our eyes, ticks at a constant (unchanging) rate.
6. What is time in special relativity? Hint: in a given system, time is a process that, in the eyes of an observer who lives in this system, seems to tick at a constant rate (say, a radioactive decay).
7. Draw the arrow of time.
8. Does it point to the future? Why not to the past?
9. What is the direction of time?
10. Who says time flows (or progresses) forward, not backward?

11. What is entropy? Hint: in a specific kind of energy, entropy is the amount of randomness and disorder that make the energy hard to use.
12. Give an example of a kind of energy with low entropy. Hint: potential nuclear energy in the atom.
13. Is this energy of high quality? Hint: yes, it could convert to electricity.
14. Give an example of a kind of energy with high entropy. Hint: heat.
15. Is this energy of high quality? Hint: no, it is too random, disordered, and hard to use.
16. What happens to entropy in time? Hint: in a closed system, the total entropy must increase, not decrease. In other words, although the energy remains constant, it gets less and less useful.
17. What is the correct direction of time? Hint: towards the future! After all, this is the direction of increasing entropy.
18. What was the amount of entropy at the singularity? Hint: at its absolute minimum.
19. What happens in nuclear fusion? Hint: two small atoms, with only a few protons, neutrons, and electrons, merge into one heavy atom.
20. Look at the original small atoms. What energy do they have? Hint: potential nuclear energy.
21. Is this energy of high quality? Hint: yes, this is useful energy, of low entropy.
22. Now, look at the new heavy atom. What energy does it have? Hint: potential nuclear energy.
23. More or less? Hint: more than the original atoms together.
24. It seems that entropy has descreased. After all, there is now more good energy, with low entropy. How come? Hint: in the process, a lot of heat is also generated. This is useless energy, with high entropy. Thus, in a closed system, the total entropy has increased, as in the second law of thermodynamics.

7

Binary Trees
and Chaos Theory

To study stability in calssical mechanics, we use a simple tool from discrete math and set theory: a multilevel hierarchy, or a tree. This may help not only to solve practical problems but also to develop a suitable theory in the first place. To make things clearer, we take a geometrical point of view.

From the dawn of civilization, people used to ask: how to measure weight, distance, or time? This is indeed a most practical problem. For example, you could ask your grocer to have 1700 grams of apples, and pay 400 cents. But it would make more sense to ask 1.7 kilogram, and pay 4 dollars. By grouping 1000 grams in one kilogram, and 100 cents in one dollar, one could move from a too fine scale to a coarser (and more practical) unit of measurement.

The above example shows how often we use multilevel in day-to-day life to group small elements in one composite unit. This way, we "climb" up the hierarchy, from an elementary object to a more complex object, containing many small objects.

In this chapter, we look at a few useful multilevel hierarchies. In particular, we focus on binary trees, with their applications in chaos theory. In this vein, the Cantor set will play a major role: it will help design fractals, and look at chaos theory from a geometrical point of view.

Thanks to trees, we can also look at stability and instability in dynamical systems. This is indeed the start of modern physics: initial data are not always good enough to predict the future. In fact, due to instability, even a tiny error may grow exponentially, and contaminate the results in the long run. We must therefore be more humble, and always have a little doubt. Things are not always as they seem: we can never understand everything. This will be emphasized even more in quantum mechanics and relativity, later on in the book.

7.1 Induction and Deduction

7.1.1 Induction

To show how useful multilevel could be, we start with an important example from the philosophy of science. Thanks to multilevel, this will take a geometrical face, easy to comprehend.

Induction and deduction are probably the most fundamental analytical tools introduced by the ancient Greeks. To help comprehend the deep insight behind them, let's put them in a new two-level scheme.

Assume that you are given a concrete engineering problem: say, to build a road between two particular cities. For this purpose, you are also given the relevant data about resources, costs, topography, etc.

Unfortunately, the specific details may be too tedious and technical. They may cloud the issue, and obscure the entire problem area. To see the light, better drop them.

Furthermore, tomorrow you may be given a slightly different problem: say, to build yet another road between two other cities. Should you then start all over again from scratch? Why not extend the original problem to a more general one, and introduce more general guidelines, to help solve every problem of the kind?

This is indeed induction: extend the concrete problem into a more general one, by introducing new concepts and terminology. The extended problem may seem harder, but is not. On the contrary: the new general terms may help identify the main ingredients, and lead to a new theory, useful in other fields as well.

This way, induction contributes not only to solving the practical problem but also to a deeper understanding of the theoretical components in it. Furthermore, induction may enrich our language and vocabulary, introducing new terms and concepts not only for the present problem but also for many others to come.

7.1.2 Deduction

Once the general problem has been solved efficiently, deduce a specific solution to your concrete problem. For this purpose, just feed the original data to your solver, to specify the relevant circumstances explicitly. This will adapt your well-written code to the original special case, as required. This is indeed deduction.

To help understand the entire process as a whole, let's put it in a two-level scheme.

7.1.3 Two-Level Tree

To help visualize the induction–deduction process geometrically, let's model it in a two-level tree (Figure 7.1). The original problem is placed at the bottom-left. Induction is then used to extend it into a more general problem at the top level. Once the required theory has been developed to help solve the problem in most general terms, deduction is used to help solve practical instances at the bottom-right.

7.1.4 V-Cycle

To visualize this in yet another way, reverse the picture upside-down. This gives the so-called V-cycle (Figure 7.2).

Where did this name come from? Well, the picture is like the Latin letter 'V,' which has two legs: induction at the left leg, to "slide" to the general case at the bottom, and deduction at the right leg, to "climb" up to the top-right, where the general model is applied to the original case.

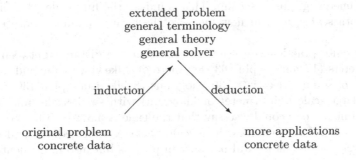

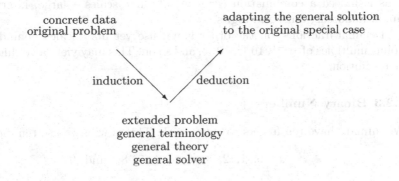

Fig. 7.1. The induction–deduction tree: the original problem at the bottom-left is extended into a more general problem. To help solve it, a general theory is developed at the top level. From this, concrete applications are then deduced at the bottom-right.

Fig. 7.2. The V-cycle – the upside-down tree: the original problem at the top-left is rewritten in most general terms at the bottom. General theory is then developed to form the general solution, which is then adapted to the original data at the top-right.

So far, we've only used a two-level tree. Let's go ahead and introduce more levels, to have a more complicated structure: multilevel. To help motivate this, let's see how common this is in practice.

7.2 Multiscale

7.2.1 How Did Multiscale Develop?

From the dawn of history, people faced a common practical question: how to measure? The answer was simple: multiscale! In the big (coarse) scale, use big units: meter, kilogram, or hour. This helps measure most of the amount. To measure the rest, on the other hand, we can no longer use big units. We must turn to smaller units: centimeter, gram, or second. This is indeed the fine scale. By combining (or adding) both scales, we can now measure the entire amount accurately and efficiently.

This is quite practical: one can now say that the distance between two points is 17.63 meters (17 meters plus 63 centimeters). Likewise, one could also say that the weight of some object is 8.130 kilograms (8 kilograms plus 130 grams). This combines large-scale units (meter or kilogram) with small-scale units (centimeter or gram). Finally, one could also say that the time is 5:31:20 (5 hours, 31 minutes, and 20 seconds). This combines three scales: coarse, intermediate, and fine.

Moreover, multiscale is vital not only in physics but also in elementary math.

7.2.2 Decimal Numbers

To write a natural number, we must use multiscale. For example,

$$178 = 100 + 70 + 8 = 1 \cdot 100 + 7 \cdot 10 + 8 \cdot 1 = 1 \cdot 10^2 + 7 \cdot 10^1 + 8 \cdot 10^0.$$

This is indeed a combination (or sum) of three scales – large, intermediate, and small: hundreds, tens, and digits.

Decimal fractions, on the other hand, use yet finer scales behind the decimal point: multiples of 10^{-1}, 10^{-2}, 10^{-3}, and so on. This may yield a yet higher accuracy or resolution.

7.2.3 Binary Numbers

We humans have ten fingers to count with. This is why we use ten digits:

$$0, \ 1, \ 2, \ 3, \ 4, \ 5, \ 6, \ 7, \ 8, \ \text{and} \ 9$$

to help wirte a decimal number in base 10, as above. The computer, on the other hand, has just two "fingers:" 0 and 1. This is why it uses binary numbers, in base 2.

A binary number, written in base 2, uses powers of 2 rather than 10. A binary number like 1101.01 is interpreted as

$$1 \cdot 2^3 + 1 \cdot 2^2 + 0 \cdot 2^1 + 1 \cdot 2^0 + 0 \cdot 2^{-1} + 1 \cdot 2^{-2}.$$

This representation combines (or sums) six different scales: from 2^3 (the largest or coarsest scale) to 2^{-2} (the smallest or finest scale). The coefficients (either 0 or 1) are stored in the original binary number: 1101.01.

7.3 Dimension

7.3.1 Dimension: A New Definition

Let's use multiscale to define dimension. So far, we've interpreted the concept of dimension geometrically: an interval has dimension 1, a square has dimension 2, and so on. But what is the deeper meaning of dimension?

Fig. 7.3. To cover the unit interval, use ten intervals, of length 1/10 each.

Well, consider the (closed) unit interval

$$[0, 1] \equiv \{x \text{ is a real number} \mid 0 \leq x \leq 1\}.$$

Let's cover it with smaller intervals, of length 1/10 each. How many are needed? Clearly, ten (Figure 7.3).

Now, let's move on to the next finer scale, and use yet smaller intervals, of length 1/100 each. How many are needed? Clearly, one hundred. In general, we need 10^n intervals of length 10^{-n} each $(n \geq 0)$. In this case, we say that our original interval has dimension 1.

7.3.2 Two Dimensions

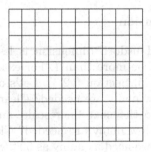

Fig. 7.4. To cover the unit square, on the other hand, use 100 cells, of size $(1/10) \times (1/10)$ each.

Consider now the (closed) unit square:

$$[0, 1]^2 \equiv \{(x, y) \mid 0 \leq x, y \leq 1\}.$$

Let's cover it with small cells, of size $(1/10) \times (1/10)$ each. How many are required? Clearly, one hundred (Figure 7.4).

Now, let's move on to the next finer scale, and use yet smaller cells, of size $(1/100) \times (1/100)$ each. How many are needed? Clearly, $10,000$. In general, to cover the original square with cells of size $10^{-n} \times 10^{-n}$, we need as many as 10^{2n}. This is much more than in Section 7.3.1. We therefore say that the unit square has dimension 2.

What's so good about this new interpretation? Well, it may help define a new kind of geometry, with a strange dimension: no longer an integer number, but a fraction.

7.4 Fractals and Hausdorff Dimension

7.4.1 Fractal and Its Dimension

Look at Cantor's set (in base 10) (Chapter 5, Section 5.2.2). Let's cover it with small intervals, of length $1/10$ each. How many are needed? Clearly, nine (Figure 5.3).

Now, let's move on to the next finer scale, and use yet smaller intervals, of length $1/100$ each. How many are needed? Clearly, $9 \cdot 9 = 81$ (Figure 5.4). In general, we need 9^n intervals of length 10^{-n} each ($n \geq 0$). This is substantially less than in Section 7.3.1:

$$9^n < 10^n.$$

For this reason, Cantor's set is a fractal: it has a fractional dimension, smaller than 1. This is called Hausdorff dimension.

7.4.2 Hausdorff Dimension

In the above example, Hausdorff dimension is less than 1. But this is now the only option: it could also be more than 1. For example, look at the Cantor set in two dimensions (Chapter 5, Section 5.2.3). Let's cover it with small cells, of size $(1/10) \times (1/10)$ each. How many are needed? Clearly, 99 (Figure 5.5).

Now, let's move on to the next finer scale, and use yet smaller cells, of size $(1/100) \times (1/100)$ each. How many are needed? Clearly, $99 \cdot 99$. In general, to use cells of size $10^{-n} \times 10^{-n}$, we need as many as 99^n ($n \geq 0$). This number lies strictly in between

$$10^n < 99^n < 10^{2n}.$$

Therefore, this kind of Cantor set is not really two-dimensional. On the contrary: its dimension must be a fraction in between 1 and 2. This is a fractal as well.

7.4.3 Self Similarity

How does a fractal look like? Well, it has a nice geometrical property: it is similar to itself. For example, look at Cantor's set, restricted to just one subinterval in Figure 5.4. It looks just like the entire Cantor set in the entire interval. Likewise, look at just one subsquare in Figure 5.6. In it, Cantor's set looks just like the entire Cantor set in the entire square.

This is not just a mathematical artifact: it also appears in nature. For example, look at the coast of Norway. Now, look at a small portion of it: It could be as complicated as the entire coast. This is indeed self-similarity.

Of course, nature is finite. Therefore, at a very fine scale, self-similarity breaks down. Indeed, if you stand at the beach, then you'd see a nearly straight line. We don't talk about such a small scale. In every reasonable scale, as in a photograph taken from a satellite, the coast always looks like a fractal: similar to itself.

Later on, we'll see self-similarity in an infinite binary tree as well. Before going into this, let's see self-similarity in its simplest form: a unary tree.

7.5 Mathematical Induction

7.5.1 Unary Tree

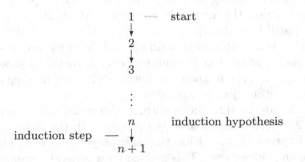

Fig. 7.5. Mathematical induction as a unary tree: at the top, make sure that 1 has the property. For $n = 1, 2, 3, 4, \ldots$, assume that n has the property (the induction hypothesis), and prove that $n + 1$ has the same property as well (the induction step).

So far, we've seen a few examples of multiscale, used in elementary math and physics. Now, let's look at multilevel from a geometrical point of view.

In the simplest kind of multilevel, each level has just one item in it. This could be modeled in a unary tree: from each level, issue just one branch, pointing downwards, to the next lower level underneath. As a matter of fact, we've already met this model, and used it implicitly in mathematical induction.

The ancient Greeks introduced the principle of induction as a powerful thinking tool. From this wide principle, we can now obtain mathematical induction as a formal version.

Indeed, mathematical induction is similar, but much more technical. Assume that we already know that 1 has some desirable property. Assume that we also have a mechanism to prove that, if some number has the property, then its successor has the same property as well. How to show that 1000 has the same property as well?

For this purpose, we must first show that 999 has the property. For this, we must first show that 998 has the property, and so on. This is indeed recursion. It goes backward, as in reverse engineering. Why not go forward, from 1 onward, as in standard engineering? This is indeed the other face of the same process: mathematical induction.

Unfortunately, the above recursion is a long and tedious process: we must use the same mechanism over and over again, as many as 999 times! To avoid this, use a powerful thinking tool: induction. To apply it, extend the original task into a yet more general task: prove the property not only for 1000 but also for larger numbers as well. This seems more difficult, but is not. After all, what's so special about 1000? Better use the same mechanism not only up to 1000 but also beyond! For this purpose, prove one thing only: if $1, 2, \ldots, n$ have the property (the induction hypothesis), then $n + 1$ has it as well (the induction step).

That's it: we are done! We can now deduce that 1000 indeed has the property, as required. After all, in theory, we could always start from 1, and march step by step to $2, 3, 4, \ldots, 1000$, proving that all of them have the property as well. Fortunately, we don't need to do this explicitly any more. After all, we've already established that this is possible in theory.

To visualize this better, draw a simple multilevel hierarchy: a unary tree (Figure 7.5). First, place 1 at the top. For this purpose, prove that 1 indeed has the property. Then, place $2, 3, 4, \ldots, n$ in lower and lower levels, each in a separate level. This way, the nth level contains just one number: n.

To do this, you need to prove nothing. After all, the induction hypothesis says that levels $2, 3, \ldots, n$ indeed have the property. In the induction step, we use this hypothesis to prove that $n+1$ has the same property as well. This allows us to place $n + 1$ in the next lower level. This is done for a general (unspecified) n. In fact, the same could be done for each and every n. That's it: we're done!

We can now use our work, and deduce a special case: $n = 1000$. For this purpose, no more work is needed. After all, in theory, one could always start from 1, and "jump" down from level to level, until reaching the 1000th level, as required.

Mathematical induction could be used not only to establish a particular property but also to define the natural numbers in the first place. Later on, we'll use it for our own goal: to define a tree.

7.5.2 Example: Finite Power Series

A finite (geometric) power series can be written in two equivalent ways:

$$1 + q + q^2 + q^3 + \cdots + q^n = \sum_{i=0}^{n} q^i,$$

where $n \geq 0$ is a given integer number, and $q \neq 1$ is a given parameter. Here, i is just an index that runs from 0 to n.

Let's use mathematical induction to prove that the sum is

$$\sum_{i=0}^{n} q^i = \frac{q^{n+1} - 1}{q - 1}.$$

Indeed, for $n = 0$, this is easy to prove. After all, in this trivial case, the series is very short: it contains one term only – the leading term $q^0 = 1$:

$$\sum_{i=0}^{n} q^i = \sum_{i=0}^{0} q^i = q^0 = 1 = \frac{q^1 - 1}{q - 1} = \frac{q^{n+1} - 1}{q - 1}.$$

This proves the above formula for $n = 0$.

Let us now consider a more interesting case: $n \geq 1$. Assume that our formula holds for a slightly shorter series:

$$\sum_{i=0}^{n-1} q^i = \frac{q^n - 1}{q - 1}.$$

This is indeed the induction hypothesis: it is obtained from the original formula by substituting $n - 1$ for n. Thanks to it, we can now prove the original formula. For this purpose, split the original series into two parts. The first part contains most of the terms: the n leading terms. Fortunately, their sum has just been written down. The second part, on the other hand, contains one term only – the final term:

$$\sum_{i=0}^{n} q^i = \left(\sum_{i=0}^{n-1} q^i \right) + q^n$$
$$= \frac{q^n - 1}{q - 1} + q^n$$
$$= \frac{q^n - 1 + q^n(q - 1)}{q - 1}$$
$$= \frac{q^{n+1} - 1}{q - 1},$$

as required. This completes the induction step. This completes the mathematical induction, proving that

$$\sum_{i=0}^{n} q^i = \frac{q^{n+1} - 1}{q - 1}$$

for every integer number $n \geq 0$, as required.

This is just algebra: after all, we only used numbers, and arithmetic operations between them. Let's move on to calculus: the limit process. For this purpose, we need a new tool: infinity, or ∞.

7.5.3 Infinite Power Series

Consider now the infinite power series:

$$1 + q + q^2 + q^3 + \cdots = \sum_{i=0}^{\infty} q^i.$$

What is the meaning of this? We must distinguish between a few different cases. Let's start with a few strange cases. If $q = 1$, then this is just the constant series:

$$1 + 1 + 1 + \cdots = \sum_{i=0}^{\infty} 1.$$

Later on, we'll see that this series approaches infinity, or converges to ∞.

A more interesting case, on the other hand, is $q \neq 1$. From the previous section, we already know how to sum the $n + 1$ leading terms, to produces the partial sum S_n:

$$S_n \equiv \sum_{i=0}^{n} q^i = 1 + q + q^2 + \cdots + q^n = \frac{q^{n+1} - 1}{q - 1}.$$

We can now deal with another degenerate case: $q = -1$. In this case, there are alternating signs, so each term cancels the previous one:

$$1 - 1 + 1 - 1 + 1 - \cdots = (1 - 1) + (1 - 1) + 1 - \cdots.$$

In other words, each pair of two consecutive terms sums to 0, contributing nothing to the total sum. As a result, we must distinguish between two cases: an even n and an odd n:

$$S_n = \begin{cases} 1 & \text{if } n \text{ is even} \\ 0 & \text{if } n \text{ is odd.} \end{cases}$$

Fortunately, this is also in agreement with our formula:

$$S_n = \frac{q^{n+1} - 1}{q - 1} = \frac{(-1)^{n+1} - 1}{-2} = \begin{cases} 1 & \text{if } n \text{ is even} \\ 0 & \text{if } n \text{ is odd.} \end{cases}$$

So, in this case, S_n can never make up its mind. As n grows, S_n never converges: it always switches from 1 to 0 and back to 1 and so on, with no limit. We then say that the original series diverges. Likewise, if $q < -1$, then the series diverges as well.

Another strange case is $q > 1$. In this case, as n grows, q^n gets as large as you like:

$$q^n \to_{n \to \infty} \infty.$$

Therefore, the power series "converges" to infinity, or approaches ∞:

$$S_n = \frac{q^{n+1} - 1}{q - 1} \to_{n \to \infty} \infty.$$

This is often written as

$$\lim_{n \to \infty} S_n = \infty,$$

or simply

$$\sum_{i=0}^{\infty} q^i = \infty.$$

Finally, we arrive at the really interesting case: $|q| < 1$. In this case, the power series really converges: not to infinity, but to a finite limit. Indeed, in this case, as n grows, $|q^n|$ gets as small as you like:

$$q^n \to_{n \to \infty} 0.$$

Thus,

$$S_n = \frac{q^{n+1} - 1}{q - 1} \to_{n \to \infty} \frac{-1}{q - 1} = \frac{1}{1 - q}.$$

This is often written as

$$\sum_{i=0}^{\infty} q^i = \lim_{n \to \infty} S_n = \frac{1}{1 - q}.$$

These results will be useful later on.

7.6 Trees

7.6.1 General Tree

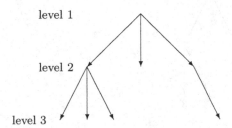

level 1

level 2

level 3

Fig. 7.6. A three-level tree. From the head, three branches issue. The left and right branches point to a new two-level subtree. The middle branch, on the other hand, points to a leaf.

The unary tree is rather degenerate and boring, and hardly deserves to be called a tree. Indeed, in it, each level contains just one branch, pointing to an equally boring unary subtree. A general tree, on the other hand, is much more interesting: a complex multilevel hierarchy, which develops and grows level by level.

To define a general tree, use mathematical induction. Start from a degenerate one-level tree: just one dangling node (or leaf), to serve as a head (or root).

Now, for $n = 1, 2, 3, 4, \ldots$, look at a fixed n. Assume that we already know how to construct any k-level tree ($1 \le k \le n$). This is indeed the induction hypothesis.

Now, how to construct a bigger $(n+1)$-level tree? First, form the head (or root), and place there a new node. From it, issue a few new edges (or branches), each pointing to a new k-level subtree (for some $1 \le k \le n$).

Thanks to the induction hypothesis, we already know how to construct these subtrees. In this construction, at least one of them must have $k = n$ levels. The rest, on the other hand, may contain $k \le n$ levels. This way, the entire tree indeed has

$n + 1$ levels, as required. Still, it is not necessarily uniform (symmetric): different subtrees may have a different number of levels, smaller than or equal to n. (Later on, we'll also design more symmetric trees.)

From the general tree, we can now deduce a few examples. The small two-level tree in Figure 7.1 is easy enough to construct. For this purpose, set $n = 2$. From the head, issue just two branches. At the end of each branch, stick a trivial one-level tree: just a leaf.

The three-level tree in Figure 7.6, on the other hand, is a little more complicated. To construct it, set $n = 3$. From the head, issue three branches. To the middle branch, attach a leaf. To the left and right branches, on the other hand, attach a new two-level subtree.

7.6.2 Arithmetic Expression

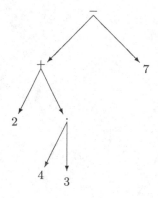

Fig. 7.7. Storing the arithmetic expression $2 + 4 \cdot 3 - 7$ in a four-level tree, and calculating its value, bottom to top. The top-priority arithmetic operation, $4 \cdot 3$, is carried out at the bottom. The next operation, $2 + (4 \cdot 3)$, is carried out at the second level. Finally, the least prior operation, $(2 + 4 \cdot 3) - 7$, is carried out at the top.

What is the tree good for? Well, it could be used to store an arithmetic expression (Figure 7.7). In the leaves, store the operands: some numbers. In the rest of the nodes, on the other hand, store arithmetic symbols like '+', '−', '·', or '/'.

Where to place the arithmetic symbols? Well, this depends on their priority. In the head, place the symbol of *least* priority: either '+' or '−'. If your expression contains many such symbols, then pick the latter instance, and place it in the head. After all, it is least prior. This splits the original expression into two subexpressions: one on the left (stored recursively in the left subtree), and the other on the right (stored recursively in the right subtree). This is indeed mathematical induction. After all, the induction hypothesis tells us how to store a shorter subexpression in its own subtree.

Where to store the subtrees? For this purpose, issue two branches from the head. Attach the left subtree to the left branch, and the right subtree to the right branch.

If, however, there is no '+' or '−' in the entire expression, then do the same with higher-priority symbols: '·' or '/', and so on. This completes the entire tree.

7.6.3 How to Calculate?

How to calculate the numerical value of the arithmetic expression? First, calculate the values of each subexpression recursively. This way, you obtain two values. Apply to them the arithmetic operation at the head of the tree. That's it: you're done!

7.6.4 How to Check Correctness?

How do we know that this is correct? By mathematical induction, of course! After all, each subtree contains less levels than the entire tree. Thus, the induction hypothesis tells us that the subexpressions have been calculated properly. Finally, check that the symbol in the head was applied properly.

This highlights an important principle: mathematical induction mirrors recursion, and is actually the same thing! We'll come back to this later.

For example, the arithmetic expression

$$2 + 4 \cdot 3 - 7$$

is stored in a four-level tree (Figure 7.7). In this expression, what operation is least prior? Clearly, this is the subtraction at the end. Place the '$-$' symbol at the head, to be applied last.

What operation has higher priority? This is the addition at the beginning. Place the '$+$' symbol at the next lower level.

Finally, what operation has the highest priority? This is the multiplication at the middle. Place the '$\cdot$' symbol at the next lower level, to be applied to the numbers in the leaves. This way, the calculation will be carried out bottom to top, as required.

7.6.5 Virtual Tree

How does the computer calculate such an arithmetic expression? Fortunately, it uses a tree only virtually and implicitly.

Indeed, in practice, no concrete tree is needed. Instead, the computer uses a recursive algorithm (list of instructions):

1. Scan the arithmetic expression backwards: from right to left, as in Hebrew.
2. Look for the arithmetic symbol of least proiroty: either '$+$' (addition) or '$-$' (subtraction), in between two subexpressions. For example, in $2 + 4 \cdot 3 - 7$, the minus symbol is found first.
3. Once such an arithmetic symbol has been found,
   a) use it to split the original expression into two disjoint subexpressions.
   b) Calculate each of them recursively, using the same algorithm itself. This gives two subresults.
   c) Apply to them the relevant arithmetic operation. (In the above example, subtract them from each other.)
   d) This is your result: you are done.
4. If, however, there is no addition or subtraction in the entire expression, then start from the beginning, and do the above scanning once again, only this time look for arithmetic symbols of higher priority: either '$\cdot$' (multiplication) or '$/$' (division).

5. If, however, no such arithmetic symbol can be found either, then look at the first symbol in the expression (on the left).
6. If this is '−', then this is no subtraction, but just the negative of the number that follows.
7. Finally, if there are no arithmetic symbols at all, then this must be a number. So, we're at the innermost recursion, and this number is our result. Later on, we'll see that a number is often written as a polynomial: sum of powers of 10. We'll also see a few ways to calculate it.

Fortunately, this algorithm mirrors the underlying tree, without constructing it explicitly. Indeed, each recursive call opens a new virtual subtree.

What operation is carried out first? Well, first of all, the algorithm carries out the innermost recursive call: just evaluate an individual number, as required.

Let's go ahead and mirror this in a Boolean expression as well.

7.6.6 Boolean Expression

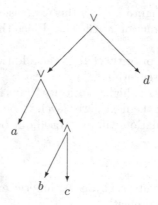

Fig. 7.8. Storing the Boolean expression $a \vee b \wedge c \vee d$ in a four-level tree. The calculation is made bottom to top. The top-priority Boolean operation, $b \wedge c$, is made at the bottom. The next operation, $a \vee (b \wedge c)$, is made at the second level. Finally, the least prior operation, $(a \vee b \wedge c) \vee d$, is made at the top.

Boolean expressions mirror arithmetic expressions: they have the same form and structure, with just a different interpretation. Here, the "and" operation (denoted by '$\wedge$') is prior to the "or" operation (denoted by '$\vee$'). The "not" operation, on the other hand, is prior to both. This operation is different: it is a unary operation, applied to one operand only. In this respect, it mirrors the minus sign in arithmetic.

Instead of numbers, we now have Boolean variables, which may take just two possible values: either 1 for true, or 0 for false. As a result, the value of the entire expression must also be either 1 (true) or 0 (false). As before, it is calculated recursively, bottom to top. Indeed, the example in Figure 7.8 mirrors the example in Figure 7.7.

7.7 Binary Tree

7.7.1 Full Binary Tree

So far, we looked at a general tree: from each node, issue as many branches as you like. Now, let's look at a (full) binary tree: from each node, issue either zero or two branches.

What does "full" mean here? It means that all nodes must have the same number of branches. In our case, this number is two. Indeed, all nodes have two branches. Only leaves have a different number: zero. As a matter of fact, we've already seen two such examples in Figures 7.7–7.8. In what follows, we deal with full trees only, so there is no need to keep writing "full."

How could a (full) binary tree look like? Well, there are two options. It could be really small: no branches at all – just one dangling leaf, and that's it. On the other hand, it could look like a real tree. For this, it must have two branches, issuing from the head. To the left branch, attach a binary subtree, defined recursively. To the right branch, on the other hand, attach yet another binary subtree, which may have a different number of levels. This way, the entire binary tree may be uneven and nonsymmetric: deeper on one side than on the other.

What have we done above? Just mathematical induction! Indeed, we've defined a one-level binary tree: just a single node, and that's it. Now, for $n = 1, 2, 3, 4, \ldots$, look at a fixed n. Assume that we already know how to define a k-level binary tree ($1 \le k \le n$). This is indeed the induction hypothesis.

How to define a slightly bigger binary tree, with $n + 1$ levels? First, place a new node at the head. From it, issue two new branches. At the end of one branch, stick a new n-level binary subtree. At the end of the other branch, stick a new k-level binary subtree, for some $1 \le k \le n$. Because k might be smaller than n, the entire binary tree may be uneven and nonuniform.

We've already seen two examples of a (full) binary tree (Figures 7.7–7.8). Why didn't we mention this at the time? Because arithmetic and Boolean expressions should better be stored in a general tree, not a full tree. After all, in many cases, they may need to use a unary operation: the minus sign, or the "not" operation.

As discussed above, the binary tree is not necessarily even or uniform: it may contain leaves not only at the bottom level but also at an intermediate level. In Figures 7.7–7.8, for example, there are leaves not only in the fourth level but also in the third level (the left node), and even the second level (the right node). A perfect tree, on the other hand, is much more even and symmetric.

7.7.2 Perfect Binary Tree

In the trees discussed so far, all levels may contain leaves. This is why they are uneven and nonuniform: deeper on one side than on the other. A perfect tree, on the other hand, is much more uniform: only one level may contain leaves: the bottom level. In Figure 7.9, for example, there are leaves in the fourth level only.

How to define a perfect binary tree? As befores, use mathematical induction. A one-level tree is very small: just one dangling node, and that's it. Next, let's define more interesting trees. For $n = 1, 2, 3, 4, \ldots$, assume that we already know how to define an n-level perfect binary tree. With this knowledge at hand, we can now go

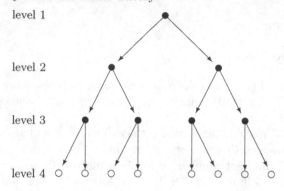

level 1

level 2

level 3

level 4

Fig. 7.9. A four-level perfect binary tree. The arrows stand for branches. The bullets stand for nodes at intermediate levels. The circles, on the other hand, stand for leaves at the bottom.

ahead and define a new $(n+1)$-level perfect binary tree as well: place a new node at the head, and issue two branches from it. At the end of the left branch, stick an n-level perfect binary subtree. At the end of the right branch, on the other hand, stick yet another n-level perfect binary subtree. This way, the new $(n+1)$-level tree is indeed even and symmetric, as required. To prove this, use mathematical induction once again.

7.7.3 Mathematical Induction and Recursion

What is the difference between mathematical induction and recursion? Well, they mirror each other, and can be viewed as two sides of the same coin. Mathematical induction is the theoretical principle. It goes forward, and engineers the process from scratch onwards. Recursion, on the other hand, puts it into practice. It works the other way around: it goes backwards, and reverse-engineers the process.

To construct a perfect binary tree, for example, the induction step uses two n-level subtrees. But how is a subtree constructed in the first place? Recursively, of course! For this purpose, one must use two $(n-1)$-level subtrees, and so on. This is indeed recursion.

Although recursion does the hard work, mathematical induction gets all the glory. After all, it is useful to prove theorems. Our perfect binary tree is a good example for this. Thanks to its multilevel structure, it has some nice properties:

- As in a general tree, the bottom level contains leaves only.
- The rest of the levels, on the other hand, contain no leaf any more. On the contrary: every node in them has exactly two branches.
- This is indeed perfect: even and uniform, as in Figure 7.9.
- The number of nodes doubles from level to level: in the first level, there are $2^0 = 1$ nodes. In the next lower level, there are $2^1 = 2$ nodes. In the ith level, there are 2^{i-1} nodes ($1 \le i \le n$).
- The total number of nodes is, thus,

$$\sum_{i=1}^{n} 2^{i-1} = \sum_{i=0}^{n-1} 2^i = \frac{2^n - 1}{2 - 1} = 2^n - 1$$

(Section 7.5.2). To prove these properties, use mathematical induction. In the exercises below, we'll also calculate the total number of branches.

7.8 Application in Chaos Theory

7.8.1 Exponential Growth

Thus, the number of nodes increases exponentially: it doubles from level to level. This is a very rapid growth: at the 11th level, for example, there are as many as

$$2^{11-1} = 2^{10} = 1024 > 1000 = 10^3$$

nodes. The index i, on the other hand, grows much more slowly: linearly. Thus, while the index grows moderately from i to $i+1$, the number of nodes doubles from 2^{i-1} to 2^i.

Such a sharp jump is not always benign. For example, an algorithm (or method) that requires exponential time or storage is often considered as impractical. This is indeed a "bad" exponential growth: it leaves us with no control. This is indeed in the heart of chaos theory.

7.8.2 Relative vs. Absolute Error

To introduce chaos, let's consider a simple algebraic process. It starts from a given parameter: $v_0 \neq 0$. How does it continue? Step by step. At each step, it multiplies by yet another given parameter: $\alpha \neq 0$. In other words, for $n = 0, 1, 2, \ldots$, define

$$v_{n+1} \equiv \alpha v_n.$$

As a matter of fact, this is just mathematical induction.

This is called a linear process: in each step, v_n is multiplied by a constant number: α. Later on, we'll also see a nonlinear process, in which v_n is multiplied by itself as well.

Thanks to mathematical induction, v_n also has a more explicit form:

$$v_n = \alpha v_{n-1} = \alpha^2 v_{n-2} = \cdots = \alpha^n v_0.$$

Assume now that v_n is some physical quantity, measured to some finite precision only. So, the v_n's can never be known precisely, but only within a small error.

As a matter of fact, the error may appear already in the initial data. To account for this, assume that v_0 is available only approximately, as

$$v_0 + \delta_0,$$

where the error δ_0 is small in magnitude relative to v_0:

$$|\delta_0| \ll |v_0|.$$

Under these circumstances, we can no longer compute v_n exactly, but only approximately: with an error δ_n. Indeed, mathematical induction can still be used to define the approximations

$$\begin{aligned}
v_n + \delta_n &\equiv \alpha(v_{n-1} + \delta_{n-1}) \\
&= \alpha^2(v_{n-2} + \delta_{n-2}) \\
&= \cdots = \alpha^n(v_0 + \delta_0) \\
&= \alpha^n v_0 + \alpha^n \delta_0 \\
&= v_n + \alpha^n \delta_0.
\end{aligned}$$

Thus, the error δ_n undergoes the same linear process as well:

$$\delta_n = \alpha^n \delta_0.$$

If $|\alpha| > 1$, then the absolute error $|\delta_n|$ may grow exponentially with n. Fortunately, what matters is not the absolute but the relative error, which doesn't grow at all:

$$\left| \frac{\delta_n}{v_n} \right| = \left| \frac{\alpha^n \delta_0}{\alpha^n v_0} \right| = \left| \frac{\delta_0}{v_0} \right|.$$

This is indeed a stable process: the relative error remains small, so $v_n + \delta_n$ approximate v_n well. For this reason, $v_n + \delta_n$ is good enough in practice. In a nonlinear process, on the other hand, this is not necessarily the case.

7.9 Stability vs. Instability

7.9.1 Unstable Quadratic Process

To describe a nonlinear process, we need yet another parameter: $\beta \neq 0$. With it, we can now define a quadratic process:

$$v_{n+1} \equiv \alpha v_n + \beta v_n^2.$$

Let's write this as

$$v_{n+1} - v_n = y(v_n),$$

where

$$y(x) \equiv (\alpha - 1)x + \beta x^2$$

is the quadratic function that tells us how the process progresses.

Let us distinguish between two major cases. Assume first that

$$\alpha \leq 1.$$

For simplicity, assume also that

$$\beta > 0.$$

(Otherwise, just work with $-v_n$ and $-\beta$.) This way, the graph of $y(x)$ is a straight parabola, as in Figure 7.10.

Do the v_n's converge to any limit? Well, let v_∞ be such a hypothetical limit. It must satisfy

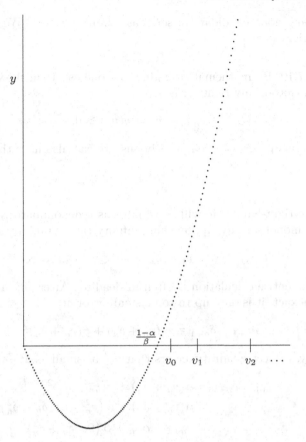

Fig. 7.10. An unstable quadratic process. The v_n's increase monotonically towards infinity, away from $(1 - \alpha)/\beta$. The calculation is unstable and meaningless.

$$0 = v_\infty - v_\infty$$
$$= \lim_{n\to\infty} v_n - \lim_{n\to\infty} v_n$$
$$= \lim_{n\to\infty} v_{n+1} - \lim_{n\to\infty} v_n$$
$$= \lim_{n\to\infty} (v_{n+1} - v_n)$$
$$= \lim_{n\to\infty} y(v_n)$$
$$= y(v_\infty).$$

Thus, v_∞ must solve the quadratic equation

$$y(x) = 0.$$

We already have one solution: $x = 0$. Still, there is a more interesting solution:

$$v_\infty \equiv \frac{1 - \alpha}{\beta} \geq 0.$$

This is indeed a good candidate to serve as a limit. Or is it? Well, assume that we start to the right of v_∞:

$$v_0 > v_\infty,$$

as in Figure 7.10. By mathematical induction, one can then show that the process progresses monotonically rightwards:

$$v_{n+1} - v_n = y(v_n) > 0.$$

How rapid is this progress? Well, as a bonus, we can also have the ratio

$$1 \le \frac{v_{n+1}}{v_n} = \alpha + \beta v_n.$$

(This positive right-hand side will serve later as a denominator.) In summary, the v_n's increase monotonically, approaching infinity rather rapidly, away from v_∞:

$$v_\infty < v_0 < v_1 < v_2 < \cdots < v_n \to_{n\to\infty} \infty.$$

Yet worse: the entire calculation is often misleading. After all, in practice, the calculation is inexact: it is only up to some small error δ:

$$v_{n+1} + \delta_{n+1} \equiv \alpha(v_n + \delta_n) + \beta(v_n + \delta_n)^2.$$

For simplicity, neglect quadratic errors that are as small as δ^2:

$$\begin{aligned}
v_{n+1} + \delta_{n+1} &= \alpha(v_n + \delta_n) + \beta(v_n + \delta_n)^2 \\
&= \alpha(v_n + \delta_n) + \beta\left(v_n^2 + 2v_n\delta_n + \delta_n^2\right) \\
&= v_{n+1} + \alpha\delta_n + \beta\left(2v_n\delta_n + \delta_n^2\right) \\
&\doteq v_{n+1} + \alpha\delta_n + 2\beta v_n\delta_n \\
&= v_{n+1} + (\alpha + 2\beta v_n)\,\delta_n.
\end{aligned}$$

Thus, the new error is approximately

$$\delta_{n+1} \doteq (\alpha + 2\beta v_n)\,\delta_n.$$

Usually, $\delta_n \ne 0$, so we can also divide:

$$\frac{\delta_{n+1}}{\delta_n} \doteq \alpha + 2\beta v_n.$$

In calculus, this is actually the derivative of v_{n+1} at v_n, denoted by $v'_{n+1}(v_n)$. It tells us how the error grows with n.

Thanks to the above, we can also go ahead and approximate the relative error as well:

$$\frac{\delta_{n+1}}{v_{n+1}} \doteq \frac{(\alpha + 2\beta v_n)\,\delta_n}{v_{n+1}} = \frac{(\alpha + 2\beta v_n)\,\delta_n}{\alpha v_n + \beta v_n^2} = \frac{\alpha + 2\beta v_n}{\alpha + \beta v_n} \cdot \frac{\delta_n}{v_n}.$$

Thus, upon advancing from n to $n+1$, the relative error multiplies by the approximate factor

$$\frac{\alpha + 2\beta v_n}{\alpha + \beta v_n} = 2 - \frac{\alpha}{\alpha + \beta v_n} \ge \begin{cases} 2 & \text{if } \alpha \le 0 \\ 2 - \frac{\alpha}{\alpha + \beta v_0} & \text{if } \alpha > 0. \end{cases}$$

This is bad news. Indeed, in either case, the relative error grows exponentially fast:

$$\left| \frac{\delta_n}{v_n} \right| \geq \left(\frac{|\alpha| + 2\beta v_0}{|\alpha| + \beta v_0} \right)^n \left| \frac{\delta_0}{v_0} \right|.$$

Thus, the computed value $v_n + \delta_n$ is no good: it is no approximation to the desirable value v_n at all. Let's move on to a more meaningful case.

7.9.2 The Logistic Equation

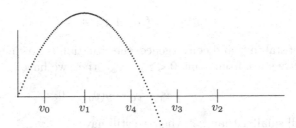

Fig. 7.11. A stable quadratic process. The v_n's approach v_∞. The calculation is now stable and meaningful.

So far, we've assumed that $\alpha \leq 1$. Unfortunately, the result was an unstable process, which can never be calculated in practice.

Let us now move on to a fundamentally different case [35]. For this purpose, assume now that

$$\alpha > 1.$$

For simplicity, assume also that

$$\beta < 0.$$

(Otherwise, just work with $-v_n$ and $-\beta$.) This way, the graph of $y(x)$ is now an upside-down parabola, as in Figure 7.11. As before, it still meets the x-axis at

$$v_\infty \equiv \frac{1-\alpha}{\beta} \geq 0.$$

This solves the so-called logistic equation:

$$y(v_\infty) = (\alpha - 1)v_\infty + \beta v_\infty^2 = 0.$$

In terms of stability and convergence, the situation is now much better than before. Indeed, if we start from some $0 < v_0 < v_\infty$, then we have

$$v_1 - v_0 = y(v_0) > 0.$$

If v_1 is still smaller than v_∞, then we still have

$$v_2 - v_1 = y(v_1) > 0.$$

But v_2 could be too large: it could skip v_∞, as in Figure 7.11. In this case, we have

$$v_3 - v_2 = y(v_2) < 0,$$

so v_3 may again approach v_∞, from the right.

This is not the end of the story: v_4 may skip v_∞ once again, and be on its left again. Fortunately, v_5 may still approach v_∞ from the left, and so on. In summary, the process may approach v_∞ from alternating sides.

7.9.3 Control Parameter

By now, it seems that our v_n's do approach their limit: v_∞. Or do they? To check on this, assume that they are already pretty close to v_∞, within a new error δ. This new δ is not the same as before: it stands now for the distance from the limit. We can now write

$$
\begin{aligned}
v_\infty + \delta_{n+1} = v_{n+1} \\
&= \alpha v_n + \beta v_n^2 \\
&= \alpha(v_\infty + \delta_n) + \beta(v_\infty + \delta_n)^2 \\
&= \alpha(v_\infty + \delta_n) + \beta\left(v_\infty^2 + 2v_\infty\delta_n + \delta_n^2\right) \\
&= v_\infty + \alpha\delta_n + \beta\left(2v_\infty\delta_n + \delta_n^2\right) \\
&\doteq v_\infty + \alpha\delta_n + 2\beta v_\infty\delta_n \\
&= v_\infty + \alpha\delta_n + 2(1-\alpha)\delta_n \\
&= v_\infty + (2-\alpha)\delta_n.
\end{aligned}
$$

In summary,

$$\delta_{n+1} \doteq (2 - \alpha)\delta_n.$$

Assuming that $\delta_n \neq 0$, we can also divide:

$$\frac{\delta_{n+1}}{\delta_n} \doteq 2 - \alpha.$$

In the language of calculus, this is the derivative of v_{n+1} at v_∞, denoted by $v'_{n+1}(v_\infty)$ (Chapter 24).

Is the process stable? Well, it depends on the original parameter α: if it is not too large,

$$1 < \alpha \leq 3,$$

then the error (the distance from the limit) remains small:

$$|\delta_{n+1}| \doteq |(2 - \alpha)\delta_n| = |2 - \alpha| \cdot |\delta_n| \leq |\delta_n|.$$

In this case, the relative error remains small as well:

$$\left| \frac{\delta_{n+1}}{v_\infty} \right| \leq \left| \frac{\delta_n}{v_\infty} \right|,$$

so the calculation is indeed meaningful. This is independent of β: α alone tells us whether the process is stable or not. This is why α is called the control parameter. Still, β is also important: its sign tells us how the parabola looks like.

7.10 Chaos: Find The Hidden Order!

7.10.1 Bifurcation Points

We can now discuss different kinds of α's. In particular, the special point

$$\alpha_0 \equiv 3$$

distinguishes "good" α's from "bad" α's. This is why α_0 is called a bifurcation point. For α's larger than α_0, there is no convergence any more.

Still, even for these "bad" α's, there is still some hope. In fact, there is yet another bifurcation point:

$$\alpha_1 \doteq 3.45.$$

For those α's in between

$$\alpha_0 < \alpha < \alpha_1,$$

although the original process no longer converges, it still splits into two subprocesses that do converge: the even-numbered items converge to one limit,

$$v_0, v_2, v_4, v_6, \ldots, v_{2n} \to_{n \to \infty} v_{2\infty},$$

whereas the odd-numbered items converge to another limit:

$$v_1, v_3, v_5, v_7, \ldots, v_{2n+1} \to_{n \to \infty} v_{2\infty+1}.$$

Here, $v_{2\infty}$ and $v_{2\infty+1}$ are just convenient notations for two different numbers. And what happens beyond α_1? Well, the next bifurcation point is

$$\alpha_2 \doteq 3.55.$$

For those α's in between

$$\alpha_1 < \alpha < \alpha_2,$$

there are four disjoint subprocesses, converging to four different limits:

$$v_{4n} \to_{n\to\infty} v_{4\infty}$$
$$v_{4n+1} \to_{n\to\infty} v_{4\infty+1}$$
$$v_{4n+2} \to_{n\to\infty} v_{4\infty+2}$$
$$v_{4n+3} \to_{n\to\infty} v_{4\infty+3},$$

and so on.

7.10.2 Bifurcation Tree

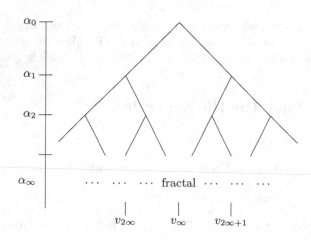

Fig. 7.12. The bifurcation tree.

This goes on: one could go on defining infinitely many bifurcation points:

$$\alpha_0 < \alpha_1 < \alpha_2 < \alpha_3 < \alpha_4 < \cdots < \alpha_m \to_{m\to\infty} \alpha_\infty \doteq 3.57.$$

What happens at $\alpha = \alpha_\infty$? Chaos! No subsequence converges any more. In other words, the original process has no accumulation point any more (Chapter 4, Section 4.4.4).

Fortunately, even at chaos, there is still some hope. Indeed, from deep chaos, a beautiful geometrical pattern emerges: a new bifurcation tree (Figure 7.12). At its "bottom," it approaches a new fractal, of dimension less than 1.

7.10.3 Binary Tree and Its Fractal

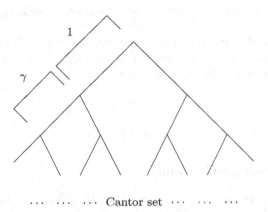

Fig. 7.13. At its bottom, the binary tree approaches a Cantor set, of dimension less than 1.

How does the bifurcation tree look like from below? To visualize this, let's turn to a simpler model: a binary tree (Figure 7.13). Assume that it has infinitely many levels: at the bottom, the levels "shrink," and get more and more crowded. In other words, upon advancing from one level to the next lower level, the branches get shorter by the constant factor

$$0 < \gamma < \frac{1}{2}.$$

How does this look like? Look at the left edge. The top branch, issuing from the head, is of length 1. The next lower branch, on the other hand, is of length γ. The next lower branch, on the other hand, is of length γ^2, and so on (Figure 7.13).

These are the leftmost branches. Together, they make the left edge of the entire tree: from the head at the top, down to the lower-left corner. What is the total length of the entire edge? Tthis is an infinite power series (Section 7.5.3):

$$1 + \gamma + \gamma^2 + \gamma^3 + \cdots = \sum_{i=0}^{\infty} \gamma^i = \frac{1}{1-\gamma}.$$

By now, we've measured the entire left side of the tree: from the head to the lower-left corner. Now, let's drop the forst (top-left) branch, and start from the second one. Without the first branch, the length is now

$$\gamma + \gamma^2 + \gamma^3 + \cdots = \sum_{i=1}^{\infty} \gamma^i = \gamma \sum_{i=0}^{\infty} \gamma^i = \frac{\gamma}{1-\gamma}.$$

This is γ times as short. This splits the entire left side into two parts: a long part (one branch only), and a shorter part (the rest). The bottom could now split in the same proportion (as in Figure 5.1). Thus, the bottom is indeed as sparse and transparent as the original Cantor set.

To see this, cover the entire bottom with small intervals (as in Section 7.4.1). There is just a slight change: start with intervals of length γ (times the distance from the lower-left corner to the lower-right corner). How many are needed? Clearly, two.

Now, turn to the next finer scale, and use intervals γ times as short. How many are needed? Clearly, four. This goes on and on: with intervals γ^n times as short, we need 2^n times as many. In summary, since $2\gamma < 1$,

$$2^n < \left(\gamma^{-1}\right)^n,$$

so this is indeed a fractal, of dimension less than 1.

7.10.4 Feigenbaum's Constant

In the above binary tree, the levels "shrink" by a constant factor: γ. Indeed, upon advancing from one level to the next lower level, the branch gets γ times as short. The original bifurcation tree in Figure 7.12 is nearly the same: its levels "shrink" by a nearly constant factor. Indeed,

$$\frac{\alpha_m - \alpha_{m-1}}{\alpha_{m+1} - \alpha_m} \to_{m \to \infty} \delta \doteq 4.669.$$

Here, δ is a new constant: Feigenbaum's constant. It governs many phenomena in physics and other sciences.

In summary, for $\alpha \geq \alpha_\infty$, the quadratic process gets chaotic: it makes a new fractal. This is indeed the aim of chaos theory: to find some pattern even in a complete disorder.

This is quite practical. Consider, for example, weather forecast. For this purpose, a new field was developed: computational fluid dynamics. It offers a few complex models to help make a reliable and stable forecast. This is quite tricky: there are highly nonlinear interactions between unknowns like velocity, temperature, and pressure.

Due to its chaotic nature, weather can be predicted for a very short time: five days ahead, or so. For more than that, no prediction can be made: there is too much sensitivity to initial data. The difficulty is mathematical, not computational: even a supercomputer would fail. This is why Edward Lorenz said: "the flap of a butterfly wing in Brazil may set off a tornado in Texas!" This is what happens when one attempts to predict the future...

7.11 Some Philosophical Remarks

7.11.1 Can We Predict The Future?

In his science-fiction trilogy "Original Foundation," Isaac Asimov describes the state of humanity a few centuries ahead in time. Mankind is no longer limited to the Earth: there are many settlements on other planets as well, where people are better off. People who stay here on the Earth, on the other hand, are not so well off. This drives a social gap, and eventually a severe social crisis.

Fortunately, the great sociologist is about to give a lecture about this. Everyone awaits the rare opportunity to learn some new hints or perhaps even a solution to the acute problem.

Of course, the great sociologist is long gone. Fortunately, before his death, he left a sealed video cassette. It is time to open it, and listen to what he has to say.

Upon opening the cassette, everyone is shocked. It turns out that the great sociologist failed to foresee the future. He talks about a completely different state, in which the entire humanity is still confined to the Earth, and never left. How could his thoughts be relevant?

Fortunately, some people in the audience still manage to benefit from the talk. Although the great sociologist didn't predict the present state, his speech still offers some deep insight and comprehension in broad terms. After all, the detailed circumstances are not so important. What is important is the fundamental principles that characterize and motivate sociological processes.

In the ancient (and middle) ages, people used to look for answers in the Bible or Plato's or Aristotle's scripts. Fortunately, in the Bible, the prophets rarely predict the concrete future. More often, they just set moral standards and guidelines, relevant at any time and age. Thanks to this advice, one could have more peace of mind, and make better decisions.

Likewise, Asimov's sociologist doesn't know all. On the contrary: he only knows what's before his eyes. From this, he could never predict a concrete disaster, say a global pandemic. Still, he may well foresee a more general event that may force us to change our way of life. To cope with this, he uses his own wisdom and intuition, and offers us a general theory. This is indeed induction. It is even better than a specific solution: it leaves us some work to do. Indeed, it is up to us to read between the lines, examine the advice we hear, and act upon it here and now. This is indeed deduction. (Sections 7.1.1–7.1.2).

7.11.2 The Phase Plane

Our society is a highly complex system. This is too complicated. To simplify things, let's focus on one simple phenomenon in physics, economics, sociology, or any other field. How to model it geometrically? Well, there may still be a lot of unknown variables that control or affect or govern it. For simplicity, assume that there are just two unknowns: p and q.

So far, we've only considered a scalar process: v. Here, on the other hand, v is a two-dimensional vector, with two components, or degrees of freedom:

$$v \equiv (p, q)$$

(Figure 7.14). During the process, the scalar unknowns p and q may interact nonlinearly with each other, and change. This produces a nonlinear curve (trajectory) of states: start from the initial state ($v_0 \equiv (p_0, q_0)$), and advance gradually to slightly different states that may take place in the future.

How could one say anything about the future? Well, for this purpose, a prophet (or an academic) could only use the information available at the present: p_0 and

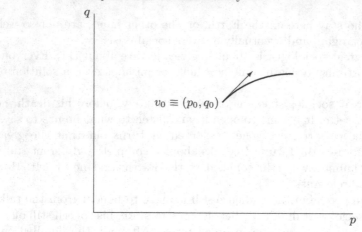

Fig. 7.14. A stable nonlinear process. From the initial state $v_0 \equiv (p_0, q_0)$, issue a smooth curve of new states. The tangent vector points at the direction of the best prediction available.

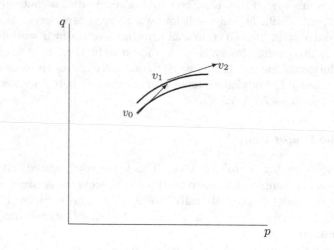

Fig. 7.15. The discrete process $v_0, v_1, v_2, \ldots$ approximates the smooth nonlinear curve. At v_0, a tangent vector is used to advance to v_1. At v_1, on the other hand, a new tangent vector is used to advance to v_2, and so on.

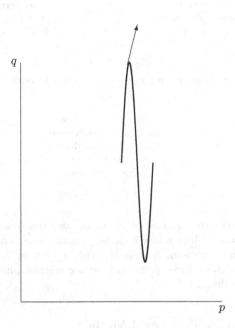

Fig. 7.16. A highly nonlinear (unstable) process. The curve changes too sharply, so linear prediction is useless.

q_0. To predict the future, one must linearize: look straight ahead, along the tangent vector in Figure 7.14.

Still, the tangent vector is inaccurate: it only approximates the true nonlinear curve. Even the best prediction must contain an error, due to uncertainty. After all, only a vague idea is available: the first derivative, or the first-order approximation.

How strong is the nonlinear interaction between p and q? Well, if it is moderate, then the curvature is rather small, so the relative error is kept small as well. In this case, one may step along a discrete path: $v_0, v_1, v_2, \ldots$, approximating the original curve rather well (Figure 7.15).

In the highly nonlinear case, on the other hand, the curve may change too sharply: no discrete path is useful any more (Figure 7.16), This means that the original phenomenon is too uncertain, chaotic, and unstable: no linearization could work any more.

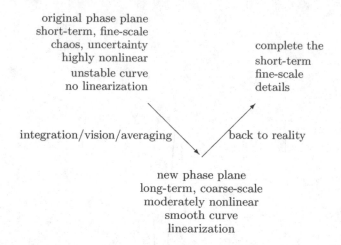

original phase plane
short-term, fine-scale
chaos, uncertainty
highly nonlinear
unstable curve
no linearization

complete the
short-term
fine-scale
details

integration/vision/averaging   back to reality

new phase plane
long-term, coarse-scale
moderately nonlinear
smooth curve
linearization

Fig. 7.17. The V-cycle: the original phase plane uses too many degrees of freedom: the scale is too fine. Average it into a new (coarser) phase plane. There, pick a suitable initial state, with a smooth curve issuing from it, with a tangent vector that points towards the desired long-term state. Finally, go back to the original phase plane, to work out the remaining short-term details.

7.11.3 Leadership: A Two-Level Model

Could the leader impose his/her own will upon the future, and shape it in his/her own vision? Could he/she focus on the long term, and work out the short-terms details later, without spoiling the long-term goal?

To do this, better work with two models: fine and coarse (Figure 7.17). The fine model is just our original phase plane. Now, average on it. This gives a new model: a new (coarser) phase plane, with a new (smoother) curve, suitable for linearization. Only after working in it should one go back to the original model, to work out the short-term details, without interacting with the coarse scale too much, and without spoiling the work done so far.

So, vision is not enough. To realize his/her view, the leader could also nominate a hierarchy of reliable managers to carry out his/her original plan, and make sure that the fine details are kept well under control, and don't undermine the entire project.

In a way, the leader uses a longer time scale: a spacial kind of "glasses," to look ahead into the deep future, disregarding any detail that may obscure the big picture. In this new (approximate) phase plane, the leader can pick a suitable initial state, with a sufficiently smooth curve issuing from it. This could indeed be linearized: the tangent vector may then point in the right direction – towards the desired state. All that is left to do is to go back to the fine model, to sort out the short-term details, without introducing any instability that could harm the entire goal. This could be viewed as an induction–deduction tree. (See Figure 7.17, and compare also to Figure 7.2.)

7.11.4 Feedback and Instability

In the above, we have just two unknowns: p and q. Now, one could also introduce many more unknowns, and work not only in a plane but also in a (multidimensional) hyperspace:

$$v \equiv (p, q, \ldots).$$

Like the sociologist and the prophet, the leader can only set guidelines: long-term ideas or principles. Geometrically, this means averaging on reality: producing a coarse (inexact) phase plane, where curves may get smoother, and easier to linearize, with less instability. Once this has been done, we may be well on track towards our ultimate goal, at least in broad terms. Still, the details must be worked out carefully, to fit in the same direction.

So, can the leader shape the future? This is discussed in a series of talks by Gershon Hacohen. He says that, by setting a clear vision, the leader may generate just enough faith, motivation, and power of will in the hearts of the people, to get them right on track to overcome any difficulty, and complete the original task successfully. After all, faith is the key to making the initial step in the right direction.

Still, there must be some symmetry: a constant interaction and communication between the leader and his/her people. After the initial step, the people may indeed give an important feedback to the leader, to update the original plan, if necessary. After all, the fine short-term details, however tiny, may still interact with the long-term process, and affect it quite substantially.

Geometrically, this is illustrated in the curvature. If it is moderate, then it could be smoothed out in the coarse scale. This is indeed stability. In this case, the feedback could be positive, and even build up: generate more and more motivation and will, to follow in the original plan, or even improve on it.

7.12 Zorn's Lemma in Quantum Mechanics

7.12.1 The Uncertainty Principle

How does an atom look like? In a naive model, the nucleus is at the center, and the electrons orbit it, round and round. Quantum mechanics, on the other hand, tells us a different story: more stochastic, and less deterministic. In fact, the principle of uncertainty says that one can never know all. At a given time, one could measure one quantity only: either the position of the electron, or its momentary momentum, but not both. Thus, the electron has no definite place: it only has some probability (or likelihood) to occupy a certain place at a given time. We'll discuss this later in more detail.

7.12.2 Tree of Universes

To model this, one could design a tree of potential universes (Figure 7.18). In each moment, the universe splits into many different universes, each assumes that the electron is at a different place in the atom. Together, these new universes make a new level in the tree. From each universe in this level, many more universes could emerge in the next moment, each at a certain probability. Together, they form the

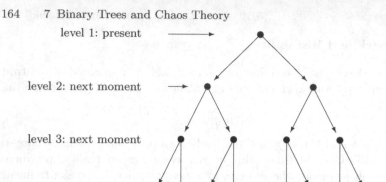

level 1: present

level 2: next moment

level 3: next moment

level 4: next moment

Fig. 7.18. Tree of potential universes: at the head, there is our familiar universe, at the present. At the next moment, it splits into two (or more) potential universes. Each has its own probability to be picked as the correct (real) universe. At the next moment, it will also split into two (or more) new universes, and so on.

next level, and so on, moment by moment, producing more and more potential universes, each with its own (small) probability. Together, all these levels make an infinite tree of many potential universes that may appear later.

As a matter of fact, the same could be done not only for one particular electron but also for each and every electron in the world. Fortunately, the total number of electrons is still finite (Chapter 6). For simplicity, we consider one electron only. This is particularly easy to illustrate in our tree.

7.12.3 Entropy: Partial Order

In our tree, define a new (strict) partial order: a particular universe is "smaller" than those new universes that may emerge from it later. Is this a legitimate order? Well, let's check:

- It is clearly transitive.
- It is anti-symmetric: there is no going back to an old universe. After all, entropy must increase, as in the second law of thermodynamics (Chapter 6, Section 6.3.5).
- It is anti-reflexive: no universe is static.

7.12.4 Experiment or Observation

At each individual moment, only one particular universe appears in reality. To pick it, there is a need to make an experiment, and observe the exact location of the electron in the atom. At each level, this picks just one (correct and real) universe. By repeating this moment by moment, we get an infinite path, level by level, down the tree, along those universes that really appear. All others remain science fiction.

7.12.5 Parallel Universes

This models the theory of parallel universes [8]. In it, you must use your imagination: you only *imagine* that you are in this world, here and now. As a matter of fact, you also have many duplicate copies in other potential universes (at the same level), who also think that they are the only ones!

7.12.6 Zorn's Lemma

But time is unlimited: there are infinitely many moments, each with its own level in the tree. How to form the infinite path down the tree? At each moment, which universe should be picked from the level?

Fortunately, the tree must contain no dead-end: from each universe, at least one new universe emerges. In terms of our partial order, this means no maximal universe.

Thanks to this property, such a path indeed exists. This is indeed Zorn's lemma: there is a chain with no upper bound. This way, we can always advance along the path, down the tree. This can be done for arbitrarily long time. This may help "solve" Zeno's paradox: time never stops, but always restarts.

7.13 Exercises: Complexity

7.13.1 Binary Tree and Its Complexity

1. Show that, in a general tree, every node but the head has exactly one incoming branch (or edge) pointing at it.
2. Conclude that the total number of edges is smaller by 1 than the total number of nodes. Hint: map each node (but the head) to the incoming branch leading to it.
3. Show that, in a perfect binary tree, the ith level contains 2^{i-1} nodes ($i \geq 1$). Hint: use mathematical induction on $i = 1, 2, 3, \ldots$.
4. Conclude that, in an n-level perfect binary tree, there are

$$\sum_{i=1}^{n} 2^{i-1} = \sum_{i=0}^{n-1} 2^i = \frac{2^n - 1}{2 - 1} = 2^n - 1$$

   nodes.
5. Prove this in yet another way. Hint: use mathematical induction on $n = 1, 2, 3, \ldots$.
6. Conclude that, in an n-level perfect binary tree, there are $2^n - 2$ branches (edges). Hint: we've already established that the total number of edges is smaller by 1 than the totla number of nodes, even in a general tree.
7. Prove this in yet another way. Hint: use mathematical induction on $n = 1, 2, 3, \ldots$.

8. Write an algorithm that embeds a given arithmetic expression in a tree, and calculates its value. Hint: use recursion. The solution can be found in Chapter 1 in [44].

9. Write an algorithm that embeds a given Boolean expression in a tree, and calculates its true value (1 or 0).

10. Write an algorithm that transforms a given natural number from its decimal representation (in base 10) to its binary representation (in base 2).

11. How many arithmetic operations are needed?

12. This is the cost. It grows with the total number of digits in the original number. How fast is this growth? Exponential or linear?

13. Write an algorithm that transforms a given natural number from its binary representation back to its decimal representation.

14. How many arithmetic operations are needed?

15. This is the cost. It grows with the total number of digits in the original number. How fast is this growth? Exponential or linear?

16. Improve both algorithms, and make sure that their cost grows only linearly, not exponentially. Hint: use recursion. The solution can be found in Chapter 1 in [44].

17. In Figure 7.14, where is the time variable? Why is it missing? Hint: this is a phase plane. In it, time is implicit: it is just a parameter that "pushes" along the curve. To see this more vividly, make a movie that shows the advance from state to state along the curve.

The Binomial Formula
and Quantum Statistical Mechanics

Binomial and Quantum Statistical Mechanics

As discussed above, a binary tree grows exponentially: the next level contains twice as many nodes. This is quite a rapid growth. This is why it is useful to model instability.

There is, however, a more moderate growth. Pascal's triangle, for example, grows only linearly. Indeed, the next level contains just one more cell. Still, even in this moderate growth, there is a hidden exponential growth. Indeed, in Pascal's triangle, one could draw all sorts of geometrical paths. Together, these paths make a familiar binary tree. This will lead to the (extended) binomial formula, useful in quantum statistical mechanics.

In this part, we introduce the binomial formula from two points of view: algebraic and geometrical. Indeed, thanks to mathematical induction, we get to learn a lot about the binomial coefficients and their algebraic properties. Still, they also have a geometrical face: they are mirrored by geometrical paths in Pascal's triangle. This leads to an interesting application in statistical physics: Brownian motion.

Furthermore, in its extended form, the binomial formula is useful in quantum mechanics as well. Indeed, thanks to conservation of mass and energy, we can go ahead and estimate the number of particles in each energy level. This leads to three types of distributions: molecules are distributed in Maxwell-Boltzmann statistics, bosons in Bose-Einstein statistics, and electrons in Fermi-Dirac statistics.

8

Newton's
Binomial and Trinomial
Formulas

So far, we've used binary trees to introduce chaos theory from a geometrical point of view. We've seen that the (infinite) binary tree is transparent from below. Indeed, at its bottom, it approaches a fractal: a Cantor set, of dimension less than 1. This helps model a bifurcation tree in chaos theory.

Here, on the other hand, we use binary trees for yet another purpose: count geometrical paths in Pascal's triangle. This leads to two important formulas: the binomial and trinomial formulas. Later on, we'll extend them even more, and use them in physics.

What is a binary tree? It is a good example of a multilevel hierarchy, with an exponential growth. Indeed, the next level contains twice as many nodes (and branches).

In this chapter, on the other hand, we study a more modest growth: linear. Indeed, in Pascal's triangle, the next lower level contains just one more cell. Still, we soon get our good old binary tree back again: in Pascal's triangle, one could draw all sorts of geometrical paths. Together, they make a new binary tree, with its familiar exponential growth. This way, we can sort these paths, and count them. This leads to an important combinatorial result: Newton's binomial formula. To show how useful it is, we use it in a stochastic process in statistical physics.

This approach gives a lot of geometrical insight. Still, there is also a more algebraic proof, using mathematical induction only. This approach is more general: it helps extend the original binomial coefficient, and develop yet more advanced formulas. Later on, we'll use them in quantum mechanics and relativity.

8.1 Geometrical Point of View

8.1.1 Pascal's Triangle

The binomial formula is most useful in quantum mechanics and relativity. To introduce it properly, we need some geometrical-combinatorial background. For this purpose, we start from Pascal's triangle.

Pascal's triangle is made of oblique cells, listed row by row (Figure 8.1). Later on, each cell will contain a number (entry).

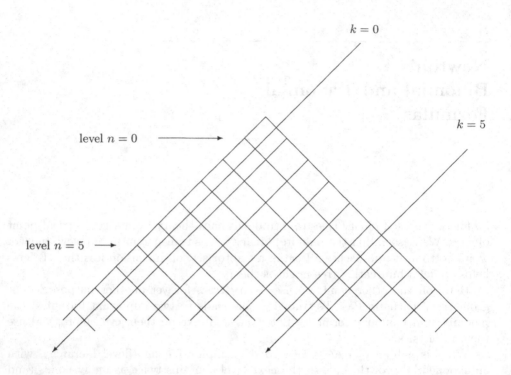

Fig. 8.1. Pascal's triangle: to index the levels (or rows), use $n = 0, 1, 2, \ldots$. To index the individual cells in the nth level, use $k = 0, 1, 2, \ldots, n$.

How are the cells indexed? By two indices: n and k. The rows (or levels) are indexed by $n = 0, 1, 2, 3, \ldots$, top to bottom. For example, the zeroth level (at the top) contains just one cell (and one entry in it). The next lower level, on the other hand, contains two cells, indexed by $k = 0, 1$. The next lower level, on the other hand, contains three cells, indexed by $k = 0, 1, 2$, and so on.

How to define the entries in the cells? By mathematical induction, of course! In the top cell, place the entry 1 (Figure 8.2). This fills the zeroth level, as required. Now, for $n = 0, 1, 2, 3, \ldots$, assume that the nth level has already been filled with entries, as required. (This is the induction hypothesis.) How to fill the $(n+1)$th level just below it?

This is done as follows. In each cell in the $(n+1)$th level, place the sum of two entries: one from the upper-left cell, and the other from the upper-right cell.

Clearly, there are two exceptions. The leftmost cell has no upper-left neighbor. Therefore, the entry in it must be the same as the upper-right entry: 1. Likewise, the rightmost cell has no upper-right neighbor. Thus, the entry in it must be the same as the upper-left entry: 1.

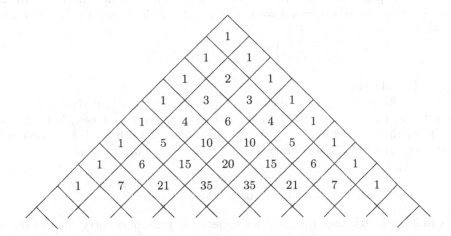

Fig. 8.2. The entries in Pascal's triangle: each entry is the sum of the upper-left and upper-right entries.

So, at all levels, the leftmost and rightmost entries are just 1. The intermediate entries, on the other hand, are greater than 1. This can be proved easily by mathematical induction.

The above definition is constructive: it actually offers a practical algorithm to fill Pascal's triangle with entries: recursively, level by level. Still, what if we only need one particular entry? Is it possible to calculate it on its own, without filling the entire triangle?

8.1.2 The Binomial Coefficient

How to calculate an individual entry, without calculating all the others? For this purpose, recall that the factorial function is defined recursively by

$$n! \equiv \begin{cases} 1 & \text{if } n = 0 \\ (n-1)! \cdot n & \text{if } n > 0. \end{cases}$$

In Pascal's triangle, how are the cells indexed? Recall: the levels are indexed top to bottom: $n = 0, 1, 2, 3, \ldots$. In each particular level, the cells are indexed left to right: $k = 0, 1, 2, 3, \ldots, n$. This way, each cell is indexed by two indices: n and k ($n \geq k \geq 0$). Let's use these indices.

In the nth level, let's focus on the kth entry (from the left). How does it look like algebraically? Let's prove that it looks like this:

$$\binom{n}{k} \equiv \frac{n!}{k!(n-k)!}.$$

This is the binomial coefficient.

How to prove this? By mathematical induction, of course! Indeed, at the zeroth level, where $n = 0$, we must have $k = 0$ as well, so the formula is indeed correct:

$$\binom{0}{0} = \frac{0!}{0! \cdot (0-0)!} = \frac{1}{1 \cdot 1} = 1,$$

as in Pascal's triangle.

Now, for $n = 1, 2, 3, \ldots$, assume that we already know that the above formula is correct at the $(n-1)$th level. (This is the induction hypothesis.) Is it also correct at the nth level just below it? Well, let's look at the nth level, and check it entry by entry. For the leftmost entry, for which $k = 0$, we indeed have

$$\binom{n}{0} = \frac{n!}{0!(n-0)!} = \frac{n!}{1 \cdot n!} = 1,$$

as in Pascal's triangle. Likewise, let's check the rightmost entry, for which $k = n$:

$$\binom{n}{n} = \frac{n!}{n!(n-n)!} = \frac{n!}{n! \cdot 1} = 1,$$

as in Pascal's triangle. Finally, let's check those intermediate entries, for which $0 < k < n$. Fortunately, we already know that the formula holds at the previous level just above. After all, this is the induction hypothesis. Now, in the nth level, how was the kth entry defined in the first place? It was defined as the sum of the upper-left and upper-right entries:

$$\begin{aligned}
\binom{n-1}{k-1} + \binom{n-1}{k} &= \frac{(n-1)!}{(k-1)!(n-k)!} + \frac{(n-1)!}{k!(n-k-1)!} \\
&= \frac{k}{n} \cdot \frac{n!}{k!(n-k)!} + \frac{n-k}{n} \cdot \frac{n!}{k!(n-k)!} \\
&= \frac{k+n-k}{n} \cdot \binom{n}{k} \\
&= \binom{n}{k},
\end{aligned}$$

as the formula indeed says. This shows that the formula is correct at the nth level as well. This completes the induction step.

Thus, the formula is true in general: each individual entry is the same as the corresponding binomial coefficient.

8.1.3 Geometrical Paths in Pascal's Triangle

In Pascal's triangle, consider a path: start from the top, and "jump" from level to level, down the triangle. In each "jump," move to the next lower level: either down-left, or down-right (Figures 8.3–8.4).

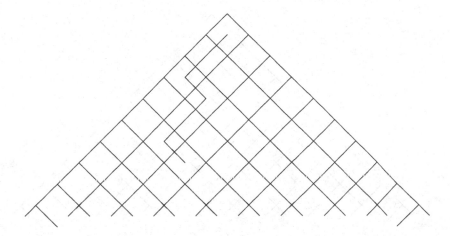

Fig. 8.3. A path leading to cell $k = 2$ in level $n = 6$ could be modeled by the binary vector $(0, 0, 1, 0, 0, 1)$: the third and sixth moves are down-right, and the rest are down-left.

Now, in the nth level, focus on the kth cell from the left $(0 \leq k \leq n)$. How many paths lead to it? Well, the answer is already written in the cell! To prove this, use mathematical induction, and the original definition of the entries. Indeed, to reach our cell, the path must approach it, either from the upper-left, or from the upper-right.

How many paths approach from the upper-left, and how many from the upper-right? Fortunately, thanks to the induction hypothesis, the answers are already written in the previous level. So, we just need to sum the upper-left and upper-right entries (if exist), to obtain the correct answer. Fortunately, this is exactly what is already written in the cell.

8.2 Combinatorial Point of View

8.2.1 Binary Vectors

Not convinced? Let's prove this in yet another way: not only geometrically, but also combinatorially. A path as above contains n moves: either down-right, or down-left. Let's index them from 1 to n:

$$v_1, \ v_2, \ v_3, \ \ldots, \ v_n.$$

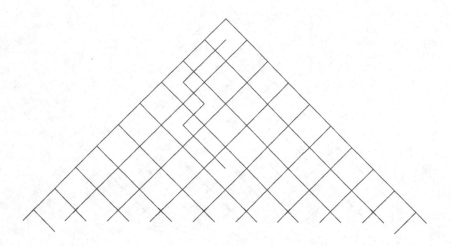

Fig. 8.4. A path leading to cell $k = 3$ in level $n = 6$ could be modeled by the binary vector $(0, 0, 1, 0, 1, 1)$: the third, fifth, and sixth moves are down-right, and the rest are down-left.

For $1 \leq i \leq n$, the ith move is modeled by

$$v_i \equiv \begin{cases} 1 & \text{if the } i\text{th move is down-right} \\ 0 & \text{if the } i\text{th move is down-left.} \end{cases}$$

This way, the entire path is modeled by a binary n-dimensional vector: a zero component stands for a down-left move, whereas 1 stands for a down-right move (Figures 8.3–8.4).

There is one degenerate case: for $n = 0$, the empty vector $\emptyset$ models the static "path," with no move at all. Let's look at more interesting cases, with $n > 0$.

8.2.2 Subsets

Let's get some more insight from set theory. Indeed, what is this binary vector? It could be interpreted as a characteristic function, which maps n natural numbers, either to 0 or to 1:

$$v : \{1, 2, 3, \ldots, n\} \rightarrow \{0, 1\}.$$

Which numbers are mapped to 1? In other words, what is the origin of 1? This makes a new subset of the form

$$S \subset \{1, 2, 3, \ldots, n\},$$

containing those i's for which $v_i \neq 0$:

$$i \in S \Leftrightarrow v_i = 1, \quad 1 \leq i \leq n.$$

In Figure 8.3, for example,

$$v = (0, 0, 1, 0, 0, 1),$$

so

$$S = \{3, 6\} \subset \{1, 2, 3, 4, 5, 6\}.$$

In Figure 8.4, on the other hand,

$$v = (0, 0, 1, 0, 1, 1),$$

so

$$S = \{3, 5, 6\} \subset \{1, 2, 3, 4, 5, 6\}.$$

Our original question was geometrical: how many paths lead to the kth cell in the nth level? Thanks to the above, it is now combinatorial: how many binary vectors have k 1's and $n - k$ 0's? Or, in set-theory language, what is the cardinality (size) of the set

$$\{v \equiv (v_1, v_2, \ldots, v_n) \in \{0, 1\}^n \mid v_1 + v_2 + \cdots + v_n = k\}?$$

Or, in terms of subsets, what is the cardinality of the set

$$\{S \subset \{1, 2, 3, \ldots, n\} \mid |S| = k\}?$$

8.2.3 Subsets and The Binomial Coefficient

In Section 8.1.3, we've already studied Pascal's triangle, and counted the geometrical paths in it. Thanks to the language of subsets, we can now do this in yet another way, and obtain the same result once again: the binomial coefficient.

In fact, our original question takes now a pure combinatorial form: how many different subsets of the form

$$S \equiv \{s_1, s_2, s_3, \ldots, s_k\}$$

could be extracted from $\{1, 2, 3, \ldots, n\}$? Well, to design such a subset, we need to pick k different natural numbers, not larger than n.

How to do this? Well, let's pick s_1. It could be either 1, or 2, or 3, or ..., or n. So, there are n options to pick s_1.

For each such an option, there are now $n - 1$ options to pick s_2. After all, s_2 could be either smaller or larger than s_1, but not the same. So, in total, there are $n(n - 1)$ options to pick s_1 and s_2.

Or are there? After all, in Figure 8.3,

$$\{3, 6\} = \{6, 3\}$$

are the same subset. There is no need to count it twice.

So, the above count contains some duplication. To avoid this, we must divide by 2. This means that we adopt a new convention: write $\{s_1, s_2\}$ in a unique writing style, in which the elements are written from small to large:

$$\{3,6\} \quad \text{rather than} \quad \{6,3\}.$$

Thanks to this convention, there is now no duplication any more, so there are only

$$\frac{n(n-1)}{2} = \binom{n}{2}$$

different options to pick

$$s_1 < s_2,$$

to help design each subset of the form $\{s_1, s_2\}$ only once, not twice.

For each such an option, we now have $n-2$ options to pick s_3. After all, s_3 could be either smaller than s_1, or larger than s_2, or in between s_1 and s_2, but mustn't be equal to s_1 or s_2.

Unfortunately, there is still some duplication. For example, to design the path in Figure 8.4, one should pick

$$\{s_1, s_2, s_3\} = \{3,6,5\} = \{3,5,6\} = \{5,6,3\}.$$

In other words, one could pick

$$s_1 = 3 \quad \text{and} \quad s_2 = 6$$

(as in Figure 8.3), and then

$$s_3 = 5.$$

But this is not the only way: one could equally well pick

$$s_1 = 3, \ s_2 = 5, \quad \text{and} \quad s_3 = 6,$$

or

$$s_1 = 5, \ s_2 = 6, \quad \text{and} \quad s_3 = 3.$$

To avoid this kind of duplication, stick to our convention: use the unique option in which

$$s_1 < s_2 < s_3,$$

and drop the other two. With no duplication, there are just

$$\frac{n(n-1)}{2} \cdot \frac{n-2}{3} = \frac{n!}{(n-3)!} \cdot \frac{1}{3!} = \binom{n}{3}$$

different options to design $\{s_1, s_2, s_3\}$.

By repeating this process (or by mathematical induction on $k = 1, 2, 3, \ldots, n$), we have that there are

$$\binom{n}{k-1} \cdot \frac{n-k+1}{k} = \frac{n!}{(k-1)!(n-k+1)!} \cdot \frac{n-k+1}{k} = \frac{n!}{k!(n-k)!} = \binom{n}{k}$$

different options to pick

$$s_1 < s_2 < s_3 < \cdots < s_k,$$

to help design each subset of the form $\{s_1, s_2, \ldots, s_k\}$ exactly once: uniquely, as required.

8.2.4 Monotonically Increasing Sequences

By now, we've counted how many sequences of the form

$$s_1 < s_2 < s_3 < \cdots < s_k$$

could be extracted from $\{1, 2, 3, \ldots, n\}$. Still, these strong inequalities are a little inconvenient: the s_i's must lie in different intervals:

$$1 \leq s_1 \leq n - k + 1$$
$$2 \leq s_2 \leq n - k + 2$$
$$3 \leq s_3 \leq n - k + 3$$
$$\vdots$$
$$k \leq s_k \leq n.$$

How to transform them to new numbers that do lie in the same interval? For this purpose, let's define the new numbers

$$t_1 \equiv s_1 - 1$$
$$t_2 \equiv s_2 - 2$$
$$t_3 \equiv s_3 - 3$$
$$\vdots$$
$$t_k \equiv s_k - k.$$

Fortunately, the t_i's do lie in the same interval:

$$0 \leq t_1 \leq n - k$$
$$0 \leq t_2 \leq n - k$$
$$0 \leq t_3 \leq n - k$$
$$\vdots$$
$$0 \leq t_k \leq n - k.$$

Furthermore, they make a weakly-increasing sequence:

$$0 \leq t_1 \leq t_2 \leq t_3 \leq \cdots \leq t_k \leq n - k.$$

Fortunately, this transformation is invertible. Indeed, given a weakly-increasing sequence like this, we can always produce a strongly-increasing sequence from it, by defining

$$s_1 \equiv t_1 + 1$$
$$s_2 \equiv t_2 + 2$$
$$s_3 \equiv t_3 + 3$$
$$\vdots$$
$$s_k \equiv t_k + k.$$

To simplify, let's introduce a new number:

$$m \equiv n - k.$$

(What is geometrical nature of m? To see this, see the exercises below.) This way, both m and k have the same status. Indeed, since $n \geq k \geq 0$, both m and k must be nonnegative:

$$m, k \geq 0.$$

Thanks to this new definition, our weakly-increasing sequence lies in the new interval $[0, m]$:

$$0 \leq t_1 \leq t_2 \leq t_3 \leq \cdots \leq t_k \leq m.$$

How many such sequences could be extracted from $\{0, 1, 2, 3, \ldots, m\}$? We already know the answer:

$$\binom{m + k}{k}.$$

Does this formula work even in degenerate cases? Yes, it does! For $k = 0$, there is just one possible sequence: the empty sequence $\emptyset$. This goes well with the formula. For $m = 0$, there is just one possible sequence:

$$0 \leq 0 \leq 0 \leq \cdots \leq 0.$$

This agrees hand-in-hand with the formula as well.

8.3 Geometrical Problem

8.3.1 Discrete Simplex

By now, we've managed to avoid strong (or strict) inequalities, and replace them by weak inequalities. This way, our new t_i's lie in a more uniform interval: $[0, m]$. Still, by now, the discussion got too combinatorial. How to make it geometrical again, easy to illustrate and visualize?

To answer this question, let's split it into two subquestions. First, how to get rid of the inequalities altogether? Second, our t_i's are still extracted from an existing set: $\{0, 1, 2, 3, \ldots, m\}$. How to get rid of this requirement altogether?

For this purpose, let's define new differences:

$$d_1 \equiv t_1$$
$$d_2 \equiv t_2 - t_1$$
$$d_3 \equiv t_3 - t_2$$
$$\vdots$$
$$d_k \equiv t_k - t_{k-1}.$$

What algebraic conditions must the d_i's satisfy? First, they must be nonnegative:

$$d_i \geq 0, \quad 1 \leq i \leq k.$$

Next, their sum mustn't exceed m. After all, it is equal to t_k:

$$\sum_{i=1}^{k} d_i = d_1 + d_2 + d_3 + \cdots + d_k = t_k \leq m.$$

Thus, the t_i's have been transformed to the new d_i's. This is an invertible transformation. After all, given d_i's that satisfy the above conditions, one could always calculate their partial sums, and obtain a new weakly-increasing sequence:

$$t_1 \equiv d_1$$
$$t_2 \equiv d_1 + d_2$$
$$t_3 \equiv d_1 + d_2 + d_3$$
$$\vdots$$
$$t_k \equiv d_1 + d_2 + d_3 + \cdots + d_k.$$

In summary, we got what we wanted: instead of our original t_i's, we can now work with the new d_i's, which are no longer extracted from any existing set, but just satisfy two geometrical onditions:

$$d_i \geq 0, \quad 1 \leq i \leq k,$$

and

$$\sum_{i=1}^{k} d_i \leq m.$$

In what sense are these conditions geometrical? To see this, just place the d_i's in a new k-dimensional vector:

$$(d_1, d_2, d_3, \ldots, d_k) \in \left(\mathbb{Z}^+\right)^k,$$

where

$$\mathbb{Z}^+ \equiv \mathbb{N} \cup \{0\}$$

contains the nonnegative integer numbers.

Where is this new vector? Well, it is in the k-dimensional infinite grid. Furthermore, its components are nonnegative, and their sum is at most m. In summary, our vector is in the discrete simplex of size m:

$$\left\{ (d_1, d_2, \ldots, d_k) \in \left(\mathbb{Z}^+\right)^k \mid \sum_{i=1}^{k} d_i \leq m \right\}.$$

8.3.2 Discrete Triangle

How does the discrete simplex look like? For $k > 3$, this is not easy to see. For $k = 2$, on the other hand, this is just a discrete triangle. In Figure 8.5, we illustrate a discrete triangle of size $m = 3$: each edge contains just four grid points.

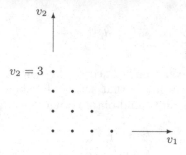

Fig. 8.5. The case $k = 2$: a discrete triangle of size $m = 3$. It is placed in the two-dimensional v_1-v_2 grid. Inside the triangle, we have $v_1 \geq 0$, $v_2 \geq 0$, and $v_1 + v_2 \leq m$.

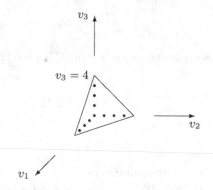

Fig. 8.6. The case $k = 3$: a discrete tetrahedron of size $m = 4$. It is placed in the three-dimensional v_1-v_2-v_3 grid. Inside the tetrahedron, we have $v_1 \geq 0$, $v_2 \geq 0$, $v_3 \geq 0$, and $v_1 + v_2 + v_3 \leq m$.

8.3.3 Discrete Tetrahedron

For $k = 3$, on the other hand, this is a discrete tetrahedron. In Figure 8.6, we illustrate a discrete tetrahedron of size $m = 4$: each edge contains five grid points.

Let's go ahead and count the points in a discrete triangle or tetrahedron. Better yet, let's consider a discrete simplex of just any dimension, and count the points in it.

8.3.4 How Many Points?

How many points are there in the discrete k-dimensional simplex of size m? We already know the answer:

$$\left| \left\{ (d_1, d_2, \ldots, d_k) \in \left(\mathbb{Z}^+ \right)^k \ \middle| \ \sum_{i=1}^k d_i \le m \right\} \right| = \binom{m+k}{k}.$$

Does this formula work even in degenerate cases? Let's check: for $k = 0$, there is just one possible vector: the empty vector $\emptyset$, with no components at all. This goes well with the formula. For $m = 0$, there is just one possible vector – the origin:

$$(0, 0, 0, \ldots, 0).$$

This agrees with the formula as well.

8.3.5 Subsimplex of Lower Dimension

Thanks to the discrete simplex, we have a geometrical insight back again. Why is this so important? Because it gives us not only a better intuition but also a practical bonus. This is relevant for every nontrivial simplex, in one dimension or more: $k \ge 1$.

In our discrete simplex, the d_i's must sum to m or less. Here, on the other hand, they sum to m exactly:

$$d_1 + d_2 + d_3 + \cdots + d_k = m.$$

What does this mean geometrically? Well, instead of looking at the entire simplex, look at its upper face only.

Is this a new simplex in its own right? Yes, it is! Indeed, what are we doing here? We just assume that the last coordinate, d_k, can no longer be picked freely. On the contrary: it must be defined as

$$d_k \equiv m - \sum_{i=1}^{k-1} d_i.$$

Fortunately, we are still free to pick $d_1, d_2, \ldots, d_{k-1}$. These should satisfy two geometrical conditions:

$$d_i \ge 0, \quad 1 \le i \le k - 1,$$

and

$$\sum_{i=1}^{k-1} d_i \le m.$$

In what sense are these conditons geometrical? Well, they make a familiar geometrical structre: a discrete $(k-1)$-dimensional simplex of size m. In summary, all we did here amounts to reducing the dimension from k to $k-1$.

How many points are there in this new subsimplex? Well, we already know the answer:

$$\left| \left\{ (d_1, d_2, \ldots, d_k) \in (\mathbb{Z}^+)^k \ \Big| \ \sum_{i=1}^{k} d_i = m \right\} \right|$$

$$= \left| \left\{ (d_1, d_2, \ldots, d_{k-1}) \in (\mathbb{Z}^+)^{k-1} \ \Big| \ \sum_{i=1}^{k-1} d_i \leq m \right\} \right|$$

$$= \binom{m + k - 1}{k - 1}.$$

8.3.6 Splitting The Discrete Simplex

Thanks to the above bonus, we can now split (or slice) the original discrete simplex into $m + 1$ disjoint subsimplices (or oblique slices), and obtain a new combinatorial formula:

$$\binom{m + k}{k}$$

$$= \left| \left\{ (d_1, d_2, \ldots, d_k) \in (\mathbb{Z}^+)^k \ \Big| \ \sum_{i=1}^{k} d_i \leq m \right\} \right|$$

$$= \left| \cup_{j=0}^{m} \left\{ (d_1, d_2, \ldots, d_k) \in (\mathbb{Z}^+)^k \ \Big| \ \sum_{i=1}^{k} d_i = j \right\} \right|$$

$$= \sum_{j=0}^{m} \left| \left\{ (d_1, d_2, \ldots, d_k) \in (\mathbb{Z}^+)^k \ \Big| \ \sum_{i=1}^{k} d_i = j \right\} \right|$$

$$= \sum_{j=0}^{m} \binom{j + k - 1}{k - 1}.$$

This is indeed how our new geometrical insight is used to develop a new interesting combinatorial formula. In the exercies below, we'll prove it in yet another way. Now, let's see a few applications in physics.

8.4 Application in Statistical Physics

8.4.1 Brownian Motion

The binomial coefficients, and the geometrical paths associated with them, can be used in yet another interesting application: Brownian motion. Consider a particle confined to the real axis, moving step by step. In each step, the particle "jumps" along the real axis: either rightwards or leftwards.

In the beginning, the particle starts from the origin: 0. In each step, it moves exactly one unit: either rightwards (from some l to $l + 1$) or leftwards (from l to $l - 1$).

The process is random (or stochastic, or nondeterministic): in each individual step, we don't know in what direction the particle will jump, or where it will land. Fortunately, we still do know how *likely* it is to move in each direction. To move leftwards, the probability is a (for some fixed $0 \leq a \leq 1$). To move rightwards, on the other hand, the probability is b (for some fixed $0 \leq b \leq 1$). Since there is no other option, the probabilities must sum to 1:

$$a + b = 1.$$

After all, the particle must move, either rightwards, or leftwards. We'll come back to this later, in our introduction to quantum mechanics.

The interesting question is: where would the particle be after n steps? Of course, we can never tell this for sure. Fortunately, we can still tell how likely it is to be at any specific point on the real axis.

In n steps, the particle must make k leftward moves and $n - k$ rightward moves, for some $0 \leq k \leq n$. Thus, after n steps, the particle could reache the point

$$-k + (n - k) = n - 2k$$

on the real axis. How likely is the particle to get there? We already know the answer: there are $\binom{n}{k}$ distinct ways to pick k leftward moves and $n - k$ rightward moves (Section 8.1.3). Furthermore, since the steps are independent of each other, all these ways have the same probability: $a^k b^{n-k}$. Thus, after n steps, the particle would reach $n - 2k$ at probability

$$\binom{n}{k} a^k b^{n-k}.$$

These probabilities must sum to 1. After all, after n steps, the particle must lie somewhere. Later on, we'll also prove this algebraically.

8.4.2 Symmetric Diffusion

Consider, for example, the symmetric case, in which the particle is equally likely to move rightwards or leftwards:

$$a = b = \frac{1}{2}.$$

In this case, the above process models a symmetric one-dimensional diffusion.

After n steps, how likely is the particle to reach $n - 2k$? The above formula already tells us the answer:

$$\binom{n}{k} \left(\frac{1}{2}\right)^k \left(\frac{1}{2}\right)^{n-k} = \frac{1}{2^n} \binom{n}{k}.$$

This distribution is illustrated in three instances: $n = 5$, 6, and 7 (Figures 8.7–8.9).

The nonsymmetric case $a < b$, on the other hand, models convection-diffusion, with a slight wind, blowing rightwards. Likewise, the case $a > b$ models a wind blowing leftwards.

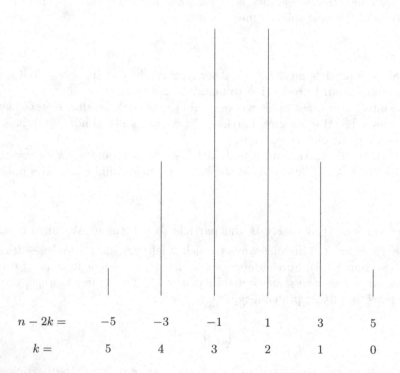

Fig. 8.7. Symmetric Brownian motion ($a = b = 1/2$): distribution after $n = 5$ steps. The columns in the diagram stand for the probability of the particle to reach the point $n - 2k$ ($0 \leq k \leq n$) after $n = 5$ steps. (This requires k leftward and $n - k$ rightward moves.)

8.4.3 Closed System: Entropy Must Increase

In a closed physical system, entropy measures the amount of uniformity: to what extent everything is the same? For example, in the above model, assume that all particles are initially concentrated at the origin.: 0 This means minimum entropy. Indeed, there is no uniformity at all: the origin, on one hand, is full of matter, whereas the rest of the real axis is completely empty.

Afterwards, thanks to diffusion, things get better: the particles start to move, rightwards or leftwards, independent of each other. The distribution of the particles gets gradually more and more spread out, as in Figures 8.7–8.9. This illustrates how entropy indeed increases monotonically: the diagrams get more and more spread out, uniform, and flat.

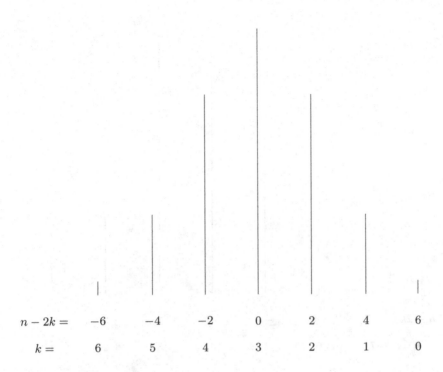

Fig. 8.8. Symmetric Brownian motion: distribution after $n = 6$ steps. The columns stand for the probability to reach the point $n - 2k$ after $n = 6$ steps.

8.4.4 Big Bang and Inflation

This is indeed the second law of thermodynamics: on average, entropy must always increase. At the very beginning of the universe, at the big bang, entropy was at its minimum: all matter and energy had been concentrated at one singular point. Since then, on average, entropy has only increased.

In one theory, in the inflation that followed, the universe had expanded exponentially fast. Afterwards, matter and energy continued to spread out, increasing entropy even more.

Still, not everything expanded. Thanks to gravity, when the universe cooled down, stars and galaxies formed. Still, this didn't decrease entropy. On the contrary: this generated more and more heat, spreading energy around, and increasing entropy even more on average.

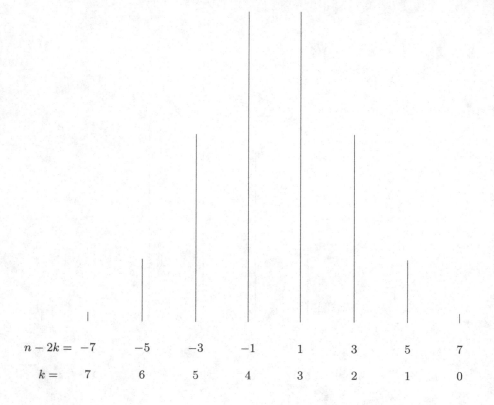

Fig. 8.9. Symmetric Brownian motion: distribution after $n = 7$ steps. The columns stand for the probability to reach the point $n - 2k$ after $n = 7$ steps.

8.5 The Binomial Formula: Mathematical Induction

8.5.1 Newton's Binomial Formula

By now, we've used the binomial coefficients for a combinatorial purpose: counting increasing sequences (or subsets). This task also has a geometrical version: counting geometrical paths in Pascal's triangle, or points in a discrete simplex. This leads to our main objective: the binomial formula.

Let n be a nonnegative integer number. Let a and b be some numbers. Consider the nth power of $a + b$:

$$(a + b)^n \equiv (a + b)(a + b)(a + b) \cdots (a + b) \qquad (n \text{ times}).$$

In this product, there are n factors of the same form: $a + b$. In other words, the same factor $a + b$ multiplies n times.

How to open parentheses? From each factor of the form $a + b$, pick either a or b. This way, you must pick k a's and $n - k$ b's (for some $0 \leq k \leq n$). Upon opening parentheses, this produces a new product: $a^k b^{n-k}$.

How many times is this product produced? In other words, how many different ways are there to pick k a's from k factors of the form $a + b$? Well, each such way can be modeled by a distinct n-dimensional binary vector: 1 stands for picking a, and 0 for picking b. As we've seen above, there are

$$\left| \left\{ v \in \{0,1\}^n \mid \sum_{i=1}^{n} v_i = k \right\} \right| = \binom{n}{k}$$

such vectors.

Thus, once the parentheses open, each product of the form $a^k b^{n-k}$ appears $\binom{n}{k}$ times in the resulting sum:

$$(a + b)^n = \sum_{k=0}^{n} \binom{n}{k} a^k b^{n-k}.$$

This is Newton's binomial formula. Not convinced? In a moment, We'll prove this more directly.

How to calculate these binomial coefficients in practice? Easy: just fill Pascal's triangle, level by level, until the nth level. For this purpose, you only need to add numbers, and never multiply. This is rather cheap and efficient.

8.5.2 Mathematical Induction in The Binomial Formula

To prove Newton's binomial formula, we've studied the binomial coefficients, gaining a lot of insight about their geometrical and combinatorial nature. Still, there is yet another approach. After all, we are now much more experienced in mathematical induction. Why not use it directly?

Indeed, for $n = 0$, the binomial formula is true:

$$(a + b)^0 = 1 = \binom{0}{0} a^0 b^0 = \sum_{k=0}^{0} \binom{0}{k} a^k b^{0-k}.$$

Now, for $n = 1, 2, 3, \ldots$, assume that the induction hypothesis holds:

$$(a + b)^{n-1} = \sum_{k=0}^{n-1} \binom{n-1}{k} a^k b^{n-1-k}.$$

Using this hypothesis and the formula in Section 8.1.2, we can now make the induction step:

$$(a + b)^n = (a + b)(a + b)^{n-1}$$

$$= (a + b) \sum_{k=0}^{n-1} \binom{n-1}{k} a^k b^{n-1-k}$$

$$= a \sum_{k=0}^{n-1} \binom{n-1}{k} a^k b^{n-1-k} + b \sum_{k=0}^{n-1} \binom{n-1}{k} a^k b^{n-1-k}$$

$$= \sum_{k=0}^{n-1} \binom{n-1}{k} a^{k+1} b^{n-(k+1)} + \sum_{k=0}^{n-1} \binom{n-1}{k} a^k b^{n-k}$$

$$= \sum_{k=1}^{n} \binom{n-1}{k-1} a^k b^{n-k} + \sum_{k=0}^{n-1} \binom{n-1}{k} a^k b^{n-k}$$

$$= a^n + \sum_{k=1}^{n-1} \binom{n-1}{k-1} a^k b^{n-k} + \sum_{k=1}^{n-1} \binom{n-1}{k} a^k b^{n-k} + b^n$$

$$= a^n + \sum_{k=1}^{n-1} \left(\binom{n-1}{k-1} + \binom{n-1}{k} \right) a^k b^{n-k} + b^n$$

$$= a^n + \sum_{k=1}^{n-1} \binom{n}{k} a^k b^{n-k} + b^n$$

$$= \sum_{k=0}^{n} \binom{n}{k} a^k b^{n-k},$$

as required. This completes the induction step, and indeed the entire proof.

8.6 Extended Formulas

8.6.1 Extended Formula: Mathematical Induction

Let's extend the binomial formula yet further. For this purpose, let a and n be two nonnegative integer numbers:

$$a, n \geq 0.$$

(This new a has nothing to do with the old a used in $a + b$ above.) Define a ratio of two factorials:

$$C_{a,n} \equiv \begin{cases} \frac{a!}{(a-n)!} & \text{if } a \geq n \\ 0 & \text{if } a < n. \end{cases}$$

How to calculate this? You could follow the original definition: first, calculate the numberator. Then, calculate the denominator. Finally, divide.

But isn't this a bit silly? After all, why calculate the entire numerator explicitly? After all, it contains a lot of factors that appear in the denominator as well, and are soon going to cancel out anyway!

How to avoid this? The original definition should be used for theoretical purposes only. In practice, on the other hand, better use a new (recursive) definition:

$$C_{a,n} \equiv a(a-1)(a-2) \cdots (a-(n-1)) = \begin{cases} 1 & \text{if } n = 0 \\ a C_{a-1,n-1} & \text{if } n \geq 1. \end{cases}$$

This way, we also get a bonus: a could now be just any number, integer or not, positive or not. Furthermore, for every two numbers a and b, we can now mirror Newton's binomial formula:

$$C_{a+b,n} = \sum_{k=0}^{n} \binom{n}{k} C_{a,k} C_{b,n-k}.$$

To prove this, use mathematical induction. Indeed, the formula is clearly true for $n = 0$. Now, for $n \geq 1$, assume that the induction hypothesis holds: for every a and b,

$$C_{a+b,n-1} = \sum_{k=0}^{n-1} \binom{n-1}{k} C_{a,k} C_{b,n-1-k}.$$

Using this hypothesis and the formula in Section 8.1.2, we can now make the induction step:

$$
\begin{aligned}
C_{a+b,n} &= (a+b)C_{a+b-1,n-1} \\
&= aC_{(a-1)+b,n-1} + bC_{a+(b-1),n-1} \\
&= a\sum_{k=0}^{n-1} \binom{n-1}{k} C_{a-1,k} C_{b,n-1-k} + b\sum_{k=0}^{n-1} \binom{n-1}{k} C_{a,k} C_{b-1,n-1-k} \\
&= \sum_{k=0}^{n-1} \binom{n-1}{k} C_{a,k+1} C_{b,n-(k+1)} + \sum_{k=0}^{n-1} \binom{n-1}{k} C_{a,k} C_{b,n-k} \\
&= \sum_{k=1}^{n} \binom{n-1}{k-1} C_{a,k} C_{b,n-k} + \sum_{k=0}^{n-1} \binom{n-1}{k} C_{a,k} C_{b,n-k} \\
&= C_{a,n} + \sum_{k=1}^{n-1} \binom{n-1}{k-1} C_{a,k} C_{b,n-k} + \sum_{k=1}^{n-1} \binom{n-1}{k} C_{a,k} C_{b,n-k} + C_{b,n} \\
&= C_{a,n} + \sum_{k=1}^{n-1} \left(\binom{n-1}{k-1} + \binom{n-1}{k} \right) C_{a,k} C_{b,n-k} + C_{b,n} \\
&= C_{a,n} + \sum_{k=1}^{n-1} \binom{n}{k} C_{a,k} C_{b,n-k} + C_{b,n} \\
&= \sum_{k=0}^{n} \binom{n}{k} C_{a,k} C_{b,n-k},
\end{aligned}
$$

as required. This completes the proof.

8.6.2 The Extended Binomial Coefficient

How to use this formula fully? Well, in Section 8.1.2, the binomial coefficient has been defined with integer numbers only. Let's extend it to every number a, integer or not, positive or not:

$$\binom{a}{k} \equiv \frac{C_{a,k}}{k!}.$$

As in the original definition, k still has to be a nonnegative integer number. a, on the other hand, can now be just any number: even complex!

Now, look at this new definition. Is it consistent with the old one? Let's check: what happens in the special case in which a happens to be integer? Well, there are two options: if $a \geq k$, then the new definition indeed agrees with the old one. Indeed, both give the same familiar binomial coefficient. If, on the other hand, $0 \leq a < k$, then there is no old definition at all. Fortunately, the new definition tells us that, in this case, the (extended) binomial coefficient vanishes.

Now, look again at the formula proved in Section 8.6.1. To make it look better, divide it by $n!$. This gives

$$\binom{a+b}{n} = \frac{1}{n!}C_{a+b,n}$$

$$= \frac{1}{n!}\sum_{k=0}^{n}\binom{n}{k}C_{a,k}C_{b,n-k}$$

$$= \sum_{k=0}^{n}\frac{1}{k!(n-k)!}C_{a,k}C_{b,n-k}$$

$$= \sum_{k=0}^{n}\frac{C_{a,k}}{k!}\cdot\frac{C_{b,n-k}}{(n-k)!}$$

$$= \sum_{k=0}^{n}\binom{a}{k}\binom{b}{n-k}.$$

This formula mirrors the binomial formula: it gives $\binom{a+b}{n}$ as a sum rather than a product. Moreover, in the special case in which $a = b = n$, we have

$$\binom{2n}{n} = \sum_{k=0}^{n}\binom{n}{k}\binom{n}{n-k} = \sum_{k=0}^{n}\binom{n}{k}^2.$$

This way, the middle binomial coefficient (the highest column in Figure 8.8) is written as a sum rather than a product.

8.6.3 The Trinomial Formula

In the binomial formula, there are two numbers: a and b. In the trinomial formula, on the other hand, there is yet another number: c.

To obtain the trinomial formula, let's apply the binomial formula twice:

$$(a+b+c)^n = ((a+b)+c)^n$$

$$= \sum_{k=0}^{n}\binom{n}{k}(a+b)^k c^{n-k}$$

$$= \sum_{k=0}^{n}\binom{n}{k}\left(\sum_{l=0}^{k}\binom{k}{l}a^l b^{k-l}\right)c^{n-k}$$

$$= \sum_{k=0}^{n}\sum_{l=0}^{k}\binom{n}{k}\binom{k}{l}a^l b^{k-l}c^{n-k}$$

$$= \sum_{k=0}^{n}\sum_{l=0}^{k}\frac{n!}{k!(n-k)!}\cdot\frac{k!}{l!(k-l)!}a^l b^{k-l}c^{n-k}$$

$$= \sum_{k=0}^{n}\sum_{l=0}^{k}\frac{n!}{l!(k-l)!(n-k)!}a^l b^{k-l}c^{n-k}$$

$$= \sum_{0 \le l,j,m \le n,\ l+j+m=n} \frac{n!}{l!j!m!} a^l b^j c^m.$$

In the latter sum, we just introduced two new indices: $j \equiv k - l$, and $m \equiv n - k$. This way, the new indices l, j, and m have the same status. This is as in the transformation in Section 8.3.1.

This is the trinomial formula. Let's go ahead and mirror it, to produce a yet more advanced formula. For this purpose, let's apply the formula in Section 8.6.1 twice:

$$C_{a+b+c,n} = C_{(a+b)+c,n}$$

$$= \sum_{k=0}^{n} \binom{n}{k} C_{a+b,k} C_{c,n-k}$$

$$= \sum_{k=0}^{n} \binom{n}{k} \left(\sum_{l=0}^{k} \binom{k}{l} C_{a,l} C_{b,k-l} \right) C_{c,n-k}$$

$$= \sum_{k=0}^{n} \sum_{l=0}^{k} \binom{n}{k} \binom{k}{l} C_{a,l} C_{b,k-l} C_{c,n-k}$$

$$= \sum_{k=0}^{n} \sum_{l=0}^{k} \frac{n!}{k!(n-k)!} \cdot \frac{k!}{l!(k-l)!} C_{a,l} C_{b,k-l} C_{c,n-k}$$

$$= \sum_{k=0}^{n} \sum_{l=0}^{k} \frac{n!}{l!(k-l)!(n-k)!} C_{a,l} C_{b,k-l} C_{c,n-k}$$

$$= \sum_{0 \le l,j,m \le n,\ l+j+m=n} \frac{n!}{l!j!m!} C_{a,l} C_{b,j} C_{c,m}.$$

To make this formula look better, divide it by $n!$. This gives

$$\binom{a+b+c}{n} = \sum_{0 \le l,j,m \le n,\ l+j+m=n} \binom{a}{l} \binom{b}{j} \binom{c}{m}.$$

In the latter sum, how many terms are there? Thanks to Section 8.3.5, we already know the answer:

$$\binom{n+3-1}{3-1} = \binom{n+2}{2} = \frac{(n+1)(n+2)}{2}.$$

8.7 Exercises: Pascal's Triangle

8.7.1 Oblique Lines in Pascal's Triangle

1. Show that, for every nonnegative integer number $n \ge 0$,

$$\binom{n}{0} = \binom{n}{n} = 1.$$

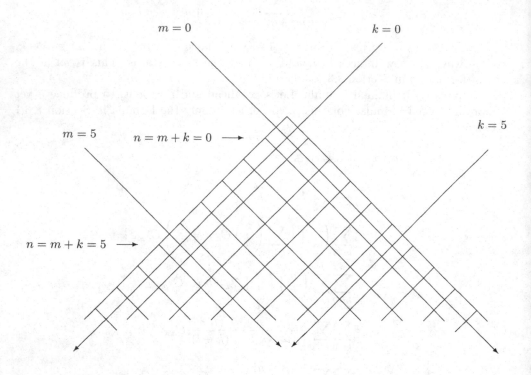

Fig. 8.10. Pascal's triangle: an oblique m-k grid. To index the oblique down-left lines, use $k = 0, 1, 2, \ldots$, as before. To index the oblique down-right lines, on the other hand, use the new index $m = 0, 1, 2, \ldots$. Finally, to index the horizontal levels, use $n = m+k = 0, 1, 2, \ldots$, as before.

2. Show that, for every two integer numbers $n \geq k \geq 0$, the binomial coefficient is symmetric in the sense that

$$\binom{n}{k} = \binom{n}{n-k}.$$

3. What does this mean geometrically? Hint: Figures 8.7–8.9 are indeed symmetric.
4. Show that, for every two natural numbers $n > k > 0$,

$$\binom{n}{k} = \binom{n-1}{k} + \binom{n-1}{k-1}.$$

Hint: see Section 8.1.2.
5. In Pascal's triangle, draw the lines $n =$ const. How do they look like geometrically? Hint: these are horizontal rows (Figure 8.1).
6. Next, draw the lines $k =$ const. How do they look like geometrically? Hint: these are oblique lines, pointing down-left (Figures 8.1 and 8.10).

7. Define the new index $m \equiv n - k$. What is its geometrical meaning? Hint: it indexes the oblique down-right lines.

8. Use this new index to draw the lines $m =$const. How do they look like geometrically? Hint: these are oblique lines, pointing down-right (Figure 8.10).

9. View Pascal's triangle as an oblique (infinite) m-k grid.

10. In this oblique grid, draw the lines $m + k =$const. How do they look like geometrically? Hint: these are horizontal rows (Figure 8.10).

11. Use the result in Section 8.1.2 recursively, time and again:

$$\binom{n}{k} = \binom{n-1}{k-1} + \binom{n-1}{k} = \binom{n-1}{k-1} + \binom{n-2}{k-1} + \binom{n-2}{k} = \cdots .$$

12. How does this process look like geometrically? Hint: in Pascal's triangle, look at the oblique line $k - 1 =$const. March along it, cell by cell, and sum the entries up. Stop just before hitting the nth level. On your left, you'd then see the same sum.

13. To help simplify this, avoid the old index n. Instead, use the new index $m = n - k$.

14. Use the above to prove once again the combinatorial formula in Section 8.3.6. Hint: use mathematical induction on $m = n - k = 0, 1, 2, \ldots$.

15. Write this mathematical induction as a recursive algorithm.

16. What is the geometrical meaning of this process? Is this familiar? Hint: we've already seen it before. In Pascal's triangle, look at the oblique line $k - 1 =$const. March along it, cell by cell, and sum the entries up. Stop just before hitting the nth level. On your left, you'd then see the same sum.

17. In terms of a discrete simplex, what is the geometrical meaning of this formula? Hint: look at the discrete k-dimensional simplex of size m. Count the points in it, face by face.

18. Conclude that it was indeed a good idea to view Pascal's triangle as an infinite m-k grid.

8.7.2 Binary Vectors and Characteristic Functions

1. Define the set of n-dimensional binary vectors:

$$V \equiv V(n) \equiv \{(v_1, v_2, \ldots, v_n) \mid v_i = 0 \text{ or } v_i = 1, \ 1 \le i \le n\}.$$

2. Note that each binary vector could also be viewed as a characteristic (or binary) function on the set $\{1, 2, 3, \ldots, n\}$. Hint: define the associated characteristic function simply as $v(i) \equiv v_i$.

3. Write V as the set of all characteristic (binary) functions on the set $\{1, 2, 3, \ldots, n\}$. Hint: in the notation of set theory,

$$V = V(n) \equiv \{0, 1\}^{\{1,2,\ldots,n\}}.$$

4. Note that V depends on n. Still, we often drop the '(n).'

5. What happens in the degenerate case $n \equiv 0$? How to define $V(0)$? Hint: $V(0)$ contains just one "vector:" the empty "vector" $\emptyset$, which contains no component at all:

$$V(0) \equiv \{\emptyset\}.$$

6. What is the cardinality of $V(n)$? In other words, how many vectors are there in $V(n)$?

7. Use mathematical induction on $n \geq 0$ to show that $V \equiv V(n)$ contains

$$|V| \equiv |V(n)| = 2^n$$

distinct vectors.

8. Let $n \geq 1$ be fixed. Let $V_k \subset V$ contain those vectors with exactly k nonzero components:

$$V_k \equiv \left\{ v \in V \mid \sum_{i=1}^{n} v_i = k \right\}.$$

9. Use mathematical induction on $k = 0, 1, 2, \ldots, n$ to show that V_k contains

$$|V_k| = \binom{n}{k}$$

distinct vectors. Hint: see Section 8.2.3.

10. Show that, for $k \neq l$, V_k and V_l are disjoint:

$$V_k \cap V_l = \emptyset.$$

11. Conclude that V could be written as the union of n disjoint subsets:

$$V = \cup_{k=0}^{n} V_k.$$

12. Conclude that

$$2^n = |V| = \sum_{k=0}^{n} |V_k| = \sum_{k=0}^{n} \binom{n}{k}.$$

13. Does this agree with the binomial formula?

14. In Pascal's triangle, how many distinct paths lead from the top cell (the zeroth level) to the kth cell in the nth level ($0 \leq k \leq n$)? Hint: see Section 8.1.3.

15. Why must this number be the same as $|V_k|$? Hint: model each path by a unique vector in V_k.

16. Prove more directly (by mathematical induction on n rather than k) that V_k indeed contains

$$|V_k| = \binom{n}{k}$$

distinct vectors. Hint: split V_k into two disjoint subsets: the first containing those vectors ending with 0, and the second containing those vectors ending with 1. Then, use the induction hypothesis, and the formula in Section 8.1.2.

17. In the algebraic expression $(a + b)^n$, open the parentheses. The result is a sum. In this sum, look at one particular term: $a^k b^{n-k}$ (where k is a fixed number between 0 and n). How many duplicate copies of this term appear in the above sum? Hint: upon opening parentheses, to obtain this term, pick exactly k a's from exactly k factors of the form $a + b$. To model this pick, use a unique vector in V_k.

18. Conclude that there are

$$|V_k| = \binom{n}{k}$$

duplicate copies of that term.

19. Conclude that, in Newton's binomial formula, the term $a^k b^{n-k}$ must indeed have the coefficient

$$\binom{n}{k}.$$

Applications
in Quantum
Statistical Mechanics

The binomial formula is particularly useful in statistical mechanics. In Chapter 8, Sections 8.4.1–8.4.3, we've already seen one example: Brownian motion. Here, we see more. For this purpose, we need to extend the binomial and trinomial formulas, to help model a system of molecules. This leads to the Maxwell-Boltzmann statistics.

Molecules are distinguishable from each other. Bosons, on the other hand, are not. Fortunately, thanks to our original study of discrete simplex, we can now model a system of bosons as well. This leads to the Bose-Einstein statistics.

Electrons are indistinguishable as well. Still, they are different from bosons: they obey Pauli's exclusion principle. Still, thanks to our earlier study in discrete math, we can now model them as well. This leads to the Fermi-Dirac statistics [7, 19, 29, 40]. In summary, discrete math helps model three different kinds of physical systems.

9.1 The Quadrinomial Formula

9.1.1 The Trinomial Coefficient

To obtain the present applications in quantum statistical mechanics, recall a few results from discrete math. The trinomial formula tells us that

$$(a + b + c)^n = \sum_{0 \le l,j,m \le n,\ l+j+m=n} \frac{n!}{l!j!m!} a^l b^j c^m$$

(Chapter 8, Section 8.6.3). What is the combinatorial meaning of this? Well, upon opening parentheses on the left-hand side, we obtain a sum of many products. Each product may contain l a's, j b's, and m c's, where l, j, and m are some nonnegative integers that sum to n:

$$l + j + m = n.$$

This produces a product of the form

$$a^l b^j c^m,$$

as in the right-hand side.

How many duplicate copies of this product are there? In other words, how many different ways are there to pick l a's, j b's, and m c's? The answer is the trinomial coefficient:

$$\frac{n!}{l!j!m!}.$$

Let's go ahead and extend this.

9.1.2 Extending The Trinomial Formula

Let's extend the above yet further:

$$(a+b+c+d)^n = \sum_{0 \le l,j,m,k \le n, \; l+j+m+k=n} \frac{n!}{l!j!m!k!} a^l b^j c^m d^k.$$

This is the quadrinomial formula. What is its combinatorial meaning? Well, upon opening parentheses on the left-hand side, we obtain a sum of products of the form

$$a^l b^j c^m d^k,$$

where l, j, m, and k are some nonnegative integers that sum to n:

$$l + j + m + k = n.$$

To obtain such a product, how many different ways are there to pick l a's, j b's, m c's, and k d's? The answer is the quadrinomial coefficient:

$$\frac{n!}{l!j!m!k!}.$$

Let's go ahead and use this in physics.

9.2 Mass and Energy

9.2.1 Closed System: Constant Energy

To use the above in physics, we need four new indices:

$$n_1 \equiv l$$
$$n_2 \equiv j$$
$$n_3 \equiv m$$
$$n_4 \equiv k.$$

Consider now a system of n molecules. Assume that the system is closed: isolated and insulated. This way, the system has constant energy.

9.2.2 Energy Levels

In quantum mechanics, energy is not continuous but discrete: it comes in discrete energy levels:

$$0 < E_1 < E_2 < E_3 < E_4 < E_5 < \cdots.$$

Only these E_i's are allowed: the system can never have any other energy in between. In this sense, energy is different from position and time, which remain continuous: at each individual time, the molecule can still lie in any position in space.

9.2.3 Conservation of Mass

Now, what are the a's? Let's interpret them as molecules of energy E_1. Likewise, let's interpret the b's, the c's, and the d's as molecules of energy E_2, E_3, and E_4, respectively. This way, in our original system of n molecules, n_i molecules have energy E_i ($1 \leq i \leq 4$). This splits the original system into four disjoint subsets:

$$\sum_{i=1}^{4} n_i = n_1 + n_2 + n_3 + n_4 = n.$$

This is actually conservation of mass: uniting the subsets produces the original system back again.

9.2.4 Configuration – Distribution

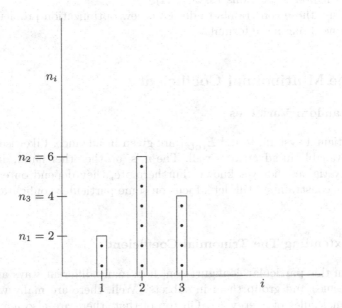

Fig. 9.1. Distribution of molecules: there are n_1 molecules of energy E_1, n_2 molecules of energy E_2, and so on.

As a matter of fact, why stop at 4? After all, there could be many more subsets of yet higher energy levels: n_5 molecules of energy E_5, n_6 molecules of energy E_6, and so on. This way, the above sum takes the new form

$$\sum_{i=1}^{\infty} n_i = n.$$

This is not really an infinite sum. After all, only a few n_i's are nonzero, and all the rest vanish. Still, because we don't know which i's are redundant, we leave them all in.

The n_i's make a new configuration or distribution. This is illustrated in a column diagram (Figure 9.1): the ith column contains n_i molecules of energy E_i. This is how the original n molecules split into separate energy levels.

9.2.5 Conservation of Energy

So far, we've used one constraint only: conservation of mass. This law is more mathematical than physical. After all, the total number of molecules must be n. Still, we also have a more fundamental physical law – conservation of energy:

$$\sum_{i=1}^{\infty} n_i E_i = E_{\text{total}}$$

(the total energy in the entire system). This is indeed the first law of thermodynamics (Chapter 1, Sections 1.6.2–1.7.4).

Let's use these constraints to design a new optimization problem. For this purpose, we need one more formula.

9.3 The Multinomial Coefficient

9.3.1 Random Variables

In our original system, n and E_{total} are given in advance. Likewise, the individual E_i are available in advance as well. The n_i's, on the other hand, are not: they are random variables, not yet known. Furthermore, they depend on each other, as in the above constraints. Still, let's focus on some particular configuration, with some specific n_i's.

9.3.2 Extending The Trinomial Coefficient

To design this particular configuration, how many different ways are there to pick the molecules and group them in subsets? Well, there are many ways to pick the first n_1 molecules of energy E_1. On top of that, there are also many ways to pick n_2 molecules of energy E_2, and so on. In total, there are

$$n! \frac{1}{n_1!} \cdot \frac{1}{n_2!} \cdot \frac{1}{n_3!} \cdots$$

different ways. This is just a finite product. After all, only a few n_i's are nonzero. Still, this product could be quite long. This is indeed the multinomial coefficient.

9.3.3 Degeneracy

In our configuration, there are n_i molecules of energy E_i $(i \geq 1)$. Still, they are not necessarily in the same state. On the contrary: they split into d_i different states (or degeneracies), where d_i is a constant natural number. This is illustrated in Figure 9.2: the ith column splits into d_i subsquares.

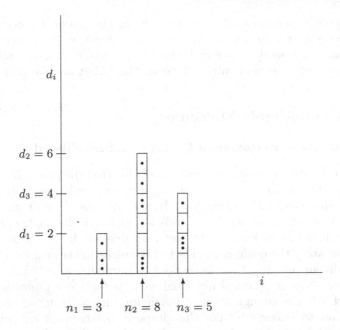

Fig. 9.2. At energy level E_i, the original n_i molecules split now into d_i different states (degeneracies), each containing a few molecules.

How many different ways are there to split the original n_i molecules into d_i disjoint subsets? Clearly, the first molecule could be placed in either of the d_i subsets. On top of that, the second molecule could be placed in either of the d_i subsets as well, and so on. Thus, the answer is

$$\left| \{1, 2, 3, \ldots, d_i\}^{\{1,2,3,\ldots,n_i\}} \right| = d_i^{n_i}.$$

Thus, the product in Section 9.3.2 should be modified to read

$$n! \frac{d_1^{n_1}}{n_1!} \cdot \frac{d_2^{n_2}}{n_2!} \cdot \frac{d_3^{n_3}}{n_3!} \cdots.$$

This is the total number of different ways to design our configuration.

9.3.4 Configuration and Its Entropy

This product is called the entropy of the configuration. It tells us how likely the configuration is to take place: the higher it is, the more options the configuration has to form, and the more likely it is to appear in reality. As time goes by, the system may change, but only in one direction: our system could only "jump" from one particular configuration to yet "better" configuration, with more entropy, and a larger probability to occur. Indeed, thanks to the second law of thermodynamics, entropy can only increase, but never decrease. (This is discussed throughout our study in discrete math: in Chapter 6, Section 6.3.5, Chapter 7, Section 7.12.3, and Chapter 8, Section 8.4.3.) As a result, in the physical process in our closed system, a configuration can only transform into a new configuration of a yet higher entropy and likelihood. In the end, the process will approaches an equilibrium: a configuration of maximum entropy (among those that are allowed).

9.4 Distinguishable Molecules

9.4.1 Equilibrium: Maximum Entropy and Likelihood

Let's use the entropy to obtain a new useful quantity: the probability of the configuration. For this purpose, we only need to normalize: make sure that all probabilities of all configurations indeed sum to 1. How to do this? Easy: sum up all entropies of all possible configurations. Then, look at one particular configuration. To obtain its probability, just look at its entropy, and divide it by the above sum. This will tell us how likely the configuration is to take place in the real world.

Actually, in practice, there is no need to normalize at all. After all, this makes no difference to our practical question: what is the most probable configuration? Clearly, this is the configuration of maximum entropy (among those that are allowed). How to uncover it? For this purpose, better work not with the entropy itself, but with its natural logarithm. After all, the logarithm is a monotonically-increasing function (Figure 24.17). Thus, maximizing it is the same as maximizing the original entropy.

We must also work under two constraints: conservation of mass and energy. This way, we'll obtain our solution: the configuration that maximizes entropy among those configurations that obey conservation of mass and energy. This will indeed be equilibrium, approached by the physical process in our closed system. This is called the Maxwell-Boltzmann distribution (or statistics).

9.4.2 Maxwell-Boltzmann Statistics

To obtain the probability of a specific configuration, we looked at its entropy, and divided it by the sum of all entropies of all possible configurations. This was only implicit: in practice, it is unnecessary.

As time goes by, the system will eventually arrive at its equilibrium: the optimal configuration. How does it look like? Well, this is just a column diagram: at energy level i, there are n_i molecules of energy E_i (Figure 9.1). In other words, this is a discrete distribution. To make it continuous, we must first normalize. This, however, is a different kind of normalization. (After all, by now, we already have only one configuration, not many.) This new normalization has one purpose only: to make the columns sum to 1, to make a proper probability function.

Thus, in the Maxwell-Boltzmann distribution, look at the ith column, and normalize it. In other words, look at the number of molecules of energy E_i, and divide by the total number of molecules. This defines a new probability function:

$$p(E_i) \equiv \frac{n_i(E_i)}{n}.$$

Now, if you pick a molecule from the system at random, how likely is it to have energy E? We now have the answer: the probability for this is $p(E)$. Fortunately,

$p(E)$ doesn't depend on the discrete index i any more, but on the (continuous) energy E only.

Later on, in quantum mechanics, we'll see that energy is not really continuous, but only discrete: it comes in discrete lumps only. Still, these lumps are so tiny that they could make a continuous graph, to illustrate the behavior of p as a function of E. For this purpose, however, p must renormalize, to make sure that the area under the graph is 1, as required in a continuous distribution.

9.5 Indistinguishable Bosons

9.5.1 Indistinguishable Bosons

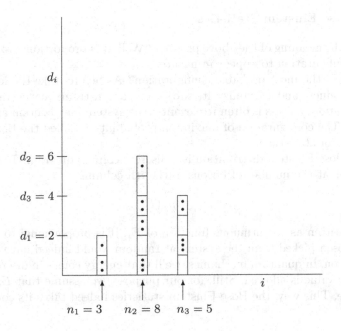

Fig. 9.3. Bosons, on the other hand, are indistinguishable: it doesn't matter which n_1 bosons to place in the first energy level, whcih n_2 bosons to place in the next energy level, and so on.

Instead of molecules, consider now n bosons, in one particular configuration. These are indistinguishable particles: there is no way to distinguish one boson from another. Therefore, it doesn't matter which n_1 bosons are placed in the first energy level. Likewise, it doesn't matter which n_2 bosons are placed in the second energy level, and so on.

Still, there are many ways to split the first n_1 bosons among the d_1 possible states in the first energy level (Figure 9.3). How many? Well, to split n_1 bosons into d_1 disjoint states, we need to decide how many are in state 1, how many in state

2, and so on. (After all, since the bosons are indistinguishable, it doesn't matter which ones, but only how many.) For this purpose, we need to pick d_1 nonnegative integers that sum to n_1. How many ways are there to do this? We already know the answer:

$$\binom{n_1 + d_1 - 1}{d_1 - 1}$$

(Chapter 8, Sections 8.3.5–8.3.6). The same could be done in the other energy levels as well. In summary, to have our original configuration, there are

$$\binom{n_1 + d_1 - 1}{d_1 - 1}\binom{n_2 + d_2 - 1}{d_2 - 1}\binom{n_3 + d_3 - 1}{d_3 - 1}\cdots$$

possible ways.

9.5.2 Bose-Einstein Statistics

What is the meaning of the above product? Well, it is proportional to the probability of our configuration to appear in nature.

What is the most probable configuration? As before, take the logarithm of the above product, and maximize it, subject to a constraint: conservation of energy. (Conservation of mass is often irrelevant – the system may contain arbitrarily many bosons.) The configuration of maximum probability is called the Bose-Einstein distribution (or statistics).

The Bose-Einstein distribution is a discrete column diagram. To make it continuous, look at the number of bosons in the ith column:

$$n_i\left(E_i\right).$$

This is written as a continuous function of E_i. It is proportional to the probability that a boson picked from the system at random would indeed have energy E_i.

Later on, in quantum mechanics, we'll that energy comes in discrete lumps only: not every value is allowed. Still, for our purpose, we assume that E_i may take just any value. This way, the Bose-Einstein statistics indeed takes its continuous face.

9.6 Indistinguishable Electrons

9.6.1 Pauli Exclusion Principle

Instead of bosons, let us now consider n electrons, arranged in our original configuration. Like bosons, electrons are indistinguishable. Furthermore, they must also obey Pauli's exclusion principle: in each energy level, each state may contain one electron at most. For this reason, the number of electrons in this level can never exceed the number of states:

$$n_i \le d_i, \quad i \ge 1.$$

Under these circumstances, how to split n_i electrons into d_i disjoint states? For this purpose, look at the ith column in Figure 9.4. It contains d_i subsquares. Out

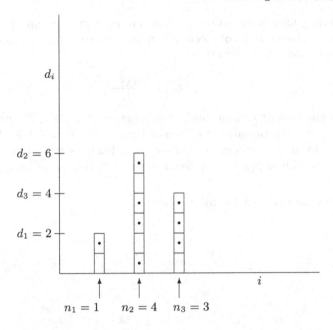

Fig. 9.4. Electrons, on the other hand, must obey Pauli's exclusion principle: each state (degeneracy) may contain one electron at most, so $n_i \leq d_i$.

of these, pick just n_i subsquares to contain an electron, leaving $d_i - n_i$ subsquares empty. (After all, because electrons are indistinguishable, it doesn't matter which electrons are picked, but only which states are picked.) How many ways are there to do this? We already know the answer:

$$\binom{d_i}{n_i}.$$

Thus, to produce our original configuration, there are

$$\binom{d_1}{n_1} \binom{d_2}{n_2} \binom{d_3}{n_3} \cdots$$

different ways.

9.6.2 Fermi-Dirac Statistics

What is the meaning of the above product? Well, it is proportional to the probability of our configuration to take place in reality.

What is the most probable configuration? As before, take the logarithm of the above product, and maximize it, subject to two constraints: conservation of mass and energy. The configuration of maximum probability is called the Fermi-Dirac distribution (or statistics).

The Fermi-Dirac distribution is a discrete column diagram. To make it continuous, look at the number of electrons in the ith column, and normalize it. This defines a new probability function:

$$p(E_i) \equiv \frac{n_i(E_i)}{n}.$$

This could also be written as a continuous function of E_i. (For this purpose, however, there is a need to renormalize, as discussed below.) How to use it? Well, if you pick an electron from the system at random, how likely is it to have energy E? The probability for this is $p(E)$. Let's see how this function looks like.

9.6.3 Fermi Level and Fermi Energy

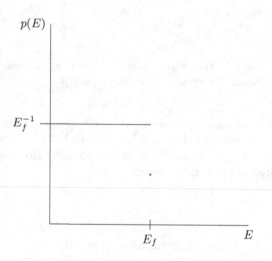

Fig. 9.5. As $T \to 0^+$, p may take two constant values only: below the Fermi energy, p is a positive constant (full levels). Above the Fermi energy, on the other hand, $p = 0$ (empty levels). For this reason, the Fermi energy is the maximal energy in the entire system: no electron may have more energy.

It turns out that p has a closed analytic form [19], in terms of the exponent function:

$$p(E) = \frac{E_f^{-1}}{\exp\left(\frac{E - E_f}{K_B T}\right) + 1},$$

where

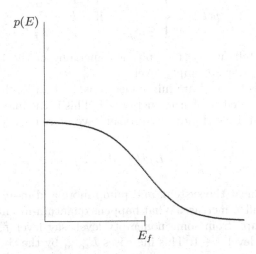

Fig. 9.6. As T increases, E_{total} increases as well, so some electrons may get enough energy to "jump" from a level just below the Fermi level to a higher level just above the Fermi level. This way, p remains symmetric around E_f.

- f is a fixed natural number: the Fermi level,
- E_f is the Fermi energy: the energy at the Fermi level,
- T is the temperature (in Kelvin's scale),
- and K_B is Boltzmann's constant.

This probability function should be used only weakly, not strongly. In other words, it should never be interpreted pointwise, but only in some interval. For instance, let $b > a \geq 0$ be two nonnegative numbers. Now, pick an electron from the system at random. How likely is the electron to have its energy in between a and b? The probability for this is

$$\int_a^b p(E)dE.$$

This is no longer a sum, but an integral. Therefore, a new kind of normalization is needed, to make sure that the total area under the integral is 1:

$$\int_{-\infty}^{\infty} p(E)dE = 1,$$

as required in a continuous distribution.

How does p look like? Let's start from the easy case, in which p is uniform. This is the limit case, in which E_{total} is minimal, and T approaches the absolute zero:

$$T \to 0^+ = (-273°C).$$

At this limit,

$$p(E) \to \begin{cases} E_f^{-1} & \text{if} \quad E < E_f \\ 0 & \text{if} \quad E > E_f \\ E_f^{-1}/2 & \text{if} \quad E = E_f \end{cases}$$

(Figure 9.5). This tells us the true physical meaning of the Fermi level: the first nonfull level, or the last nonempty level.

In fact, all levels below f are full of electrons, and all levels above f are empty. The energy at f is called the Fermi energy: E_f. This is the maximal possible energy in the entire system. Indeed, no electron can have more energy:

$$\int_0^{E_f} p(E)dE = E_f^{-1}E_f = 1.$$

Now, let's start to heat the system, and pump more and more energy into it. This way, both E_{total} and T increase. What happens quantum-mechanically? Well, some electron must "jump" from some nonempty level, say level f, to the next higher (nonfull) level, say level $f + 1$. This increases E_{total} by the tiny bit of

$$E_{f+1} - E_f > 0.$$

As T increases further, more electrons may "jump" as well, say from level $f - 1$ to level $f + 2$. This way, $p(E)$ will always remain symmetric around E_f, as in Figure 9.6.

9.7 Exercises: Spin

9.7.1 Spin-Up and Spin-Down

1. Extend the trinomial formula, and obtain the quadrinomial formula:

$$(a + b + c + d)^n = \sum_{0 \leq l,j,m,k \leq n,\ l+j+m+k=n} \frac{n!}{l!j!m!k!} a^l b^j c^m d^k.$$

Hint: extend the proof in Chapter 8, Section 8.6.3.
2. How many terms are there in this sum? Hint:

$$\binom{n+4-1}{4-1} = \binom{n+3}{3} = \frac{(n+1)(n+2)(n+3)}{6}.$$

(Chapter 8, Sections 8.3.5–8.3.6).
3. In the above sum, look at the quadrinomial coefficient:

$$\frac{n!}{l!j!m!k!}.$$

What is its combinatorial meaning? Hint: this is the total number of different ways to pick l a's, j b's, m c's, and k d's.

4. What is its physical meaning? Hint: this is the total number of different ways to pick l molecules of energy E_1, j molecules of energy E_2, m molecules of energy E_3, and k molecules of energy E_4 from a system of $n = l + j + m + k$ molecules.

5. What would happen if, instead of molecules, we had bosons? Hint: unlike molecules, bosons are indistinguishable. Therefore, all the above ways would coincide, and become one and the same.

6. What would happen if, instead of bosons, we had electrons? Hint: electrons are indistinguishable as well. Thus, there would be no change: all the above ways still coincide.

7. Consider a system of n electrons. In the first energy level, there are just two possible states (degeneracies): either spin-up, or spin-down. In other words,

$$d_1 = 2.$$

How many ways are there to have $n_1 = 2$ electrons in this energy level? Hint: there is only one way: one electron must have spin-up, and the other must have spin-down.

8. Is this as expected from our formula? Hint: yes. Indeed, in this case, there is only

$$\binom{d_1}{n_1} = \binom{2}{2} = 1$$

way.

9. One student gave a different answer: "there are two different ways:
- either the first electron has spin-up and the second has spin-down,
- or the first electron has spin-down and the second has spin-up."

What's wrong with this answer? Hint: these ways coincide, because there is no way to distinguish between the "first" and "second" electrons.

10. How many ways are there to have $n_1 = 1$ (just one electron in the first energy level)? Hint: two: either spin-up, or spin-down.

11. Is this as expected from our formula? Hint: yes. Indeed, in this case, there are

$$\binom{d_1}{n_1} = \binom{2}{1} = 2$$

ways.

12. Conclude that $n_1 = 1$ is more likely than $n_1 = 2$.

13. Does this mean that, in the Fermi-Dirac distribution, the first energy level must always contain one electron only? Hint: no! The Fermi-Dirac distribution must obey two constraints: conservation of mass and energy:

$$E_{\text{total}} = n_1 E_1 + n_2 E_2 + n_3 E_3 \cdots$$

(for some given parameter E_{total}). Under these constraints, if the most probable configuration has $n_1 = 2$, then the first level is full:

$$n_1\,(E_1) = 2.$$

In practice, this is often the case: the first energy level is usually full (Figures 9.5–9.6).

14. Consider now a system of bosons rather than electrons. In the first energy level, there are now three possible states (degeneracies): either spin-up, or spin-down, or spin-zero. In other words,

$$d_1 = 3.$$

How many ways are there to have n_1 bosons in this energy level? Hint:

$$\binom{n_1 + 3 - 1}{3 - 1} = \binom{n_1 + 2}{2} = \frac{(n_1 + 1)(n_1 + 2)}{2}.$$

15. Later on, in the context of Lie algebras, we'll learn more about bosons and electrons and their spin.

Part IV

Introduction
to Relativity

Introduction to Relativity

A physical phenomenon should better be independent of the mathematical coordinates used to describe it. After all, the laws of nature must be consistent. You must always observe the same laws, no matter what coordinates you use to carry out your experiment.

But what if you are on a spaceship, travelling away from the Earth at a constant speed? In your spaceship, you may have a lamp. Look at its light, and measure its speed. The result must be 300, 000 kilometer per second.

I on the Earth, on the other hand, also want to calculate the speed of light. But I'm a bit naive: I take the speed of light that you've just measured, and add to it yet another speed: the speed of the spaceship with respect to the Earth. So, I think that light is faster than 300, 000 kilometer per second!

Fortunately, we know better: light travels at the same speed in all systems, moving or not. Indeed, this was already proved in many experiments. How to solve the above paradox? Well, speeds should not be added linearly, but nonlinearly.

This is how special relativity improves on Newton's original laws. Fortunately, in practical engineering problems, this effect is usually negligible, so Newtonian mechanics is good enough.

Still, in modern technology, this can no longer be ignored. To account for this, relativity theory designs a new geometry: it takes the time axis, and sticks it to space, to form a new four-dimensional hyperspace: spacetime. In it, we get to see two new effects most vividly: time dilation, and length contraction. Later on, these effects will be extended to general relativity as well.

In general relativity, spacetime takes a new geometrical form: a four-dimensional differentiable manifold. How to design it? Around each event in spacetime, draw a little chart: a local t-x-y-z coordinate system, to help carry out physical experiments locally.

Thanks to Zorn's lemma, there exists a maximal atlas of charts. Furthermore, thanks to the axiom of choice, at each event in spacetime, we can pick one coordinate system, to help describe our local "lab." This is also a good preparation work for Lie groups, used often in quantum mechanics.

Thanks to this, we can envelope spacetime with tangent spaces all over. These are Riemann normal coordinates: local, not global. This is indeed Einstein's equivalence principle, in its geometrical face. This was indeed the starting point of general relativity.

10

Introduction
to Special Relativity:
Momentum-Energy and Mass

To describe physics, Euclidean geometry is too poor: it only studies static shapes in the two-dimensional plane. To help model motion, Newton introduced a new time axis, perpendicular to the plane. This may help model a new force, applied to an object from the outside, to accelerate its original speed, and make it move faster.

This fits well in Plato's philosophy: to think about a general concept, we must introduce a new word in our language, to represent not only one concrete instance but also the "godly" spirit behind all possible instances. Likewise, a Newtonian force acts from the outside, to give "life" to a static object.

Einstein, on the other hand, threw the time dimension back into the very heart of geometry. This way, time is not much different from any other spatial dimension. Once the time axis is united with the original spatial axes, we have a new four-dimensional manifold: spacetime.

This is more in the spirit of Aristotle's philosophy. A word takes its meaning not from the outside but from the very inside: the deep nature of the general concept it stands for.

Special relativity [11, 58] improves on Newtonian mechanics. It teaches us how to add speeds more accurately: not linearly, but nonlinearly. As a result, we also have a new (relativistic) definition of energy and momentum. This is how physics should indeed be introduced: invariant at all (inertial) reference frames (or coordinate systems), traveling at a constant speed, at a constant direction.

10.1 The Michelson-Morley Experiment

10.1.1 Light: An Electromagnetic Wave

In the 19th century, Maxwell introduced his equations that united electricity and magnetism into one physical phenomenon: an electromagnetic field. The common solution to these equations is an electromagnetic wave, traveling at the speed of light: $c \equiv 300,000$ kilometer per second.

Nowadays, we know that light is actually a special kind of electromagnetic wave: while traveling forward, it also oscillates at a frequency that we can see. X-rays, on the other hand, make a different kind of radiation, too frequent for us to see. Radio waves, on the other hand, make yet another kind of wave, not frequent enough for us to see. Still, all three kinds of waves are in principle the same: an oscillating electromagnetic wave. They only occupy different parts of the spectrum of possible frequencies.

10.1.2 Wave in Its Medium

In Maxwell's day, everybody assumed that a wave must oscillate in some medium. In what medium could the electromagnetic wave oscillate? After all, it could travel and oscillate even in vacuum!

To answer this, everyone believed that the entire space around us must be full of some new invisible substance: the (so-called) luminiferous aether. This must be the medium in which light travels.

10.1.3 Aether: An Absolute Reference Frame

For this purpose, the aether must stand still. The entire Earth may rotate, day and night, but the aether doesn't: it stands still, and makes an absolute reference frame.

The entire Earth rotates eastwards: this is why the sun rises from the east. Assume that this rotation is at velocity $v > 0$ with respect to the aether. Still, we humans look at things the other way around: we think that we stand still, and that the aether moves at velocity v westwards. After all, we look at things from our own private perspective: our own reference frame. Actually, we should feel a slight wind, blowing at our faces: the aether wind, blowing westwards at velocity v. Fortunately, the aether is to fine for us to feel.

10.1.4 Inertial Reference Frames

In Maxwell's day, the aether was accepted as an absolute reference frame. With respect to it, the Earth makes an inertial reference frame, moving at the constant velocity v eastwards. For simplicity, assume that this motion makes a straight line. This way, these are inertial reference frame: they never accelerate with respect to one another, and never change direction. This is the type considered in special relativity. Acceleration will be discussed later, in general relativity.

10.1.5 The Michelson-Morley Experiment

To help measure v, Michelson and Morley carried out a new experiment (Figure 10.1). In the middle, they placed an oblique half silvered mirror. From the left, they fired a light beam, straight eastwards. This way, once the light beam arrived at the center, it splits: only half of it transmits straight eastwards, reflects at the far right, and returns to the center, where half of it turns southwards, towards the screen. The other half, on the other hand, doesn't transmit at the center, but turns northwards, and reflects all the way southwards to the screen.

Now, assume that the northern mirror is at distance d from the center. How much time is required to pass this distance? Well, look at things from the perspective of the Earth. On its journey northwards on the Earth, the half light beam feels a slight wind, coming from the east at velocity v: the aether wind. To overcome it, the half

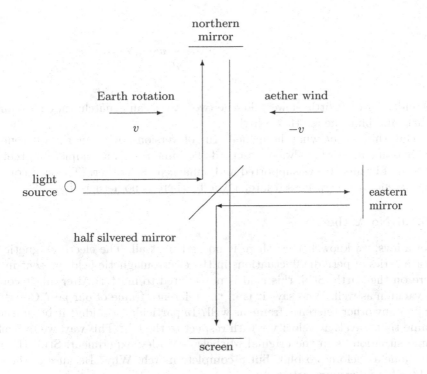

Fig. 10.1. The Michelson-Morley experiment. The light beam comes from the left, and splits at the center: half transmits straight rightwards, reflects on the far right, and returns to the center, where it turns southwards. The other half, on the other hand, turns northwards at the center, reflects on the far north, and returns all the way southwards, through the center, to the screen at the far south. It turns out that both halves match, unaffected by any "aether wind."

light beam must actually travel a little obliquely eastwards. Thanks to Pythagoras' theorem, its net velocity northwards is only

$$\sqrt{c^2 - v^2} < c.$$

Thus, the round trip from the center to the northern mirror and back should last

$$\Delta t = \frac{2d}{\sqrt{c^2 - v^2}}$$

seconds (distance divided by velocity).

How about the other half, transmitting straight eastwards? It first travels eastwards, against the aether wind. Then, it reflects on the far right, and returns westwards, with the aether wind at its back. Assuming that velocity is additive, this should last

$$\frac{d}{c - v} + \frac{d}{c + v} = d\frac{c + v + c - v}{(c - v)(c + v)}$$

$$= 2d\frac{c}{c^2 - v^2}$$

$$= \frac{2d}{\sqrt{c^2 - v^2}} \cdot \frac{c}{\sqrt{c^2 - v^2}}$$

$$> \frac{2d}{\sqrt{c^2 - v^2}}$$

seconds. This is a little slower. So, the two halves can't match: they must miss each other, and blur one another a little.

But this is *not* what happened. In all versions, with many different angles, Michelson and Morley always observed the same result: a complete match! In fact, both half beams always supported and enhanced each other. (This is a constructive interference, to be discussed later.) Maybe there is no aether after all?

10.1.6 No Aether!

Nowadays, we know better: there is no aether at all. The electromagnetic wave is just a series of periodic fluctuations in the electromagnetic field, present in the lab, here on the Earth. Still, this field is not related to matter. After all, it could exist in vacuum as well. Who says it is in the reference frame of our lab? Couldn't it be in just any other reference frame as well? In particular, couldn't it be in a reference frame that travels at velocity v with respect to the lab? This way, we're back to the same situation as in the original Michelson-Morley experiment. Still, the result is the same as before: no blur, but a complete match. Why? Because c, the speed of light, obeys a strange equation:

$$c \text{ "}+\text{" } v = c \text{ "}-\text{" } v = c.$$

10.1.7 Relativistic Time

Sounds nonsense! How could this be? Well, the quotation marks tell us the answer: this is a nonstandard "addition:" not linear, but nonlinear.

To understand this, we must realize that time is not absolute but relativistic: each reference frame has its own time scale, measured in its private clock.

10.1.8 Einstein's Law: Constant Speed of Light

In this reference frame, the new (relativistic) time scale matches its own length scale: they still have the same ratio as before. This way, the speed of light remains c. This is true in all (inertial) reference frames. This is Einstein's law: the speed of light is constant, independent of the reference frame (and coordinates) used to measure it.

10.1.9 Synchronized Reference Frames

To see this, our (inertial) reference frames should better be synchronized: both clocks should start ticking at the same initial time: zero. At this time, both should also share the same reference point: the origin, where distance is measured from.

Later on, time and space will combine into a new geometrical structure: space-time. Here, for a start, we take an algebraic point of view.

10.2 Adding Velocities Accurately

10.2.1 How to Add Velocities?

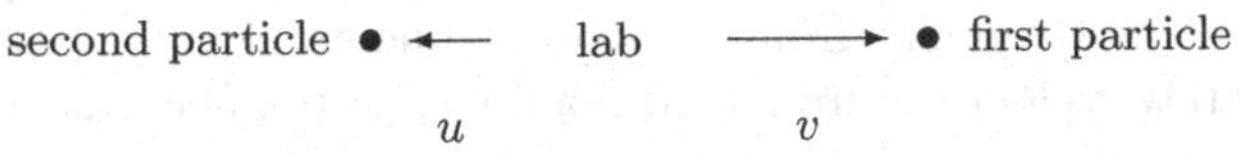

second particle • ◄—— lab ———► • first particle

u v

Fig. 10.2. In our lab, the first particle moves rightwards at velocity v, while the second particle moves leftwards at velocity u.

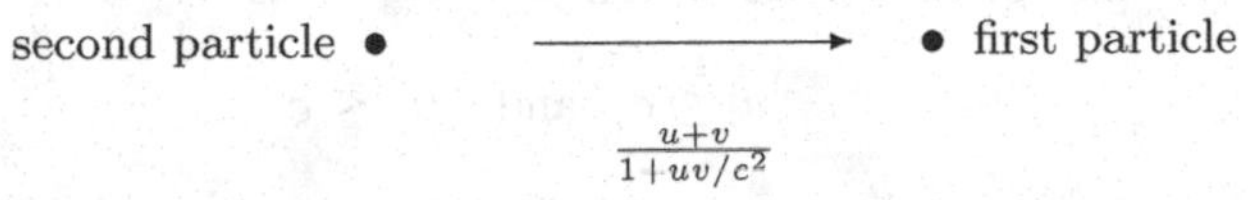

second particle •   ————————► • first particle

$$\frac{u+v}{1+uv/c^2}$$

Fig. 10.3. How fast does the first particle get away from the second particle? Well, the answer is not quite $u + v$, but a little different: $(u + v)/(1 + uv/c^2)$.

In Newtonian mechanics, velocities are added linearly. Consider, for example, a particle that travels at the constant velocity v at some constant direction. At the same time, another particle travels at the constant velocity u at the opposite direction (Figure 10.2). These velocities are not absolute: they are only relative to our lab.

Still, our lab is not an absolute reference frame: it is not static, but dynamic as well. After all, it is on the Earth, which orbits the sun, which orbits the center of our galaxy: the Milky Way. Fortunately, this underlying motion could be disregarded.

After all, we don't feel it at all. In fact, we are only interested in the velocities of the particles with respect to our lab. For this purpose, we may assume that our lab is at rest.

Better yet, disregard the entire lab, and focus on the relation between the particles: how fast do they get away from one another? This way, we actually look at one particle from the perspective of the other. After all, this is the true physical quantity here: the lab is not really a part of the game.

So, how fast does the first particle get away from the second one? In Galileo-Newton's theory, the velocities add up, so the answer is simply $u + v$. For small velocities, this indeed makes sense. But what happens when v is as large as the speed of light c? In this case, the sum $c + u$ would exceed the speed of light, which is impossible, as is evident from many experiments.

Fortunately, velocities are added nonlinearly: the accurate answer is not $u + v$ but

$$\frac{u + v}{1 + \frac{uv}{c^2}}$$

(Figure 10.3). Later on, we'll see how plausible this really is, both algebraically and geometrically. Before going into this, better look at a few examples and special cases.

Clearly, so long as both u and v are moderate, this is nearly the same as $u + v$. This is why Galileo-Newton's theory works well in practical engineering problems. Still, strictly speaking, Newtonian mechanics is inaccurate. At high velocities, this subtle inaccuracy becomes crucial, and can't be ignored any more.

10.2.2 Never Exceed The Speed of Light!

Thanks to the above formula, we can now be consistent, and keep a universal rule: never exceed the speed of light! Indeed, assume that, with respect to the lab, both particles don't exceed the speed of light:

$$|u| \leq c \quad \text{and} \quad |v| \leq c.$$

Now, with respect to each other, could the particles ever exceed the speed of light? No, they can't! To prove this algebraically, we must first scale the velocities properly. For this purpose, just divide them by the speed of light, and define

$$\beta_u \equiv \frac{u}{c} \quad \text{and} \quad \beta_v \equiv \frac{v}{c}.$$

In these new terms, what does Einstein's law tell us? Well, it says that c is the same in all systems. Furthermore, no speed could ever exceed c:

$$|\beta_u| \leq 1 \quad \text{and} \quad |\beta_v| \leq 1.$$

Now, in these new terms, what do we assume? Well, assume first that both u and v are strictly smaller (in magnitude) than the speed of light:

$$|u| < c \quad \text{and} \quad |v| < c,$$

so

$$|\beta_u| < 1 \quad \text{and} \quad |\beta_v| < 1.$$

In this case,

$$1 - \beta_u - \beta_v + \beta_u\beta_v = (1 - \beta_u)(1 - \beta_v) > 0,$$

so

$$\beta_u + \beta_v < 1 + \beta_u\beta_v,$$

or

$$\frac{\beta_u + \beta_v}{1 + \beta_u\beta_v} < 1.$$

Moreover, if both u and v change sign, then the above is still valid. Therefore, we also have

$$\left|\frac{\beta_u + \beta_v}{1 + \beta_u\beta_v}\right| = \frac{|\beta_u + \beta_v|}{1 + \beta_u\beta_v} < 1.$$

By multiplying both sides of this inequality by c, we also have

$$\left|\frac{u + v}{1 + \frac{uv}{c^2}}\right| < c.$$

This is good: our new formula for adding velocities is consistent! Indeed, with respect to each other, the particles never exceed the speed of light, as required.

10.2.3 Particle as Fast as Light

So far, we've assumed that no particle is as fast as light. What happens when one of them is? In this case, things may get strange. Consider, for instance, the following extreme case:

$$v = c \quad \text{and} \quad -c < u \leq c.$$

In this case,

$$\frac{u + c}{1 + \frac{uc}{c^2}} = \frac{u + c}{1 + \frac{u}{c}} = c.$$

What is the physical meaning of this? Well, the first particle travels rightwards as fast as light, leaving the lab behind. Being so fast, it can no longer distinguish between the lab and the second particle. It views them as one and the same thing, left behind at the speed of light.

10.2.4 Singularity: Indistinguishable Particles

Still, there is a yet stranger case:

$$v = c \quad \text{and} \quad u = -c.$$

In this (singular) case, the second particle is as fast: it follows the first particle rightwards, at the speed of light too. In these extreme circumstances, our rule is no good any more: it divides zero by zero.

Fortunately, this case is not very interesting. After all, at the initial time of $t = 0$, both particles are at the origin, and have the same speed: c.

Later on, in quantum mechanics, we'll see that this is nonphysical. Still, in the present terms, knowing nothing about quantum mechanics, what can we say about such particles? Well, they are actually one and the same: indistinguishable. In this sense, although the mathematical formula fails, physics still tells us the whole story. Later on, in quantum chemistry, we'll learn more about indistinguishable particles.

10.3 Systems and Their Time

10.3.1 How to Measure Time?

A reference frame should better have its own coordinates, to help describe it geometrically. In our lab, use the Cartesian coordinates x, y, and z. Assume that, at the initial time $t = 0$, the first particle lies at the origin: $(0,0,0)$. From there, it starts moving rightwards, towards $(1,0,0)$.

Because there is no external force, the particle never accelerates or changes direction: it keeps traveling in the x-direction all the time. This could model any kind of (inertial) motion: not only in the x-direction but also in any other fixed direction. After all, when you design your coordinate system in the first place, you could always pick your x-axis to match the original direction of motion. Thus, this is actually a one-dimensional motion: the y and z coordinates play no part in it at all, and remain the same in all systems. Therefore, they could be ignored, and even dropped: we actually "live" in the x-t plane only.

How should the time t be measured? Well, in a standard clock, a time unit is often measured in terms of length. For example, in one second, the long hand of the clock makes an angle of $6°$: $1/60$ of a complete circle.

In the present context, on the other hand, we might want to use a linear (rather than circular) clock. For this purpose, just use a light beam. Of course, this is not very practical. Still, this is a nice theoretical "clock:" in each second, the light beam advances one more light second. This tells us that one more second passed.

What's so good about this clock? Well, in this clock, time is scaled by c. Instead of the original time variable t, measured in seconds, we actually use now the length variable ct, measured in light seconds. After all, while t increases by one second, ct advances by one light second: $300,000$ kilometer.

This way, instead of the original t-axis, we now have the new ct-axis. Why is this attractive? Because ct is a length variable, just like x. This way, instead of the original x-t plane, we can now use the new x-ct plane.

10.3.2 The Self System

The original coordinates x, y, and z tell us the position in the lab. Similarly, t tells us the time, as measured in the lab. Later on, we'll see that these measurements are relevant in the lab only. In other (moving) labs, on the other hand, time and space could be measured differently.

For example, the first particle also has a self system that travels with it (at the same direction, and at the same speed: v). This system has new (primed) coordinates: x' and t'. Don't get confused: this prime has nothing to do with differentiation. In fact, t' just tells us how much time passed since the initial time $t' = 0$ (measured by a tiny clock carried by the self system, in the "pocket" of the particle). Likewise, x' tells us the distance from the origin $x' = 0$ (measured by a tiny ruler carried by the self system, in the same pocket). If we are to the right of the particle, then $x' > 0$. If, on the other hand, we are to the left of the particle, then $x' < 0$. In between, at $x' = 0$, we have the particle itself. As before $y' \equiv y$ and $z' \equiv z$ remain irrelevant, and could be ignored.

How to measure the time in the self system? As before, better use not t' but ct'.

10.3.3 Synchronization

As before, assume that the systems are synchronized: initially, at $t' = t = 0$, the particle lies at the origin in both systems:

$$\begin{pmatrix} x \\ y \\ z \\ ct \end{pmatrix} = \begin{pmatrix} x' \\ y' \\ z' \\ ct' \end{pmatrix} = \begin{pmatrix} 0 \\ 0 \\ 0 \\ 0 \end{pmatrix}.$$

This is the initial event in spacetime. At any later time, on the other hand, the systems may start to differ, not only in position but also in time: t' may differ from t. This way, time is indeed relativistic: it depends on the system where it is measured.

10.4 Lorentz Transformation and Matrix

10.4.1 Coordinates and Their Equal Status

In terms of the self system, the particle is static. Indeed, its position is always

$$x' = 0.$$

In terms of the lab, on the other hand, the particle is dynamic: it moves at speed v. Therefore, at time t, it is at

$$x = \frac{dx}{dt}t = vt = \frac{v}{c}ct = \beta_v ct,$$

where

$$\beta_v \equiv \frac{v}{c} \equiv \frac{dx}{d(ct)}$$

is the scaled velocity, obtained from differentiation with respect to the scaled time ct.

Consider now a more general point at a fixed distance x' from the particle. In the self system, this marks a fixed position: x'. If $x' > 0$, then this is to the right of the particle. If, on the other hand, $x' < 0$, then this is to the left of the particle. In either case, this remains static relative to the particle.

In the lab, on the other hand, the above point is dynamic: it moves at speed v. At the initial time of $t = 0$, it starts from $x = x'$. Then, at time $t \geq 0$, it advances to

$$x = x' + vt = x' + \beta_v ct.$$

In other words,

$$x' = x - \beta_v ct.$$

Still, this is not the whole story: x' and t' should make a new pair. Together, x' and t' must preserve the same speed of light: c. After all, the speed of light is a true physical quantity, and mustn't depend on the coordinates, which are mathematical artifacts only.

This way, x' and t' will indeed take equal status, and mirror x and t well. Later on, we'll see this geometrically (Figure 11.10). Here, on the other hand, we focus on algebra: to make sure that x' and t' have the same status, we better transform them by a symmetric matrix. Later on, we'll put this in the wider context of Lie algebras.

10.4.2 Lorentz Transformation

Let's use the above to transform the original x-ct lab coordinates to the new x'-ct' self coordinates. This transformation must be insensitive to interchanging x and ct. After all, both measure distance: x measures the distance from the lab's origin, whereas ct measures the distance made by the light beam in our linear clock (Section 10.3.1).

Once the dummy coordinates $y' \equiv y$ and $z' \equiv z$ are dropped, things get simpler: we obtain the new Lorentz transformation that transforms x and ct into x' and ct':

$$\begin{pmatrix} x \\ ct \end{pmatrix} \rightarrow \begin{pmatrix} x' \\ ct' \end{pmatrix} \equiv \gamma\left(\beta_v\right) \begin{pmatrix} 1 & -\beta_v \\ -\beta_v & 1 \end{pmatrix} \begin{pmatrix} x \\ ct \end{pmatrix},$$

where

$$\gamma \equiv \gamma\left(\beta_v\right) \equiv \frac{1}{\sqrt{1 - \beta_v^2}} \geq 1$$

is picked to make sure that the determinant is 1. (For the definition of the determinant, see Figure 2.11.)

This is the Lorentz transformation: it gives the self coordinates in terms of the lab coordinates. Algebraically, it is given in terms of a symmetric 2×2 matrix. This way, time and space indeed relate symmetrically to each other. In principle, time is not much different from any spatial dimension. Later on, we'll illustrate this geometrically. Here, on the other hand, we take an algebraic point of view.

10.4.3 Lorentz Matrix and The Infinity Point

Lorentz transformation preserves area: thanks to the coefficient γ, the Lorentz matrix has determinant 1.

What happens when $|v|$ is as large as c? In this case, γ is no longer a number: it is the infinity point ∞. Still, it is assumed that

$$0 \cdot \infty = 0.$$

We'll return to this point later.

Let's look at another extreme case: $v = 0$. In this case, Lorentz matrix is just the 2×2 identity matrix:

$$I \equiv \begin{pmatrix} 1 & 0 \\ 0 & 1 \end{pmatrix}.$$

This may look silly, but is not: it will help rewrite the Lorentz matrix better. For this purpose, define the new 2×2 matrix

$$J \equiv \begin{pmatrix} 0 & 1 \\ 1 & 0 \end{pmatrix}.$$

The original Lorentz matrix can now be written as

$$\gamma\left(\beta_v\right)\left(I - \beta_v J\right).$$

This will be useful later.

10.4.4 Interchanging Coordinates

Fortunately, the Lorentz matrix commutes with J:

$$J\gamma\left(\beta_v\right)\left(I - \beta_v J\right) = \gamma\left(\beta_v\right)\left(I - \beta_v J\right)J.$$

Therefore, the Lorentz transformation is insensitive to interchanging x and ct:

$$\begin{pmatrix} ct \\ x \end{pmatrix} = J\begin{pmatrix} x \\ ct \end{pmatrix}$$

$$\rightarrow \gamma\left(\beta_v\right)\left(I - \beta_v J\right)J\begin{pmatrix} x \\ ct \end{pmatrix}$$

$$= J\gamma\left(\beta_v\right)\left(I - \beta_v J\right)\begin{pmatrix} x \\ ct \end{pmatrix}$$

$$= J\begin{pmatrix} x' \\ ct' \end{pmatrix}$$

$$= \begin{pmatrix} ct' \\ x' \end{pmatrix}.$$

In other words, the Lorentz transformation is blind (completely insensitive) to interchanging x and ct, as required.

10.4.5 Composition of Lorentz Transformations

Thanks to the Lorentz transformation, we can now prove the rule of adding velocities (Section 10.2.1). For this purpose, note that every two Lorentz matrices commute with each other.

Consider the composition of two Lorentz transformations. What does this mean geometrically? Well, let's look at our first particle, moving rightwards at speed v. The second particle, on the other hand, moves leftwards at speed u, or rightwards at speed $-u$. These velocities are with respect to our lab, which is assumed to be at rest.

Now, look at things the other way around: from the perspective of the second particle, the entire lab moves rightwards at speed u. This way, the second particle is now assumed to be at rest. How does the first particle travel away from it?

To calculate this, we need to compose two motions: the motion of the first particle with respect to the lab, on top of the motion of the entire lab away from the second particle. In other words, we need to compose two Lorentz transformations, or just multiply two Lorentz matrices. Since

$$J^2 = I,$$

the product is

$$\gamma\left(\beta_u\right)\left(I-\beta_u J\right)\gamma\left(\beta_v\right)\left(I-\beta_v J\right) = \gamma\left(\beta_u\right)\gamma\left(\beta_v\right)\left(I-\beta_u J\right)\left(I-\beta_v J\right)$$
$$= \gamma\left(\beta_u\right)\gamma\left(\beta_v\right)\left(I-\beta_u J-\beta_v J+\beta_u\beta_v J^2\right)$$
$$= \gamma\left(\beta_u\right)\gamma\left(\beta_v\right)\left(\left(1+\beta_u\beta_v\right)I-\left(\beta_u+\beta_v\right)J\right)$$
$$= \gamma\left(\beta_u\right)\gamma\left(\beta_v\right)\left(1+\beta_u\beta_v\right)\left(I-\frac{\beta_u+\beta_v}{1+\beta_u\beta_v}J\right)$$
$$= \gamma\left(\frac{\beta_u+\beta_v}{1+\beta_u\beta_v}\right)\left(I-\frac{\beta_u+\beta_v}{1+\beta_u\beta_v}J\right)$$

(because each Lorentz matrix has determinant 1, and their product must have determinant 1 as well). This composition describes the total motion of the first particle away from the second one. The total velocity is, thus, not $u+v$ but

$$c\frac{\beta_u+\beta_v}{1+\beta_u\beta_v} = \frac{u+v}{1+\frac{uv}{c^2}},$$

as asserted in Section 10.2.1.

10.4.6 The Inverse Transformation

In particular, look at the special case in which $u=-v$: the second particle coincides with the first one. In this case, the above composition takes the form

$$\gamma\left(\beta_{-v}\right)\left(I-\beta_{-v}J\right)\gamma\left(\beta_v\right)\left(I-\beta_v J\right) = \gamma\left(\frac{\beta_{-v}+\beta_v}{1+\beta_{-v}\beta_v}\right)\left(I-\frac{\beta_{-v}+\beta_v}{1+\beta_{-v}\beta_v}J\right)$$
$$= \gamma(0)I$$
$$= I.$$

This gives us the inverse transformation. Indeed, it is represented by the inverse matrix:

$$\gamma\left(\beta_{-v}\right)\left(I-\beta_{-v}J\right),$$

or

$$\gamma\left(\beta_v\right)\left(I+\beta_v J\right).$$

This is a legitimate Lorentz matrix as well: it has determinant 1 as well, as required. What does it mean geometrically? It transforms back to the lab. Indeed, with respect to the particles (which are now one and the same), the entire lab moves leftwards at speed v, or rightwards at speed $-v$. In other words, the inverse transformation works the other way around: it considers the particles to be at rest, and the entire lab as moving at speed $-v$ away from them. This is why the inverse transformation only picks a minus sign: it uses $-v$ rather than v.

The inverse Lorentz matrix could also be obtained from Cramer's formula (Chapter 13, Section 13.8.3):

$$\begin{pmatrix} 1 & -\beta_v \\ -\beta_v & 1 \end{pmatrix}^{-1} = \frac{1}{1-\beta_v^2}\begin{pmatrix} 1 & \beta_v \\ \beta_v & 1 \end{pmatrix} = \gamma^2\left(\beta_v\right)\begin{pmatrix} 1 & \beta_v \\ \beta_v & 1 \end{pmatrix}.$$

Indeed, just divide both sides by $\gamma(\beta_v)$, and obtain the same inverse matrix as before.

In summary, the set of Lorentz matrices could actually be mirrored by the open interval $(-1,1)$:

$$\gamma\left(\beta_v\right)\left(I-\beta_v J\right) \leftrightarrow \beta_v \in (-1,1).$$

10.5 Proper Time in The Self System

10.5.1 Proper Time: Invariant

In its self system, the particle is static: it is always at $x' = 0$. In its "pocket," it carries a tiny clock. What time does it show in the self system? This is the proper time of the particle:

$$s \equiv t'.$$

Once s is determined in the self system, it could also be calculated from any other system (including the lab), and always give the same result. Below, we'll see how this calculation is made. Thus, the proper time is not relativistic but invariant. This will be useful later.

In the x'-t' self system, the tiny clock is at

$$(x', t') = (0, s).$$

Let's use this to form the matrix

$$\begin{pmatrix} ct' & x' \\ x' & ct' \end{pmatrix} = \begin{pmatrix} cs & 0 \\ 0 & cs \end{pmatrix} = csI.$$

Clearly, this matrix has determinant $c^2 s^2$. Let's apply the inverse Lorentz matrix to it. Do this column by column: start with the second column: the inverse Lorentz matrix transforms it back to the x-t lab coordinates:

$$\begin{pmatrix} x' \\ ct' \end{pmatrix} \to \gamma(\beta_v)(I + \beta_v J) \begin{pmatrix} x' \\ ct' \end{pmatrix} = \begin{pmatrix} x \\ ct \end{pmatrix}.$$

Fortunately, the inverse Lorentz matrix commutes with J (Section 10.4.4). Therefore, the first column transforms in a similar way:

$$\begin{pmatrix} ct' \\ x' \end{pmatrix} = J \begin{pmatrix} x' \\ ct' \end{pmatrix} \to J\gamma(\beta_v)(I + \beta_v J) \begin{pmatrix} x' \\ ct' \end{pmatrix} = J \begin{pmatrix} x \\ ct \end{pmatrix} = \begin{pmatrix} ct \\ x \end{pmatrix}.$$

In summary, the entire matrix transforms to

$$\begin{pmatrix} ct & x \\ x & ct \end{pmatrix} = \gamma(\beta_v)(I + \beta_v J) \begin{pmatrix} ct' & x' \\ x' & ct' \end{pmatrix} = cs\gamma(\beta_v)(I + \beta_v J).$$

Since the inverse Lorentz matrix has determinant 1, the above matrix still has determinant $c^2 s^2$:

$$c^2 t^2 - x^2 = c^2 s^2.$$

So, we've managed to calculate the proper time not only from the self system but also from the lab:

$$s \equiv \sqrt{t^2 - \frac{x^2}{c^2}}.$$

In these new terms, what is x? this is the location of the particle in the lab. On the other hand, what is t? This is the proper time of the lab: the time read from a static clock in the lab. After all, this is how t was defined in the first place. Still, it has yet another (mathematical) meaning: the maximal proper time of any particle

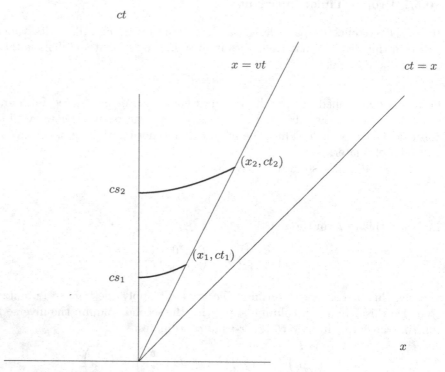

Fig. 10.4. The proper time of the lab is just t. Indeed, this is what a static clock shows in the lab. At discrete times, it ticks like $t = t_1, t_2, t_3, \ldots$. This is the maximal time. A clock moving at the constant speed of $x/t = v$, on the other hand, has a smaller (or slower) proper time: $t_1' = s_1 < t_1$, then $t_2' = s_2 < t_2$, and so on.

(Figure 10.4). After all, every moving particle would have a shorter proper time. Indeed, at any time t, the particle would be at $x = vt$, so its proper time would be

$$s \equiv \sqrt{t^2 - \frac{x^2}{c^2}} = \sqrt{t^2 - \beta_v^2 t^2} = \frac{t}{\gamma(\beta_v)} \leq t.$$

In summary, if you want to think that a lot of time has passed, then you should better look at your own static clock. Just hold it in your hand, and look at it. This way, it will tick fast, telling you that many seconds have passed. If, on the other hand, it moved towards or away from you, then it would tick more slowly, telling you that less seconds have passed.

This is your "penalty" for looking at someone else's clock... Later on, we'll refer to this as time dilation. It leads to the twin paradox.

10.5.2 The Twin Paradox

Suppose that I live in the system of the lab. With me, I have a static clock to show me my proper time: t. I also have a twin brother who lives in the system of the

Fig. 10.5. The twin paradox: I live in the lab system. My twin, on the other hand, lives in the system of the particle, getting away at speed v. I say: "my time ticks faster, so I'm older!" My twin, on the other hand, sees things the other way around, and says: "my time ticks faster, so I'm older!" Who is right?

particle, moving at the constant velocity v away from me (Figure 10.5). With him, he carries his own clock that shows him his own proper time: s.

As discussed above, I think that my own proper time ticks faster, so I'm older. My twin, on the other hand, views things the other way around. He thinks that he is static, and that I travel at velocity $-v$ away from him. Therefore, he believes that his own proper time ticks faster, and that he is older. Who is right?

Here is a possible answer. Later on, we'll also discuss yet another effect: length contraction. From my own point of view, distances in the moving system are shorter. Therefore, I'm bigger than my twin: my veins are longer, and my blood must flow a longer distance to get to my entire body. This requires more time. Fortunately, thanks to time dilation, I do have more time. This way, both effects cancel each other. As a result, both twins have the same metabolism, and age at the same rate.

Why did we have a paradox in the beginning? Because we looked at a single position: $x' = 0$. Still, in the x'-t' plane, $x' = 0$ is not just a single point, but a complete line: the entire t'-axis. Once transformed from system to system, this axis scales differently, leading to the twin paradox. To avoid this, better transform not just a thin line but a wider area. This should scale right: after all, as a transformation from $\mathbb{R}^2$ to $\mathbb{R}^2$, the Lorentz transformation preserves area, as required. Later on, we'll illustrate this geometrically in spacetime (Figure 11.10).

10.5.3 Hyperbolic Geometry: Minkowski Space

In our lab, look at a fixed time $t = t_0 > 0$. At t_0, where is the particle? Well, this depends on its velocity: to reach x, the velocity must have been $v = x/t_0$. Let's look at all those x's that could have been reached by any particle, traveling at any possible speed v, not exceeding the speed of light: $|v| \leq c$. This makes a horizontal line segment in the x-t plane:

$$\{(x, t_0) \mid |x| \leq ct_0\}.$$

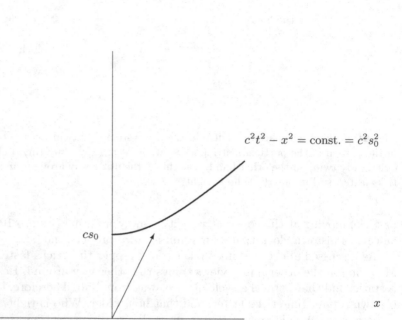

Fig. 10.6. A level set of s: a hyperbola in the original x-t lab coordinates. More specifically, (x, t) is on the hyperbola if x could be reached at time t by a particle moving at speed $v = x/t$ with respect to the lab. In the self system of the particle, on the other hand, this will happen at proper time s_0.

This is a level set of t. Indeed, in it, t is constant: $t = t_0$. Still, now we know better: because it moves at its own speed v, the particle also has its own proper time s. Therefore, we should better look at a level set of s: the hyperbola

$$\left\{ (x, t) \mid c^2 t^2 - x^2 = c^2 s_0^2 \right\},$$

where s_0 is constant.

The motion of such a particle is modeled by the arrow in Figure 10.6. Once the arrow hits the hyperbola, the particle arrives at x. The tiny clock carried in its pocket will then show its proper time: s_0.

Finally, let's look at all possible s_0's. Together, these level sets make a new two-dimensional manifold: the x-s Minkowski space.

10.5.4 Length Contraction

In its self system, the particle is always at $x' = 0$. In the lab, on the other hand, it is at x. Where is this geographically? Easy: just start from 0 in the lab, measure x rightwards, and you arrive at x.

Thus, the location has no meaning on its own, but only relative to a reference point: 0. Only the distance between two locations is meaningful.

Consider, for example, a horizontal stick that moves at velocity v with respect to our lab. In its self system, the stick is at rest: one endpoint at x_1', and the other at x_2'. Thus, in its own system, its length is

$$\Delta x' \equiv x_2' - x_1'.$$

What is the view from the lab? Well, let's use Lorentz transformation:

$$\begin{pmatrix} x_1' \\ ct_1' \end{pmatrix} = \gamma\,(\beta_v) \begin{pmatrix} 1 & -\beta_v \\ -\beta_v & 1 \end{pmatrix} \begin{pmatrix} x_1 \\ ct_1 \end{pmatrix}$$

and

$$\begin{pmatrix} x_2' \\ ct_2' \end{pmatrix} = \gamma\,(\beta_v) \begin{pmatrix} 1 & -\beta_v \\ -\beta_v & 1 \end{pmatrix} \begin{pmatrix} x_2 \\ ct_2 \end{pmatrix}.$$

Now, subtract the former equation from the latter:

$$\begin{pmatrix} \Delta x' \\ c\Delta t' \end{pmatrix} = \gamma\,(\beta_v) \begin{pmatrix} 1 & -\beta_v \\ -\beta_v & 1 \end{pmatrix} \begin{pmatrix} \Delta x \\ c\Delta t \end{pmatrix}.$$

To measure the length of the moving stick, a viewer who sits in the lab has no access to the self system: he/she must use the x-t lab coordinates. For this purpose, he/she must have both endpoints x_1 and x_2 at the same time $t_1 = t_2$:

$$\begin{pmatrix} \Delta x' \\ c\Delta t' \end{pmatrix} = \gamma\,(\beta_v) \begin{pmatrix} 1 & -\beta_v \\ -\beta_v & 1 \end{pmatrix} \begin{pmatrix} \Delta x \\ 0 \end{pmatrix}$$

$$= \gamma\,(\beta_v) \begin{pmatrix} \Delta x \\ -\beta_v \Delta x \end{pmatrix}.$$

We only need the top equation:

$$\Delta x' = \gamma\,(\beta_v)\,\Delta x,$$

or

$$\Delta x = \frac{\Delta x'}{\gamma\,(\beta_v)}.$$

Since $\gamma \geq 1$, $|\Delta x| \leq |\Delta x'|$. This is called length contraction. From its own self system, the stick looks longer than from any other system (such as our lab). As a matter of fact, we've already seen this effect, and used it to "solve" the twin paradox (Section 10.5.2).

Moreover, the above length Δx (observed from the lab) decreases monotonically as $|v|$ increases. In the extreme case of $|v| = c$ and $\gamma = \infty$, for example, $\Delta x = 0$. This means that, from the lab, the stick seems like one point, passing as fast as light.

This confirms what was already said in Section 10.2.4: two particles that follow each other at the speed of light are indistinguishable, and could be considered as one and the same. This is indeed what happens here to the endpoints of our stick: they shrink to just one point. This is also why, in a particle as fast as light, no change could ever be observed. It looks like its proper time is stuck, and doesn't tick at all!

10.5.5 Simultaneous Events

In the above, in the lab, both endpoints of the stick are measured at the same time $t_1 = t_2$. These are simultaneous events. Still, in the x-t plane, they are not identical. After all, they take place in two different locations: $x_1 \neq x_2$.

In the self system, on the other hand, these events are no longer simultaneous. Indeed, in the equation in Section 10.5.4, look now at the bottom:

$$c\Delta t' = -\gamma(\beta_v)\beta_v\Delta x = -\beta_v\Delta x' \neq 0.$$

Thus, the events are simultaneous in the lab only. In every other system, on the other hand, they are not.

10.5.6 Time Dilation

Thus, in spacetime, two events could differ in time or location or both. So far, we discussed simultaneous events that happen at the same time. Next, let's consider events that take place at the same location, but at different times.

Consider again our particle that moves at velocity v with respect to the lab (Figure 10.4). Consider also a clock, moving at the same speed. This way, in the self system, the clock is static.

Now, in this clock, consider two successive times: $t_2' > t_1'$. (In both ticks, the clock is still at the same location in the self system: $x_2' = x_1'$.) This way, in the self system, the time difference is

$$\Delta t' \equiv t_2' - t_1'.$$

(This difference is in terms of proper time, as in Section 10.5.1.) A viewer who looks at the moving clock from the lab, on the other hand, measures the time difference

$$\Delta t \equiv t_2 - t_1.$$

Is this the same? Well, thanks to the inverse Lorentz transformation,

$$\begin{pmatrix} x_1 \\ ct_1 \end{pmatrix} = \gamma(\beta_v) \begin{pmatrix} 1 & \beta_v \\ \beta_v & 1 \end{pmatrix} \begin{pmatrix} x_1' \\ ct_1' \end{pmatrix}$$

and

$$\begin{pmatrix} x_2 \\ ct_2 \end{pmatrix} = \gamma(\beta_v) \begin{pmatrix} 1 & \beta_v \\ \beta_v & 1 \end{pmatrix} \begin{pmatrix} x_2' \\ ct_2' \end{pmatrix}.$$

Let's subtract the former equation from the latter:

$$\begin{pmatrix} \Delta x \\ c\Delta t \end{pmatrix} = \gamma(\beta_v) \begin{pmatrix} 1 & \beta_v \\ \beta_v & 1 \end{pmatrix} \begin{pmatrix} \Delta x' \\ c\Delta t' \end{pmatrix}$$

$$= \gamma(\beta_v) \begin{pmatrix} 1 & \beta_v \\ \beta_v & 1 \end{pmatrix} \begin{pmatrix} 0 \\ c\Delta t' \end{pmatrix}$$

$$= \gamma(\beta_v) \begin{pmatrix} \beta_v c\Delta t' \\ c\Delta t' \end{pmatrix}.$$

We only need the bottom equation:

$$\Delta t = \gamma \left(\beta_v\right) \Delta t'.$$

Since $\gamma \geq 1$, $\Delta t \geq \Delta t'$. This is time dilation.

Be careful: on its own, this could lead to the twin paradox. Fortunately, by now, we also have a counter-effect: length contraction. Together, these effects make perfect sense: they cancel (or balance) each other, and "solves" the twin paradox (Section 10.5.2).

Now, let $|v|$ get greater and greater. This way, time dilation gets stronger and stronger. In other words, as $|v|$ increases, the time difference Δt (observed from the lab) increases as well. In the extreme case of $|v| = c$ and $\gamma = \infty$, $\Delta t = \infty$ as well. This is why, in a particle as fast as light, no change could ever be observed: even a tiny change seems to last forever.

10.6 Spacetime and Velocity

10.6.1 Doppler's Effect

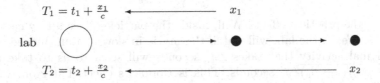

Fig. 10.7. Doppler's effect (a view from the lab). At time $t_1 > 0$, the particle is at $x_1 = vt_1$. At this time, a signal issues back to the lab, to arrive at $T_1 = t_1 + x_1/c$. At $t_2 > t_1$, on the other hand, the particle is already at $x_2 = vt_2$. At this time, another signal issues back to the lab, to arrive at $T_2 = t_2 + x_2/c$.

In the above, the particle gets away from the lab at speed v. In its "pocket," the particle has a tiny clock, to show its proper time. We, in the lab, can also read the time from this clock, although with some time dilation. This is still quite theoretical: how could this be done in practice? After all, this information must travel from the clock back to the lab, at a finite speed, not exceeding the speed of light!

In its own self system, the tiny clock shows time t'. Once read from the lab, this time must transform to t. For example, as things look from the lab, at time $t_1 > 0$, the particle gets as far as $x_1 = vt_1$. At this time, a signal as fast as light issues from the particle, to carry the news all the way back to the lab. To arrive, it needs some more time: $x_1/c = vt_1/c$. (Here, we assume for simplicity that $v > 0$, as in Figure

10.7.) Denote the arrival time by T_1. Later on, at time $t_2 > t_1$, the next signal will issue as well, to arrive at $T_2 > T_1$.

How to write the arrival-time difference in terms of the real-time difference? In other words, how to write $T_2 - T_1$ in terms of the original time difference $t_2' - t_1'$, read in the self system itself? Well, thanks to time dilation (Section 10.5.6) and the above discussion,

$$
\begin{aligned}
\Delta T &\equiv T_2 - T_1 \\
&= t_2 + \frac{x_2}{c} - \left(t_1 + \frac{x_1}{c}\right) \\
&= t_2 + \frac{vt_2}{c} - \left(t_1 + \frac{vt_1}{c}\right) \\
&= t_2 - t_1 + \frac{v}{c}(t_2 - t_1) \\
&= (\Delta t)(1 + \beta_v) \\
&= (\Delta t')\gamma(\beta_v)(1 + \beta_v) \\
&= (\Delta t')\frac{1 + \beta_v}{\sqrt{1 - \beta_v^2}} \\
&= (\Delta t')\frac{1 + \beta_v}{\sqrt{1 - \beta_v}\sqrt{1 + \beta_v}} \\
&= (\Delta t')\sqrt{\frac{1 + \beta_v}{1 - \beta_v}}.
\end{aligned}
$$

What is the practical effect? Well, inside the particle, place a tiny camera, to take a movie. We, in the lab, will watch the movie in slow motion. Indeed, since $v > 0$, an original activity that takes $\Delta t'$ seconds will seem to us to take as many as $\sqrt{(1 + \beta_v)/(1 - \beta_v)}\,\Delta t'$ seconds. This is Doppler's effect. Later on, we'll see how this affects color.

10.6.2 Velocity in Spacetime

How do things look like from the second particle? Recall that this particle travels at speed $-u$ with respect to the lab. To describe its self system, let's use now the x-t coordinates (with no prime). This will be our new spacetime.

So, we work now the other way around. Unlike before, our lab will use the primed x'-t' coordinates, traveling at speed u with respect to our new spacetime. This will prove useful below.

In the x'-t' lab coordinates, how does the first particle move? Well, it moves at speed v:

$$
\frac{dx'}{dt'} = v,
$$

or

$$
\frac{dx'}{cdt'} = \frac{v}{c} = \beta_v.
$$

This could also be written in terms of a two-dimensional column vector:

$$
\begin{pmatrix} dx' \\ cdt' \end{pmatrix} = \begin{pmatrix} \beta_v \\ 1 \end{pmatrix}.
$$

Indeed, just divide the top component by the bottom component, and get the same thing as before. (This is as in projective geometry: see Chapter 6 in [45].)

But how do things look like from our new spacetime? To see this, we need to transform the above vector back to our new spacetime: the x-t self system of the second particle. For this purpose, apply an inverse Lorentz matrix:

$$\begin{pmatrix} dx \\ cdt \end{pmatrix} = \gamma\left(\beta_u\right) \begin{pmatrix} 1 & \beta_u \\ \beta_u & 1 \end{pmatrix} \begin{pmatrix} dx' \\ cdt' \end{pmatrix}$$

$$= \gamma\left(\beta_u\right) \begin{pmatrix} 1 & \beta_u \\ \beta_u & 1 \end{pmatrix} \begin{pmatrix} \beta_v \\ 1 \end{pmatrix}$$

$$= \gamma\left(\beta_u\right) \begin{pmatrix} \beta_v + \beta_u \\ \beta_u\beta_v + 1 \end{pmatrix}.$$

Now, in this vector, divide the top component by the bottom component:

$$\frac{dx}{cdt} = \frac{\beta_v + \beta_u}{\beta_u\beta_v + 1},$$

or

$$\frac{dx}{dt} = c\frac{\beta_v + \beta_u}{\beta_u\beta_v + 1}, = \frac{u + v}{1 + \beta_u\beta_v}.$$

Isn't this familiar? This is our rule of adding velocities (Section 10.2.1). This is indeed how fast the first particle gets away from the second one.

10.6.3 Moebius Transformation

The Lorentz transformation transforms two-dimensional vector to two-dimensional vector. The Moebius transformation, on the other hand, works more directly: it transforms ratio to ratio: the old velocity of the first particle (as observed from the lab) to its new velocity (as observed from the second particle):

$$\beta_v = \frac{dx'}{cdt'} \rightarrow \frac{dx}{cdt} = \frac{\beta_u + \beta_v}{1 + \beta_u\beta_v}.$$

Let's use the same approach to calculate the perpendicular velocity as well.

10.6.4 Perpendicular Velocity

Let us now consider a two-dimensional motion: not only in the x' but also in the y' spatial direction. For this purpose, our lab still uses primes in its coordinates: x', y', and t'.

Assume now that the first particle moves at an oblique velocity $(v_{x'}, v_{y'})$ with respect to the lab: velocity $v_{x'}$ in the x' direction, and also velocity $v_{y'}$ in the perpendicular y' direction (Figure 10.8). (Don't get confused: these are not partial derivatives, but just velocity components.) The second particle, on the other hand, still moves at velocity $(-u, 0)$ in the x' direction only. This is how things look like from the lab.

How do things look like from the second particle? Well, this is now our new spacetime: the x-y-t self system of the second particle. To observe things from there, let's use the same approach as before, with a few changes.

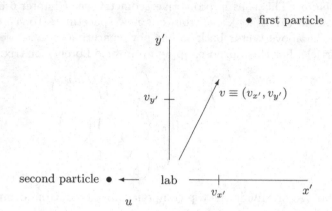

Fig. 10.8. View from the lab: the first particle travels at velocity $(v_{x'}, v_{y'})$, while the second particle travels at velocity $(-u, 0)$.

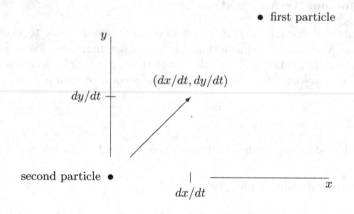

Fig. 10.9. View from the second particle: the first particle gets away at a new velocity: $(dx/dt, dy/dt)$ in the x-y-t system.

In its own self system, the second particle is at rest, while the entire lab moves at velocity $(u, 0)$. In these terms, how does the first particle move?

To transform back to the self system of the second particle, we must now apply an extended 3×3 Lorentz matrix, which leaves the second component unchanged:

$$
\begin{pmatrix} x' \\ y' \\ ct' \end{pmatrix} \to \begin{pmatrix} x \\ y \\ ct \end{pmatrix} = \begin{pmatrix} \gamma\left(\beta_u\right) & & \\ & 1 & \\ & & \gamma\left(\beta_u\right) \end{pmatrix} \begin{pmatrix} 1 & & \beta_u \\ & 1 & \\ \beta_u & & 1 \end{pmatrix} \begin{pmatrix} x' \\ y' \\ ct' \end{pmatrix}.
$$

(As usual, blank spaces stand for zero matrix elements.) Let's use this matrix to calculate the velocities:

$$
\begin{aligned}
\begin{pmatrix} dx \\ dy \\ cdt \end{pmatrix} &= \begin{pmatrix} \gamma\left(\beta_u\right) & & \\ & 1 & \\ & & \gamma\left(\beta_u\right) \end{pmatrix} \begin{pmatrix} 1 & & \beta_u \\ & 1 & \\ \beta_u & & 1 \end{pmatrix} \begin{pmatrix} dx' \\ dy' \\ cdt' \end{pmatrix} \\
&= \begin{pmatrix} \gamma\left(\beta_u\right) & & \\ & 1 & \\ & & \gamma\left(\beta_u\right) \end{pmatrix} \begin{pmatrix} 1 & & \beta_u \\ & 1 & \\ \beta_u & & 1 \end{pmatrix} \begin{pmatrix} \beta_{v_{x'}} \\ \beta_{v_{y'}} \\ 1 \end{pmatrix} \\
&= \begin{pmatrix} \gamma\left(\beta_u\right) & & \\ & 1 & \\ & & \gamma\left(\beta_u\right) \end{pmatrix} \begin{pmatrix} \beta_{v_{x'}} + \beta_u \\ \beta_{v_{y'}} \\ \beta_u \beta_{v_{x'}} + 1 \end{pmatrix}.
\end{aligned}
$$

We can now go ahead and divide by dt, to obtain the new velocities dx/dt and dy/dt, as observed from the second particle. (This way, we actually eliminate the x'-y'-t' lab coordinates, and drop them altogether.) As observed from the second particle, the x-velocity of the first particle is still

$$
\frac{dx}{dt} = c\frac{\beta_u + \beta_{v_{x'}}}{1 + \beta_u \beta_{v_{x'}}} = \frac{u + v_{x'}}{1 + \beta_u \beta_{v_{x'}}},
$$

as in Sections 10.6.2–10.6.3. The y-velocity, on the other hand, is now

$$
\frac{dy}{dt} = c\frac{\beta_{v_{y'}}}{\gamma\left(\beta_u\right)\left(1 + \beta_u \beta_{v_{x'}}\right)} = \frac{v_{y'}}{\gamma\left(\beta_u\right)\left(1 + \beta_u \beta_{v_{x'}}\right)}.
$$

Thus, as observed from the second particle, the first particle indeed follows the arrow in Figure 10.9. To draw this arrow, we now have the new velocities dx/dt and dy/dt in terms of three known parameters: u, $v_{x'}$, and $v_{y'}$.

10.7 Relativistic Momentum and Its Conservation

10.7.1 Invariant Mass

Consider now a new particle, traveling rightwards at a new velocity: not v, but u (with respect to the lab). Assume that it has mass m. This mass is not relativistic, but invariant: once measured in one system, it could also be measured in any other system (including the self system) in the same way, and also yield the same result: m. Still, to obtain m, the measurement must be carried out without changing the velocity. In the lab, for example, the mass must be measured while the particle is at motion, at velocity u.

10.7.2 Momentum: Old Definition

before the explosion:

after the explosion:

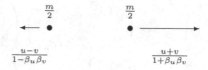

Fig. 10.10. Conservation of momentum: after the original particle (top picture) breaks into two subparticles (bottom picture), the total momentum is still $mu\gamma(\beta_u)$, as before.

Suddenly, without supplying any force or energy from the outside, the particle explodes, and splits into two equal subparticles (Figure 10.10). How should they fly away, with respect to the original particle? Well, thanks to symmetry, they should fly sideways: one subparticle flies rightwards at the extra velocity v, while the other flies leftwards at the extra velocity v.

In Newtonian mechanics, the momentum is defined as mass times velocity (Chapter 2, Section 2.3.1). This is the linear momentum in the x dimension. In these terms, the momentum indeed remains the same:

$$\frac{m}{2}(u+v) + \frac{m}{2}(u-v) = \frac{m}{2}(u+v+u-v) = mu.$$

10.7.3 Relativistic Momentum

Thanks to special relativity, we already know better: this is *not* the correct way to add velocities (Section 10.2.1). To fix this, let's redefine momentum in a more accurate way.

In the lab, the initial velocity is

$$u \equiv \frac{dx}{dt}.$$

Still, this could never be relativistic. After all, this is the ratio between two relativistic quantities: space and time. Better differentiate with respect to an invariant (nonrelativistic) time: the proper time s. This will define a new relativistic momentum p:

$$p \equiv m\frac{dx}{ds} = m\frac{dx}{dt} \cdot \frac{dt}{ds} = mu\gamma\left(\beta_u\right).$$

The location of the particle at proper time s can now be uncovered simply by integrating over the constant p:

$$mx(s) = m\int_0^s \frac{dx}{ds}ds = \int_0^s pds = ps.$$

Since s is invariant, it could be specified in the lab, and remain the same in all inertial systems as well. For this reason, $x(s)$ and p keep the same proportion in all inertial systems.

This defines the momentum not absolutely, but only relative to the lab, as required. In the rest of this chapter, when talking about momentum, we mean relativistic momentum.

10.7.4 Relativistic Momentum and Energy

So far, we've seen that x and ct make a relativistic pair, transformed together by the Lorentz transformation. Why? Because they are proportional to one another:

$$x = \frac{dx}{dt}t = \frac{dx}{cdt}ct = \frac{u}{c}ct = \beta_u ct.$$

Later on, we'll see that momentum is proportional to energy$/c$ in the same way:

$$p = mu\gamma\left(\beta_u\right) = \frac{u}{c}mc\gamma\left(\beta_u\right) = \beta_u\frac{mc^2\gamma\left(\beta_u\right)}{c} = \beta_u\frac{\text{energy}}{c}.$$

This will be proved later (Einstein's formula). Thus, momentum and energy$/c$ will also make a relativistic pair, transformed together by the Lorentz transformation.

In summary, momentum mirrors space. In fact, momentum could replace space, and play the same role, in a more fundamental way. This is why momentum is often constant in space: insensitive to any spatial translation. Likewise, energy could replace time, and play the same role, in a more fundamental way. After all, energy is often conserved and time invariant: unchanged under any time translation. This marks a new symmetry line in spacetime. We'll come back to this in quantum and Hamiltonian mechanics.

10.7.5 Rest Mass vs. Relativistic Mass

m is also called the rest mass. Later on, we'll see why. Still, this could be a bit confusing: better use just mass. As discussed above, the mass m is invariant.

To define the relativistic momentum, however, we used not m but the product

$$m\gamma\left(\beta_u\right).$$

This is often called the relativistic mass. It plays the same role as the inertial mass in classical mechanics: it resists any force applied from the outside in an attempt to change the momentum without changing m.

Still, if sufficient energy is available to overcome this resistance and apply such a force, then u could get very high, and even close to the speed of light (without changing m). This way, the relativistic mass gets high as well, and resists the force even more. In this sense, the particle gets "heavier:" not in terms of rest mass, but only in terms of relativistic mass. This is why a particle of a positive mass $m > 0$ could never reach the speed of light: its relativistic mass would get too big:

$$m\gamma(\beta_u) \to_{u \to c} \infty.$$

It would resist the outer force ever so strongly, and win!

Still, rest mass is more elementary than relativistic mass. In the rest of this chapter, when talking about mass, we mean rest mass, not relativistic mass.

10.7.6 Moderate Velocity

The new definition of the relativistic momentum is indeed a natural extension of the old definition. After all, for a moderate velocity $|u| \ll c$, we also have $\beta_u \ll 1$ and $\gamma(\beta_u) \sim 1$, so

$$mu\gamma(\beta_u) \sim mu.$$

So, little has changed. For a large velocity u, on the other hand, the new definition is a substantial improvement.

10.7.7 Closed System: Lose Mass – Gain Motion

Indeed, to make the explosion happen, some energy is required, which must come from somewhere. Now, our system is isolated: no force or energy could come from the outside. Thus, the extra energy must come from mass: at the explosion, the subparticles must lose some of their original mass. Indeed, before the explosion, each subparticle has mass of $m/2$. After the explosion, on the other hand, each subparticle has mass of only

$$\frac{m}{2\gamma(\beta_v)} < \frac{m}{2\gamma(0)} = \frac{m}{2}.$$

(This will be discussed more later.) What is the physical meaning of this? Well, the mass after the explosion must be less than before (when the subparticle was still inside the original particle, and had velocity $v = 0$ with respect to it). Still, no mass was lost for nothing: it supplied the extra energy required to make the explosion happen.

10.7.8 The Momentum Matrix

Let's define the (so-called) momentum matrix: mass times Lorentz matrix. (Later on, we'll focus on just one matrix element: the relativistic momentum.) For this purpose, we need to look at things from the lab, as in Figure 10.10. For each subparticle, we need to compose two Lorentz matrices, as in Section 10.4.5:

mass · Lorentz matrix

$$= \frac{m}{2\gamma\,(\beta_v)}\gamma\left(\frac{\beta_u+\beta_v}{1+\beta_u\beta_v}\right)\left(I+\frac{\beta_u+\beta_v}{1+\beta_u\beta_v}J\right) + \frac{m}{2\gamma\,(\beta_v)}\gamma\left(\frac{\beta_u-\beta_v}{1-\beta_u\beta_v}\right)\left(I+\frac{\beta_u-\beta_v}{1-\beta_u\beta_v}J\right)$$

$$= \frac{m}{2\gamma\,(\beta_v)}\gamma\,(\beta_v)\,(I+\beta_v J)\,\gamma\,(\beta_u)\,(I+\beta_u J) + \frac{m}{2\gamma\,(\beta_v)}\gamma\,(\beta_v)\,(I-\beta_v J)\,\gamma\,(\beta_u)\,(I+\beta_u J)$$

$$= \frac{m}{2}\,(I+\beta_v J + I - \beta_v J)\,\gamma\,(\beta_u)\,(I+\beta_u J)$$

$$= m\gamma\,(\beta_u)\,(I+\beta_u J).$$

Thus, after the explosion, the momentum matrix remains the same (with respect to the lab). This is indeed conservation of momentum, in its matrix form.

10.7.9 Momentum and Its Conservation

The above is a matrix equation: it actually contains four scalar equations. Let's focus on one of them, say the upper-right one:

$$\frac{m}{2\gamma\,(\beta_v)}\gamma\left(\frac{\beta_u+\beta_v}{1+\beta_u\beta_v}\right)\frac{\beta_u+\beta_v}{1+\beta_u\beta_v} + \frac{m}{2\gamma\,(\beta_v)}\gamma\left(\frac{\beta_u-\beta_v}{1-\beta_u\beta_v}\right)\frac{\beta_u-\beta_v}{1-\beta_u\beta_v} = m\gamma\,(\beta_u)\,\beta_u.$$

To simplify, multiply this by c:

$$\frac{m}{2\gamma\,(\beta_v)}\frac{u+v}{1+\beta_u\beta_v}\gamma\left(\frac{\beta_u+\beta_v}{1+\beta_u\beta_v}\right) + \frac{m}{2\gamma\,(\beta_v)}\frac{u-v}{1-\beta_u\beta_v}\gamma\left(\frac{\beta_u-\beta_v}{1-\beta_u\beta_v}\right) = mu\gamma\,(\beta_u).$$

This is indeed conservation of momentum: after the explosion, with respect to the lab, mass times velocity times γ still sums to the same: $mu\gamma(\beta_u)$.

10.8 Relativistic Energy and Its Conservation

10.8.1 Force: Derivative of Momentum

So far, we've defined the relativistic momentum, and made sure it is indeed conserved. What about relativistic energy? To help define it, we'll need to differentiate $\gamma(\beta_v)$:

$$\gamma'\,(\beta_v) \equiv \left(\left(1-\beta_v^2\right)^{-1/2}\right)'$$

$$= -\frac{1}{2}\left(1-\beta_v^2\right)^{-3/2}(-2\beta_v)$$

$$= \beta_v\left(1-\beta_v^2\right)^{-3/2}.$$

Let's use this to differentiate $\gamma(\beta_v)$ as a composite function of v:

$$\frac{d}{dv}\gamma\,(\beta_v) = \gamma'\,(\beta_v)\,\frac{d\beta_v}{dv}$$

$$= \frac{1}{c}\gamma'\,(\beta_v)$$

$$= \frac{1}{c}\beta_v\left(1-\beta_v^2\right)^{-3/2}.$$

Let's use this to differentiate the product $v\gamma(\beta_v)$:

$$\frac{d}{dv}\left(v\gamma\left(\beta_v\right)\right) = \gamma\left(\beta_v\right) + v\frac{d}{dv}\gamma\left(\beta_v\right)$$

$$= \gamma\left(\beta_v\right) + \frac{v}{c}\beta_v\left(1-\beta_v^2\right)^{-3/2}$$

$$= \gamma\left(\beta_v\right) + \beta_v^2\left(1-\beta_v^2\right)^{-3/2}$$

$$= \gamma^{-2}\left(\beta_v\right)\gamma^3\left(\beta_v\right) + \beta_v^2\left(1-\beta_v^2\right)^{-3/2}$$

$$= \left(1-\beta_v^2\right)\left(1-\beta_v^2\right)^{-3/2} + \beta_v^2\left(1-\beta_v^2\right)^{-3/2}$$

$$= \left(1-\beta_v^2\right)^{-3/2}.$$

We are now ready to define relativistic energy. This definition will be accurate not only for small but also for large velocities.

Consider a particle of mass m that is initially at rest in the x-t coordinates in the lab. Then, an external force F is applied to it from time 0 until time $q > 0$, to increase both its momentum and energy (without changing m). Thanks to the increase in its energy, the particle doesn't have to lose any mass any more: it may remain with the same mass m throughout the entire time interval $[0, q]$.

To calculate the force, we need to differentiate the relativistic momentum (Section 10.7.3) with respect to time. This will help define the relativistic energy:

$$\int_{x(0)}^{x(q)} F(x)dx = \int_0^q F(x(t))dx(t)$$

$$= \int_0^q F(x(t))\frac{dx(t)}{dt}dt$$

$$= \int_0^q F(x(t))v(t)dt$$

$$= m\int_0^q \frac{d}{dt}\left(v\gamma\left(\beta_v\right)\right)v(t)dt$$

$$= m\int_0^q \frac{d}{dv}\left(v\gamma\left(\beta_v\right)\right)\frac{dv}{dt}v(t)dt$$

$$= m\int_{v(0)}^{v(q)} \frac{d}{dv}\left(v\gamma\left(\beta_v\right)\right)vdv$$

$$= m\int_{v(0)}^{v(q)} \left(1-\beta_v^2\right)^{-3/2}vdv$$

$$= mc^2\int_{v(0)}^{v(q)} \left(1-\beta_v^2\right)^{-3/2}\frac{1}{c}\beta_v dv$$

$$= mc^2\left(\gamma\left(\beta_{v(q)}\right) - \gamma\left(\beta_{v(0)}\right)\right)$$

$$= mc^2\left(\gamma\left(\beta_{v(q)}\right) - 1\right).$$

10.8.2 Relativistic Energy: Kinetic plus Potential

This is indeed the new kinetic energy that the external force has introduced into the particle from time 0 until time q.

The potential (nuclear) energy stored in the particle, on the other hand, is not relativistic, but invariant. This is the amount subtracted in the above formula:

$$E_{\text{potential}} \equiv E(0) \equiv mc^2.$$

Together with this potential energy, we now have the total energy (as a smooth function of v):

$$E(v) \equiv E_{\text{potential}} + E_{\text{kinetic}}(v) = mc^2 + mc^2 \left(\gamma\left(\beta_v\right) - 1\right) = mc^2 \gamma\left(\beta_v\right).$$

This is indeed the new relativistic energy: kinetic plus potential. This new definition is closely related to the momentum matrix

$$m\gamma\left(\beta_v\right)\left(I + \beta_v J\right).$$

Indeed, just look at the lower-right entry, and multiply by c^2. In the rest of this chapter, when talking about energy, we mean relativistic energy.

10.8.3 Moderate Velocity

The new definition improves on the old one: it is accurate not only for small but also for large velocities. In fact, for a moderate velocity $|v| \ll c$, it is nearly the same as the classical (inaccurate) formula. Indeed, from the Taylor expansion around 0, we have that, for $\beta_v^2 \ll 1$,

$$\gamma\left(\beta_v\right) = \left(1 - \beta_v^2\right)^{-1/2} \sim 1 + \frac{1}{2}\beta_v^2.$$

Therefore, for $|v| \ll c$, the new definition nearly agrees with the classical one:

$$E(v) = mc^2 \gamma\left(\beta_v\right) \sim mc^2 \left(1 + \frac{1}{2}\beta_v^2\right) = mc^2 + \frac{1}{2}mv^2.$$

10.8.4 Tradeoff: Kinetic Energy vs. Mass

If, however, no external force has been applied to it, could the particle still have a nonzero velocity $v \neq 0$? Well, it could. Still, in this case, its kinetic energy must come from somewhere: from its original potential energy (before it started to move). For this, there is a price to pay: the particle must lose some mass. Only a static particle ($v = 0$) could keep its entire original (maximal) mass m. This is why it is also called rest mass.

10.8.5 Energy Is Conserved – Mass Is Not

In Section 10.8.2, we've assumed that an external force is applied to the particle, to increase both its momentum and energy (without changing m). This is why its mass remains m all the time. The system is not isolated, but open to an outer influence.

In Figure 10.10, on the other hand, the explosion takes place in a closed (isolated) system: no external force is allowed. For this reason, not only the total momentum but also the total energy remains unchanged. But mass is not.

During the explosion, where do the subparticles get their extra kinetic energy from? Well, it must come from somewhere: from the original potential (nuclear) energy, stored in the original particle and its mass. There is a price to pay: to make the explosion happen, some mass must be lost.

As a matter of fact, this is true not only for an exploding particle but also for any particle that starts moving. Indeed, at the initial time, the particle accelerates: it changes its own velocity from 0 to $v \neq 0$ (while preserving its total energy). Its new kinetic energy must come from somewhere: from its potential nuclear energy. As a price, its mass must decrease from $m(0) = m$ to

$$m(v) \equiv \frac{m(0)}{\gamma\left(\beta_v\right)} < m(0) = m.$$

This way, its total energy remains the same as at rest:

$$E(v) = m(v)c^2 \gamma\left(\beta_v\right) = \frac{m(0)}{\gamma\left(\beta_v\right)} c^2 \gamma\left(\beta_v\right) = m(0)c^2 = E(0).$$

This is indeed conservation of energy in a closed system.

10.8.6 Rest Mass – Invariant

This is why $m(0)$ is also called the rest mass: the original mass measured in the lab, before the particle started to move.

In the lab, this is $m(0) = m$, as discussed above. At any other (inertial) reference frame, on the other hand, the particle may travel, and the measurement must be carried out while it is at motion, giving m again. But be careful: in the other reference frame, this mass is no longer measured at rest. Thus, the term rest mass is a little confusing: better use just mass.

The lab is the system we're really interested in. Why? Because our physical process takes place in the lab. In fact, we're not interested in any fictitious motion due to a transformation to any other system, but only in a real motion, coming from converting energy from potential to kinetic. Once the particle started to move in our lab, the new mass $m(v)$ gets smaller:

$$m(v) = \frac{m(0)}{\gamma\left(\beta_v\right)} < m(0) = \frac{m(0)}{\gamma(0)} = m.$$

10.8.7 Closed System: Energy Can Only Convert

Like the explosion studied before, the above decrease from $m(0)$ to $m(v)$ takes place in an isolated system. This is essential: since no external force is welcome, the total energy must remain constant. The mass $m(v)$, on the other hand, decreases as $|v|$ increases. Thus, energy is never lost, but only converts from potential to kinetic. In this process, the total energy remains the same:

$$E = E_{\text{potential}}(v) + E_{\text{kinetic}}(v) = \frac{m(0)}{\gamma\left(\beta_v\right)} c^2 + \frac{m(0)}{\gamma\left(\beta_v\right)} c^2 \left(\gamma\left(\beta_v\right) - 1\right) = m(0)c^2.$$

This is Einstein's famous formula. It tells us how powerful the original nuclear energy really is: it may produce a lot of kinetic energy.

10.8.8 Open System

This is quite different from the situation in Section 10.8.2, in which the system is open, and welcomes an external force from the outside. In that case, the kinetic energy increases, while the mass remains constant. This was necessary to help define the kinetic energy obtained from the work done by the external force.

Mass, on the other hand, may change even in a closed system. In fact, in a particle that starts to move, as $|v|$ increases, its mass $m(v)$ decreases. If v grows and grows and reaches $|v| = c$, then we have $\gamma = \infty$, so the particle has no mass at all any more: all its potential energy has already been exploited, and converted into kinetic energy.

Still, even with no mass at all, the particle is not gone: it still has nonzero momentum. This is why momentum and energy are more fundamental than mass: they are relativistic, not absolute.

10.9 Momentum–Energy and Their Transformation

10.9.1 New Invariant Mass

In their new definitions (Sections 10.7.3–10.8.2), both energy and momentum are relativistic: they depend on the velocity, which may change from system to system. In the lab, in particular, the particle travels rightwards at velocity v. As discussed above, its mass is $m(0)/\gamma(\beta_v)$. Still, who cares about its history? Who cares whether it was initially at rest or not? Better redefine m as

$$m \leftarrow \frac{m(0)}{\gamma(\beta_v)}.$$

This is the new mass, measured when the particle is at rest not in the lab but in another system, traveling at velocity $\pm v$ with respect to the lab. Alternatively, m could also be measured in the lab, while the particle is at motion. This mass is now used to calculate both energy and momentum in the lab.

10.9.2 Spacetime

To describe the lab, let's use again the x'-t' coordinates (as in Section 10.6.2). (This prime has nothing to do with differentiation – it just reminds us that both x' and t' are measured in the lab.) Thus, in the lab, the momentum of the particle is primed as well:

$$p' \equiv mv\gamma(\beta_v).$$

Likewise, in the lab, the total energy of the particle is primed as well:

$$E' \equiv mc^2\gamma(\beta_v).$$

Why are we using primes here? Because we are not really interested in the lab. We are more interested in a second particle, traveling leftwards at velocity u with respect to the lab. To describe its self system, we use the x-t coordinates (as in Section 10.6.2). This will be our spacetime. In it, what are the energy E and the momentum p of the first particle?

10.9.3 A Naive Approach

To transform things to our new spacetime, we could take a naive approach:

- Use the rule of adding velocities to calculate the velocity of the first particle away from the second one. This would give us the velocity of the first particle in our new spacetime.
- Use this new velocity to calculate the new momentum and energy in our new spacetime, as required.

Still, this would require a lot of calculations. Is there a more direct approach?

10.9.4 The Momentum–Energy Vector

Fortunately, there is. For this purpose, note that p' and E'/c have a familiar ratio:

$$\frac{p'}{E'/c} = \frac{mv\gamma\left(\beta_v\right)}{mc\gamma\left(\beta_v\right)} = \frac{v}{c} = \beta_v.$$

So, let's put them in a new column vector, proportional to the column $(\beta_v, 1)^t$. In fact, this is just the second column in the momentum matrix (times c):

$$\begin{pmatrix} p' \\ E'/c \end{pmatrix} \equiv cm\gamma\left(\beta_v\right)\left(I + \beta_v J\right) \begin{pmatrix} 0 \\ 1 \end{pmatrix}.$$

10.9.5 The Momentum Matrix in Spacetime

How do things look like from our new spacetime? Well, from the perspective of the second particle, the entire lab travels rightwards at velocity u. On top of this, the first particle also travels in the lab at velocity v. Thus, to transform the old momentum matrix to our new spacetime, we need to apply an inverse Lorentz matrix:

$$m\gamma\left(\beta_u\right)\left(I + \beta_u J\right)\gamma\left(\beta_v\right)\left(I + \beta_v J\right).$$

This is the new momentum matrix in spacetime.

10.9.6 Lorentz Transformation on Momentum–Energy

To have the energy and momentum of the first particle from the perspective of the second one, just look at the second column of this new momentum matrix, and multiply by c:

$$\begin{pmatrix} p \\ E/c \end{pmatrix} = cm\gamma\left(\beta_u\right)\left(I + \beta_u J\right)\gamma\left(\beta_v\right)\left(I + \beta_v J\right) \begin{pmatrix} 0 \\ 1 \end{pmatrix}$$

$$= \gamma\left(\beta_u\right)\left(I + \beta_u J\right) \begin{pmatrix} p' \\ E'/c \end{pmatrix}.$$

So, we got what we wanted: to drop the primes from the old vector $(p', E'/c)^t$, just apply the inverse Lorentz transformation (Section 10.4.6). This is indeed direct: you work with the momentum-energy vector only, avoiding the explicit transformation of the entire x'-t' system back to the x-t system. As a result, both the energy and the momentum of the first particle are now available not only with respect to the lab but also with respect to the second particle, with no need to add the velocities u and v explicitly any more.

10.10 Energy and Mass

10.10.1 Invariant Nuclear Energy

In the above, we put the energy and the momentum in the second column of the momentum matrix. We could work the other way around, and put them in the first column too:

$$\begin{pmatrix} E'/c & p' \\ p' & E'/c \end{pmatrix} = cm\gamma\,(\beta_v)\,(I + \beta_v J).$$

Fortunately, a Lorentz matrix must have determinant 1:

$$\begin{aligned}
\frac{E'^2}{c^2} - p'^2 &= \det\left(\begin{pmatrix} E'/c & p' \\ p' & E'/c \end{pmatrix}\right) \\
&= \det\left(cm\gamma\,(\beta_v)\,(I + \beta_v J)\right) \\
&= c^2 m^2 \det\left(\gamma\,(\beta_v)\,(I + \beta_v J)\right) \\
&= m^2 c^2.
\end{aligned}$$

In the right-hand side, we have an invariant constant, independent of system. Therefore, in the left-hand side too, the primes could probably drop. How to make them drop? We already know how: apply another Lorentz matrix (Section 10.9.6). Fortunately, this has no effect on the determinant:

$$\begin{aligned}
\frac{E^2}{c^2} - p^2 &= \det\left(\begin{pmatrix} E/c & p \\ p & E/c \end{pmatrix}\right) \\
&= \det\left(\gamma\,(\beta_u)\,(I + \beta_u J)\begin{pmatrix} E'/c & p' \\ p' & E'/c \end{pmatrix}\right) \\
&= \det\left(\begin{pmatrix} E'/c & p' \\ p' & E'/c \end{pmatrix}\right) \\
&= m^2 c^2.
\end{aligned}$$

Thus, the determinant is invariant: it doesn't depend on u, and doesn't change from system to system. To simplify, multiply by c^2:

$$E^2 - c^2 p^2 = m^2 c^4 = \left(mc^2\right)^2.$$

Is this familiar? This is indeed the squared nuclear energy stored in the particle. This is not relativistic, but invariant: once fixed in one system, it is the same at all systems. Indeed, you could put the primes back on if you like.

The above equation relates two relativistic quantities to an invariant one. Is this familiar? we've already seen this structure before:

$$c^2 t^2 - x^2 = c^2 s^2,$$

where s is invariant: the proper time (Section 10.5.1).

In the exercises below, we'll extend this to a three-dimensional setting. In this model, the momentum p will be not just a scalar, but a complete three-dimensional vector. In this case, p^2 should be replaced by the inner product $\|p\|^2 = (p, p)$.

10.10.2 Invariant Mass

In the above, we saw once again that the mass m is invariant: once fixed in one system, it remains the same in all systems, and never changes under any Lorentz transformation. In our lab, m was measured while the particle is at motion, at velocity v (Section 10.9.1). In another system, the mass will be the same, even though the particle moves at a different velocity.

In fact, the above formula actually gives us a new method to calculate m.

- First, calculate the relativistic momentum and energy in your system.
- This will give you p and E (or p' and E').
- Fortunately, these primes don't matter at all: in either case, use the above formula to calculate the same m, as required.

This will be useful below.

10.10.3 Einstein's Formula

In particular, why not calculate m in the self system of the first particle itself? After all, in this system, there is no velocity or momentum or kinetic energy at all, so the above formula simplifies to read

$$E^2_{\text{potential}} = m^2 c^4,$$

or

$$E_{\text{potential}} = mc^2.$$

This is Einstein's formula back again.

10.11 Center of Mass

10.11.1 Collection of Subparticles

In the lab, assume now that the velocity v of the first particle is unknown. Fortunately, it could still be obtained in terms of the momentum p' and the energy E':

$$v = \frac{mv\gamma\left(\beta_v\right)}{mc^2\gamma\left(\beta_v\right)}c^2 = \frac{p'}{E'}c^2.$$

When is this useful? When the momentum and energy are available, but the velocity is not. This is quite practical: p' and E' are more fundamental than v, which is often missing.

Throughout this chapter, the second particle could be just theoretical, with no size or mass at all. After all, it only serves as a reference point for the first particle and its motion.

The first particle, on the other hand, is more real and physical. To emphasize this, let's break it into a collection of $k \geq 1$ subparticles, each with its own velocity v_i, momentum p'_i, and energy E'_i with respect to the lab ($1 \leq i \leq k$).

What are the total momentum and energy? Well, as fundamental (and conserved) quantities, they sum up:

$$p' \equiv \sum_{i=1}^{k} p_i' \quad \text{and} \quad E' \equiv \sum_{i=1}^{k} E_i'.$$

The velocity of the entire collection, on the other hand, is not necessarily the sum of the v_i's. After all, the subparticles may have different masses, which are not always available. As a matter of fact, some subparticles may even have no mass at all (those that are as fast as light, and have $|v_i| = c$). To define the total velocity properly, better use the fundamental relativistic quantities: momentum and energy.

10.11.2   Center of Mass

We are now ready to define the velocity of the entire collection:

$$v \equiv \frac{p'}{E'} c^2 = c^2 \frac{\sum_{i=1}^{k} p_i'}{\sum_{i=1}^{k} E_i'}.$$

This new velocity describes the motion of no concrete physical object, but only a theoretical object: the center of mass of the entire collection. Where is this? To tell this, let's use the second particle.

The above velocity is in terms of the lab. Next, let's look at things from our new spacetime: the self system of the second particle, which travels in the lab at velocity $-u$. In this spacetime, what are the new momentum and energy of the collection? We already know what they are:

$$\begin{pmatrix} p \\ E/c \end{pmatrix} = \gamma(\beta_u)(I + \beta_u J) \begin{pmatrix} p' \\ E'/c \end{pmatrix}$$

(Section 10.9.6). In this formula, look at the top equation. Assume also that the second particle follows at the same speed:

$$u \equiv -v, \quad \text{so} \quad \beta_u = -\beta_v = -\frac{p'}{E'/c}.$$

This way, the top equation simplifies to read

$$p = 0.$$

Thus, with respect to the second particle, the entire collection has no momentum at all: it is at a complete rest. This is why the second particle marks the center of mass itself.

10.11.3   Mass of the Collection

What is the mass m of the entire collection? Is it the same as the sum of the individual masses? To find out, we better work with more fundamental quantities: momentum and energy.

In our spacetime (the self system of the second particle, traveling in the lab at velocity $-u = v$), the collection has no momentum at all: $p = 0$. Therefore, the formula in Section 10.10.1 tells us that

$$E^2 = m^2 c^4,$$

or

$$m \equiv \frac{E}{c^2}.$$

This is indeed a proper definition of the total mass of the entire collection.

What is the physical meaning of this? Well, we are now in our spacetime: both p and E have no prime any more. This is also the self system of the collection (or its center of mass). In it, the collection is at a complete rest, with no momentum or kinetic energy at all. Thus, the above formula actually defines its mass m (which is also its rest mass) in terms of its total energy E. Being relativistic and conserved, E can also be written as a sum:

$$m \equiv \frac{E}{c^2}$$

$$= \frac{1}{c^2} \sum_{i=1}^{k} E_i,$$

where E_i is the energy of the ith subparticle in our spacetime: kinetic and potential alike. Indeed, to sum up, E_i must be relativistic: not only potential but also kinetic. The total mass, on the other hand, is neither relativistic nor conserved: it is *not* the sum of individual masses.

10.12 Oblique Force and Momentum

10.12.1 Oblique Momentum in x'-y'

As in Section 10.6.4, assume again that the lab is described by the x'-y'-t' co-ordinates. This way, in the lab, the entire collection moves obliquely, at the new velocity $v \equiv (v_{x'}, v_{y'})^t$. This means that its speed is $v_{x'}$ in the x'-direction, and $v_{y'}$ in the perpendicular y'-direction. In this setting, the momentum is oblique as well: $p' \equiv (p'_{x'}, p'_{y'})^t$, proportional to v. This is the view from the lab (Figure 10.11). (From the second particle, on the other hand, things look different – Figure 10.12.)

Don't get confused: there is no differentiation here. The above primes mean no derivative, but only remind us that we are in the lab. Likewise, the subscripts $_{x'}$ and $_{y'}$ mean no partial derivative, but only spatial coordinates.

Thus, in the formula in Section 10.10.1, p'^2 should be replaced by the inner product

$$\|p'\|^2 = (p', p') = p'^2_{x'} + p'^2_{y'}.$$

After all, in theory, we could always redefine x' to align with v and p' (see exercises below). Fortunately, there is no need to do this explicitly.

10.12.2 View From Spacetime

The second particle, on the other hand, still moves in the x'-direction only: at velocity $(-u, 0) \neq (0, 0)$ with respect to the lab. To transform from the lab to our

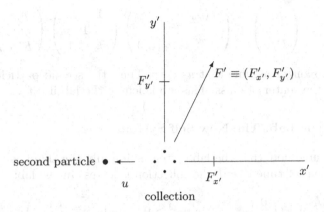

Fig. 10.11. View from the lab: initially, at time $t' = 0$, the collection is still at rest at $(x', y') = (0, 0)$. At $t' = 0$, an oblique external force $F' = (F'_{x'}, F'_{y'})$ starts to act upon it, to increase its momentum and kinetic energy (without changing its mass).

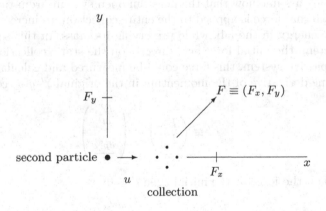

Fig. 10.12. View from the second particle: the force that acts on the collection at the initial time $t = t' = 0$ is the same in the x-direction, but seems weaker in the perpendicular y-direction.

new spacetime (the self system of the second particle), we must now use an extended 3×3 Lorentz matrix:

$$\begin{pmatrix} p_x \\ p_y \\ E/c \end{pmatrix} = \begin{pmatrix} \gamma\,(\beta_u) & & \\ & 1 & \\ \gamma\,(\beta_u) & & \end{pmatrix} \begin{pmatrix} 1 & & \beta_u \\ & 1 & \\ \beta_u & & 1 \end{pmatrix} \begin{pmatrix} p'_{x'} \\ p'_{y'} \\ E'/c \end{pmatrix}.$$

Here, we no longer assume that $u = -v$. Thus, the second particle no longer coincides with the center of mass. This job is left to the lab itself.

10.12.3 The Lab: The New Self System

Indeed, assume now that the lab system is initially the same as the self system of the collection: at time $t' = 0$, the collection is at rest in the lab:

$$p' \equiv \begin{pmatrix} p'_{x'} \\ p'_{y'} \end{pmatrix} = \begin{pmatrix} 0 \\ 0 \end{pmatrix}, \quad E' = mc^2 > 0, \quad \text{and} \quad v \equiv \begin{pmatrix} v_{x'} \\ v_{y'} \end{pmatrix} \equiv \frac{p'}{E'}c^2 = \begin{pmatrix} 0 \\ 0 \end{pmatrix}.$$

This is true at time $t' = 0$ only. At $t' > 0$, on the other hand, things may change, due to an external force.

10.13 Force in an Open System

10.13.1 Force in an Open Passive System

Unlike before, assume now that the lab is an open system: from time $t = t' = 0$ onward, an external force is applied to the entire collection, to increase its momentum and kinetic energy in the lab, while preserving its mass. In this sense, the lab is a passive system: the initial force acts directly on the static collection in it.

In the passive system, this force could be measured and calculated. In fact, it is just the time-derivative of the momentum in the original x'-y'-t' coordinates:

$$F' \equiv \begin{pmatrix} F'_{x'} \\ F'_{y'} \end{pmatrix} \equiv \begin{pmatrix} \frac{dp'_{x'}}{dt'} \\ \frac{dp'_{y'}}{dt'} \end{pmatrix}.$$

Let's focus on the force at the initial time $t' = 0$:

$$F' \equiv \begin{pmatrix} F'_{x'} \\ F'_{y'} \end{pmatrix} \equiv \begin{pmatrix} \frac{dp'_{x'}}{dt'}(0) \\ \frac{dp'_{y'}}{dt'}(0) \end{pmatrix}.$$

(After all, each fixed time $t' > 0$ could in theory be shifted back to zero.) How does this force look like from the second particle? In other words, how to transform the force to our new spacetime (the x-y-t self system of the second particle)?

10.13.2 How to Have The Force in Spacetime?

Of course, we could take a naive approach: transform the momentum and energy to our new spacetime, and define the force F there by differentiating the new momentum with respect to t. Still, this requires a lot of calculations. Is there a more direct approach?

10.13.3 Proper Time in The Lab

Fortunately, there is. To differentiate the momentum with respect to time, let's use the trick in Section 10.6.4.

To start, let's differentiate t with respect to t'. This seems easy: after all, in the lab, t' is the proper time, isn't it? So, from our new spacetime, it should look slower (Section 10.5.1):

$$t' = \frac{t}{\gamma\left(\beta_u\right)},$$

shouldn't it? Indeed, it should undergo time dilation (Section 10.5.6):

$$\Delta t' = \frac{\Delta t}{\gamma\left(\beta_u\right)},$$

shouldn't it?

Unfortunately not. After all, t' could be proper only in an isolated lab, which welcomes no external force. In our open lab, on the other hand, t' is only nearly proper: only at $t' = 0$, before the force had time to act, does t' behave like a proper time.

10.13.4 Nearly Proper Time in The Lab

To see this, transform the lab back to our spacetime, using the inverse Lorentz transformation:

$$ct = \gamma\left(\beta_u\right)\left(\beta_u x' + ct'\right),$$

or

$$t = \gamma\left(\beta_u\right)\left(\frac{\beta_u}{c}x' + t'\right).$$

Differentiate this with respect to t':

$$\frac{dt}{dt'} = \gamma\left(\beta_u\right)\left(\frac{\beta_u}{c}\cdot\frac{dx'}{dt'} + 1\right) = \gamma\left(\beta_u\right)\left(\frac{\beta_u}{c}v_{x'} + 1\right).$$

At $t' = 0$, in particular, $v_{x'} = 0$, so this simplifies to read

$$\frac{dt}{dt'} = \gamma\left(\beta_u\right).$$

Thus, at $t' = 0$, t' is indeed nearly proper: it behaves just like a proper time. Let's use this to look at the force from the second particle as well (Figure 10.12).

10.14 Perpendicular Force

10.14.1 Force: Time-Derivative of Momentum

Let's start with the perpendicular force in spacetime: F_y. Thanks to the above 3×3 matrix, in the y-direction, the momentum is still the same:

$$p_y = p'_{y'}.$$

Thus, the differentiation is simple:

$$
\begin{aligned}
F_y &\equiv \frac{dp_y}{dt} \\
&= \frac{dp'_{y'}}{dt} \\
&= \frac{dp'_{y'}}{\gamma\left(\beta_u\right) dt'} \\
&= \frac{1}{\gamma\left(\beta_u\right)} \cdot \frac{dp'_{y'}}{dt'} \\
&= \frac{1}{\gamma\left(\beta_u\right)} F'_{y'}.
\end{aligned}
$$

10.14.2 Passive System – Strong Perpendicular Force

This is weaker! Thus, the passive system (our lab) feels the maximal perpendicular force. Any other system, on the other hand, would feel a weaker perpendicular force. In particular, our spacetime (the self system of the second particle) feels a perpendicular force that is $\gamma(\beta_u)$ times as weak.

10.15 Nonperpendicular Force

10.15.1 Force: Time-Derivative of Momentum

What about the force in the x-direction? Does it also feel weaker? Well, let's use the same trick:

$$
\begin{aligned}
F_x &= \frac{dp_x}{dt} \\
&= \frac{d\left(\gamma\left(\beta_u\right)\left(p'_{x'} + \beta_u \frac{E'}{c}\right)\right)}{\gamma\left(\beta_u\right) dt'} \\
&= \frac{d\left(p'_{x'} + \beta_u \frac{E'}{c}\right)}{dt'} \\
&= \frac{dp'_{x'}}{dt'} + \beta_u \frac{dE'}{c\,dt'} \\
&= F'_{x'} + \frac{\beta_u}{c} \cdot \frac{dE'}{dt'}.
\end{aligned}
$$

10.15.2 Energy in an Open System

To simplify this, let's look at the latter term, and show that it contributes nothing. For this purpose, let's look at the original equation

$$E'^2 = c^2 (p', p') + m^2 c^4.$$

This equation comes from the original definition of E' and p' in the lab. Therefore, it holds for every time $t' \geq 0$ (with the same m, but possibly with a different E', p', and v). In a moment, we'll come back to the initial time $t' = 0$.

10.15.3 Open System – Constant Mass

In this equation, the latter term remains constant. After all, thanks to the external force, the potential energy (and the mass) remain unchanged. So, upon differentiating both sides with respect to t', the latter term drops:

$$2E'\frac{dE'}{dt'} = 2c^2 \left(p', \frac{dp'}{dt'} \right) = 2c^2 (p', F').$$

10.15.4 Nearly Constant Energy in The Lab

Recall again that we're particularly interested in the initial time $t' = 0$, when the momentum was zero:

$$2E'\frac{dE'}{dt'} = 2c^2(0,0)F' = 0.$$

Since $E' > 0$, we must therefore have

$$\frac{dE'}{dt'} = 0,$$

as required.

10.15.5 Nonperpendicular Force: Same at All Systems

In summary, at $t = t' = 0$,

$$F_x = F'_{x'} + \frac{\beta_u}{c} \cdot \frac{dE'}{dt'} = F'_{x'}.$$

Thus, unlike the perpendicular force, the nonperpendicular force is the same at all systems.

10.15.6 The Photon Paradox

What is light? It has two faces. On one hand, it is a wave. On the other hand, it is also a particle: a massless photon, traveling at speed c with respect to any (inertial) reference frame.

The photon is in a kind of singularity. Indeed, from our perspective, it can have no size at all (due to length contraction, as in Section 10.5.4). Still, this is only from our own (subjective) point of view. The photon may disagree: in its own self system, it is at rest, with positive mass and size, while he entire universe shrinks to a singular point, traveling as fast as light in the opposite direction. Who is right?

The answer is that the photon has no *inertial* self system. After all, Einstein's law tells us that the speed of light is c in all inertial systems. In its own self system, on the other hand, the photon has speed 0, not c. This is why this system can never be inertial. This is also why it can never be transformed to our system: the Lorentz matrix gets singular.

10.16 Exercises: Special Relativity in 3-D

10.16.1 Lorentz Matrix and Its Determinant

1. Show that the determinant of a 2×2 matrix is the same as the area of the parallelogram made by its column vectors. Hint: see Figure 2.11.
2. Show that the determinant of a 2×2 matrix is the same as the area of the parallelogram made by its row vectors.
3. Consider two 2×2 matrices. Show that the determinant of their product is the same as the product of their determinants. Hint: use the original definition in Figure 2.11. Alternatively, see Chapter 13, Section 13.6.7.
4. Show that the Lorentz matrix has determinant 1.
5. Consider a 2×2 matrix. Multiply it by a Lorentz matrix. What is the effect on the determinant? Hint: no effect at all (thanks to the previous exercises).
6. Conclude that the Lorentz transformation preserves area in the two-dimensional Cartesian plane
7. Calculate the inverse Lorentz matrix. Hint: just change sign: replace v by $-v$ (Section 10.4.6).
8. What is the physical meaning of this? Hint: from the moving system, our lab seems to move in the opposite direction.
9. Conclude that the inverse Lorentz matrix has determinant 1 as well.
10. Conclude that the inverse transformation preserves area as well.
11. Use Cramer's formula (Chapter 13, Section 13.8.3) to calculate the inverse Lorentz matrix directly.
12. Do you get the same result?

10.16.2 Motion in Three Dimensions

1. Let

$$v \equiv \begin{pmatrix} v_1 \\ v_2 \\ v_3 \end{pmatrix} \in \mathbb{R}^3$$

be some nonzero three-dimensional real vector. Define the 3×3 matrix O_v, whose columns are v (normalized), a vector that is orthogonal to v (normalized as well), and their vector product:

$$O_v \equiv \left(\frac{v}{\|v\|} \ \middle| \ \frac{v^\perp}{\|v^\perp\|} \ \middle| \ \frac{v \times v^\perp}{\|v\| \cdot \|v^\perp\|} \right).$$

2. Show that O_v is an orthogonal matrix. Hint: see Chapter 26, Section 26.3.4.

3. Show that O_v has determinant 1. Hint: use the Euler angles at the end of Chapter 2. Alternatively, see Chapter 13, Section 13.6.8.

4. Consider a particle that moves at velocity $v \in \mathbb{R}^3$ with respect to the lab. In other words, the particle moves at direction $v/\|v\|$ at speed $\|v\|$. Let (x', y', z', ct') denote the lab coordinates, and (x'', y'', z'', ct'') the self coordinates of the particle. We are now ready to define the more general Lorentz transformation

$$
\begin{pmatrix} x' \\ y' \\ z' \\ ct' \end{pmatrix} \rightarrow \begin{pmatrix} x'' \\ y'' \\ z'' \\ ct'' \end{pmatrix} = L_v \begin{pmatrix} x' \\ y' \\ z' \\ ct' \end{pmatrix},
$$

where L_v is the following 4×4 Lorentz matrix:

$$
L_v \equiv \begin{pmatrix} O_v & \\ & 1 \end{pmatrix} \begin{pmatrix} \gamma\left(\beta_{\|v\|}\right) & & & 1 \\ & 1 & & \\ & & 1 & \\ \gamma\left(\beta_{\|v\|}\right) & & & \end{pmatrix} \begin{pmatrix} 1 & & & -\beta_{\|v\|} \\ & 1 & & \\ & & 1 & \\ -\beta_{\|v\|} & & & 1 \end{pmatrix} \begin{pmatrix} O_v^t & \\ & 1 \end{pmatrix}.
$$

(As usual, blank spaces stand for zero matrix elements.)

5. Show that this indeed transforms the lab system to the self system of the particle.

6. Show that L_v has determinant 1.

7. Consider also a second particle, moving at velocity $-u \in \mathbb{R}^3$ with respect to the lab. Denote its self coordinates by (x, y, z, ct). This will be our new spacetime.

8. With respect to this spacetime, how does the entire lab move? Hint: it moves at velocity $u \in \mathbb{R}^3$.

9. How could this spacetime transform to the lab? Hint: the transformation is

$$
\begin{pmatrix} x \\ y \\ z \\ ct \end{pmatrix} \rightarrow \begin{pmatrix} x' \\ y' \\ z' \\ ct' \end{pmatrix} = L_u \begin{pmatrix} x \\ y \\ z \\ ct \end{pmatrix}.
$$

10. Consider the composite Lorentz transformation

$$
\begin{pmatrix} x \\ y \\ z \\ ct \end{pmatrix} \rightarrow \begin{pmatrix} x'' \\ y'' \\ z'' \\ ct'' \end{pmatrix}
$$

from our spacetime (the self system of the second particle) to the self system of the first particle. Show that it is represented by the matrix product $L_v L_u$:

$$
\begin{pmatrix} x'' \\ y'' \\ z'' \\ ct'' \end{pmatrix} = L_v \begin{pmatrix} x' \\ y' \\ z' \\ ct' \end{pmatrix} = L_v L_u \begin{pmatrix} x \\ y \\ z \\ ct \end{pmatrix}.
$$

11. Show that $L_v L_u$ has determinant 1 as well. Hint: see Chapter 13, Section 13.6.7.

12. Does L_u commute with L_v? Hint: only if u is a scalar multiple of v.

13. Consider the inverse Lorentz transformation

$$\begin{pmatrix} x'' \\ y'' \\ z'' \\ ct'' \end{pmatrix} \rightarrow \begin{pmatrix} x \\ y \\ z \\ ct \end{pmatrix}$$

from the self system of the first particle back to our spacetime (the self system of the second particle). Show that it is represented by the inverse matrix

$$(L_v L_u)^{-1} = L_u^{-1} L_v^{-1} = L_{-u} L_{-v}.$$

14. Conclude that the last column in $L_{-u} L_{-v}$ describes the motion of the first particle away from the second one. Hint: see the exercises below.

15. Show that, in its own self system, the first particle is at rest:

$$x'' = y'' = z'' = 0.$$

16. Conclude that t'' is its proper time. Hint: see Section 10.5.1.

17. Write the above equation in its vector form. Hint:

$$\begin{pmatrix} x'' \\ y'' \\ z'' \\ ct'' \end{pmatrix} = \begin{pmatrix} 0 \\ 0 \\ 0 \\ ct'' \end{pmatrix}.$$

18. Write the above equation in its differential form. Hint:

$$\begin{pmatrix} dx'' \\ dy'' \\ dz'' \\ cdt'' \end{pmatrix} = \begin{pmatrix} 0 \\ 0 \\ 0 \\ 1 \end{pmatrix}.$$

19. What is the meaning of this? Hint: this is just a proportion. After all, upon dividing the left-hand side by t'', you get zero velocity, as is indeed the case in a static particle in its own self system.

20. Transform this back to our spacetime. Hint:

$$\begin{pmatrix} dx \\ dy \\ dz \\ cdt \end{pmatrix} = L_{-u} L_{-v} \begin{pmatrix} dx'' \\ dy'' \\ dz'' \\ cdt'' \end{pmatrix} = L_{-u} L_{-v} \begin{pmatrix} 0 \\ 0 \\ 0 \\ 1 \end{pmatrix}.$$

21. How to do this efficiently? Hint: up to a scalar multiple,

$$L_{-v} \begin{pmatrix} 0 \\ 0 \\ 0 \\ 1 \end{pmatrix} = \begin{pmatrix} v \\ c \end{pmatrix}.$$

Therefore, we only need to calculate

$$\begin{pmatrix} dx \\ dy \\ dz \\ cdt \end{pmatrix} = L_{-u} \begin{pmatrix} v \\ c \end{pmatrix}.$$

22. Divide the left-hand side by dt, and obtain the velocity in space:

$$\begin{pmatrix} dx/dt \\ dy/dt \\ dz/dt \end{pmatrix} \in \mathbb{R}^3.$$

23. What is its physical meaning? Hint: it tells us how the first particle gets away from the second one.
24. Consider a special case, in which u aligns with the x-axis:

$$u \equiv \begin{pmatrix} \|u\| \\ 0 \\ 0 \end{pmatrix}.$$

25. Show that, in this case, one could design

$$O_u \equiv I$$

(the 3×3 identity matrix).
26. Show that, in this case,

$$L_u \equiv \begin{pmatrix} \gamma\left(\beta_{\|u\|}\right) & & & 1 \\ & 1 & & \\ & & 1 & \\ 1 & & & \gamma\left(\beta_{\|u\|}\right) \end{pmatrix} \begin{pmatrix} 1 & & & -\beta_{\|u\|} \\ & 1 & & \\ & & 1 & \\ -\beta_{\|u\|} & & & 1 \end{pmatrix}.$$

27. Show that, in this case,

$$L_{-u} \equiv \begin{pmatrix} \gamma\left(\beta_{\|u\|}\right) & & & 1 \\ & 1 & & \\ & & 1 & \\ 1 & & & \gamma\left(\beta_{\|u\|}\right) \end{pmatrix} \begin{pmatrix} 1 & & & \beta_{\|u\|} \\ & 1 & & \\ & & 1 & \\ \beta_{\|u\|} & & & 1 \end{pmatrix}.$$

28. Show that this is not just a special case, but a most general case. Hint: for a general u, pick the x-axis to align with u in the first place.
29. Interpret L_u as a projective mapping in the real projective space. Hint: see Chapter 6 in [45].
30. Interpret the above method as the three-dimensional extension of the methods in Sections 10.6.2–10.6.4.
31. Likewise, in Section 10.6.4, interpret the inverse Lorentz transformation back to spacetime as a projective mapping in the real projective plane.
32. What does this mapping do? Hint: it maps the original velocity $(dx'/dt', dy'/dt')$ of the first particle in the lab (Figure 10.8) to the new velocity $(dx/dt, dy/dt)$ of the first particle away from the second one (Figure 10.9).

33. In Figures 10.8–10.9, where is the time variable? Why is it missing? Hint: these are phase planes, as in Figures 1.8 and 7.14. They only show us the position that the particle may reach, but not when. Time remains implicit: it is just a parameter that "pushes" the particle along the arrow. To see this motion more vividly, make a movie that shows how the particle travels along the arrow.

34. In these figures, how did we get rid of time? Hint: we just divided by dt. This way, the time variable was eliminated, leaving just the velocity arrow in space.

11

Towards General Relativity: Spacetime and Its Local Coordinates

What is spacetime? So far, spacetime was a four-dimensional hyperspace, spanned by the x, y, z, and t axes. Here, on the other hand, spacetime is more general: a four-dimensional manifold. Furthermore, it is also differentiable: at each event in spacetime, we have a local "lab:" a little chart, made of three-dimensional time levels. This makes a new t-x-y-z coordinate system to help carry out physical experiments locally.

Thanks to Zorn's lemma, there exists a maximal atlas of charts. Moreover, thanks to the axiom of chice, at each event in spacetime, we can pick one individual chart: some coordinate system, to help describe our experiment in our local "lab." This is also a good preparation work for Lie groups, used often in quantum mechanics.

We can now go ahead and envelope spacetime with tangent spaces all over. These are Riemann normal coordinates: local, not global. This is indeed Einstein's equivalence principle in its geometrical face. This was indeed the starting point of general relativity.

In Chapter 7, we've already used a binary tree to model a bifurcation tree in chaos theory. From below, this tree is completely transparent: it looks like Cantor's null set.

In general relativity, on the other hand, a binary tree is no longer good enough. From each node in the tree, we need to issue not only two but four branches. This makes a quaternary tree, useful to model a tensor.

Thanks to their recursive nature, quaternary trees are suitable to model big tensors with arbitrarily many indices, as required in numerical relativity. Indeed, it is now easy to implement all sorts of algebraic operations between tensors: addition, contraction, etc., no matter how many indices are used. This is indeed a god example of using a multilevel hierarchy efficiently.

Tensors are particularly useful in general relativity. Thanks to them, one can model the curvature in spacetime. The key to the curvature is the metric that tells us how spacetime stretches, and in what direction [2, 4, 9, 38].

How does the metric do this? At each event in spacetime, the metric specifies four directions, orthogonal to each other. Three of these directions are space-like: to each of them, the metric assigns a positive weight. The fourth direction, on the other hand, is time-like: to it, the metric assigns a negative weight. Together, these weights tell us the (local) possible speed: how much one could advance in space, compared to the advance in time (Chapter 1, Section 1.1.1).

Unfortunately, the metric is not yet known. Fortunately, Einstein equations tell us an important thing about the metric: it can never come from nothing. On the contrary: it must come from a physical source: the stress tensor. This produces the Einstein equations, whose solution is the metric itself. They will be introduces elsewhere (Chapter 15 in [45]). Here, on the other hand, we are more interested in their geometrical meaning. For this, we'll use tensors for yet another important purpose: to discuss Riemann normal coordinates in spacetime. Thanks to these (theoretical, implicit, local) coordinates, we'll be able to see Einstein's equivalence principle geometrically.

11.1 Some Background

11.1.1 Special Relativity – No Gravity

In special relativity, one assumes no gravity at all. This way, an object can fly undisturbed at a constant speed, and never accelerate or change direction.

In real systems such as the solar system, on the other hand, gravity can no longer be ignored. On the contrary: it must be explained by a new mathematical model, independent of the coordinates that happen to be used. After all, the coordinates are just mathematical artifacts to help model nature. As such, they should never affect nature or the laws that govern it.

To put general relativity in its proper context, let's start with some historical background.

11.1.2 Flat Geometry

The ancient Greeks introduced Euclidean geometry for one main purpose: to model static shapes in the two-dimensional plane – triangles, circles, and so on. Later on, this theory was also extended to the three-dimensional space. This was quite useful to calculate volume, surface area, and more.

This was still rather static. Dynamics was still missing.

11.2 Physics and Philosophy

11.2.1 Newton and Plato

Newton, on the other hand, introduced a new time axis on top, to help model not only static but also dynamic shapes. This was indeed a breakthrough: a new force can now be applied to the original shape from the outside, to accelerate its original velocity, and even change its direction.

This fits well in Plato's philosophy. To refer to a geometrical shape (or just any general object), we must introduce a new word in our language, to represent not only one concrete instance but also the "godly" spirit behind all possible instances. This way, the word stands behind the deep concept fully. Likewise, in physics, force stands behind the motion, and affects it from the outside.

11.2.2 Einstein and Aristotle

Newton viewed time as an auxiliary parameter, which makes a new (nonphysical) axis, perpendicular to the (physical) phase-space, where the original motion takes place. Einstein, on the other hand, threw the time dimension back into the very heart of geometry. This way, time is not different from any other spatial dimension. Once the time axis is united with the original three-dimensional space, we have a new four-dimensional manifold: spacetime.

This is more in the spirit of Aristotle's philosophy. A word in our language takes its meaning not from the outside but from the very inside: the deep nature of the general concept it stands for.

To describe spacetime and gravity, Einstein used differential geometry. Here is some background.

11.3 Differentiable Manifold

11.3.1 Coordinates on The Sphere: Longitude and Latitude

For a start, consider a two-dimensional manifold: a sphere. How to refer to a particular point on it? For this purpose, the point must have a "name" or label: two coordinates.

Consider, for example, the face of the Earth. On it, consider a particular location. How to refer to it? By two coordinates: latitude and longitude. These are just two numbers that label the location, and give it a name. The longitude is, say, between $0°$ and $180°$, and the latitude between $-90°$ and $90°$. (Later on, we'll exclude the Greenwich longitude, and give it a special treatment.)

11.3.2 The Equator: A Level Set

Around each location on the face of the Earth, we can draw a chart: say, horizontal and vertical lines. The equator, for example, makes a horizontal line. On it, the latitude is constant: $0°$. This is the zero level set of the latitude. The Greenwich longitude, on the other hand, makes a vertical line. On it, the longitude is constant: $0°$. This is the zero level set of the longitude.

11.3.3 Singularity at The Pole

Unfortunately, this chart is not entirely global. Indeed, it fails at two points: the north and south poles. After all, at the pole, longitude can never be defined uniquely any more. With respect to this chart, the pole is singular. Around it, we must use an alternative chart.

11.3.4 Regular Chart in The Arctic

How to design it? This could be done in many different ways. The present approach has an advantage: it sticks to the standard coordinate for as long as possible.

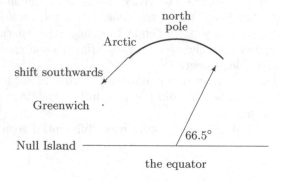

Fig. 11.1. The northern hemisphere of the Earth: a view from the east. From the Arctic Circle and south, use the standard coordinates: latitude and longitude. In the Arctic itself, on the other hand, use new coordinates: shift the entire Arctic southwards, until the north pole hits the Null Island. There, use longitude and latitude as a chart for the shifted Arctic. Finally, shift it back north, with its new chart.

For a start, look at the Arctic Circle, at latitude 66.5° (Figure 11.1). To the north of it, we have the Arctic, with the north pole at the middle. How to draw a new (regular) chart there?

For this purpose, shift the entire Arctic southwards. In this process, the north pole will shift southwards as well, through Greenwich, until it hits the equator at the origin $(0,0)$ (latitude 0°, and longitude 0°). This is the Null Island, in the gulf of Guinea, in west Africa. Around it, we already have a good chart: the familiar longitude and latitude. Use them as a chart for the shifted Arctic. Finally, shift it back north, together with its new chart. This way, the Arctic is now covered with a new regular chart, as required.

11.3.5 Local Coordinates

In the new chart in the Arctic, what are the coordinates of the north pole? Well, they are inherited from the Null Island: $(0,0)$. These are local coordinates: they label the north pole uniquely in terms of the new chart in the Arctic. Likewise, each point in the Arctic inherits its new coordinates from the surroundings of the Null Island. Again, these are just local coordinates: they "live" in the Arctic only, not outside it.

Thus, the new coordinates (inherited from the Null Island and its surroundings) map the entire Arctic continuously into $\mathbb{R}^2$, labeling each individual point by two local coordinates. Could this be done in the entire sphere as well, in a consistent way?

11.3.6 Compatible Charts on The Arctic Circle

subtropics   subtropics

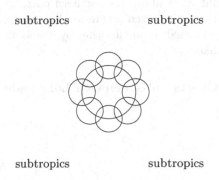

subtropics   subtropics

Fig. 11.2. The northern hemisphere: a view from above. The big circle is the Arctic Circle. It is covered by a list of small overlapping circles, round and round.

subtropics   subtropics

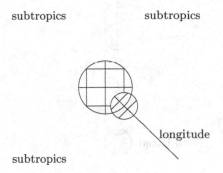

longitude

subtropics

Fig. 11.3. The northern hemisphere: a view from above. The Arctic has now a nonstandard chart, borrowed from the Null Island and its surroundings. The small circle on the bottom-right, on the other hand, still sticks to the standard chart: oblique longitude lines (directed to the north pole), and perpendicular latitude lines. In its northern part, where it overlaps with the Arctic, both charts are compatible: smoothly transferable from one to another.

Unfortunately, there is still a problem: what to do on the Arctic Circle itself? To its south, we still use the standard chart: longitude and latitude. To its north, on the other hand, we use our new chart. How to make them match?

For this purpose, let's cover the Arctic Circle with a list of small overlapping circles, round and round (Figure 11.2). (Each circle is open: only its interior, without its boundary.) In each small circle, let's keep the original coordinates: longitude and latitude.

Now, look at one small circle (Figure 11.3). In its southern part, there is no problem: it matches. Likewise, it also matches with its neighbors: the small circles on its east and west. But what about its northern part, where it overlaps with the Arctic? There, the charts don't match any more. Fortunately, they are still compatible: they could be transformed to one another by a smooth mapping, differentiable as many times as you like.

11.3.7 Compatible Charts on Greenwich Longitude

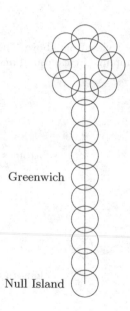

Fig. 11.4. The Greenwich longitude is also covered by a list of small overlapping circles. This accounts for the jump in longitude: from 1°, 2°, 3°, ... on the east, to 179°, 178°, 177°, ... on the west.

We are not done yet. There is still one more problem: what to do on the Greenwich longitude? After all, there is a discontinuity there: on the east, the longitude is numbered 1°, 2°, 3°, ... On the west, on the other hand, it is numbered 179°, 178°, 177°, ... How to fix this?

We already know how. After all, we already had the same problem in the small circles that cover the Arctic Circle. Indeed, look at the bottom circle in Figure 11.2. In it, we want to use longitude and latitude. Still, to guarantee continuity, we must also allow negative longitude: ..., −3°, −2°, −1°, 0°, 1°, 2°, 3°, ... This matches on the east, but not on the west. Fortunately, it is still compatible. After all, adding 180° is a very smooth mapping.

The same trick could work on the entire Greenwich longitude as well: cover it with a list of small overlapping circle (Figure 11.4). On each of them, allow negative longitude as well. This way, they match on the east, and also on the west (up to an additive constant). Thus, we still have compatibility on the west as well, as required.

11.3.8 Atlas of Compatible Charts

The above design could be mirrored in the southern hemisphere as well. This covers the entire sphere with compatible charts all over. This is called an atlas. Thanks to it, the sphere is about to become a differentiable manifold.

The above is just one example: there are many other legitimate atlases. Which one to pick?

11.3.9 Maximal Atlas

Our atlas is still rather small: it contains only a few charts. We can now extend it into a bigger atlas: just introduce a new chart around some point on the sphere. In this step, compatibility must be preserved: the new chart must be compatible to all previous ones. This could be repeated time and again, producing a chain of bigger and bigger atlases, which include one another in a row.

Is there a maximal atlas that can't be extended any more? To answer this, let's use a powerful tool from set theory: Zorn's lemma.

11.3.10   Using Zorn's Lemma

To use Zorn's lemma, consider a chain of bigger and bigger atlases that include one another in a row. Does it have an upper bound? In other words, is there a yet bigger atlas that includes all atlases in the chain? Easy: just look at their union. Is it a legitimate atlas? To check on this, pick two charts from it. Are they compatible to one another? They must be. After all, they must belong to some atlas in the chain.

Thus, every chain of atlases indeed has an upper bound: their union. Thanks to Zorn's lemma (Chapter 7, Section 7.12.6), there exists a maximal atlas. It has two attractive properties: on one hand, all charts in it are compatible to one another. On the other hand, no new chart could be introduced without violating compatibility.

11.3.11   Differentiable Manifold

Thanks to the maximal atlas, we can now consider not only our original coordinates, but also any new system of coordinates, obtained by a smooth invertible transformation. This is benign: the new coordinates could be differentiated with respect to the old ones, and vice versa. Together with this maximal atlas, the sphere indeed makes a legitimate differentiable manifold.

11.3.12 Using The Axiom of Choice

Our original atlas contained only a few charts. The maximal atlas, on the other hand, may be much bigger, and even contain infinitely many charts. In fact, each point on the sphere may be contained in many overlapping charts (compatible with each other). Which one to pick?

Fortunately, in set theory, we have the axiom of chice to help do this. Still, the axiom of choice is implicit: it offers no explicit method to pick one suitable chart. This job is left to us. Here is one attractive suggestion.

11.3.13 Riemann Normal Coordinates

For each individual point on the sphere, pick a little chart around it. Then, shift the chart a little, so that the point is now labeled by $(0,0)$ in its new (private) chart. These are Riemann normal coordinates. They are only local, not global. Still, they are particularly useful to design a vector field.

11.4 Vector Field and Its Basis

11.4.1 Continuous and Smooth Real Functions

Consider the space of continuous real functions:

$$C^0(\mathbb{R}) \equiv \{f \mid f \text{ is continuous in } \mathbb{R}\}.$$

In it, consider a smaller subspace of functions: those that are not only continuous but also smooth:

$$C^\infty(\mathbb{R}) \equiv \{f \mid f, \ f', \ f'', \ f''', \ \dots \ \text{are continuous in } \mathbb{R}\}.$$

What is so good about this function space? Well, each function in it is differentiable as many times as you like. This is quite useful in physics.

Now, consider a very simple function: a constant function that never changes:

$$f_1 \equiv 1.$$

f_1 will tell us that we are on a straight line: the real axis. Still, as a constant function, f_1 can never distinguish between different points on this line. This job is left to its (nonconstant) coefficient.

11.4.2 How to Span The Real Functions?

Consider now a new real function g. How to "span" it, or rewrite it in terms of f_1? For this purpose, we must do a silly thing: multiply f_1 by a "new" (nonconstant) coefficient, which is actually the same as g:

$$g(x) = \lambda_1(x)f_1,$$

where

$$\lambda_1(x) \equiv g(x).$$

What's good about this? It splits our jobs: λ_1 takes now the job of distinguishing between different x's, whereas f_1 takes a more geometrical job: to tell us that we have just one function, not two. This sounds silly, but is not. To see how important this is, let's move on to a higher dimension.

11.4.3 Vector Field or Function

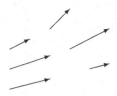

Fig. 11.5. A vector field: to each point in the Cartesian plane, attach a small arrow, issuing from it, and pointing in some direction.

Let's move on to a more interesting geometry: the two-dimensional Cartesian plane $\mathbb{R}^2$. In this case, we consider a vector field: a pair of two scalar functions, both defined in the Cartesian plane. Together, they make a new arrow, or direction vector (Figure 11.5).

As a matter of fact, this is not only a vector field but also a vector *function*, or an actual mapping: each point in $\mathbb{R}^2$ is mapped to the arrow issuing from it, which could be a completely different vector in $\mathbb{R}^2$. Still, this is not always the case: later on, we'll see a vector fields that is *not* a function.

Consider now a subspace of vector fields (or functions). Denote it by T:

$$T \subset \left(\mathbb{R}^2\right)^{\left(\mathbb{R}^2\right)} \equiv \left\{f \mid f : \mathbb{R}^2 \to \mathbb{R}^2\right\}.$$

What is T? For now, we only know that it contains some vector fields. How does a vector field look like? Well, to each point in $\mathbb{R}^2$, it attaches a small arrow, issuing from this point, and specifying a new direction vector. Still, T contains only "good" vector fields.

11.4.4 Continuous Vector Field

To deserve to belong to T, the vector field must be continuous: if the original point changes a little, then the arrow issuing from it could change as well, but only a little. For this purpose, the vector field must be made of two continuous (scalar) functions:

$$T \subset C^0\left(\mathbb{R}^2\right) \times C^0\left(\mathbb{R}^2\right) = \left(C^0\left(\mathbb{R}^2\right)\right)^2.$$

11.4.5 Smooth Vector Field

Better yet, T should contain only smooth vector fields: as the original point changes, the arrow should change smoothly (Figure 11.6). For this purpose, the vector field must be made of two smooth (real) functions:

Fig. 11.6. A smooth vector field: if the original point changes a little, then the arrow issuing from it may change as well, but this change must be small and smooth in terms of both length and direction.

$$T \equiv C^\infty\left(\mathbb{R}^2\right) \times C^\infty\left(\mathbb{R}^2\right) = \left(C^\infty\left(\mathbb{R}^2\right)\right)^2.$$

This kind of vector field is most useful in physics.

11.4.6 Basis of Vector Fields

How to span these vector fields? Again, we better split two jobs. Our first job is to take care of our new two-dimensional geometry. For this purpose, we need two constant "functions:"

$$f_1 \equiv \begin{pmatrix} 1 \\ 0 \end{pmatrix} \quad \text{and} \quad f_2 \equiv \begin{pmatrix} 0 \\ 1 \end{pmatrix}.$$

This reflects well the two-dimensional nature of the arrows in our vector field. Indeed, f_1 spans the first spatial dimension, whereas f_2 spans the second. Together, they form a basis for $\mathbb{R}^2$. This is why f_1 is often denoted by $\partial/\partial x$, and f_2 by $\partial/\partial y$.

11.4.7 How to Span a Vector Field?

Still, as constant functions, f_1 and f_2 can never distinguish between different points in $\mathbb{R}^2$. For this job, they must have new (nonconstant) coefficients. This way, a given (smooth) vector field $g \in T$ can now be represented uniquely as

$$g = \lambda_1 f_1 + \lambda_2 f_2,$$

where λ_1 and λ_2 are new (scalar) functions, defined in $\mathbb{R}^2$, and dependent on g. This way, they take the job of distinguishing between different points in $\mathbb{R}^2$. This frees the constants f_1 and f_2 to focus on their geometrical job: to distinguish between the horizontal and vertical dimensions in each arrow.

In the above example, $\mathbb{R}^2$ is spanned by an elegant geometrical basis: f_1 and f_2. In a more complex geometry, on the other hand, there may be no such basis at all.

11.5 Vector Field on a Sphere

11.5.1 Riemann Normal Coordinates: Local and Tangent

Consider now a more complicated case: the original points are no longer in the Cartesian plane, but on a sphere. In other words, from each point on the sphere, issue a little arrow, tangent to the sphere. This way, in the tangent plane, we can now draw two axes, to make two local coordinates, fitting with our local chart.

In theory, we could draw them as we like. Still, in the tangent plane, they should better be perpendicular to each other. The x-direction is denoted by $(1,0)$ or $\partial/\partial x$, and the y-direction by $(0,1)$ or $\partial/\partial y$. In terms of these axes, the original point lies at the origin $(0,0)$, and the arrow leads to (x,y) on the tangent plane (outside of the sphere).

These are Riemann normal coordinates. How to specify them more clearly? This will be our next job.

11.5.2 Tangent Plane – Local Chart

Thus, $\partial/\partial x$ is a vector field in its own right. After all, from each individual point on the sphere, it points in some direction, tangent to the sphere. This could be the horizontal direction. Still, this is not a must.

Likewise, $\partial/\partial y$ is a vector field as well. Usually, it is perpendicular to $\partial/\partial x$ in the tangent plane. Still, this is not a must.

Together, $\partial/\partial x$ and $\partial/\partial y$ span the tangent plane: $\partial/\partial x$ points horizontally (aligning with the latitude), and $\partial/\partial y$ is perpendicular to it (aligning with the longitude). There could be many other choices, but this choice makes a lot of sense. This spans the tangent plane: the geometrical realization of our local chart.

So far, we only have vector *fields*: each arrow "loves" in its own tangent plane (or local chart). Now, is it possible to identify all tangent planes with each other, and produce not only a vector field but also a vector *function*?

11.5.3 Vector Field: Not a Function!

Let's focus on $\partial/\partial x$. So far, it is only defined locally, in each individual chart, tangent to the sphere. This makes it a legitimate vector field. Still, is it also a legitimate vector *function*?

For this purpose, consider all tangent planes as one and the same. In this case, could $\partial/\partial x$ be defined not only locally but also globally, everywhere on the sphere at the same time? In other words, is

$$\frac{\partial}{\partial x} \in (C^\infty (\text{sphere}))^2?$$

No! Indeed, at the poles, this global definition would lead to singularity, and even discontinuity:

$$\frac{\partial}{\partial x} \notin (C^0 (\text{sphere}))^2.$$

How to fix this? How to smooth it out at the poles?

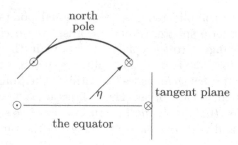

Fig. 11.7. The northern hemisphere of the Earth: a view from the side. On the right, there is a vertical tangent plane at the equator. On the upper-left, on the other hand, there is an oblique tangent plane at latitude $0 < \eta < \pi/2$.

11.5.4 Tangent Planes at The Equator

Let's focus again on $\partial/\partial x$. At the equator, it could be viewed as a legitimate vector function (Figure 11.7). Indeed, in each tangent plane, it indeed points horizontally, along the equator. This could be viewed as a constant function, just like f_1 in Section 11.4.6. As a constant function, it is indeed continuous and smooth, as required. In fact, it spans the first Riemann normal coordinate: the horizontal coordinate. This helps envelope the entire equator with tangent horizontal arrows, round and round. (Together with $\partial/\partial y$, this could even make a rotating coordinate system, spanning the tangent planes that envelope the Earth all over.)

The equator is a level set: the latitude is $\eta = 0$. Consider now a higher latitude of $0 < \eta < \pi/2$. At this latitude, look at a horizontal cross section of the Earth (Figure 11.7). The same approach works here as well: $\partial/\partial x$ is still constant, and helps envelope the entire cross section with horizontal arrows, round and round.

11.5.5 Singularity at The North Pole

But what happens as $\eta \to \pi/2$, approaching the north pole? In this case, $\partial/\partial x$ is still valid in the local chart, tangent to the north pole. As such, it still makes a legitimate vector field. However, it can no longer be viewed as a vector function. Indeed, as such, it would becomes discontinuous. Indeed, as $\eta \to \pi/2$, the entire cross section shrinks to just one point: the north pole itself. This singular point could be approached from either side: Upon approaching the north pole from the east (along longitude 90°), $\partial/\partial x$ points in one direction. Upon approaching from the west (along longitude 270°), on the other hand, $\partial/\partial x$ points the other way around. How to smooth this out?

11.6 Generating System

11.6.1 How to Smooth Out?

To smooth this out, let's define a new vector function f_1 that decrease monotonically at higher and higher latitudes, and vanishes smoothly at the north pole:

$$f_1 \equiv \exp\left(-\tan^2(\eta)\right)\frac{\partial}{\partial x}.$$

This new f_1 is still horizontal. Still, at higher and higher latitudes, the arrows get shorter and shorter, until vanishing completely at the north pole. The same is mirrored at the south pole as well. Thus, f_1 makes not only a vector field but also a legitimate vector function, smooth on the entire sphere.

11.6.2 Horizontal Level Sets

For our new f_1, the equator is still a level set. Indeed, on the equator, where $\eta = 0$, f_1 is the same as $\partial/\partial x$: $(1,0)$.

   Likewise, at a fixed altitude $0 < \eta < \pi/2$, the cross section makes a level set for f_1 as well. After all, f_1 is just a scalar multiple of $\partial/\partial x$ there. As such, it still points horizontally, tangent to the cross section, enveloping it with little horizontal arrows, round and round.

   Unfortunately, as $\eta \to \pm\pi/2$, f_1 is not very useful any more: it approaches zero. As a matter of fact, the north pole itself makes a degenerate level set: the zero level set. To fix this, we need yet another vector function, which will "live" at the poles as well. Furthermore, it will complete f_1: it will point not horizontally but vertically.

11.6.3 Oblique Tangent Direction

How to do this? Let W be the rotation that rotates the entire sphere counterclockwise (Figure 11.8). This way, the north pole "slides" southwards along longitude $270°$, until hitting the equator at the Galapagos islands. In this process, each point p on the sphere is mapped to Wp, carrying its own arrow $f_1(p)$ with it. In its new place, the arrow gets a new name:

$$f_2(Wp) \equiv f_1(p).$$

This way, f_2 points obliquely, not horizontally. Furthermore, its level sets are now vertical loops, not horizontal.

11.6.4 Vertical Level Sets

How does a level set of f_2 look like? In Figure 11.8, look at a fixed angle κ. This marks a vertical cross section of the Earth. Around it, f_2 makes a vertical loop of arrows, enveloping it with tangent arrows all over, round and round, from south to north and back.

   Often, the new level sets of f_2 cross those of f_1. This way, at each individual point on the globe, they could be used to span the tangent plane, in terms of two new local coordinates. Unfortunately, this could break.

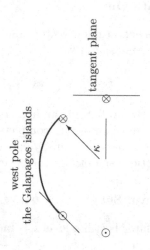

Fig. 11.8. The rotation W rotates the entire sphere counterclockwise. This way, the north pole slides along longitude 270°, until it hits the equator at the Galapagos islands. In this process, each point carries with it the horizontal arrow f_1. This makes a new arrow f_2.

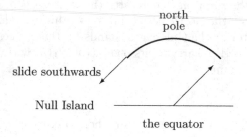

Fig. 11.9. The rotation Q rotates the entire sphere: the north pole slides along the Greenwich longitude, until it hits the equator at the Null Island. In this process, each point carries with it the original arrow f_1. In its new place, it is called f_3.

11.6.5 A Third Tangent Direction

Unfortunately, along longitudes 90° and 270°, there is still a problem: f_2 aligns with f_1: they are both horizontal. So, we still have only one tangent direction. How to design another one?

For this purpose, let Q be yet another rotation of the entire sphere. Here, however, the north pole slides along the Greenwich longitude, until hitting the equator at the Null Island (Figure 11.9). In this process, each point p on the sphere is mapped to Qp, carrying its own arrow $f_1(p)$ with it. In its new place, the arrow gets a new name:

$$f_3(Qp) \equiv f_1(p).$$

Like f_2, f_3 points obliquely, and its level sets are vertical loops. We're particularly interested in longitudes 90° and 270°. This is indeed a vertical level set of f_3: it loops the Earth with tangent arrows, round and round, from north to south and back. This aligns neither with f_1 nor with f_2. On the contrary: it is perpendicular to them there. Thus, this supplies the missing local coordinate in the tangent planes there.

Together, f_1, f_2, and f_3 make a generating system, ready to represent just any vector function. Let's see how.

11.6.6 Generating System

By now, at each point on the globe, we already have three level sets. Of these, at least two don't align with one another, and can serve as local coordinates, to help draw the local chart, and span the tangent plane.

Usually, three are too many. After all, in a two-dimensional manifold, two local coordinates should be enough to span the two-dimensional tangent plane. Still, to be on the safe side, we better have three. After all, two of them may align with each other, and point in the same direction, producing one and the same coordinate. Furthermore, at the poles, a level set may even shrink and degenerate completely. In such a case, the third one comes to our aid. Indeed, it is perpendicular to the two others there, and can therefore supply the missing coordinate.

Thus, two local coordinates are good enough locally, to span the local chart and the tangent plane, and indeed represent any given vector field. A vector *function*, on the other hand, is trickier: it is defined not only locally but also globally, in the entire sphere. To represent it, we must therefore have three vector functions.

Indeed, consider a given vector function

$$g \in (C^\infty (\text{sphere}))^2.$$

It can now be represented (albeit not uniquely) as

$$g = \lambda_1 f_1 + \lambda_2 f_2 + \lambda_3 f_3,$$

where

$$\lambda_1, \ \lambda_2, \ \lambda_3 \ \in C^\infty (\text{sphere})$$

are three scalar functions. These are the coefficients, which depend on g. Their job is to distinguish between different points on the globe.

This frees the vector functions f_1, f_2, and f_3 (which don't depend on g) to take care of the special geometry of the sphere and its tangent planes. Together, they form a new generating system.

11.6.7 Generating System: Not a Basis

Could we drop one of them? Unfortunately not. Indeed, to be smooth on the entire sphere, each of them must vanish somewhere. There, we must also have two others. In summary, the third one is necessary to supply the missing tangent vector wherever the two others align with each other (or degenerate).

The Cartesian plane is simple. This is why it has a basis of two (constant) vector functions (Section 11.4.6). The sphere, on the other hand, is more complicated. This is why it must have a generating system that is no longer a basis: it contains not only two but three vector functions. Together, they take care of the geometry, not only locally but also globally: they can span just any given vector function, once its coefficients are defined properly.

Below, however, we'll focus on vector fields that are *not* functions: they are defined only locally, not globally. To represent these, we don't need three: two are enough.

11.7 Time: Can It Ever Stop?

11.7.1 Zeno's Paradox

Why did we talk about the sphere so much? Because it could help model a slippery term: time.

To do this, consider again the northern hemisphere (Figure 11.9). On it, consider a particle that travels northwards: from the equator (at latitude $\eta = 0$), to the north pole (at latitude $\eta = \pi/2$).

Assume that the particle moves really fast: it can reach the north pole in just one second. Still, before doing this, it must first make half the way, and reach latitude $\eta = \pi/4$. This should take half a second. After doing this, it must still do one more quarter of the way, and reach latitude $\eta = 3\pi/8$. This should take another quarter of a second, and so on.

So, we have a paradox here: the particle must carry out infinitely many steps, each shorter than the previous ones. This is an infinite list. How could the particle ever complete the entire way, and reach the north pole in a finite number of steps? This is Zeno's paradox (Chapter 1, Sections 1.1.3–1.1.6).

11.7.2 Are You Afraid of Infinity?

Fortunately, by now, we're no longer afraid of infinity. After all, we can now use an infinite power series to calculate the total time that the particle needs:

$$\frac{1}{2} + \frac{1}{4} + \frac{1}{8} + \frac{1}{16} + \cdots = 1.$$

Thus, the total time is indeed one second, as expected.

11.7.3 Towards a Black Hole

But what if the particle slows down? For instance, assume that, at each step, the particle gets twice as slow. In this case, the total time would be as large as

$$1 + 1 + 1 + 1 + \cdots = \infty,$$

and the particle would never reach!

When could this happen? When our hemisphere models a more complicated process: a particle traveling towards a black hole. Later on, we'll discuss this a little more, including an interesting phenomena: in our eyes, the particle can never arrive!

11.7.4 The Nonphysical Event Horizon

The black hole is surrounded by a (theoretical) sphere: the event horizon. The particle can only approach it from the outside, but never arrive. To reach it physically, it needs an infinite time.

Why? Because, as the particle approaches the event horizon, its time gets slower and slower. This is gravitational time dilation, to be discussed later.

This is only in our eyes. In its own self system, the particle feels no change at all: its time keeps ticking as usual. Thus, in its own self system, it'll eventually pass the event horizon smoothly, and fall into the black hole. After all, the event horizon is not physical, but only mathematical.

11.7.5 Zeno's Paradox Comes True!

We, on the other hand, who look at the particle from here, see it slowing down, and never reaching the event horizon. Zeno's paradox comes true!

This also works the other way around: no particle can ever leave the event horizon and travel to our eyes in any finite time.

11.7.6 The Dark Event Horizon

Even a photon can never leave the event horizon and reach our eyes. This is why the event horizon looks dark.

Although it is as fast as light, the photon has not enough time to complete the entire journey. This is why, from our perspective, it remains stuck in the event horizon forever.

In its own self system, on the other hand, its time ticks as uniformly as ever. From its own point of view, it advances towards us, as fast as light. After all, it has no idea that it can never leave the event horizon in any finite time.

11.7.7 Singularity at The Event Horizon: Time Stops!

To us, the event horizon makes a singularity. On the way towards it, time gets slower and slower. For us, at the event horizon itself, time stops!

To make things a little more "plausible," let's see how this could help in science fiction.

11.7.8 Honey, I Shrunk The Kids!

In this movie (by Disney's studios), there is an absent-minded scientist who builds a new machine that can create a new environment with the same effects as in a very high speed: time slows down, and length shrinks.

Unfortunately, by mistake, the machine starts working with no control, and shrinks the kids! They were thrown into a strange part of spacetime, where both time and space shrink by the same factor. Below, we'll illustrate this geometrically.

There, the kids are in a totally different system, completely detached from here and now. In their jump to the new system, they feel or see no change at all. After all, their time and length keep the same ratio as before. This way, they still have the same speed of light. We, on the other hand, see them shrinking, and their time slowing down.

These systems are inherently disjoint, and could never unite or even meet. Still, in the movie, they do interact, for the sake of more drama and fun... This trick was already used in *The Wonderful Adventure of Nils Holgersson* by Selma Lagerlioff.

Let's discuss spacetime a little more, and use geometry to understand it better.

11.8 Spacetime in Special Relativity

11.8.1 Our Coordinate System

In special relativity, spacetime is particularly easy to draw [11, 58]. After all, everything is linear: there is no gravity, nor any other force, nor any curvature.

For simplicity, assume that our space is one-dimensional. In our system, this makes the x-axis. Together with the time axis on top, we have the familiar x-t system.

For even more simplicity, assume also that x is measured in light-seconds. This way, light travels at speed 1: one light-second per second. In spacetime, this draws an oblique line of angle 45° (Figure 11.10).

This is nice and symmetric: space and time take equal status. This way, a light ray could split spacetime obliquely and evenly. This works well not only in our coordinate system but also in any other (inertial) coordinate system, moving at a constant velocity with respect to ours.

11.8.2 Traveling Coordinate System

Now, consider a spaceship, traveling at a constant speed v away from us. (Again, v is measured in light-seconds per second.) Clearly, v can never exceed the speed of light:

$$|v| < 1.$$

In the spaceship, you carry your own clock and ruler. In them, you can measure your own time t' and length x'. These make two new axes: the new x'-t' coordinate system. In it, light still has the same speed: 1. Therefore, light must still draw an oblique line in between.

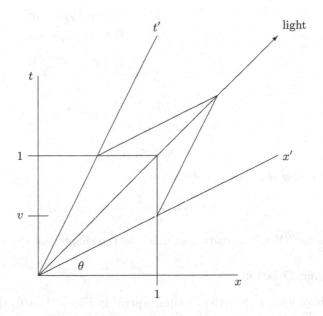

Fig. 11.10. Spacetime, illustrated in two different coordinate systems. The x-t axes span our own system here. The x'-t' axes, on the other hand, span another system, traveling rightwards at the constant speed v (away from us). In both systems, light travels at the same speed, drawing an oblique line in spacetime.

11.8.3 Area in The New Coordinates

Look again at Figure 11.10. In it, what is the angle between the x- and x'-axes? This is θ, satisfying

$$\tan(\theta) = v.$$

Thanks to symmetry, this is also the angle between the t- and t'-axes. After all, in special relativity, time and space take equal status.

Now, the x'-t' axes make a new diamond. What is its lower-left angle? This is just

$$\frac{\pi}{2} - 2\theta.$$

This is also the upper-right angle.

How long is the edge of the diamond? Thanks to Pythagoras theorem, the answer is

$$\sqrt{1 + v^2}.$$

Thanks to Euclidean geometry and the formula in Chapter 24, Section 24.4.1, we are now ready to calculate the area of the entire diamond:

$$\begin{aligned}
\text{area of diamond} &= \left(1 + v^2\right) \sin\left(\frac{\pi}{2} - 2\theta\right) \\
&= \left(1 + v^2\right) \cos(2\theta) \\
&= \left(1 + v^2\right) \left(\cos^2(\theta) - \sin^2(\theta)\right) \\
&= \left(1 + v^2\right) \left(1 - 2\sin^2(\theta)\right) \\
&= \left(1 + v^2\right) \left(1 - \frac{2v^2}{1 + v^2}\right) \\
&= \left(1 + v^2\right) \frac{1 - v^2}{1 + v^2} \\
&= 1 - v^2 \\
&= \gamma^{-2},
\end{aligned}$$

where γ is a new parameter, defined by

$$\gamma \equiv \frac{1}{\sqrt{1 - v^2}} \geq 1.$$

Is this familiar? We've already met this γ in the chapter on special relativity.

11.8.4 Time Dilation

Thus, to have area 1 (like the original square in Figure 11.10), the diamond must get bigger. For this purpose, the t'-axis must stretch by factor γ: from here, each second measured in the spaceship looks as long as γ seconds (Figure 11.11). This is called time dilation. Is this familiar? We've already seen this before. Here, however, it is more vivid: not only algebraic, but also geometrical.

11.8.5 Length Contraction

For the sake of symmetry, the x'-axis must stretch by the same factor: γ. This means that, from here, your ruler (that you carry on the spaceship) looks γ times as short. Why?

Well, assume that your ruler is as long as a light-second, so it ends at $x' = 1$. Now, in the spaceship, the ruler remains static, ending at $x' = 1$ all the time, and drawing the oblique line

$$x' \equiv 1$$

(the short oblique line in Figure 11.11). What slope does it have? Well, it is parallel to the t'-axis, so it makes angle θ with the t-axis. Thanks to triangle similarity, it hits the x-axis at

$$\gamma - \frac{(\gamma v)^2}{\gamma} = \gamma\left(1 - \frac{(\gamma v)^2}{\gamma^2}\right) = \gamma\left(1 - v^2\right) = \gamma \cdot \gamma^{-2} = \gamma^{-1}.$$

This is why, from here, the ruler indeed looks γ times as short. This is called length contraction.

Isn't this familiar? We've already seen it before. This is indeed special relativity: no gravity is assumed. What happens in general relativity? Well, gravity is now allowed. As a result, we may have the same effects as above. After all, at a place of high gravity, objects may fall at a very high speed.

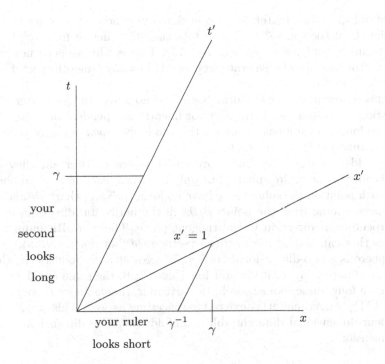

Fig. 11.11. The stretched x'-t' system of the spaceship. From here, one second of their clock looks as long as $\gamma \geq 1$ seconds. This is time dilation, illustrated by the short horizontal line. Furthermore, from here, their meter looks as short as $\gamma^{-1} \leq 1$ meters. This is indeed length contraction, illustrated by the short oblique line.

11.8.6 Singularity

What happens when the spaceship is nearly as fast as light:

$$v \to 1?$$

In this limit case, the above diamond gets degenerate: too narrow, with no width at all. This is a singularity: space and time shrink to zero. Time is too lazy to move, so every motion must be degenerate: too small to notice at all. Energy and momentum, on the other hand, get infinite. This looks like a complete chaos: time stops!

In spacial relativity, this is only theoretical: a limit case. In general relativity, on the other hand, this may really happen at the center of a black hole. There, only the laws of quantum mechanics could save us.

11.9 Spacetime in General Relativity

11.9.1 From The Sphere to Spacetime

To understand spacetime better, let's go back to our sphere for another look. What is so special about the sphere? Well, it is topologically different from $\mathbb{R}^2$: it has two "sides:" the inner, and the outer (Figure 3.15). This is why we need not only two but three vector functions to generate (represent) just any (smooth) vector function on it (Section 11.6.6).

Still, this representation has a drawback: it is no longer unique. After all, three basis functions are often too many, and not linearly independent any more. This is why they no longer form a basis. In fact, they no longer *span* but only *generate* the given vector function (Section 11.6.6).

Vector fields, on the other hand, are easier to span. After all, they are not defined globally in the entire sphere, but only locally, in each individual chart. For example, each point on the sphere may have its local (private) chart around it, with two local axes issuing from the point: $\partial/\partial x$ in the horizontal direction, and $\partial/\partial y$ in the perpendicular direction in the tangent plane. These are Riemann's normal coordinates that can help span any given vector *field* (but not *function*).

The sphere is a two-dimensional surface. Indeed, at each point on it, there are just two coordinates: say, latitude and longitude. Still, the same could be done in three (or even four) dimensions as well. In particular, thanks to set theory (Sections 11.3.9–11.3.11), a maximal atlas exists for spacetime as well. This way, spacetime makes a four-dimensional differentiable manifold: the time dimension, and three spatial dimensions.

11.9.2 Curvature

In special relativity, spacetime is completely flat and linear, with no curvature at all. The sphere, on the other hand, has a constant positive curvature. This is why it is nice and symmetric. Unlike both, in general relativity, spacetime may be completely nonlinear and uneven, with a nonconstant curvature that may change sharply from event to event. Fortunately, in each individual event, in just any coordinate system around it, the curvature can still be written in terms of energy and momentum.

11.9.3 Spacetime and Its Tangent Spaces

How to linearize our sphere? Envelope it with tangent planes. Likewise, spacetime can be enveloped with tangent spaces. Still, because spacetime is four-dimensional, these spaces are four-dimensional as well.

Around each event in spacetime, in its own private chart, we could then design four new vector fields, to span the tangent space. These are often named $\partial/\partial x^0$, $\partial/\partial x^1$, $\partial/\partial x^2$, and $\partial/\partial x^3$. These could then span just any given vector field, locally in each tangent space.

11.9.4 Basis of Local Tangent Vector Fields

Around each individual event in spacetime, these four would then be tangent to the original level sets in the original chart, which define the four local coordinates. In particular, if the original event has coordinates $(0,0,0,0)$, then these are Riemann's normal coordinates, issuing from it.

If spacetime is linear and flat, as in special relativity, then these four could be viewed not only as vector fields but also as vector functions, defined globally and smoothly, everywhere in spacetime. These could then serve as a basis, to span just any given vector function. Indeed, in this case, spacetime is just $\mathbb{R}^4$, and the (constant) basis functions are just the standard unit vectors:

$$\frac{\partial}{\partial x^0} = \begin{pmatrix} 1 \\ 0 \\ 0 \\ 0 \end{pmatrix}, \quad \frac{\partial}{\partial x^1} = \begin{pmatrix} 0 \\ 1 \\ 0 \\ 0 \end{pmatrix}, \quad \frac{\partial}{\partial x^2} = \begin{pmatrix} 0 \\ 0 \\ 1 \\ 0 \end{pmatrix}, \quad \frac{\partial}{\partial x^3} = \begin{pmatrix} 0 \\ 0 \\ 0 \\ 1 \end{pmatrix}.$$

In general relativity, on the other hand, spacetime is not flat any more.

11.9.5 Basis vs. Generating System

If, on the other hand, spacetime has a more complex geometry, as in general relativity, then four are not enough any more. To generate just any given vector function, one needs five elementary vector functions. After all, as we'll see later, Riemann's normal coordinates are only local, not global.

Fortunately, here we are more interested in vector fields, defined only locally, not globally. To span these, we only need four tangent vector fields, to form a basis, and local coordinates.

11.10 Spacetime and Gravity

11.10.1 Curved Geometry in Spacetime

What is our sphere? It is a curved two-dimensional manifold. Furthermore, at each point on it, the curvature is a positive constant, which depends on the radius only. Spacetime, on the other hand, is much more complicated: it is a four-dimensional manifold. Indeed, it has three spatial dimensions, plus time on top.

In spacetime, the curvature is not only spatial but also temporal. In other words, it stems not only from the nonlinear gravity in space, but also from its effect on time. Indeed, gravity makes time nonuniform and nonlinear: slow near a massive star, and quick elsewhere.

How curved is spacetime, and where? In other words, what determines the curvature in spacetime?

11.10.2 Metric and Shortest Path

Thanks to Einstein, energy and momentum are not only physical but also geometrical: sources of mass, gravity, curvature, and symmetry. They determine the true metric in spacetime. This way, around each event in spacetime, you can imagine spacetime as a four-dimensional hypercube in the x^0-x^1-x^2-x^3 coordinates (picked arbitrarily in advance). Now, instead of stretching spacetime to have its true curvature, just define its metric, which tells us the true distance between two events: the length of the shortest path between them.

11.10.3 It Looks Curved, But It Is Straight

Still, because spacetime is curved, such a path may no longer look straight in the usual sense. After all, the time axis is no longer uniform: the time scale may change from place to place. Near a massive star, for example, time gets slower.

This is indeed relativity: time is no longer absolute, but only relative to the place where it is being measured. If this place contains a lot of matter, then time is slower there.

11.10.4 Gravitational Time Dilation

This is called gravitational time dilation. To understand this better, consider an atomic clock that uses some radioactive process to measure time. Assume that the clock is placed near a massive star. To us humans, a second of this clock seems slower: it may take a few seconds in our time, here on the Earth.

Assume that one second has passed there. During this time, their light made one light-second there. To us humans, this looks slow. After all, we think that a few seconds passed, so light should make a few light-seconds! Once we realize that their clock runs slow, we accept this: their light should indeed make one light-second only, without violating Einstein's Law.

11.10.5 Gravitational Redshift

The light not only travels at a constant speed but also oscillates like a wave, at a constant frequency: constant number of cycles per second (Chapter 1, Section 1.5.5). In each cycle, it covers a certain number of kilometers. This is its wave-length, which specifies its color.

Still, here on the Earth, our second is quicker. Therefore, upon arriving at the Earth, the light is less frequent: per Earth-second, it makes less cycles. As a result, it changes color: it becomes less blue, and more red.

This is called gravitational redshift: we see the light redder than it originally was upon leaving the star in the first place. In other words, the star looks redder than it really is. A very massive star may even look so red that it could hardly be seen at all.

11.11 Light Can Curve!

11.11.1 Light Near a Massive Star

What happens to a light ray that passes by a massive star? Well, as it gets closer and closer to the star, its time gets slower, so it makes a shorter distance than before. So, it must turn a little, and draw an arc around the star.

Around a star as massive and dense as a black hole, the light ray could even orbit forever, with no escape. This happens at the event horizon. If it gets any closer to the black hole, then it would spiral, and eventually fall right into the black hole. In either case, we'd never see it any more.

This is indeed why it is called black hole: it looks dark: no light could ever escape from it, and get to our eyes. This could also be explained in terms of gravitational redshift: it gets so red that it can no longer be seen at all.

11.11.2 "Straight" Line

In our eyes, this kind of spiral doesn't look straight at all. In terms of the curved geometry in spacetime, on the other hand, this is as straight as ever. Indeed, in spacetime, to be straight means to follow a "valley" where time is as slow as possible. Indeed, in a slow-time zone, the light ray follows a path that is highly stretched-out and straight (as in a degenerate diamond in Figure 11.11). In terms of its own self system, on the other hand, the light ray is static, making a vertical line, along its own self-time-axis, in its own self-spacetime.

From the Earth, the light ray may seem curved. From its own perspective, on the other hand, it is vertical. After all, its own private time (measured in its own clock) remains as fast as before. Their time looks slow only from here, not from there.

11.11.3 Light Near a Black Hole

A black hole is a special kind of star – so massive and dense that even light can't escape from its powerful gravity. Still, a light ray that approaches a black hole would "feel" nothing unusual. Why? Because its own private time still ticks at the same rate as before.

After all, the light ray remains in a free fall, feeling no new force at all. Although gravity acts upon it quite strongly, this affects it only in the eyes of an observer who watches it from here. From its own perspective, on the other hand, the light ray feels much calmer: no force or acceleration or any change to its time rate.

11.11.4 Einstein's Happy Thought

This is Einstein's happy thought (or equivalence principle), which started general relativity in the first place. This could also be viewed geometrically: at each individual event, spacetime is nearly the same as its tangent space, which is as flat as in special relativity. Therefore, at this particular event, the photon feels as at a constant velocity, which doesn't accelerate at all. At the next event, on the other hand, there will be yet another tangent space. After all, spacetime is a legitimate differentiable manifold: Riemann's normal coordinates will remain valid. This reasoning could continue on and on forever.

11.12 Metric in Spacetime

11.12.1 Spacetime – A Four-Dimensional Manifold

Where could a motion take place? In classical mechanics, it takes place in space. In general relativity, on the other hand, it takes place in spacetime. Together with the time dimension, this is a new four-dimensional differentiable manifold. Each individual point in spacetime is called an event: a four-dimensional vector, specifying not only the spatial location but also the time.

What coordinates are needed in spacetime? Pick four, as you like: say, t, x, y, and z. Fortunately, we can measure them in our lab. For the sake of uniformity, they are often denoted by an upper index:

$$x^0 \equiv t$$
$$x^1 \equiv x$$
$$x^2 \equiv y$$
$$x^3 \equiv z.$$

Here, the superscript 0, 1, 2, or 3 has nothing to do with power: it just indexes a coordinate in spacetime. This upper index is often denoted by a small Greek letter, say $\alpha = 0, 1, 2, 3$.

11.12.2 Four-Dimensional Vector

Likewise, a four-dimensional row vector contains four numbers in a row:

$$v \equiv \left(v^0, v^1, v^2, v^3\right).$$

Again, the upper index is often denoted by a small Greek letter:

$$v \equiv (v^\alpha)_{0 \leq \alpha \leq 3}.$$

Again, α is no power: it just indexes the four components in the vector. Let's extend this into a square matrix.

11.12.3 Square Matrix

The above vector contains four numbers (components), listed one by one in a row. A 4×4 matrix, on the other hand, is a table that contains 16 numbers (elements), listed row by row:

$$A \equiv (a_{\mu\nu})_{0 \leq \mu,\nu \leq 3} = \begin{pmatrix} a_{00} & a_{01} & a_{02} & a_{03} \\ a_{10} & a_{11} & a_{12} & a_{13} \\ a_{20} & a_{21} & a_{22} & a_{23} \\ a_{30} & a_{31} & a_{32} & a_{33} \end{pmatrix}.$$

(Here, $a_{\mu\nu}$ means $a_{\mu,\nu}$ – the comma is often dropped.) In A, we indeed have 16 elements of the form $a_{\mu\nu}$, indexed by two lower indices: $\mu, \nu = 0, 1, 2, 3$.

An important matrix is the metric:

$$g \equiv (g_{\mu\nu})_{0 \le \mu, \nu \le 3} = \begin{pmatrix} g_{00} & g_{01} & g_{02} & g_{03} \\ g_{10} & g_{11} & g_{12} & g_{13} \\ g_{20} & g_{21} & g_{22} & g_{23} \\ g_{30} & g_{31} & g_{32} & g_{33} \end{pmatrix}.$$

This is not a constant matrix. On the contrary: it depends on the particular event under consideration, and may change from event to event in spacetime.

The metric g encapsulates important geometrical information. At each individual event in spacetime, it tells us how spacetime "stretches," and in what direction. Unfortunately, the true matric is not yet available. To uncover it, one must solve Einstein's equations (at least numerically). This requires a new algebraic tool: tensor.

11.13 Tensor: Extended Matrix

11.13.1 Tensor

The tensor may be viewed as an extension of the matrix: it can use more indices, upper or lower. For example, the $4 \times 4 \times 4 \times 4 \times 4$ tensor

$$T \equiv \left(T^{\alpha\beta}_{\gamma\delta\epsilon}\right)_{0 \le \alpha, \beta, \gamma, \delta, \epsilon \le 3}$$

contains as many as

$$4^5 = 1024$$

numbers (entries), indexed by two upper indices (α and β), and three lower indices (γ, δ, and ϵ).

Let's go ahead and "play" with tensors algebraically. Two tensors with the same number of indices could be easily added to each other, entry by entry. Still, there is a trickier algebraic operation: contraction.

11.13.2 Contraction

In the original tensor T, take one upper index, say β, and one lower index, say δ. Rename both β and δ by one and the same index, say ρ. Then, sum over $\rho = 0, 1, 2, 3$:

$$\sum_{\rho=0}^{3} T^{\alpha\rho}_{\gamma\rho\epsilon}.$$

This gives a new tensor: the contracted tensor. In it, ρ is just a dummy index, summed over $\rho = 0, 1, 2, 3$. As such, ρ is no longer free: you can't specify it any more. The only free indices are now α, γ, and ϵ: you can still specify them as you like. As a result, the new (contracted) tensor is smaller than the original one: it contains

$$4^3 = 64$$

entries only.

11.13.3 Einstein Summation Convention

Einstein's summation convention makes our lives easier. It says: drop the '$\sum$' sign! This gives the new shorthand:

$$T^{\alpha\rho}_{\gamma\rho\epsilon} \equiv \sum_{\rho=0}^{3} T^{\alpha\rho}_{\gamma\rho\epsilon}.$$

On the left-hand side, ρ appears not only as an upper but also as a lower index. This tells us that ρ is no longer free to specify. On the contrary: it is just a dummy index, summed over $\rho = 0, 1, 2, 3$, as in the right-hand side. This shorthand is used often in general relativity.

11.14 Tensors and Trees

11.14.1 Naive Implementation

In the above, we've considered a tensor T with five indices. This, however, is not the maximum: a tensor could be yet bigger, and use as many indices as you like. How to implement a general tensor, with arbitrarily many indices?

The naive approach is geometrical: to implement a vector, use a row of four components. Besides, store one more bit: 1, to tell us that the index is upper, not lower.

Next, to implement a matrix, use four rows, to store a total of 16 elements. Besides, store two bits: 0 and 0, to tell us that both indices are lower, not upper.

Next, introduce more and more dimensions, recursively and geometrically. This way, to implement the above tensor T, use a $4 \times 4 \times 4 \times 4 \times 4$ hypercube, containing

$$4^5 = 1024$$

entries, as required. Besides, store five more bits: 1, 1, 0, 0, and 0, to tell us that T has two upper indices, followed by three lower indices.

In practice, however, the computer must "know" the dimension in advance, to allocate sufficient memory. Therefore, recursion doesn't work any more: to have a bigger tensor with one more index, one must start from scratch, and implement new algebraic operations all over again. Is there a more efficient way?

To save work, better implement all possible tensors in one go, with as many indices as you like. For this purpose, better use a truly recursive structure: a multilevel hierarchy, or a tree.

11.14.2 Quaternary Tree

In discrete math, we've already seen a perfect binary tree (Figure 7.9). Let's extend this into a quaternary tree. For this purpose, make the following change: from each node, issue not only two but four branches.

How to do this formally? As always, use mathematical induction. A one-level quaternary tree is just a dangling node. It could be used to store a degenerate "tensor," with no index at all: a single scalar (number).

Now, for $n = 1, 2, 3, \ldots$, assume that we already know how to define an n-level quaternary tree. (This is the induction hypothesis.) At its bottom, it contains 4^{n-1} leaves, ready to store 4^{n-1} numbers (entries).

11.14.3 Implementing a Vector

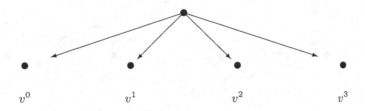

Fig. 11.12. A four-dimensional vector, implemented as a two-level tree. There are four leaves, ready to store the four components: v^0, v^1, v^2, and v^3.

For $n = 2$, for example, this is a two-level tree, with only four leaves, ready to store the components v^0, v^1, v^2, and v^3 in a given vector v (Figure 11.12). Besides, store one more bit: 1, to tell us that the index is upper, not lower.

11.14.4 Implementing a Matrix

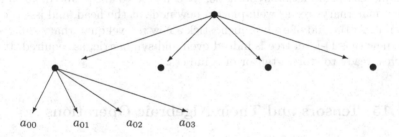

Fig. 11.13. A 4×4 matrix, implemented as a three-level tree. To reach a_{03}, pick the leftmost subtree. In it, pick the rightmost leaf. To reach a_{30}, on the other hand, work the other way around: pick the rightmost subtree. In it, pick the leftmost leaf.

For $n = 3$, on the other hand, this is already a three-level tree, with 16 leaves, ready to store the elements $a_{\mu\nu}$ in a given matrix A. In this tree, how to reach the element a_{03}, for instance? Look at Figure 11.13: pick the leftmost subtree. In it, pick the rightmost leaf. It indeed contains a_{03}, as required.

How to reach a more general element of the form $a_{\mu\nu}$ (for some $0 \le \mu, \nu \le 3$)? Look at Figure 11.14: pick the μth subtree. In it, pick the νth leaf. It indeed contains $a_{\mu\nu}$, as required.

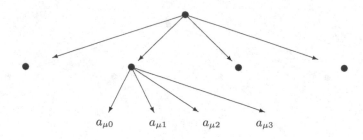

Fig. 11.14. To reach a general element of the form $a_{\mu\nu}$, pick the μth subtree. In it, pick the νth leaf.

Finally, store two more bit: 0 and 0, to tell us that both indices are lower, not upper. Next, let's extend the above: define an $(n+1)$-level quaternary tree, ready to store a tensor of n indices.

11.14.5 The Induction Step

Thanks to the induction hypothesis, we can now go ahead and define a new $(n+1)$-level quaternary tree as well: place a new node at the head, and issue four branches from it. At the end of each branch, stick a new n-level quaternary subtree. This way, the new $(n+1)$-level tree is indeed even and symmetric, as required. It contains 4^n leaves, ready to store a tensor of n indices.

11.15 Tensors and Their Algebraic Operations

11.15.1 Adding Two Tensors

Let R and Q be two tensors, with the same number of indices. How to add them to each other, entry by entry? Recursively, of course! After all, they are already implemented as quaternary trees, with the same number of levels. How many? If one, then this is easy: both R and Q are just scalars, easily added to each other. If, on the other hand, these are multilevel trees, then add them recursively, subtree by subtree.

11.15.2 Multiplying by a Scalar

Let Q be a given tensor, with as many indices as you like. Let r be a given scalar. How to multiply r times Q, entry by entry? Recursively, of course! Indeed, if Q has no index at all, then it is just a scalar, easily multiplied by r. If, on the other hand, Q is implemented as a multilevel tree, then multiply it recursively, subtree by subtree.

11.15.3   Tensor Product

Thanks to the above algorithm, we can now calculate the (outer) product of two tensors. Let R be a given tensor, with $k \geq 0$ indices (and 4^k entries). Let Q be yet another tensor, with $l \geq 0$ indices (and 4^l entries). The tensor product RQ is a new (bigger) tensor, with $k + l$ indices (and 4^{k+l} entries): each new entry is obtained as the product of an entry from R times an entry from Q.

How to calculate RQ? Recursively, of course! Indeed, if R has no index at all, then it is just a scalar, easy to multiply Q (Section 11.15.2). If, on the other hand, R is a bigger tensor, then it is implemented as a multilevel tree, with four subtrees. Each of them can now multiply Q recursively. This completes the calculation of the new tensor product RQ, as required.

11.16   Contraction

11.16.1   Substitute by a Subtree

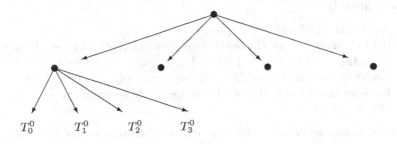

Fig. 11.15. In the algorithm Substitute(q, 1), the original tree is replaced by its own qth subtree. If $q = 0$, for instance, then the original tree is replaced by its leftmost subtree.

To help contract a tensor, we need some preparation work. In particular, we need to be able to reduce the number of levels in our tree. In other words, we want to be able to "drop" one level from our tree.

For a start, we want to be able to replace our original tree by one of its own subtrees: say, the qth subtree (for some fixed $0 \leq q \leq 3$). This is illustrated in Figure 11.15.

More than that: we may also need to replace not the entire tree but only a few subtrees in it. For example, we may need to modify the lth level (for some $l \geq 1$), and replace all subtrees in it by their own qth subsubtree. This is illustrated in Figure 11.16.

How to do this? Recursively, of course! Indeed, this is done by the following algorithm:

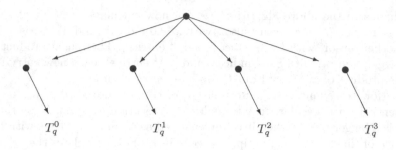

Fig. 11.16. In the algorithm Substitute($q, 2$), on the other hand, each subtree is replaced by its own qth leaf. For example, the leftmost subtree is replaced by T_q^0. In summary, the original tensor (T_ν^μ) is replaced by the new vector (T_q^μ), where q is fixed, not free.

Substitute(q, l):

1. Start from the head.
2. If $l = 1$, then replace the entire tree by its own qth subtree (Figure 11.15).
3. If, on the other hand, $l > 1$, then don't replace the entire tree. Instead, scan the four subtrees, one by one. To each subtree, apply Substitute($q, l-1$) recursively. For example, if $l = 2$, then each subtree would be replaced by its own qth subtree (Figure 11.16).

Thanks to mathematical induction, it is easy to prove that this algorithm indeed works: at the lth level, all subtrees have been replaced by their own qth subtree, as required. (See exercises below.)

What happens to the original tensor? Well, its lth index is now no longer free, but fixed: it must be the same as q. In other words, one index is gone. This will help contract the original tensor.

11.16.2 Contract The First and Second Indices

Thanks to the above algorithm, we can now contract a given tensor. For this purpose, let's contract two indices with each other: the kth index with the lth index (for some $l > k \geq 1$).

Let's start from a simple case: $k = 1$ and $l = 2$. This means that we contract the first and second indices with each other.

11.16.3 A Three-Level Tree

For instance, assume that our original tensor T has just two indices: an upper index, followed by a lower index:

$$T \equiv \left(T_\delta^\beta \right)_{0 \leq \beta, \delta \leq 3}.$$

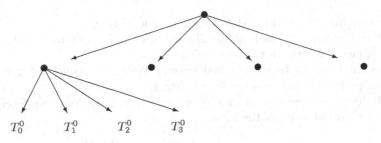

$T_0^0 \qquad T_1^0 \qquad T_2^0 \qquad T_3^0$

Fig. 11.17. To the leftmost subtree, apply Substitute$(0,1)$, to replace it by its own leftmost leaf: the scalar T_0^0.

Thanks to Einstein's summation convention, its contraction is just a scalar:

$$T_\rho^\rho \equiv T_0^0 + T_1^1 + T_2^2 + T_3^3.$$

How to calculate this number? Use the previous algorithm:

Contract$(k = 1, l = 2)$:

1. Start from the head of the original three-level tree.
2. For $i = 0,1,2,3$, look at the ith subtree, and apply to it Substitute$(i,1)$, to replace it by its ith leaf.
3. This produces a new two-level tree, with the leaves

$$T_0^0, \ T_1^1, \ T_2^2, \quad \text{and} \quad T_3^3$$

(Figure 11.17).
4. Sum them into one new leaf. This gives the required scalar: the contracted tensor.
5. Finally, use this new leaf to replace the original three-level tree.

Let's extend this algorithm further.

11.16.4 A Four-Level Tree

Let's extend this to a slightly bigger tensor, with one more (lower) index:

$$T \equiv \left(T_{\delta\epsilon}^\beta \right)_{0 \le \beta, \delta, \epsilon \le 3}.$$

Now, contract it as follows: replace both β and δ by the dummy index ρ, and sum over $\rho = 0,1,2,3$:

$$T_{\rho\epsilon}^\rho \equiv T_{0\epsilon}^0 + T_{1\epsilon}^1 + T_{2\epsilon}^2 + T_{3\epsilon}^3.$$

This leaves just one free index: ϵ.
   How to calculate this? As before:

Contract$(k = 1, l = 2)$:

1. Start from the head of the original four-level tree.
2. For $i = 0, 1, 2, 3$, look at the ith subtree, and apply to it Substitute$(i, 1)$, to replace it by its ith two-level subtree.
3. What happens to the original tensor? Well, its second index is not free any more, but the same as the first index: i.
4. Indeed, we now have a new three-level tree, with four two-level subtrees. Its leftmost subtree has the leaves

$$T^0_{00}, \ T^0_{01}, \ T^0_{02}, \quad \text{and} \quad T^0_{03}.$$

Its second subtree, on the other hand, has the leaves

$$T^1_{10}, \ T^1_{11}, \ T^1_{12}, \quad \text{and} \quad T^1_{13}.$$

Its third subtree, on the other hand, has the leaves

$$T^2_{20}, \ T^2_{21}, \ T^2_{22}, \quad \text{and} \quad T^2_{23}.$$

Finally, its rightmost subtree has the leaves

$$T^3_{30}, \ T^3_{31}, \ T^3_{32}, \quad \text{and} \quad T^3_{33}.$$

5. Add these four subtrees to each other (Section 11.15.1).
6. This produces the required two-level tree: the contracted tensor. For example, its leftmost leaf becomes now

$$T^0_{00} + T^1_{10} + T^2_{20} + T^3_{30},$$

as required.
7. Finally, use this new two-level tree to replace the original four-level tree.

We can now go ahead and introduce many more indices at the end of the original tensor: the same algorithm still works.

11.16.5 Contract With The First Index

Let's extend the above algorithm to a yet more difficult case: contract the kth and lth indices with each other, for $k = 1$ and some $l \geq 2$. This means that the first index is now being contracted with any other index (not necessarily the second one).

For example, T could now use even four indices: say, one upper, and three lower:

$$T \equiv \left(T^\beta_{\gamma\delta\epsilon} \right)_{0 \leq \beta, \gamma, \delta, \epsilon \leq 3}.$$

Thanks to Einstein's summation convention, the contracted tensor is

$$T^\rho_{\gamma\rho\epsilon} \equiv \sum_{\rho=0}^{3} T^\rho_{\gamma\rho\epsilon}.$$

(In this example, $l = 3$: the first index is contracted with the third one.)

How to calculate this? As before:

Contract$(k = 1, l \geq 2)$:

1. Start from the head of our original tree.
2. For $i = 0, 1, 2, 3$, look at the ith subtree, and apply to it Substitute$(i, l - 1)$.
3. What happens to the original tensor? Well, its lth index is now dummy: it is always the same as the first index: i.
4. From the resulting tree, pick the four subtrees, one by one, and add them to each other (Section 11.15.1).
5. This sum of subtrees is indeed the contracted tensor, as required.
6. Use it to replace the original tree.

We can now go ahead and introduce many more indices at the end of the original tensor: the same algorithm still works.

11.16.6 Contract in General

Let's extend the above algorithm to the most general case: contract the kth and lth indices with each other, for $k \geq 1$ and $l > k$. For example, T could now have even five indices: say, two upper, and three lower:

$$T \equiv \left(T^{\alpha\beta}_{\gamma\delta\epsilon} \right)_{0 \leq \alpha, \beta, \gamma, \delta, \epsilon \leq 3}.$$

Thanks to Einstein's summation convention, the contracted tensor is

$$T^{\alpha\rho}_{\gamma\rho\epsilon} \equiv \sum_{\rho=0}^{3} T^{\alpha\rho}_{\gamma\rho\epsilon}.$$

(In this example, $k = 2$ and $l = 4$, so the second index is contracted with the fourth index.)

How to calculate this? Recursively, of course:

Contract$(k \geq 1, l > k)$:

1. If $k = 1$, then use the previous version (Section 11.16.5).
2. If, on the other hand, $k > 1$, then scan the four subtrees, one by one. To each subtree, apply Contract$(k - 1, l - 1)$ recursively.

We can now go ahead and introduce many more indices at the end of the original tensor: the same algorithm still works.

11.17 Symmetric Matrix and Tensor

11.17.1 Symmetric Matrix

So far, we've used discrete math for a numerical purpose: to calculate a tensor, and its algebraic operations, with as many indices as you like. Next, let's use discrete math for yet another purpose. This will lead to an important physical consequence: Einstein equivalence principle.

To do this, let's start from a symmetric matrix. Consider a 4×4 matrix:

$$A \equiv (a_{\mu\nu})_{0 \le \mu, \nu \le 3} = \begin{pmatrix} a_{00} & a_{01} & a_{02} & a_{03} \\ a_{10} & a_{11} & a_{12} & a_{13} \\ a_{20} & a_{21} & a_{22} & a_{23} \\ a_{30} & a_{31} & a_{32} & a_{33} \end{pmatrix}.$$

Its main diagonal contains just four elements:

$$\begin{pmatrix} a_{00} & & & \\ & a_{11} & & \\ & & a_{22} & \\ & & & a_{33} \end{pmatrix}.$$

In a symmetric matrix, the main diagonal acts like a mirror: each element below it is mirrored by an element above it:

$$a_{\mu\nu} = a_{\nu\mu}, \quad 0 \le \mu, \nu \le 3.$$

For example, the following elements are the same:

$$\begin{pmatrix} & & a_{02} & \\ & & & \\ a_{20} & & & \\ & & & \end{pmatrix}.$$

For yet another example, the following elements are the same:

$$\begin{pmatrix} & & & a_{03} \\ & & & \\ & & & \\ a_{30} & & & \end{pmatrix},$$

and so on.

11.17.2 Degrees of Freedom

Thus, there is no need to store all 16 elements. It is sufficient to store ten elements only:

$$(a_{\mu\nu})_{0 \le \mu \le \nu \le 3} = \begin{pmatrix} a_{00} & a_{01} & a_{02} & a_{03} \\ & a_{11} & a_{12} & a_{13} \\ & & a_{22} & a_{23} \\ & & & a_{33} \end{pmatrix}.$$

This is the upper triangular part of A. It contains four diagonals: the main diagonal, and three more diagonals above it. Below it, on the other hand, we have three more diagonals:

$$(a_{\mu\nu})_{0 \le \nu < \mu \le 3} = \begin{pmatrix} & & & \\ a_{10} & & & \\ a_{20} & a_{21} & & \\ a_{30} & a_{31} & a_{32} & \end{pmatrix}.$$

This is the *strictly* lower triangular part of A. There is no need to store it. After all, it is mirrored by the upper triangular part, which has already been stored.

So, to define the symmetric matrix A, we have only ten degrees of freedom: we are free to pick ten elements only. By doing this, we actually specify all 16 elements, as required.

11.17.3 Counting in a Discrete Triangle

In the above, we've counted the degrees of freedom in a symmetric 4×4 matrix: ten. Let's check this. For this purpose, let's pretend we don't know the answer as yet. This will give us a more general approach, easy to extend to more complicated cases.

In our symmetric matrix, how many degrees of freedom are there? In other words, in the upper triangular part, how many elements are there? Let's mirror each such element by a point in a discrete triangle of size 3 (Figure 8.5).

More precisely, in the upper triangular part, each element has a pair of indices:

$$\mu \leq \nu.$$

How to count such pairs? In Chapter 8, Section 8.2.4, set

$$k \equiv 2$$
$$t_1 \equiv \mu$$
$$t_2 \equiv \nu$$
$$m \equiv 3.$$

Thus, the total number of such pairs is

$$\binom{m+k}{k} = \binom{3+2}{2} = \frac{5 \cdot 4}{2} = 10.$$

This is no surprise. After all, we could also count diagonal by diagonal:

$$4 + 3 + 2 + 1 = 10.$$

As a matter of fact, this is just a special case of the formula in Chapter 8, Section 8.3.6.

So, what's the big deal? To see this, let's move on to a more complicated case.

11.17.4 Symmetric Tensor

Let's extend the above to a tensor of three indices:

$$(T_{\alpha\beta\gamma})_{0 \leq \alpha, \beta, \gamma \leq 3}.$$

(For simplicity, assume that they are lower.) This way, T contains

$$4^3 = 64$$

entries. Still, in a symmetric tensor, not all of them are free to specify.

We say that T is symmetric if interchanging indices has no effect. For example, interchanging α and γ has no effect:

$$T_{\alpha\beta\gamma} = T_{\gamma\beta\alpha}.$$

Likewise, interchanging α and β (or β and γ) has no effect either. So, there is actually no need to specify all 64 entries. It is sufficient to specify only those indexed with

$$\alpha \leq \beta \leq \gamma.$$

11.17.5 Counting in a Discrete Tetrahedron

How many such triplets are there? Let's mirror each such triplet by a point in a discrete tetrahedron of size 3 (Figure 8.6).

More precisely, each triplet of the form

$$\alpha \leq \beta \leq \gamma$$

could be renamed as

$$t_1 \leq t_2 \leq t_3.$$

For this purpose, in Chapter 8, Section 8.2.4, set

$$k \equiv 3$$
$$t_1 \equiv \alpha$$
$$t_2 \equiv \beta$$
$$t_3 \equiv \gamma$$
$$m \equiv 3.$$

11.17.6 Degrees of Freedom

Thus, the total number of such triplets is

$$\binom{m+k}{k} = \binom{3+3}{3} = \frac{6 \cdot 5 \cdot 4}{6} = 20.$$

In summary, the symmetric tensor T has just 20 degrees of freedom: it is sufficient to specify just 20 independent entries, and the rest are specified automatically as well. Let's use this observation in physics.

11.18 Geometry and Physics

11.18.1 Is The Earth Flat?

In the ancient day, it was believed that the Earth was flat. After all, the land looks flat: to reach a distant place, just walk straight towards it.

Nowadays, on the other hand, we know better: the Earth is more like a ball, a little "fat" at the equator. Why does it look flat? Because this "ball" is very big relative to us.

11.18.2 Tangent Planes

In fact, at each individual place, the surface of the Earth could be approximated by a tangent plane (Figure 11.18). We humans may then think that we live on this plane, not on the round surface.

This picture uses two spatial dimensions. After all, the face of the Earth is two-dimensional: nearly a round sphere, with a (constant) positive curvature. In

Fig. 11.18. The Earth, and its tangent plane at the north pole.

particular, it is also smooth: at each individual point on it, there is a (unique) tangent plane, perpendicular to the radius.

These tangent planes are different from each other: they may have a different slope. Together, they envelope the entire Earth all over (Figures 11.7–11.8).

Let's extend this to four dimensions as well. This way, we can use the above geometrical insight in spacetime as well.

11.19 Riemann Normal Coordinates in Spacetime

11.19.1 Nearly Constant Metric

Recall that the metric is a 4×4 matrix:

$$g \equiv (g_{\mu\nu})_{0 \le \mu,\nu \le 3} = \begin{pmatrix} g_{00} & g_{01} & g_{02} & g_{03} \\ g_{10} & g_{11} & g_{12} & g_{13} \\ g_{20} & g_{21} & g_{22} & g_{23} \\ g_{30} & g_{31} & g_{32} & g_{33} \end{pmatrix}.$$

This is not a constant matrix: it depends on the event under consideration, and may change from event to event in spacetime.

The metric mustn't depend on the coordinates that happen to be used. After all, the metric is physical: it tells us how spacetime really curves. As such, it can never depend on the coordinates, which are just mathematical, not physical. Still, in each system of coordinates that describe spacetime, the metric may be written in a different way.

In the standard coordinates

$$x^0 \equiv t$$
$$x^1 \equiv x$$
$$x^2 \equiv y$$
$$x^3 \equiv z,$$

the metric takes the above form: g. In other coordinates, on the other hand, it may take a different form. Later on, we'll see how to transform from one coordinate system to another. In this context, discrete math will prove most useful.

As a matter of fact, there are no standard or favorite coordinates. After all, the coordinates are just mathematical (nonphysical) parameters, to help model spacetime.

Thus, we could actually adopt just any system of coordinates. In particular, we could adopt a new system of coordinates, in which the metric is nearly constant: its partial derivatives vanish. This way, the metric is nearly as simple as

$$\begin{pmatrix} -1 & & & \\ & 1 & & \\ & & 1 & \\ & & & 1 \end{pmatrix}.$$

This diagonal metric comes from special relativity, where no gravity is assumed at all.

11.19.2 No Self System!

These new coordinates are Riemann's normal coordinates (Sections 11.3.13 and 11.5.1). They are similar to the self coordinates, used often in special relativity. Still, there is a major difference.

In fact, Riemann's normal coordinates are only local, not global: each individual event in spacetime has its own private normal coordinates to span its own tangent space. In terms of these local coordinates, the metric is nearly constant only around this particular event, not around others. These different coordinate systems can never be "tied" together to form one global coordinate system.

11.19.3 Tangent Hyperboloids

What does this mean geometrically? Well, at each individual event in spacetime, there is a tangent four-dimensional space (Section 11.9.3). This means that spacetime is indeed smooth.

In this tangent space, gravity has no effect whatsoever. As a matter of fact, things are as simple as in special relativity. Geometrically, this means that the tangent space could split into disjoint (three-dimensional) hyperboloids. Later on, we'll discuss the physical meaning of this.

11.19.4 Constant Metric – No Curvature

In Riemann normal coordinates, the metric is nearly constant. Later on, we'll see how to design these coordinates, not globally, but only locally: around a particular event in spacetime.

Is it possible to design a coordinate system in which the metric is *exactly* constant? No! Indeed, if the metric were constant even in a tiny spot in spacetime, then there would be no curvature there at all, and no gravity either, as in special relativity. Spacetime would then be completely flat and linear, not only in this spot, but also everywhere. Indeed, just look around you. Could the Earth be flat only in a little area around you? Of course not: if it were, then it would be flat everywhere as well, which is untrue.

11.20   Transformation of Coordinates

11.20.1   Physics – Independent of Coordinates

In physics, the laws mustn't depend on the coordinates. In my lab, I may have my own time and space. You, on the other hand, may measure your own time and space in your own lab. Still, both of us must see the same nature, with the same laws.

Spacetime is real: physical, not just mathematical. Therefore, it must have the same geometry and curvature, regardless of the coordinates that are used to observe and measure them. After all, the coordinates are just mathematical artifacts, with no physical meaning whatsoever. Indeed, they are just relativistic, not absolute. For example, in my lab, I measure my own private coordinates, which seem static. From my lab, I see your coordinate system as moving (as in special relativity in Figure 11.10), or even accelerating (as in general relativity). Still, nothing is absolute: from your perspective, your system is static, while mine is moving (or even accelerating).

11.20.2   Metric – Dynamic Variables

So, the coordinates are nonphysical: they only help describe nature. What is physical? Spacetime! This is a four-dimensional manifold, whose geometry and curvature are encapsulated in the metric. Thus, the metric is physical and real: it is an integral part of spacetime, determining its shape. Still, in each coordinate system, the metric may take a different mathematical face.

The metric is a symmetric 4×4 matrix. As such, it contains ten degrees of freedom (Section 11.17.2). They are not constant: they depend on the particular event under consideration, and may change from event to event in spacetime. This is why they are called dynamic variables: they are functions of four independent variables – x^0, x^1, x^2, and x^3:

$$g_{\mu\nu} \equiv g_{\mu\nu}\left(x^0, x^1, x^2, x^3\right), \quad 0 \leq \mu \leq \nu \leq 3.$$

This is how the (nonphysical) independent variables x^0, x^1, x^2, and x^3 help describe the metric, and indeed the real physics in spacetime. Still, this is not a must: we could equally well use any other system of independent variables. The same metric would then be written in a new way: as a composite function of the new coordinates.

11.20.3   New Coordinates

How to transform? Suppose that we want to use new coordinates:

$$y^0 \equiv y^0\left(x^0, x^1, x^2, x^3\right)$$
$$y^1 \equiv y^1\left(x^0, x^1, x^2, x^3\right)$$
$$y^2 \equiv y^2\left(x^0, x^1, x^2, x^3\right)$$
$$y^3 \equiv y^3\left(x^0, x^1, x^2, x^3\right).$$

(These round parentheses are often dropped for short.) This transforms the old coordinates into new ones. We assume that this transformation is invertible: the new coordinates could be transformed back to the old ones smoothly (Sections 11.3.9–11.3.11).

11.20.4 Partial Derivatives

We can now go ahead and calculate the partial derivative of any old coordinate with respect to a new one. For example, the old time $x^0 = t$ could be differentiated with respect to the new "time" y^0:

$$\frac{\partial x^0}{\partial y^0} \equiv \frac{\partial x^0}{\partial y^0} \left(y^0, y^1, y^2, y^3 \right).$$

(These round parentheses are often dropped for short.) To obtain this partial derivative, hold y^1, y^2, and y^3 fixed, and differentiate with respect to y^0 only (Chapter 25, Sections 25.10.1 and 25.12.1). Let's go ahead and use this.

11.20.5 Transformation of The Metric

In terms of the new coordinates, how does the metric look like? For this purpose, one could write the original metric g as a composite function of the new coordinates. This, however, could still be nonlinear and complicated. Is there a more direct way?

In the new coordinates, the metric takes a new mathematical face: it is no longer called g, but h. This is a new symmetric 4×4 matrix, written as a function of the new coordinates.

To have h explicitly, let's index it in new (lower) indices – α and β. The new matrix h is then obtained from the formula

$$h_{\alpha\beta} \equiv \frac{\partial x^\mu}{\partial y^\alpha} \cdot \frac{\partial x^\nu}{\partial y^\beta} \cdot g_{\mu\nu}, \qquad 0 \le \alpha \le \beta \le 3.$$

Here, Einstein's summation convention has been used twice. This way, μ and ν are now just dummy indices, summed over $\mu, \nu = 0, 1, 2, 3$. Only α and β are free. Furthermore, they could interchange with each other, with no effect whatsoever. This proves that h is indeed a symmetric 4×4 matrix as well, and contains only ten degrees of freedom, as required (Section 11.17.2).

11.21 Riemann Normal Coordinates and Einstein's Happy Thought

11.21.1 Nearly Constant Metric

Consider a particular event in spacetime. To describe it, use the new coordinates. Around the event, draw a small spot (a little four-dimensional domain in spacetime). Could the metric be constant there? For example, could h have the constant form

$$h = \begin{pmatrix} -1 & & & \\ & 1 & & \\ & & 1 & \\ & & & 1 \end{pmatrix}$$

there? In other words, if I "lived" in the spot, could I feel no gravity at all, as in spacial relativity?

Of course not! We've already discussed this, both physically and geometrically. Let's support this algebraically as well.

Indeed, let's try and design our new coordinates in such a way that h indeed has this form. After all, we haven't specified the new coordinates as yet.

In our spot, let's focus on our original event. At this event, could we force h to have this simple form? Easy: in Section 11.20.5, the symmetric matrix h is defined in just ten equations. These are not too many. Indeed, to satisfy them, we have freedom to pick as many as 16 unspecified "unknowns:"

$$\frac{\partial x^\mu}{\partial y^\alpha}, \quad 0 \le \alpha, \mu \le 3.$$

Why are these called "unknowns?" Becuase we haven't defined the new coordinates as yet, so the partial derivatives with respect to them are still unspecified. Later on, we'll define the new coordinates cleverly, with partial derivatives that make h constant, as in the above diagonal matrix.

This is good: more unknowns than equations [36]. Thus, it is easy to specify the unknowns properly, to make h constant, as in the above diagonal form.

This is done at our event. Still, this doesn't guarantee that h has the same form in the entire spot. After all, h may change from event to event. Still, we could force h to be *nearly* constant in some little spot. For this purpose, h must have zero partial derivatives at our original event. To achieve this, at our event, differentiate the ten equations

$$h_{\alpha\beta} \equiv \frac{\partial x^\mu}{\partial y^\alpha} \cdot \frac{\partial x^\nu}{\partial y^\beta} \cdot g_{\mu\nu}, \quad 0 \le \alpha \le \beta \le 3,$$

with respect to y^δ ($0 \le \delta \le 3$). This gives 40 new equations.

How many unknowns are there? Well, we now have new unknowns: the *second* partial derivative of an old coordinate, with respect to two new coordinates:

$$\frac{\partial^2 x^\mu}{\partial y^\alpha \partial y^\delta}, \quad 0 \le \mu \le 3,\ 0 \le \alpha \le \delta \le 3.$$

Why are these called "unknowns?" Because we haven't defined our new coordinates completely as yet. After all, we've only determined how the old coordinates differentiate with respect to them. This only determines the first partial derivatives, not the second ones!

Now, these new unknowns are symmetric in terms of α and δ. Indeed, these indices could interchange, with no effect whatsoever (Chapter 25, Sections 25.10.2 and 25.12.7). So, these indices could be picked in ten different ways. On top of that, μ could be picked in four different ways. So, in total, we have 40 unknowns, just enough to solve our 40 equations.

11.21.2   Tangent Space

By picking our new coordinates cleverly, we can therefore make sure that these new unknowns are indeed as required to force h to have zero second partial derivatives at our original event. This is good: in a little neighborhood around our event, h hardly changes, and nearly keeps the same diagonal form.

11.21.3   Einstein's Happy Thought

These are Riemann normal coordinates. In them, spacetime is smooth: well-approximated by a flat linear tangent space. Thus, at the original event, things are as in special relativity: you'd feel as in a free fall, with no force or acceleration at all, as in Einstein's happy thought.

11.21.4   Riemann Normal Coordinates: Local and Implicit

There is no need to calculate Riemann normal coordinates explicitly: they can remain theoretical and implicit. Furthermore, they are not global, but only local. Indeed, in the above algorithm, they have been calculated separately for each individual event in spacetime. Thus, they depend on the event under consideration, and are defined only in a small neighborhood around it.

11.21.5   Is There a Flat Spot in Spacetime?

By now, we've got what we wanted: in terms of Riemann normal coordinates, h has zero first partial derivatives at our event, and is therefore nearly constant in a small neighborhood around it. Still, could we have more? Could we force h to have zero *second* partial derivatives as well? Not any more! Indeed, let's differentiate the above ten equations twice: not only with respect to y^δ but also with respect to y^η. Fortunately, in partial differentiation, order doesn't matter. Therefore, we only need to consider ten cases: $0 \le \delta \le \eta \le 3$. So, in total, we have 100 equations to solve.

How many unknowns are there? Well, the new unknowns are now of the form

$$\frac{\partial^3 x^\mu}{\partial y^\alpha \partial y^\delta \partial y^\eta}, \quad 0 \le \mu \le 3,\ 0 \le \alpha \le \delta \le \eta \le 3.$$

(After all, in partial differentiation, order doesn't matter.) So, we actually have here four symmetric tensors, or a total of 80 unknowns (Section 11.17.6).

This is not enough! After all, 100 is more than 80. Therefore, in general, we can't force h to be constant, not even in a tiny neighborhood of our original event.

11.21.6   Spacetime: Smooth, But Never Flat

This is no surprise: spacetime behaves just like our original example: the sphere (Section 11.18.1). Indeed, we already know a lot about the sphere: it is completely round, with positive (constant) curvature everywhere. As such, it contains no flat area at all. Still, it is smooth: each point on it has a (unique) tangent plane. In principle, spacetime is the same: smooth everywhere, but flat nowhere.

11.22   Special vs. General Relativity

11.22.1   Flat Spacetime – Constant Metric

In general, 80 unknowns are not enough to solve 100 equations. Still, in one special case, they are. When? When some of these equations are duplicate copies of others. In other words, they are not entirely independent of each other.

In this case, one could keep differentiating the above ten equations (where h is defined) more and more, use discrete math to count the equations, and make sure that they are not entirely independent of each other, so there are enough unknowns to solve them. This way, at our original event, all partial derivatives of h vanish, even the high-order ones. This means that h must be constant in some (tiny) spot around our event. What does this mean geometrically? No curvature at all in the entire spot!

Could spacetime contain such a spot? As discussed above, this would lead to a rather strange spacetime: completely flat. This could happen only theoretically in special relativity (Figure 11.10), but not in real physics: in general relativity, gravity is felt everywhere, and spacetime is curved everywhere, not flat.

11.22.2 Einstein Equivalence Principle

Thanks to Riemann normal coordinates, spacetime is smooth. Indeed, locally, spacetime is nearly flat: at each individual event, it has a tangent space.

This is a good geometrical property: in terms of Riemann normal coordinates, spacetime is nearly flat. More precisely, it is nearly hyperbolic: the metric is nearly as simple as

$$\begin{pmatrix} -1 & & & \\ & 1 & & \\ & & 1 & \\ & & & 1 \end{pmatrix}.$$

This is indeed the metric used in special relativity, where no gravity is assumed at all.

What does this mean physically? Well, if you "live" at this event, then you'd feel no force at all! After all, around this event, there is a (tiny) spot where the metric is nearly as in spacial relativity. Relative to the other events in the spot, you'd have nearly the same metric: there is hardly any change.

To you, the spot is just too big to feel any curvature. Just like you think the Earth is flat, you'd think the spot is flat as well. Physically, you are actually in a free fall.

At the event, the new system of Riemann normal coordinates encapsulates all acceleration due to gravity. This is why you feel no force at all: you are carried effortless by the new coordinates. After all, you could "live" in the new coordinates, and ignore how they were formed in the first place. Just relax, enjoy your new coordinates, and disregard how spacetime really curves or changes around you.

After all, you have no other coordinates to compare to. On the contrary: to you, Riemann normal coordinates are most natural and realistic. In fact, to you, they are as straight and linear as ever. This is why you feel as in a free fall. Why transform? Better stick to these lovely coordinates!

Now, there is nothing special about our original event. The same could be done in each and every event in spacetime. This way, around the next event, design a new tiny spot, with slightly different Riemann normal coordinates. They are still local: with respect to them, the new event lies at the origin: $(0,0,0,0)$. This could be done in each and every event in spacetime.

This is indeed Einstein equivalence principle (or happy thought): your local coordinates could take on the job of encapsulating gravity. In terms of them, you'd then feel no gravity at all!

Historically, things happened the other way around. In the beginning, Einstein used a thought experiment to discover his equivalence principle. Later, he used differential geometry to develop the complete mathematical theory of general relativity and gravity.

11.23 Anti-Symmetric Matrix and Tensor

11.23.1 Anti-Symmetric Matrix

Assume now that our matrix A is not symmetric but anti-symmetric. This means that interchanging indices picks a minus sign:

$$a_{\mu\nu} = -a_{\nu\mu}, \quad 0 \leq \mu, \nu \leq 3.$$

For this reason, each main-diagonal element must vanish. After all, it is the same as the minus of itself. Thus, it is sufficient to store the *strictly* upper triangular part:

$$(a_{\mu\nu})_{0 \leq \mu < \nu \leq 3} = \begin{pmatrix} & a_{01} & a_{02} & a_{03} \\ & & a_{12} & a_{13} \\ & & & a_{23} \\ & & & \end{pmatrix}.$$

The rest of the elements, on the other hand, don't have to be stored. After all, we already know what they are: on the main diagonal, they vanish. Below it, on the other hand, they are anti-mirrored by the (strictly) upper triangular part.

11.23.2 How Many Degrees of Freedom?

How many degrees of freedom are there? In other words, how many elements are there in the *strictly*s upper triangular part? Well, each such element is indexed by

$$\mu < \nu.$$

How many such pairs are there? To count, let's use a general method, easy to extend to more complicated cases as well. In Chapter 8, Section 8.2.3, set

$$k \equiv 2$$
$$s_1 \equiv \mu + 1$$
$$s_2 \equiv \nu + 1$$
$$n \equiv 3 + 1 = 4.$$

This way,

$$1 \leq s_1 < s_2 \leq n,$$

as required. Thus, the total number of such pairs is

$$\binom{n}{k} = \binom{4}{2} = \frac{4 \cdot 3}{2} = 6.$$

Indeed, in the *strictly* upper triangular part, there are exactly six elements:

$$a_{01}, \ a_{02}, \ a_{03}, \ a_{12}, \ a_{13}, \quad \text{and} \quad a_{23}.$$

11.23.3 Anti-Symmetric Tensor

Let's extend the above to a tensor of three indices:

$$(T_{\alpha\beta\gamma})_{0\leq\alpha,\beta,\gamma\leq3}\,.$$

(For simplicity, assume that all indices are lower.) We say that T is anti-symmetric if interchanging two indices picks a minus sign. For example, interchanging α and γ leads to

$$T_{\alpha\beta\gamma} = -T_{\gamma\beta\alpha},$$

and so on.

So, many entries must vanish: an entry that has two identical indices must be the minus of itself, or just zero. As a matter of fact, it is sufficient to specify those entries indexed with

$$\alpha < \beta < \gamma.$$

11.23.4 How Many Degrees of Freedom?

How many such triplets are there? In Chapter 8, Section 8.2.3, set

$$k \equiv 3$$
$$s_1 \equiv \alpha + 1$$
$$s_2 \equiv \beta + 1$$
$$s_3 \equiv \gamma + 1$$
$$n \equiv 3 + 1 = 4.$$

This way,

$$1 \leq s_1 < s_2 < s_3 \leq n,$$

as required. Thus, the total number of such triplets is

$$\binom{n}{k} = \binom{4}{3} = 4.$$

In summary, the anti-symmetric tensor T has just four degrees of freedom:

$$T_{012}, \ T_{013}, \ T_{023}, \quad \text{and} \quad T_{123}.$$

11.24 Towards Bundles and Lie Groups

11.24.1 Heraclitus: Dynamics and Relativity

To carry out an experiment in physics, you need a lab, with four coordinates: your own (proper) time, plus three spatial coordinates. This is indeed relativity: your lab is not static, but dynamic: it travels with the entire Earth around the sun, which travels around the center of the milky way. In fact, there is no such thing as "static:" a position is only relativistic, not absolute. As Heraclitus said, everything is moving and flowing.

11.24.2   Local Experiment in The Lab

Furthermore, your lab is only local: it focuses on a particular event, here and now, and on its immediate neighborhood in spacetime.

In your lab, how is the event specified? This is done in terms of your own coordinates: your (proper) time, and your position. Still, the result of your experiment mustn't depend on the coordinates, which are just mathematical, not physical. This is why tensors are so useful: they have a-priori rules that tell us how to transform them from one coordinate system to another. Fortunately, the results of the experiment could often be placed in a tensor, and transform in a compatible way. This way, although they may change numerically, they still have the same physical (and geometrical) meaning.

11.24.3   Metric and Shortest Path in Spacetime

In general relativity, the dynamic variables are placed in a symmetric 4×4 tensor: the metric (Section 11.20.2). This has not only physical but also geometrical meaning: it tells us what is the shortest path between two events in spacetime. The integration along the path is carried out with respect to the proper time, measured in a dynamic clock, carried along the path. To discover this optimal path, one must use the Euler–Lagrange equations [36].

The metric is then designed locally to fit these equations. Thanks to the metric, one can then define the curvature, to tell us how curved spacetime is at each individual event. Again, this is a local and nonconstant property: it depends on the particular event under consideration, and may change from event to event in spacetime.

Furthermore, thanks to the metric, one can also define the covariant derivative, to help differentiate along the curved spacetime. This helps introduce Einstein's equations, which tell us the whole story of general relativity: how the metric is related to the energy and the momentum throughout spacetime [2, 4, 9, 38, 11, 36]. This is how geometry is used fully to explain physics!

11.24.4   High-Dimensional Bundle

In [41], on the other hand, there is a yet deeper interpretation: not only local but also global. For this purpose, embed spacetime in a high-dimensional manifold: the principal bundle, mirrored by the associated bundle.

How to do this? To each individual event in spacetime, attach a complete Lie group, to model all possible (smooth) coordinate systems. This group is also mirrored by a yet simpler structure: a Lie algebra (discussed later in the book).

Still, this new mathematical structure may seem a bit artificial and unnecessary. After all, it is designed to fit the physics we already know... So, don't expect it to help in the practical calculations.

Moreover, to a physicist, this approach may look like fancy mathematics. After all, we are soon going to project the high-dimensional bundle back to spacetime, and filter out the entire Lie group, as if it had never been. This way, all different coordinate systems are lumped and united again, and considered as one and the same. So, why was all this necessary? Why not stick to good old spacetime?

11.24.5 Vertical Projection Back to Spacetime

Still, the new bundle may give us a lot of geometrical intuition: onto each individual event in spacetime, project vertically, along the "vertical" Lie group. The remaining (horizontal) component, on the other hand, aligns with the original spacetime. Along it, one can then define the curvature, to tell us how curved spacetime is at this particular event.

Furthermore, one can also differentiate in the horizontal direction: this is the covariant derivative along the curved spacetime. Upon projecting back to spacetime, we also obtain a useful field: the Young-Mills field.

11.24.6 Gauge Transformation and The Homology Group

A shift shouldn't affect differentiation. If you change the initial time, or the location of the origin, then this should have no effect on any partial derivative with respect to time or space. In other words, the coordinate system should be insensitive to any shift by a constant vector. Such a shift is often called a gauge transformation. Why? Because it helps look at things from yet another angle. After all, coordinates like time and space could be viewed as observables: they could be measured in your system, and tell you something about physics (as observed from this perspective).

Together, these shifts make a new group: the homology group. As a matter of fact, our Lie group is far bigger: it contains not only shifts but also many other (compatible) transformations of coordinates. This is indeed a more general kind of gauge transformation. Why? Because new coordinates are just a new way to look at things: physics (and indeed geometry) remains the same.

Geometrically, how does such a transformation look like? Just "jump" vertically in the bundle, and reach a new element from the Lie group: a new coordinate system. Physically, nothing is changed. This is why you are still above the same event in spacetime. Still, now you look at things from another perspective: new time, and new spatial coordinates. This is why you may get new results in your experiments. Still, this is just a numerical change, not physical or geometrical.

11.24.7 Could Bundles Really Help?

Although they gives no computational gain, bundles may still help interpret spacetime topologically better. Indeed, once the differential details are shed off, the topology takes a simple algebraic face: the holonomy group. Furthermore, thanks to the underlying Lie algebra, physics gets as simple as ever. For example, the Schur lemma helps introduce angular momentum and spin. Moreover, Weyl groups and Dynkin diagrams help sort Lie algebras, and model interesting physical phenomena.

Later on, in quantum mechanics, we'll look at the position and momentum operators [5, 7, 19, 37]. Thanks to bundles, one could even reverse-engineer the Young-Mills field, the change the coordinates properly, with no effect on physics or geometry at all.

Later on, we'll also introduce Hamiltonian mechanics. In this vein, we'll use the Poisson brackets to design a new Lie algebra, and obtain Noether's theorem for free. This may help understand conservation laws from a geometrical point of view: pick a conserved observable, and mirror it by a kinematic symmetry of the Hamiltonian: an isometric direction, in which the metric remains unchanged.

11.25 Exercises: Trees and Tensors

11.25.1 Mathematical Induction in a Tree

1. Let v be a four-dimensional vector. What tree could be used to store it? Hint: a two-level tree, with four leaves, to store the components v^0, v^1, v^2, and v^3 (Figure 11.12).

2. How to store the information that the index is upper, not lower? Hint: for this purpose, store one more bit: 1.

3. Let A be a 4×4 matrix. What tree could be used to store it? Hint: a three-level tree, with four subtrees, each being a two-level tree as above.

4. How to reach the element a_{03}? Hint: see Figure 11.13.

5. On the other hand, how to reach the element a_{30}?

6. In general, for given $0 \leq \mu, \nu \leq 3$, how to reach the element $a_{\mu,\nu}$ (or $a_{\mu\nu}$, with no comma)? Hint: see Figure 11.14.

7. How to store the information that both μ and ν are lower, not upper? Hint: for this purpose, store two bits: 0 and 0.

8. Let T be a big tensor, with five indices (Section 11.13.1). What tree could be used to store it? Hint: a six-level quaternary tree.

9. How to store the information that T has two upper indices, followed by three lower indices? Hint: for this purpose, store five bits: 1, 1, 0, 0, and 0.

10. In this T, is it allowed to contract the first and second indices with each other? Hint: no! They are both upper.

11. Furthermore, is it allowed to contract the third and fourth indices with each other? Hint: no! They are both lower.

12. On the other hand, is it allowed to contract the second and fourth indices with each other? Hint: yes! One is upper, and the other is lower, as required.

13. Use mathematical induction to show that the recursive algorithm in Section 11.15.1 indeed works, and adds two tensors (of the same number of indices) to each other. Hint: for $n = 0$, this is just a scalar addition. For $n = 0, 1, 2, 3, \ldots$, on the other hand, assume that the algorithm indeed works for n indices (or $n + 1$ levels). (This is the induction hypothesis.) Now, for $n + 2$ levels, the subtrees have $n + 1$ levels each, so they are indeed added properly.

14. Use mathematical induction to show that the recursive algorithm in Section 11.15.2 indeed works, and multiplies a tensor by a scalar. Hint: for $n = 0$, this is just scalar times scalar. For $n = 0, 1, 2, 3, \ldots$, on the other hand, assume that the algorithm indeed works for n indices (or $n+1$ levels). (This is the induction hypothesis.) Now, for $n + 2$ levels, each subtree has $n + 1$ levels, and can indeed be multiplied recursively.

15. Use mathematical induction on k to show that the recursive algorithm in Section 11.15.3 indeed works, and calculates the tensor product RQ. Hint: for $k = 0$, this is just scalar-times-tensor, which we already know how to do. For $k = 0, 1, 2, 3, \ldots$, on the other hand, assume that the algorithm indeed works when R has k indices (or $k + 1$ levels). (This is the induction hypothesis.) Now, when R has $k+2$ levels, its subtrees have $k+1$ levels each, and can therefore multiply Q recursively.

16. Use mathematical induction on l to show that the recursive algorithm in Section 11.16.1 indeed works, and sets the lth index to a fixed instance: q. Hint: for $l = 1$,

the algorithm indeed picks the qth subtree, as required. For $l = 1, 2, 3, 4, \ldots$, on the other hand, assume that the algorithm indeed works for the lth index. (This is the induction hypothesis.) Thanks to this, it can now set the $(l + 1)$st index as well: scan the four subtrees, one by one, and set their lth index to q.

17. Show that the algorithm in Section 11.16.5 indeed works, and contracts the first index with the lth index ($l \geq 2$). Hint: first, it sets the lth index to have the same value as the first index: $i = 0, 1, 2,$ or 3. Finally, it scans the four subtrees, and sums them up, as required.

18. Use mathematical induction on k to show that the recursive algorithm in Section 11.16.6 indeed works, and contracts the kth index with the lth index ($l > k \geq 1$). Hint: for $k = 1$, see the previous exercise. For $k = 1, 2, 3, 4, \ldots$, on the other hand, assume that the algorithm indeed works for some k (for every $l > k$). (This is the induction hypothesis.) Thanks to this, it also works well for $k + 1$ (for every $l + 1 > k + +1$). Indeed, it scans the foru subtrees, one by one, and contracts their own kth and lth indices with each other, as required.

19. Use the above algorithm to contract the tensor T as in Section 11.13.2. Hint: use $k = 2$ and $l = 4$ to contract the second and fourth indices with each other.

20. In the previous exercise, how does the recursion look like? Hint: in each subtree, it contracts the first and third indices with each other, using the version in Section 11.16.5.

21. How many degrees of freedom are there in a $4 \times 4 \times 4 \times 4$ tensor? Hint: all entries are free to specify: $4^4 = 256$.

22. How many degrees of freedom are there in a $4 \times 4 \times 4 \times 4$ symmetric tensor? Hint: 35 (Section 11.17.6).

23. How many degrees of freedom are there in a $4 \times 4 \times 4 \times 4$ anti-symmetric tensor? Hint: just one (Section 11.23.4).

24. What is it? Hint: T_{0123}. Indeed, all the others follow from it by interchanging indices and picking a minus sign.

25. Consider a sphere. Could it contain a flat area?

26. Consider a smooth two-dimensional surface, such as an ellipsoid or a hyperboloid. Could it contain a flat area, with no curvature at all?

27. What is the curvature in an ellipsoid? Hint: positive everywhere.

28. What is the curvature in a hyperboloid? Hint: negative everywhere.

29. What is spacetime? Hint: a four-dimensional differentiable manifold.

30. Could spacetime contain a flat spot, with no curvature at all? Hint: if it did, then the entire spacetime would have to be flat. This is science fiction: the theoretical laws of special relativity override the realistic laws of general relativity.

31. What is a vector field on the Cartesian plane? Hint: see Figure 11.6.

32. Is it also a vector function? Hint: yes, it maps $\mathbb{R}^2$ into $\mathbb{R}^2$. Indeed, to each individual point, assign one arrow.

33. What is a vector field on the sphere?

34. Is it also a function? Hint: no! The arrow are not in the same plane any more. Indeed, to each individual point on the sphere, it assigns an arrow in a different tangent plane (or local chart).

35. What are the advantages and disadvantages of our generating system? Hint: on one hand, it helps represent just any (smooth) vector function on the sphere. On the other hand, this representation is not unique.

36. In the new coordinates in Section 11.20.5, show that the metric h is still symmetric:

$$h_{\alpha\beta} = h_{\beta\alpha}, \quad 0 \le \alpha, \beta \le 3.$$

Hint: thanks to Einstein summation convention, the dummy indices μ and ν could interchange, with no effect.

Part V

Introduction
to Quantum
Physics and Chemistry

Introduction to Quantum Physics and Chemistry

In this part, we introduce quantum mechanics in simple algebraic tools: matrices, eigenvalues, and eigenvectors. This way, physical quantities can be modeled nondeterministically. Why is this important? Because, in the microscale, this is the true physical state. This way, each physical observable is mirrored by its own matrix. This is relevant to fundamental physical quantities: position, momentum, energy, angular momentum, and spin.

Why is a matrix suitable to model a physical observable? Because, in general, matrices don't commute: their commutator is nonzero. This indeed models quantum mechanics: Order does matter: observing one physical quantity before the other gives different results from observing the latter before the former. Later on, we'll put this in a wider algebraic context: Lie algebras.

This helps at a very small scale, in which physics is no longer classical but stochastic. Consider, for example, the electrons in the atom. Thanks to quantum mechanics, this is modeled by the Hartree-Fock system. For this purpose, however, we need another algebraic tool: the determinant of a matrix.

Once the determinant is used properly, although the energy is no longer known for certain, it can still be estimated nondeterministically: as a random variable. As such, we can still study its expectation and variance.

Thanks to the determinant, we can write the expected energy, and obtain the Hartree-Fock system: a pseudo-eigenvalue problem [1, 12, 18]. In this system, each (pseudo) eigenvector represents the distribution of electrons in the atom at a particular energy level.

Introduction
to Quantum Mechanics:
Energy Levels and Spin

In a very small scale, classical physics doesn't work any more. On the contrary: quantum mechanics teaches us new laws of nature. For example, order does matter: measuring position before momentum is not the same as measuring momentum before position. How to model this algebraically? We need a group with no commutative law: a non-Abelian group. Fortunately, we already have such a group: the nonsingular matrices. Indeed, two matrices often have a nonzero commutator.

In other words, in matrices, multiplying from the left is not the same as multiplying from the right. How different could this be? To measure this, we have the commutator. Later on, we'll use the commutator in Lie brackets in a new Lie algebra, to help model interesting quantum-mechanical phenomena like polarization and spin.

The commutator is thus the key algebraic tool to develop quantum mechanics. In this theory, we redefine momentum , energy, and angular momentum from scratch, in their stochastic (probabilistic) face. Thanks to the Fourier transform, we also obtain Planck's, Debroglie's, and Schrodinger's laws for free.

In Chapter 2, we've already defined angular momentum classically. This is good enough for practical engineering problems. In a very small scale, however, it must be redefined stochastically, in terms of matrices and their commutator. This may help study a few elementary particles like electrons and photons, with their new property: spin. This property is not well-understood physically, but only mathematically, in terms of matrices. Later on, we'll put this in a yet wider algebraic context: Lie algebras.

Quantum mechanics is not easy to grasp logically or intuitively. At best, it could be understood formally or mathematically, but not physically. Still, it helps make useful calculations and predictions in practice.

After learning about it, ask yourself: do I understand quantum mechanics? If you do, then something must be wrong. If, on the other hand, you are still puzzled, then you are on the right track.

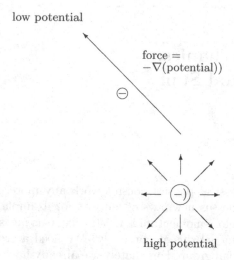

Fig. 12.1. The big charge produces a radial electrostatic field: high potential at the middle, and low potential far away. The electron tends to go to the low-potential region, to make its (positive) energy minimal.

12.1 Potential in an Electrostatic Field

12.1.1 Electrostatic Field

Classical mechanics is useful to model motion and other physical phenomena in an intermediate scale, and help solve many kinds of practical engineering problems. Still, we've already seen its limits: in a very large (astronomic) scale, it is no longer valid: relativity is necessary. Likewise, in a very small (subatomic) scale, it fails again. To see this, we need some background in electromagnetics.

Consider a static charged ball (Figure 12.1). The charge in it produces an electrostatic field, whose flux is illustrated as outgoing arrows.

At each individual point, the arrows diverge: their divergence tells us the amount of new flux produced there. Near the ball, the arrows are crowded, so the divergence is high. Far away, on the other hand, the arrows are more and more spread out, so the divergence is lower and lower. (See exercises below.)

12.1.2 Potential: Divergence of Flux

This is the potential: capacity to do work. Indeed, if we threw an electron into the field, then it would have a positive potential energy: capacity to use the electrostatic field to do work and travel farther and farther away. Like a spring that was suddenly released, the electron would travel as far as it can: to infinity. There, the potential is minimal, and the electron takes its minimal potential energy.

12.1.3 Force: Towards Minimal Energy

This could be described not only in terms of energy but also in terms of force. In these terms, the electron "feels" a new radial force, pointing away from the ball. Geometrically, this force is opposite to the gradient of the potential. Indeed, the gradient is a general form of a derivative: it points in the direction of increasing potential. The force, on the other hand, acts towards the opposite direction.

If, on the other hand, we threw not an electron but a proton into the field, then it would behave the other way around. After all, thanks to its positive charge, it would now have a negative energy, not positive. This would attract it to the ball, where the potential is high, and its (negative) potential energy is minimal.

The same principles work not only in an electric but also in a magnetic field. This time, however, our electron will have a new property: spin. Instead of electron and proton, we'll now consider spin-up and spin-down electrons. Spin-up means spinning counterclockwise around the positive part of the vertical z-axis, and spin-down means spinning clockwise.

12.2 Spin in a Magnetic Field

12.2.1 Potential and Force

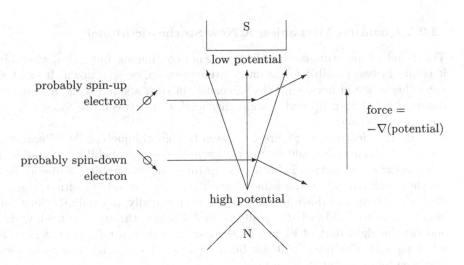

Fig. 12.2. Electrons in a nonuniform magnetic field. A spin-up electron aligns with the field, so it tends to go up (towards the low-potential region), to make its (positive) energy minimal. A spin-down electron, on the other hand, tends to go down (towards the high-potential region), to make its (negative) energy minimal.

Now, instead of the electric field, consider a new magnetic field (Figure 12.2). This time, the arrows point upwards: from the north magnetic pole below, towards the south magnetic pole above. On their way up, the arrows get more and more spread out. Thus, their divergence (or the potential) also gets lower and lower on the way up.

Now, into this field, throw tow kinds of electrons: a spin-up electron aligns with the field, so it has a positive magnetic energy. Therefore, it tends to go up, to minimize this kind of energy. In other words, it "feels" a force upwards: in the decreasing-potential direction.

A spin-down electron, on the other hand, behaves the other way around. Its magnetic energy is negative, not positive. To minimize it, it tends to go down, towards the high-potential region (Figure 12.2).

12.2.2 The Stern-Gerlech Experiment

This is indeed the Stern-Gerlech experiment: into a nonuniform magnetic field, let a stream of electrons flow from the left. Each electron may spin around its own axis of rotation that makes some angle with the positive z-direction: some electrons spin up (around an axis of rotation at angle $0°$ from the positive z-direction), some spin down (at angle $180°$), and the rest spin obliquely, at any angle in between. So, upon traveling rightwards through the field, the spin-up electrons must feel a strong force upwards, the spin-down ones must feel a strong force downwards, and the rest must feel a moderate force upwards or downwards. Thus, on the right, we expect electrons to arrive at a range of angles, say from $60°$ to $120°$.

Surprise surprise: the electrons arrive at two possible angles only: either $60°$ (spin-up electrons) or $120°$ (spin-down electrons). What happened to all others in between?

12.2.3 Quantum Mechanics: A New Stochastic Model

This is indeed quantum mechanics. Spin is not continuous, but only discrete. Indeed, it comes in two possible forms only: either spin-up, or spin-down. It can't spin in any other angle in between. Why? Because, in such a small scale, nature can only distinguish between up and down: other angles in between are too subtle to have any physical effect.

Still, the electrons in Figure 12.2 seem to spin obliquely. Why? Because spin is a random variable: it could be up, or down, at some probability. Nature is nondeterministic: each electron is both spin-up (at some probability) and spin-down (at another probability) at the same time. This is called superposition: the electron "spins" both up and down at the same time (not really, but only stochastically). It doesn't have to decide what it really is until it passes through the field. Only in the end (at the right part of Figure 12.2) must the electron tell us what it really was all along: either spin-up (if it has been found at the top-right) or spin-down (if it has been found at the bottom-right).

12.2.4 What Is Spin?

Thus, spin is nothing like classical spin (Chapter 2). We only *name* it spin because it *looks* or *behaves* like spin. Still, nobody has ever seen an electron spinning, neither up nor down. We only think of it as spinning (up or down) because this fits the place it has been found in the Stern-Gerlech experiment (either at the top-right or at the bottom-right in Figure 12.2).

We humans need a suitable term to fit our imagination and intuition. The term "spin" is indeed suitable: we can now think of the electron as spinning (up or down), to explain the results of the Stern-Gerlech experiment.

This approach is not new: we've already seen it before. For example, in Newtonian mechanics, momentum is defined as mass times velocity. Still, what about a massless photon in special relativity? How could it ever have a nonzero momentum? Well, it only has a property that *looks* and *behaves* like momentum. Therefore, it makes sense to call it momentum as well. This way, in special relativity, momentum takes a more general face: a new physical quantity, more fundamental than both mass and velocity.

Likewise, in quantum mechanics, spin is a new physical quantity, with only two possible values: either up or down. It makes sense to call it spin, because it behaves like a rigid body, spinning either up or down. Still, a rigid body could also spin at any angle in between. An electron, on the other hand, can't. It only spins obliquely in a stochastic sense, until being looked at in the lab, as in the right part of Figure 12.2. At that time, it must tell us the "truth:" it really spinned up (or down) all along.

This way, spin takes a more general and fundamental interpretation. It is no longer a continuous variable, but a discrete random variable, with only two legitimate values. Later on, we'll use algebraic tools to understand this better, geometrically and visually: Lie algebras.

12.2.5 Right-or-Left Spin

In the top-right in Figure 12.2, we have a stream of spin-up electrons. Let's focus on these electrons only: let them pass through a different Stern-Gerlech device, whose magnetic field is oriented rightwards rather than upwards. Our electrons must now make their mind once again: do they spin right or left?

But didn't we already say that they spin up? Well, this only means that they are *equally likely* to spin right or left. Therefore, the new Stern-Gerlech device will still affect them: about half of them will turn right, and the rest will turn left.

In our new stochastic model, left-or-right spin is a new random variable. With respect to it, our electrons have a neutral state: a superposition of right and left spin. In a sense, they spin both right and left at the same time. In stochastic terms, on the other hand, they are equally likely to spin right or left. This is why the second Stern-Gerlech device does affect them: it makes about half of them turn left, and the rest turn right.

Later on, we'll see that this experiment is just a special case. In general, the second magnetic field could be oriented at any angle θ away from the positive z-direction. In this case, however, our electrons are no longer neutral: they'd turn left at probability $\cos^2(\theta/2)$, or right at probability $\sin^2(\theta/2)$. This is an experimental result. Later on, we'll use Lie algebras to explain it algebraically and geometrically. Before doing this, let's see yet another paradox that helped develop quantum mechanics in the first place.

12.3 The Double-Slit Experiment

12.3.1 Nature and Chaos

In Chapters 4 and 7, we've got a taste of the important subject of stability. This marks the start of modern physics: initial data are not always sufficient to predict the future. One must look at nature with a little bit of salt: things are not always as they seem. Nature is more relative than absolute. In some cases, nature makes no sense at all. Better have a lot of doubts, and not always trust your own eyes. Nature could be quite mysterious: not quite physical, but more mathematical.

12.3.2 Light Wave and Its Polarization

For decades, people knew about two curious phenomena: electricity and magnetism. Maxwell's equations united them into one new force: electromagnetism. The typical solution to these equations is a plane wave: it propagates in one direction (say the z-direction), and oscillates (or even precesses) in the perpendicular plane (say the x-y plane). This is called polarization. (Later on, we'll model it algebraically, and illustrate it geometrically.)

Light, for example, is an electromagnetic wave. Indeed, both light and electricity travel at the same speed: $300,000$ kilometer per second, independent of the (inertial) reference frame: the system in which they are measured. This is the maximal speed known in nature.

12.3.3 Interference

As a wave, light exhibits yet another interesting phenomenon: interference. To see this, consider the double-slit experiment: a Laser beam (monochromatic light of a constant frequency) comes from the left, as in Figure 12.3. On its way, it meets a plate, with two (very narrow) horizontal slits in it. Through each slit, it passes and diffracts: restarts as a new source of light. These two light waves continue to travel rightwards, until they meet the rightmost screen, and shed their light on it.

At the middle of the screen, we can then see a horizontal strip of bright light. Why? Thanks to symmetry, both light waves (from both slits) arrive there at the same time: synchronically, in phase. This way, they enhance each other: crest hits a crest to produce a yet higher peek, and trough hits a trough to produce a yet deeper well. This is a constructive interference.

On top of the middle strip, however, symmetry breaks. The light wave from the top slit arrives a little too early: out of phase. As a result, things are not so benign any more: crest hits a trough, and trough hits a crest, canceling each other. This produces a flat (zero) wave, containing no light at all: a dark strip. This is a destructive interference.

On top of that, things get benign again. Indeed, the light from the top slit arrives yet earlier. As a result, crest hits the next crest, and trough hits the next trough. This produces a constructive interference again, and so on.

The same is mirrored at the lower half, below the middle strip. In summary, we get alternating horizontal strips: light, dark, light, dark, and so on.

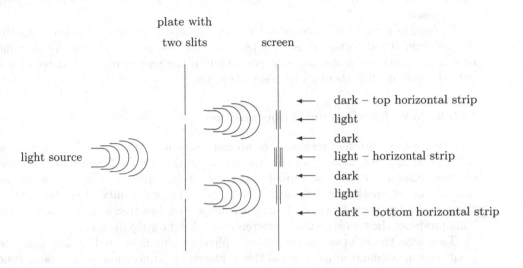

Fig. 12.3. Interference: light wave coming from the left. Through each slit, the light passes, diffracts, and restarts as a new wave. This way, each slit takes the role of a new light source, shedding light rightwards. At the screen on the right, the waves interfere with each other: either enhance or annihilate one another. This produces alternating horizontal strips: light, dark, light, dark, and so on.

12.3.4 Light: Wave or Particle?

Still, light is not only a wave but also a particle: photons. Thus, one could make the Laser beam so weak that it is actually a stream of photons, fired one at a time. This way, one might expect the photons to behave just like bullets (as in Newtonian mechanics and special relativity), and pass either through the top or bottom slit, producing just two strips of light, with no interference at all!

Surprise surprise: this is *not* what happens. On the contrary: we still see the same interference pattern as before. Why?

12.3.5 Detector: No Interference

Instead of photons, one could also fire a stream of electrons, one at a time, or even atoms that shine on the screen. The result is still the same: the electrons (or atoms) behave like waves, and interfere. Furthermore, they seem to have a "mind" of their own: they are not easily fooled.

Indeed, let's spy on them: place a detector on top of the upper slit, to tell us whether the electron passed there or not. It turns out that the electrons are too clever, and "feel" this. Indeed, they behave like bullets, not waves: there is no interference any more, but only two strips of light, across from the slits.

Even if the detector is placed to the right of the slit, so detection is delayed until after the electron "decided" where to pass, no interference appears any more. The electrons are too "shy:" to see them interfere, you must never know where exactly they passed.

In fact, to make the electrons behave like bullets, there is no need to read the results from the detector. It is sufficient to store the results chemically in some atoms: no interference appears any more. Only if the results from the detector are not recorded at all is there an interference again.

12.3.6 Wave Function and Superposition

To explain these strange results, Schrodinger came up with a new idea: nature is not unique! Instead, nature is a superposition (sum) of two different outcomes: the electron passed through the top slit (at some probability), and also through the bottom slit (at another probability). How large is the probability? This depends on the shape of the wave function. Where it is larger, the electron is more likely to be. This produces the (constructive) interference: a bright strip of light.

This way, the electron is not quite a physical thing, located at one concrete position, but a quantum-mechanical thing, placed at all positions at the same time (at some probability). The wave in Figure 12.3 is more mathematical than physical: it is just a wave function that tells us how likely the electron is to be there.

Only at the rightmost screen, where you measure the energy for the first time, does this wave function "collapses," producing a physical outcome: either constructive or destructive interference. If, however, you are too curious, and spy on it beforehand, then nature must make a decision earlier: the "collapse" must take place before any interference has a chance to form. This forces the electron to pass like a bullet, either through the top or bottom slit, with no interference at all any more.

Thanks to Schrodinger's idea, nature is more mathematical than physical. This makes sense: after all, in such a small scale, nobody can see the physics, but only imagine it mathematically. Earlier in the book, we've already made an attempt to explain this in terms of parallel universes and Zorn's lemma (Chapter 7, Section 7.12.2). Next, let's introduce the mathematical model in detail.

12.4 Nondeterminism

12.4.1 Relativistic Observation

In Newtonian mechanics, we often consider a particle, or just any physical object. Such an object must lie somewhere in space: this is its location or position.

At each individual time, the object may have a specific position. This is deterministic: we can measure the position, and tell it for sure. This way, we can also calculate how fast it changes: the (momentary) velocity. From the velocity, we can then calculate the (momentary) momentum and the kinetic energy at each individual time. This gives us a complete picture of the dynamics or motion.

In Chapter 10, on the other hand, we've seen that things are not so absolute, but more relative. The position that I measure in my lab may differ from the position that you measure in your own system. In fact, position is meaningless on its own, unless it has some fixed reference point: the origin. What is meaningful is only the difference between two positions.

The same is true for another important quantity: time. In fact, time is only relativistic and nonphysical. Indeed, the time that I see in my clock may differ from the time that you see in your own clock, particularly if you travel away from me fast.

Don't worry: no clock is wrong – both work well. Time is not absolute, but only relativistic: it depends on the perspective from which it is measured. This is why time is never defined absolutely.

Thus, in special relativity, time is just an observation. To know it, one must look and observe: make an experiment, and measure. This measurement is relativistic – it depends on the perspective from which it is made. For this reason, two observers may see a different time in their different systems. For example, the time that I see in my own clock here on the Earth may differ from the time in some other clock, placed on a satellite.

12.4.2 Determinism

So, special relativity teaches us to be more humble, and not trust our own eyes. What you see in your system is not necessarily the absolute truth: it is often different from what I see in my own (moving) system. This applies not only to time but also to other important observations, such as position. Indeed, like time, position is not absolute but only relativistic.

Together, time and position make a new pair of observations, which depend on the system where they are measured. Likewise, momentum and energy make a new pair as well. This is why momentum is more fundamental than velocity.

Fortunately, this is still deterministic. In your own system, you know what you see, with no doubt. This way, the physical quantities are still well-defined, uniquely and unambiguously. Unfortunately, this is true only in the macroscale, but not in the microscale or nanoscale, used often in molecules, atoms, and subatomic particles.

12.4.3 Nondeterminism: Observables

In quantum mechanics, things get yet worse: nothing is certain any more. After all, a particle could be so small that its position has no physical meaning any more. In this context, the position is no longer an observation, but only an observable: in theory, you could observe and measure it, but better not.

Position is no longer a physical quantity, but only a random variable. The particle *could* be somewhere at some probability: a number that tells us how likely it is to be there. Perhaps it is there, and perhaps not. We'll never know, unless we're ready to take the risk and change the original physical state forever.

Indeed, to know the position for sure, you must take a measurement. But this is not advisable: the particle is often so small that detecting its position is too hard, and requires a complicated experiment, which may change the entire physical state. This way, we may lose a lot of information about other important observables, such as momentum.

This also works the other way around: the particle is so small that measuring its momentum is too hard, and may require a complicated experiment. As a result, vital information is gone, and the original position may never be discovered any more!

Fortunately, thanks to our advanced algebraic tools, we can now model even a highly nondeterministic state like this. Indeed, to model an observable, we can now use a matrix. After all, matrices enjoy all sorts of useful algebraic properties.

This way, there is no need to look or observe as yet: this could wait until later. In the meantime, we can still "play" with our matrices, and design more and more interesting observables.

12.5 State – Wave Function

12.5.1 Physical State

In Newtonian mechanics, we often consider a particle, traveling along the x-axis. At time t, it is at position $x(t)$. Thanks to differentiation, one could also calculate the velocity $x'(t)$, and the momentum $mx'(t)$ (where m is the mass). This is the linear momentum in the x-direction. This could be done at each individual time t.

In quantum mechanics, on the other hand, this is not so easy any more. Indeed, there is no determinism any more. In "true" physics, the particle is nowhere (or everywhere). The physical state only tells us where it *could* be. Perhaps it is there, and perhaps not.

The physical state is no longer a function like $x(t)$, but a new random variable, whose probabilities are stored in a new (nonzero) n-dimensional (complex) vector $v \in \mathbb{C}^n$. In fact, this vector contains every information that nature tells us about the particle, including where it might be at time t. For this purpose, v isn't static, but dynamic: it may change in time: $v \equiv v(t)$.

12.5.2 Matrix Mechanics: The Position Matrix

So, where might the particle be? For this purpose, we have a new $n \times n$ diagonal matrix: X. On its main diagonal, you can find possible positions that the particle may take. This is also called matrix mechanics.

Consider, for example, some element on the main diagonal: $X_{k,k}$ (for some $1 \leq k \leq n$). How likely is the particle to be at position $x = X_{k,k}$? Well, the probability for this can be deduced from v: it is just $|v_k|^2$.

Clearly, the probabilities must sum to 1. For this purpose, we must assume that v has already been normalized to have norm 1. This could be written in terms of inner product, or sum of squares:

$$\|v\|^2 = (v, v) = v^h v = \bar{v}^t v = \sum_{j=1}^{n} \bar{v}_j v_j = \sum_{j=1}^{n} |v_j|^2 = 1.$$

So, what is important is only the direction of v, not its norm. For every practical purpose, one may assume that v is defined up to a scalar multiple only.

12.5.3 Dynamics: Schrodinger Picture

Unfortunately, n may be too small. After all, the particle could get farther and farther away from the origin, and reach infinitely many positions. To allow this, X must be an infinite matrix, with an infinite order.

For example, a particle could "jump" from number to number along the real axis. To model this, X must be as big as

$$X \equiv \begin{pmatrix} \ddots \\ & -3 \\ & & -2 \\ & & & -1 \\ & & & & 0 \\ & & & & & 1 \\ & & & & & & 2 \\ & & & & & & & 3 \\ & & & & & & & & \ddots \end{pmatrix}.$$

In Figure 8.7, for example the particle starts from 0, and makes exactly 5 steps, so X must have order 11 (or more). In Figure 8.9, on the other hand, the particle makes 7 steps, so X is of order 15 (or more).

To model an unlimited number of steps, on the other hand, X must be an infinite matrix: on its main diagonal, it must have all possible positions that the particle could reach: all integer numbers. In a yet more realistic model, on the other hand, X should be even bigger: on its main diagonal, it should contain not only integer but all real numbers. In this case, X is not just a matrix, but actually an operator.

This kind of dynamics is called Schrodinger's picture: X remains constant at all times, whereas v changes from time to time. This setting is more common than Heisenberg's picture, which works the other way around: v remains constant, whereas X changes in time.

For simplicity, however, we try to avoid infinite dimension. Instead, we mostly stick to our finite dimension n, and our original n-dimensional vector and $n \times n$ matrix.

12.5.4 Wave Function and Phase

How does the state v look like? Well, it may look like a discrete sine or cosine wave. This is why v is often called a wave function.

Still, v is not necessarily real: it may be a complex vector in $\mathbb{C}^n$. For example, it may be a discrete Fourier mode (Figure 12.4).

Thus, in general, each component v_k is a complex number in its own right. As such, it has its own polar decomposition – magnitude times exponent:

$$v_k = |v_k| \exp\left(i\theta_k\right),$$

where $i \equiv \sqrt{-1}$ is the imaginary number, and θ_k is the phase: the angle that v_k makes with the positive part of the real axis.

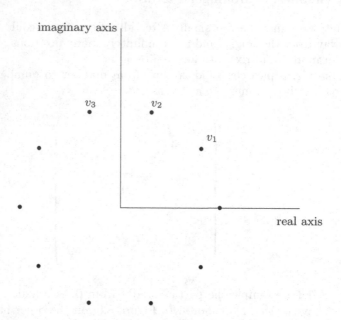

Fig. 12.4. An example of a state v that looks like a discrete Fourier mode. All components $v_1, v_2, \ldots, v_n$ are placed on the unit circle in the complex plane. Once normalized by $\sqrt{n}$, this gives the uniform probability: $|v_1|^2 = |v_2|^2 = \cdots = |v_n|^2 = 1/n$.

The magnitude $|v_k|$ tells us how likely the particle is to be at position $X_{k,k}$. More precisely, the probability for this is $|v_k|^2$. The exponent, on the other hand, is a complex number of magnitude 1. As such, it has no effect on this probability. Still, it may store important information about other physical properties.

The original component v_k is often called the (complex) amplitude. Once all components have been written in their polar decomposition, v takes a more familiar face: a wave function. This way, the original particle also has a new mathematical face: wave. As such, it enjoys an interesting physical phenomenon: interference.

12.5.5 Superposition and Interference

Two electrons cannot be in exactly the same state at the same time. This is Pauli's exclusion principle. (See exercises below.) Two photons, on the other hand, can. In this case, they have the same wave function.

Two wave functions may sum up, and produce a new wave function: their superposition (Section 12.3.6). In this process, the original states sum up, component by component. To add two corresponding components to each other, their phases are most important. If they match, then they enhance each other. This is a constructive interference. If, on the other hand, they don't match, then they may even cancel (annihilate) each other. This is a destructive interference (Figure 12.3).

Once the wave functions have been summed up and normalized, we have the new (joint) state: the superposition of the two original states. In summary, a particle actually has two mathematical faces: on one hand, it is a particle. On the other hand, it is also a wave. Each face is useful to analyze and explain different physical phenomena.

12.6 Observables Don't Commute!

12.6.1 Don't Look!

Let's return to our original particle. It only has a nondeterministic position, where it *might* be. But where is it located in fact?

Don't ask! Because, to find out, you must carry out an experiment. At probability $|v_k|^2$, you'd then discover the particle at $x = X_{k,k}$ (for some $1 \leq k \leq n$).

What happens then? Well, we now know for sure that $x = X_{k,k}$. In other words, the probability for this is now as large as 1: there is no doubt at all any more. So, our v has changed forever. After all, we now know that $|v_k|^2 = 1$. Since v has norm 1, all other components must now vanish. Indeed, the particle can no longer lie at any other position but $X_{k,k}$.

In summary, in your experiment, you spoiled the original v completely, with all the valuable information that was in it! Instead of the original interesting v, you now have a boring deterministic v: a standard unit vector $e^{(k)}$.

Here one may ask: why do we still need v at all? After all, we already got what we wanted: we discovered the true position! Still, v contained information not only about the position but also about many other physical properties.

To appreciate better the information we lost, let's study our original v once again. The probability that $x = X_{k,k}$ could also be written in terms of inner product:

$$|v_k|^2 = \left| \left(e^{(k)}, v \right) \right|^2,$$

where $e^{(k)}$ is a standard unit vector: an eigenvector of X, with eigenvalue $X_{k,k}$. Still, position is not the only observable we're interested in. We might also want to know the momentum p of the particle (in the x-direction).

12.6.2 The Momentum Matrix and Its Eigenvalues

To have the momentum, we are given yet another $n \times n$ Hermitian matrix: P. What do we mean by "Hermitian?" Well, if we take the transpose, and also the complex conjugate of each individual element, then we get the same matrix back again:

$$P^h \equiv \bar{P}^t \equiv \left(\bar{P}_{j,i} \right) = (P_{i,j}) = P.$$

This way, in P, the main diagonal could serve as a "mirror:" each element on the upper-right is mirrored by the corresponding element on the lower-left, with just one change: its imaginary part picks a minus sign.

Again, the physical state doesn't tell us the momentum for sure: it just tells us how likely the particle is to have a particular momentum. For example, let λ_k be

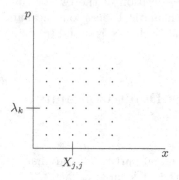

Fig. 12.5. The position-momentum grid. At each individual time t, solve for the n-dimensional state v. This way, v contains the entire physical information at time t, in terms of probability. To be at $x = X_{j,j}$, the particle has probability $|v_j|^2$. To have momentum $p = \lambda_k$, the particle has probability $|(u^{(k)}, v)|^2$ $(1 \le j, k \le n)$.

an eigenvalue of P. How likely is the momentum to be $p = \lambda_k$? The probability for this is given in terms of inner product:

$$\left| \left(u^{(k)}, v \right) \right|^2,$$

where v is our (normalized) state, and $u^{(k)}$ is the (normalized) eigenvector of P, corresponding to λ_k.

Together, X and P give us all possible position-momentum pairs $(X_{j,j}, \lambda_k)$ $(1 \le j, k \le n)$. This makes a new two-dimensional grid of such possible pairs (Figure 12.5).

The above grid mirrors the phase plane in classical mechanics, which contains all possible position-momentum pairs. Ideally, the grid should have been infinite and continuous: a two-dimensional Cartesian plane. After all, in theory, both position and momentum could take any value, not just n discrete values. To model this, n should be infinite. In this case, an inner product should be interpreted as an infinite sum, or even an integral. (We'll come back to this later: Chapter 13, Section 13.1.1.) For simplicity, however, we stick to our finite dimension n, and our original n-dimensional vectors and $n \times n$ matrices.

12.6.3 Order Matters!

Although you might be curious to know the exact position of the particle, better restrain yourself, and not look! Because, if you looked, then you'd spoil v forever, and lose the valuable probabilities of the form $|(u^{(k)}, v)|^2$ that gave you an idea about the momentum.

This also works the other way around. Better not measure the momentum, because then you'd damage v, and lose the valuable probabilities $|v_k|^2$ of the original position.

This means that order matters: measuring x and then p may give different results from measuring p and then x. In other words, applying X before P is not the same as applying P before X: they don't commute with each other.

12.6.4 Commutator

So, it was indeed a good idea to use matrices to model physical observables. After all, matrices often don't commute with each other. In our case, X and P indeed have a nonzero commutator:

$$[X, P] \equiv XP - PX \neq (0),$$

where '(0)' stands for the zero $n \times n$ matrix.

Recall that X and P are both Hermitian. Therefore, their commutator $[X, P]$ is an anti-Hermitian matrix:

$$[X, P]^h = (XP - PX)^h = P^h X^h - X^h P^h = PX - XP = [P, X] = -[X, P].$$

Later on, we'll see that this is a special case of Lie brackets, in a Lie algebra.

12.6.5 Planck Constant

To have this commutator in its explicit form, we have a new law of nature:

$$[X, P] = i\bar{h}I,$$

where $i = \sqrt{-1}$ is the imaginary number, I is the $n \times n$ identity matrix, and $\bar{h}$ is called Planck constant: a universal constant, positive, and very small.

Why is this law plausible? Well, the error due to measuring in two different orders mustn't depend on the original particle. After all, we might have the same error even if there was no particle at all. In fact, even with no particle at all, we might still have a nonzero energy (the zero-point energy, at the ground state).

Thus, $[X, P]$ should better be a constant matrix, independent of the particle under consideration. This is why it must be of the form $i\bar{h}I$: thanks to the imaginary number i, it is indeed anti-Hermitian, as required. Finally, thanks to the small constant $\bar{h}$, it is very small, and has an effect only in microscale or nanoscale, used in quantum mechanics. In macroscale, on the other hand, it has no practical effect at all. This is why it was ignored in both geometrical mechanics (Chapter 2) and special relativity (Chapter 10).

12.7 Observable and Its Expectation

12.7.1 Observable or Measurable

The matrices introduced above are called observables (or measurables, or experiments). After all, they let us observe. For example, by applying X to v, we get some idea about the nondeterministic position: its expectation at state v, as discussed below.

The actual observation, on the other hand, should better wait until later. After all, it requires an experiment, which may spoil the original state, with all the valuable probabilities that could have been deduced from it about other observables, such as momentum. In the meantime, we can still "play" with our matrices, and apply all sorts of algebraic operations to them.

Consider an $n \times n$ matrix A, not necessarily Hermitian. Let's write it as the sum of two matrices:

$$A = \frac{A + A^h}{2} + \frac{A - A^h}{2}.$$

The former term is the Hermitian part of A. It is indeed Hermitian:

$$\left(\frac{A + A^h}{2}\right)^h = \frac{A^h + A}{2} = \frac{A + A^h}{2}.$$

The latter term, on the other hand, is the anti-Hermitian part of A. It is indeed anti-Hermitian:

$$\left(\frac{A - A^h}{2}\right)^h = \frac{A^h - A}{2} = -\frac{A - A^h}{2}.$$

12.7.2 Symmetrization

A proper observable should better be Hermitian. This way, its eigenvectors are orthogonal to each other. Once normalized, they make a new orthonormal basis, which can be used to decompose just any vector. Furthermore, a Hermitian matrix has real eigenvalues (Chapter 26, Sections 26.4.4–26.4.5). This makes sense: in the nature around us, an observed value must be real, with no imaginary part.

Fortunately, the anti-Hermitian part of A is often as small as $\bar{h}$, and can be disregarded. This is called symmetrization: replacing A by its Hermitian part. Fortunately, this could wait until later. In the meantime, we can still stick to our original A, Hermitian or not.

12.7.3 Observation

So far, we've seen two observables: the position X, and the momentum P. In the context of special relativity, the time t could be viewed as an observable as well (Chapter 10). After all, to have the time, you must observe: either look at your own clock, to see the proper time, or at least look at someone else's clock, to see a new observable: a different time (Figures 10.4–10.6).

Of course, in special relativity, the scale is so big that nondeterminism has no place. In every practical sense, the observables commute with each other, and everything is deterministic.

In a very small scale, on the other hand, random effects can no longer be ignored. On the contrary: the "true" physical state is no longer deterministic. It only gives us the probability to observe something, not the actual observation. This is the "true" nature: just probability. Of course, we humans will never get to see this "truth." After all, we must make a decision: what to measure first, and what to measure later. Each choice may give us different results: the true original nature remains a mystery.

12.7.4 Random Variable

Thus, an observable is more mathematical than physical. It makes a new random variable: we can't tell for sure what its value is. Fortunately, we can still tell what its value *might* be. For example, its value could be λ: some eigenvalue of the observable. The probability for this is $|(u, v)|^2$, where v is the (normalized) state, and u is the (normalized) eigenvector corresponding to λ.

An observable like X or P must be Hermitian. To have its expectation (or average) at state v, apply it to v, and take the inner product with v. Let's do this even for a more general matrix A, Hermitian or not. This gives a new complex number:

$$(v, Av) = \left(v, \frac{A + A^h}{2}v\right) + \left(v, \frac{A - A^h}{2}v\right).$$

In this sum, the former term is real, and the latter is imaginary. Thus, in terms of absolute value, each term is smaller than (or equal to) the entire sum:

$$|(v, Av)| = \left|\left(v, \frac{A + A^h}{2}v\right) + \left(v, \frac{A - A^h}{2}v\right)\right| \geq \left|\left(v, \frac{A - A^h}{2}v\right)\right|.$$

This will be useful below.

12.7.5 Observables and Their Expectation

At state v, the expectation of X is

$$(v, Xv) = \left(\sum_{j=1}^{n} v_j e^{(j)}, X \sum_{j=1}^{n} v_j e^{(j)}\right)$$

$$= \left(\sum_{j=1}^{n} v_j e^{(j)}, \sum_{j=1}^{n} X_{j,j} v_j e^{(j)}\right)$$

$$= \sum_{j=1}^{n} X_{j,j} |v_j|^2.$$

Likewise, the expectation of P is

$$(v, Pv) = \left(\sum_{k=1}^{n} \left(u^{(k)}, v\right) u^{(k)}, P \sum_{k=1}^{n} \left(u^{(k)}, v\right) u^{(k)}\right)$$

$$= \left(\sum_{k=1}^{n} \left(u^{(k)}, v\right) u^{(k)}, \sum_{k=1}^{n} \lambda_k \left(u^{(k)}, v\right) u^{(k)}\right)$$

$$= \sum_{k=1}^{n} \lambda_k \left|\left(u^{(k)}, v\right)\right|^2$$

(thanks to orthonormality). This is indeed what expectation should be: sum of possible values, each multiplied by its own probability.

12.8 Heisenberg's Uncertainty Principle

12.8.1 Variance

The original random variable might take all sorts of possible values. How likely are they to spread out, and deviate from the average? To get some idea about this, we define the variance at state v. Again, let's do this even for a general matrix A, Hermitian or not:

$$\|(A - (v, Av) I) v\|^2.$$

Let's estimate the variances of our original observables: X and P. Fortunately, their product has a lower bound: their covariance.

12.8.2 Covariance

At state v, the expectation of X is (v, Xv), and the expectation of P is (v, Pv). Since these matrices are Hermitian, these expectations are real.

Now, at state v, look at the product of variances: that of X, times that of P. How to estimate this from below? Well, thanks to the Cauchy-Schwarz inequality (see exercises at the end of Chapter 26),

$$\|(X - (v, Xv) I) v\| \cdot \|(P - (v, Pv) I) v\| \geq |((X - (v, Xv) I) v, (P - (v, Pv) I) v)|.$$

What do we have on the right-hand side? This is the covariance of X and P at state v. To estimate it from below, recall that, although X and P are both Hermitian, their product XP is not. Fortunately, we still have the estimate in Section 12.7.4:

$$
\begin{aligned}
\|(X - (v, Xv) I) v\| \cdot \|(P - (v, Pv) I) v\| &\geq |((X - (v, Xv) I) v, (P - (v, Pv) I) v)| \\
&= |(v, (X - (v, Xv) I) (P - (v, Pv) I) v)| \\
&\geq \frac{1}{2} |(v, [X - (v, Xv) I, P - (v, Pv) I] v)| \\
&= \frac{1}{2} |(v, [X, P] v)| \\
&= \frac{1}{2} |(v, i\bar{h} I v)| \\
&= \frac{\bar{h}}{2} |(v, v)| \\
&= \frac{\bar{h}}{2}
\end{aligned}
$$

(assuming that v has already been normalized in advance).

12.8.3 Heisenberg's Uncertainty Principle

Finally, take the square of the above inequality. This gives a lower bound for the product of the variances of X and P:

$$\|(X - (v, Xv) I) v\|^2 \|(P - (v, Pv) I) v\|^2 \geq \frac{\bar{h}^2}{4}.$$

This is Heisenberg's uncertainty principle. It tells you that you can't enjoy both worlds. If you measured the precise position, then the variance of X becomes zero. Unfortunately, there is a price to pay: the variance of P becomes huge. As a result, there is no hope to measure the original momentum any more.

This also works the other way around: if you measured the precise momentum, then the variance of P would vanish. Unfortunately, in this case, there is yet another price to pay: the variance of X would become infinite. As a result, there is no hope to measure the original position any more!

This is why we better wait with the actual observation until later. In the meantime, we can still "play" with our original observables algebraically, to design all sorts of new interesting observables.

12.9 Duality: Particle as a Wave

12.9.1 Momentum: A Differentiation Operator

So far, we discussed Heisenberg's uncertainty principle in the context of probability theory, and proved it algebraically. Still, what is its physical meaning? Well, it highlights a key concept: duality. A particle like photon or electron could be matter and wave at the same time.

Indeed, assume that the position $x \in \mathbb{R}$ could take just any real value. This way, X is an infinite diagonal matrix (or operator), with all real numbers on its main diagonal. This way, X can pick just any real number as an eigenvalue, telling us where the particle really is.

Likewise, P should be an operator as well. What should P do? Well, P should look at the particle as a wave: the Fourier mode $\exp(ikx)$, where $k \in \mathbb{R}$ is a fixed real number. Like x, k is nondeterministic: a random variable. This is why it needs an operator to pick it.

How to specify the Fourier mode? In other words, how to pick k? Fortunately, P picks k as an eigenvalue. For this purpose, better define P as differentiationn:

$$P \equiv -i\bar{h}\frac{d}{dx}.$$

This way,

$$P\exp(ikx) = -i\bar{h} \cdot ik\exp(ikx) = \bar{h}k\exp(ikx).$$

This way, P indeed picks $\bar{h}k$ as an eigenvalue, telling us how the wave looks like. With a specific k at hand, one can now illustrate the original mode $\exp(ikx)$ visually.

12.9.2 Debroglie's Relation: Momentum and Wave Number

This is indeed Debroglie's relation: momentum is proportional to the wave number:

$$p = \bar{h}k.$$

This is nondeterministic: the probability to pick a concrete k is stored in the state. Heisenberg's uncertainty principle forces us to choose want we want to see: either a particle at position x, or a wave of wave number k, but not both at the same time.

The better we know x, the worse we know k and p, and vice versa. Each picture has its own advantage. The particle picture helps explain the photo-electric effect: only a photon of sufficiently high frequency and energy can knock out an electron. The wave picture, on the other hand, helps explain interference in the double-slit experiment.

12.9.3 The Commutator

With the above P, what is the commutator? Well, it is a new operator. To see what it is, let's pick a real (differentiable) function f, and apply the commutator to it:

$$
\begin{aligned}
[X, P]f &= (XP - PX)f \\
&= -(PX - XP)f \\
&= i\bar{h}((xf)' - xf') \\
&= i\bar{h}(f + xf' - xf') \\
&= i\bar{h}f.
\end{aligned}
$$

Thus, the commutator is $i\bar{h}$ times the identity operator, as required.

12.9.4 Electron in The Atom

So far, x could be just any real number. But what if x was confined to some unit circle? In other words, what if the above function f must be periodic? In this case, the above Fourier mode must be periodic as well. For this purpose, the wave number k must be integer. This is indeed quantum mechanics: nature is discrete, not continuous.

In the atom, the electron orbits the nucleus not only as a particle but also as a wave. Indeed, the above Fourier mode can be interpreted nondeterministically as a (periodic) wave function. Like the linear momentum $p = \bar{h}k$, the angular momentum takes discrete values only. Indeed, the electron must orbit at a suitable radius, where the electrostatic attraction to the nucleus supplies the required centripetal force. This is indeed Bohr's model: the nucleus is surrounded by more and more shells (energy levels), where the electron could orbit at more and more energy. Thanks to Debroglie, this is indeed stable: the periodic wave function makes a standing wave, so the electron radiates no electromagnetic wave, and loses no mass or energy.

12.10   Time-Energy Uncertainty

12.10.1   Time-Energy and Space-Momentum

In special relativity, we've already seen that space and time make a relativistic pair. Indeed, thanks to the Lorentz transformation, x and t can be transferred together to any other inertial system. Likewise, energy and momentum make a relativistic pair as well. Furthermore, x and t have the same status: the system is insensitive to interchanging their roles. In this sense, time mirrors space, and energy mirrors momentum (Chapter 10, Section 10.7.4).

In quantum mechanics too, time-energy mirror space-momentum. Thus, not only position and momentum have an uncertainty principle, but also time and energy. You can't enjoy both worlds: you must choose which of them you want to know more accurately. The better you know the energy, the worse you know what time it is, and vise versa.

Like momentum, energy looks at the particle as a wave. Indeed, look at a Fourier mode in time: $\exp(-i\omega t)$. The energy operator, the Hamiltonian H, will pick the frequency ω as an eigenvalue.

12.10.2 Schrodinger Equation

For this purpose, H mirrors P: it differentiates in time:

$$H \equiv i\bar{h}h\frac{d}{dt}.$$

This is indeed Schrodinger's equation. Later on, we'll also rewrite it with a partial derivative.

12.10.3 Planck Relation: Frequency and Energy

Thanks to this definition, the above Fourier mode is indeed an eigenfunction:

$$H\exp(-i\omega t) = i\bar{h}(-i\omega)\exp(-i\omega t) = \bar{h}\omega\exp(-i\omega t).$$

This is indeed Planck's relation: energy is proportional to frequency:

$$E = \bar{h}\omega.$$

This was used by Einstein to explain the photo-electric effect: only a photon with sufficiently high frequency has enough energy to knock out an electron.

Better yet, let's redefine both P and H in terms of a partial derivative:

$$P \equiv -i\bar{h}\frac{\partial}{\partial x}$$
$$H \equiv i\bar{h}h\frac{\partial}{\partial t}.$$

This way, P and H can now act even on a function of two variables, like the following wave:

$$\exp(-i\omega t)\exp(ikx) = \exp(i(kx - \omega t)).$$

From the left-hand side, this is indeed a standing wave: a function of x, precessing at frequency ω in time. In the atom, the true wave function is a superposition: a sum of such waves, each with its own frequency, and its own coefficient, which depends on the probability to find the electron at the relevant energy level. Next, let's consider an $n \times n$ matrix once again, and study its algebraic properties.

12.11 Eigenvalues

12.11.1 Shifting an Eigenvalue

Let's use linear algebra to study a general commutator. Let C and T be two $n \times n$ matrices, Hermitian or not. Assume also that they don't commute with each other. On the contrary: they have a nonzero commutator, proportional to T:

$$[C, T] \equiv CT - TC = \alpha T,$$

for some complex number $\alpha \neq 0$.

Let u be an eigenvector of C, with the eigenvalue λ:

$$Cu = \lambda u.$$

How to find more eigenvectors? Just apply T to u. Indeed, if Tu is a nonzero vector, then it is an eigenvector as well:

$$CTu = (TC + [C, T]) u = (TC + \alpha T) u = (\lambda + \alpha) Tu.$$

What is its eigenvalue? $\lambda + \alpha$. Thus, we've "shifted" λ by α, obtaining a new eigenvalue of C: $\lambda + \alpha$ (provided that Tu is not the zero vector).

Moreover, we can now go on applying T time and again (so long as we don't hit the zero vector). This way, we obtain more and more eigenvectors of C, with new eigenvalues, shifted more and more:

$$\lambda, \ \lambda + \alpha, \ \lambda + 2\alpha, \ \lambda + 3\alpha, \ \ldots.$$

Of course, since n is finite, this process must stop somewhere. Still, if n was infinite, then the process could continue forever.

12.11.2 Shifting an Eigenvalue of a Product

Let's use the above in a special case. For this purpose, let A and B be $n \times n$ matrices (not necessarily Hermitian). Assume that they don't commute with each other. On the contrary: they have a nonzero commutator, proportional to the identity matrix:

$$[A, B] \equiv AB - BA = \beta I,$$

for some complex number $\beta \neq 0$.

Now, let's look at the product BA. What is its commutator with B? We already know what it is:

$$[BA, B] \equiv BAB - BBA = B(AB - BA) = B[A, B] = \beta B.$$

We can now use the result in Section 12.11.1 to shift an eigenvalue of BA. For this purpose, let u be an eigenvector of BA with the eigenvalue λ:

$$BAu = \lambda u.$$

If Bu is a nonzero vector, then it is an eigenvector of BA as well:

$$BA(Bu) = (\lambda + \beta)Bu.$$

Likewise, one could also shift in the opposite direction. For this purpose, look again at the product BA. What is its commutator with A? It is just

$$[BA, A] = BAA - ABA = (BA - AB)A = -[A, B]A = -\beta A.$$

We can now use the result in Section 12.11.1 once again, to design a new eigenvector of BA. For this purpose, take the original eigenvector u, and apply A to it:

$$BA(Au) = (\lambda - \beta)Au.$$

This way, if Au is a nonzero vector, then it is indeed an eigenvector of BA as well, with the new eigenvalue $\lambda - \beta$. Moreover, we can now go on and repeat this procedure time and again (so long as we don't hit the zero vector). This way, we design more and more eigenvectors of BA, with new eigenvalues:

$$\lambda, \ \lambda \pm \beta, \ \lambda \pm 2\beta, \ \lambda \pm 3\beta, \ \ldots$$

Of course, since n is finite, this process must stop somewhere. Still, if n was infinite, then the process could continue forever.

12.11.3   A Number Operator

In the above, consider a special case, in which

$$B \equiv A^h \quad \text{and} \quad \beta \equiv 1.$$

This way, the assumption in the beginning of Section 12.11.2 takes the form

$$[A, A^h] = I.$$

Furthermore, the product studied above is now

$$BA = A^h A.$$

As a Hermitian matrix, $A^h A$ has real eigenvalues only. Furthermore, thanks to the above discussion, we can now lower or raise an eigenvalue of $A^h A$: from λ to

$$\lambda + 1, \ \lambda + 2, \ \lambda + 3, \ \ldots$$

(until hitting the zero vector), and also to

$$\lambda - 1, \ \lambda - 2, \ \lambda - 3, \ \ldots$$

(until hitting the zero vector, which will be very soon).

$A^h A$ is an important matrix: it is called a number operator. Why? Because its eigenvalues are $0, 1, 2, 3, 4, \ldots$.

12.11.4 Eigenvalue – Expectation

Indeed, let u be an eigenvector of $A^h A$, with the eigenvalue λ:

$$A^h A u = \lambda u.$$

With some effort, both λ and u could have been calculated. Better yet, there is no need to calculate them explicitly at all!

Since $A^h A$ is Hermitian, λ must be real (Chapter 26, Section 26.4.4). Could λ be negative? No! Indeed, look at the expectation of $A^h A$ at u:

$$\lambda(u, u) = (u, A^h A u) = (Au, Au) \geq 0.$$

Let's use u to design more eigenvectors.

12.11.5 Ladder Operator: Lowering an Eigenvalue

Recall that we consider now a special case: in Section 12.11.2, set $B \equiv A^h$, and $\beta \equiv 1$. This way, the above eigenvector u could be used to design a new eigenvector: Au. Indeed, if Au is a nonzero vector, then it is an eigenvector of $A^h A$ in its own right, with a lower eigenvalue: $\lambda - 1$.

In this context, the matrix A serves as a "ladder" operator. By applying it to u, we go down the ladder, to a lower eigenvalue.

What is the norm of Au? To find out, look at the expectation of $A^h A$ at u:

$$\|Au\|^2 = (Au, Au) = (u, A^h A u) = \lambda(u, u) = \lambda\|u\|^2.$$

In other words,

$$\|Au\| = \sqrt{\lambda}\|u\|.$$

Later on, we'll use this to normalize the eigenvectors.

12.11.6 Null Space

This lowering procedure can't continue forever, or we'd hit a negative eigenvalue, which is impossible (Section 12.11.4). It must stop upon reaching some eigenvector w whose eigenvalue is zero. At this stage, lowering is no longer possible. So, we see in retrospect that our λ must have been a nonnegative *integer* number: this is the only way to make sure that the lowering procedure eventually hits zero, and stops.

In summary, we must eventually reach a new vector w, for which

$$A^h A w = 0$$

(the zero vector, not scalar). How does w look like? Well, at w, $A^h A$ must have zero expectation:

$$(Aw, Aw) = (w, A^h A w) = (w, \mathbf{0}) = 0.$$

Therefore,

$$Aw = 0$$

as well. Thus, w lies in the null spaces of both A and $A^h A$ (Chapter 26, Section 26.4.2).

Fortunately, we can now go ahead and apply the reverse procedure to w, to obtain bigger eigenvalues back again.

12.11.7 Raising an Eigenvalue

For this purpose, let's use Section 12.11.2 once again (again, with $B \equiv A^h$ and $\beta \equiv 1$). This way, from our original eigenvector u, we can now form $A^h u$: a new eigenvector of $A^h A$, with a bigger eigenvalue: $\lambda + 1$. In this context, A^h serves as a new ladder operator, to help "climb" up the ladder.

What is the norm of $A^h u$? Well, since $[A, A^h] = I$,

$$
\begin{aligned}
\|A^h u\|^2 &= \left(A^h u, A^h u\right) \\
&= \left(u, A A^h u\right) \\
&= \left(u, \left(A^h A + [A, A^h]\right) u\right) \\
&= \left(u, \left(A^h A + I\right) u\right) \\
&= \left(u, (\lambda + 1)\, u\right) \\
&= (\lambda + 1)\, (u, u).
\end{aligned}
$$

In other words,

$$
\|A^h u\| = \sqrt{\lambda + 1}\|u\|.
$$

Later on, we'll use this to normalize these eigenvectors, as required.

12.12 Hamiltonian and Its Eigenvalues

12.12.1 Harmonic Oscillator and Its Hamiltonian

So far, we've studied the number operator $A^h A$ from an algebraic point of view: its eigenvalues and eigenvectors. Still, what is its physical meaning? To see this, let's model a harmonic oscillator, or a spring (Chapter 1, Sections 1.5.1–1.7.4).

For this purpose, let's use our position and momentum matrices: X and P. Let's use them to define a new matrix – the Hamiltonian:

$$
H \equiv \frac{m\omega^2}{2} \left(X^2 + \frac{1}{m^2\omega^2} P^2 \right),
$$

where m and ω are given parameters: the mass and the frequency of the harmonic oscillator. (Don't confuse ω with the vector w in Section 12.11.6, or with the angular velocity in geometrical mechanics.) This is the Hamiltonian observable. It will be used to observe the total energy in the harmonic oscillator – kinetic and potential alike. Again, this is only nondeterministic: a random variable, telling us what the total energy *could* be: an eigenvalue of H. At what probability? This already depends on the corresponding (normalized) eigenvector: take its inner product with the (normalized) state v, take the absolute value, and square it up.

12.12.2 Concrete Number Operator

How do the eigenvalues of H look like? To see this, let's define A in Section 12.11.3 more concretely:

$$
A \equiv= \sqrt{\frac{m\omega}{2\hbar}} \left(X + \frac{i}{m\omega} P \right).
$$

This way, its Hermitian adjoint is

$$A^h = \sqrt{\frac{m\omega}{2\bar{h}}}\left(X - \frac{i}{m\omega}P\right).$$

Thus, the commutator is still

$$[A, A^h] = \frac{m\omega}{2\bar{h}}\left[X + \frac{i}{m\omega}P, X - \frac{i}{m\omega}P\right] = -\frac{m\omega}{2\bar{h}}\cdot\frac{i}{m\omega}\left([X,P] - [P,X]\right) = I,$$

as in Section 12.11.3. For this reason, the above properties still hold, including raising and lowering eigenvalues. To construct normalized eigenvectors of $A^h A$, just start from some w in the null space of A, normalize it to have norm 1, apply A^h time and again, and normalize:

$$w,\ A^h w,\ \frac{1}{\sqrt{2}}\left(A^h\right)^2 w,\ \frac{1}{\sqrt{3!}}\left(A^h\right)^3 w,\ \ldots\ \frac{1}{\sqrt{k!}}\left(A^h\right)^k w,\ \ldots.$$

These are the orthonormal eigenvectors of $A^h A$. They can now help decompose just any vector (as in Chapter 26, Sections 26.4.5).

12.12.3   Energy Levels

Mathematically, we already know a lot about our concrete number operator, and its eigenvalues and eigenvectors. Still, what is its physical meaning? To see this, let's calculate it explicitly:

$$\begin{aligned}
A^h A &= \frac{m\omega}{2\bar{h}}\left(X - \frac{i}{m\omega}P\right)\left(X + \frac{i}{m\omega}P\right)\\
&= \frac{m\omega}{2\bar{h}}\left(X^2 + \frac{1}{m^2\omega^2}P^2 + \frac{i}{m\omega}[X,P]\right)\\
&= \frac{m\omega}{2\bar{h}}\left(X^2 + \frac{1}{m^2\omega^2}P^2\right) + \frac{i}{2\bar{h}}i\bar{h}I\\
&= \frac{1}{\bar{h}\omega}H - \frac{1}{2}I.
\end{aligned}$$

Thus, the Hamiltonian is strongly related to the concrete number operator:

$$H = \bar{h}\omega\left(A^h A + \frac{1}{2}I\right).$$

Thus, the total energy in the harmonic oscillator has a very simple form. After all, it is just an eigenvalue of H:

$$\frac{\bar{h}\omega}{2},\ \frac{3\bar{h}\omega}{2},\ \frac{5\bar{h}\omega}{2},\ \frac{7\bar{h}\omega}{2},\ \ldots.$$

In units as small as $\bar{h}$, this is just the frequency ω, times a (nonnegative) integer number plus one half. This is indeed quantum mechanics: energy is no longer continuous, but comes in discrete levels. Position and time, on the other hand, remain continuous: at each individual time, the particle may still lie in just any position.

Furthermore, the eigenvectors of H are the same as those of $A^h A$, designed above. Thanks to conservation of energy, each of them makes a constant state, with no dynamics at all: if your initial physical state is an eigenvector, then it must remain so forever. After all, it must preserve the same energy level – its eigenvalue. For this reason, the wave function must be a standing wave that never moves. Let's see how this looks like.

12.12.4   Ground State: Zero-Point Energy

First, let's look at w – the eigenvector that lies in the null space of A (Figure 12.6). What is its physical meaning? It represents a very strange case: no momentum, and no motion at all!

This is the minimal energy level. Indeed, w is in the null space of $A^h A$ as well:

$$A^h A w = 0.$$

For this reason, w is an eigenvector of H as well:

$$H w = \bar{h}\omega \left(A^h A + \frac{1}{2}I \right) w = \frac{\bar{h}\omega}{2} w.$$

Thus, w is indeed the ground state: even with no particle at all, there is still some minimal energy: zero-point energy (or vacuum energy). This energy is very small: in units as small as $\bar{h}$, it has just $\omega/2$ units, where ω is the frequency.

12.12.5   Gaussian Distribution

How does w look like? It makes a Gaussian distribution, with zero expectation. In Figure 12.6, we illustrate the components w_k as a function of x. To lie at $x = X_{k,k}$, the particle has probability $|w_k|^2$ (provided that $\|w\| = 1$).

12.13   Coherent States

12.13.1   Energy Levels and Their Superposition

In Schrodinger's picture, the matrices X, P, and H never change (Section 12.5.3). The dynamics is in the state, which may change in time, along with the physical information it carries: the probabilities encapsulated in it.

How can such a moving state look like? Let's try a vector we already know: a (normalized) eigenvector of H:

$$\frac{1}{\sqrt{k!}} \left(A^h \right)^k w, \quad k \geq 0.$$

But this is no good: there is no dynamics here at all! After all, can this state ever change? No, it can't! Indeed, the total energy must remain the same eigenvalue of H. So, to introduce time-dependence, the best you can do is to multiply this state by the (complex) number $\exp(i\phi t)$. But this introduces no dynamics at all. After

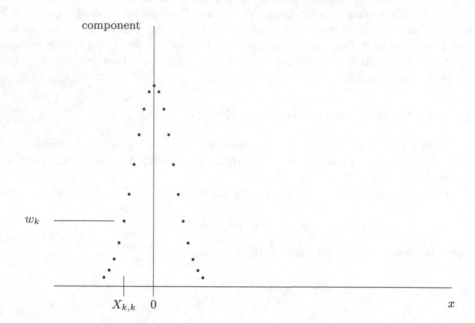

Fig. 12.6. Gaussian distribution: w lies in the null spaces of both A and $A^h A$, with zero expectation: $(Aw, Aw) = (w, A^h Aw) = 0$. To be at $x = X_{k,k}$, the particle has probability $|w_k|^2$. Thus, it is highly likely to be at $x = 0$ – the expectation.

all, this has no effect on the probabilities. Besides, the state is defined up to a scalar multiple only. Thus, this is still a standing wave function that travels nowhere.

A standing wave is rather rare: most waves travel in time. For this purpose, let's turn to a more general state v. Let's expand it in terms of the orthonormal eigenvectors of H:

$$v = \sum_{k \geq 0} \left(\frac{1}{\sqrt{k!}} \left(A^h \right)^k w, v \right) \frac{1}{\sqrt{k!}} \left(A^h \right)^k w.$$

12.13.2 Energy Levels and Their Precession

In this expansion, look at the kth coefficient. How does it change in time? It is multiplied by $\exp(-i\omega(k + 1/2)t)$:

$$v(t) = \sum_{k \geq 0} \exp\left(-i\omega \left(k + \frac{1}{2} \right) t \right) \left(\frac{1}{\sqrt{k!}} \left(A^h \right)^k w, v(0) \right) \frac{1}{\sqrt{k!}} \left(A^h \right)^k w.$$

This new exponent has no effect on the probability to be in the kth energy level, which remains

$$\left| \left(\frac{1}{\sqrt{k!}} \left(A^h \right)^k w, v \right) \right|^2 = \frac{1}{k!} \left| \left(\left(A^h \right)^k w, v \right) \right|^2$$

(assuming that $\|v(0)\| = 1$). Still, it does have an effect on the probability to be at a particular position, and to have a particular momentum. Why? Because, to calculate these probabilities, different energy levels sum up and interfere with each other. Since each energy level precesses at a different frequency, this may produce either constructive or destructive interference. We'll come back to this later, in the context of Lie algebras (Chapter 23, Section 23.12).

12.13.3 Coherent State

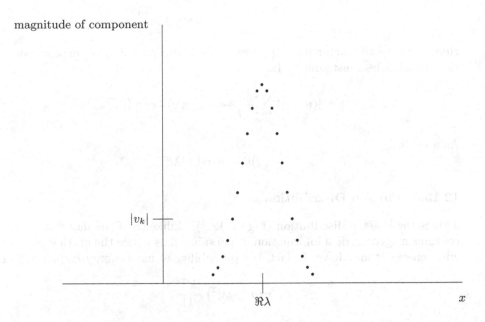

Fig. 12.7. A coherent state: a Gaussian distribution, shifted by a complex number λ. For each k, $|v_k|^2$ is the probability to be at $x = X_{k,k}$.

For example, assume that v is a coherent state: an eigenvector of A, not of $A^h A$:

$$Av = \lambda v,$$

where λ is now not necessarily real: it may have a nonzero imaginary part. After all, A is not Hermitian.

How does v look like? It makes a Gaussian distribution, shifted by the complex number λ. This is illustrated in Figure 12.7: to lie at position $x = X_{k,k}$, the particle has probability $|v_k|^2$.

12.13.4 Probability to Have Energy

In a coherent state, we've just seen the probability to be at position $x = X_{k,k}$. This is illustrated in Figure 12.7. Still, there is yet another interesting probability: what is the probability to have a particular amount of energy? Well, not every amount is allowed, but only our discrete energy levels. Fortunately, the probability to be at the kth energy level is already available in Section 12.13.2. In a coherent state, it is even simpler:

$$
\begin{aligned}
\frac{1}{k!}\left|\left(\left(A^h\right)^k w, v\right)\right|^2 &= \frac{1}{k!}\left|(w, A^k v)\right|^2 \\
&= \frac{1}{k!}\left|(w, \lambda^k v)\right|^2 \\
&= \frac{1}{k!}|\lambda|^{2k}\left|(w, v)\right|^2 \\
&= |(w, v)|^2 \frac{|\lambda|^{2k}}{k!}.
\end{aligned}
$$

Here, we have the factor $|(w, v)|^2$. How to estimate it? For this purpose, note that the probabilities must sum to 1:

$$
1 = |(w, v)|^2 \sum_{k \geq 0} \frac{|\lambda|^{2k}}{k!} \doteq |(w, v)|^2 \exp\left(|\lambda|^2\right).
$$

As a result,

$$
|(w, v)|^2 \doteq \exp\left(-|\lambda|^2\right).
$$

12.13.5 Poisson Distribution

This is the Poisson distribution (Figure 12.8). Unlike the Gaussian distribution, it contains no geometrical information: it doesn't tell us where the particle is, but only what energy it may have. In fact, the probability to have energy $\bar{h}\omega(k + 1/2)$ is

$$
\exp\left(-|\lambda|^2\right) \frac{|\lambda|^{2k}}{k!},
$$

where λ is the eigenvalue of the coherent state with respect to A.

12.13.6 Conservation of Nondeterministic Energy

Thus, our coherent state is essentially different from the ground state, or any other eigenvector of $A^h A$. Indeed, energy is no longer known for sure. On the contrary: energy is nondeterministic: a random variable, expanded as a sum (superposition). In this sum, each term stands for some energy level. The probability to be in this energy level remains the same, and never changes. This is indeed conservation of energy in its nondeterministic face.

12.14 Particle in Three Dimensions

12.14.1 Tensor Product

So far, we've seen three important observables: position, momentum, and energy. Let's design a new observable: angular momentum. For this purpose, we need more dimensions.

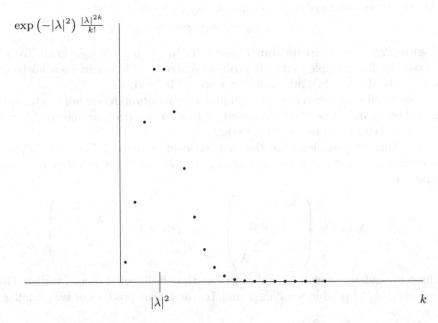

$\exp\left(-|\lambda|^2\right)\frac{|\lambda|^{2k}}{k!}$

$|\lambda|^2$

k

Fig. 12.8. The Poisson distribution. To have energy $\hbar\omega(k+1/2)$, the probability is $\exp(-|\lambda|^2)|\lambda|^{2k}/k!$, where λ is the eigenvalue of the coherent state with respect to A. The maximal probability is at $k \doteq |\lambda|^2$.

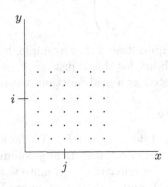

y

i

j

x

Fig. 12.9. The discrete two-dimensional grid: m horizontal rows, of m points each. Since $n = m^2$, our state v makes a (complex) grid function, defined at each grid point. Furthermore, in each particular row, P acts in the same way: it couples grid points in the same row.

Assume now that $n = m^2$, for some integer m. This way, our original state v can be viewed not only as a vector but also as a grid function:

$$v \in \mathbb{C}^n \cong \mathbb{C}^{m \times m}$$

(Figure 12.9). Let's redefine our observables in the new $m \times m$ grid. To represent x-position, for example, our new position matrix should act in each horizontal row separately: it shouldn't mix different rows in the grid.

Let x and p denote the x-position and x-momentum in one horizontal row in the grid. Let X and P be the corresponding $m \times m$ matrices, acting in this row only. How to extend them to the entire grid?

For this purpose, let I be the $m \times m$ identity matrix. Let's use X, P, and I to define extended $n \times n$ block-diagonal matrices that act in the x-direction only, x-row by x-row:

$$X \otimes I \equiv \begin{pmatrix} X & & & \\ & X & & \\ & & \ddots & \\ & & & X \end{pmatrix} \quad \text{and} \quad P \otimes I \equiv \begin{pmatrix} P & & & \\ & P & & \\ & & \ddots & \\ & & & P \end{pmatrix}.$$

These are indeed block-diagonal matrices: the off-diagonal blocks vanish. Here, the new symbol '$\otimes$' produces a bigger matrix: the tensor product of two smaller matrices.

Likewise, to act in the y-direction in the grid (y-column by y-column), define new $n \times n$ (Hermitian) matrices:

$$I \otimes X \equiv \begin{pmatrix} X_{1,1}I & & & \\ & X_{2,2}I & & \\ & & \ddots & \\ & & & X_{m,m}I \end{pmatrix} \quad \text{and} \quad I \otimes P \equiv \begin{pmatrix} P_{1,1}I & \cdots\cdots & P_{1,m}I \\ \vdots & \ddots\cdots & \vdots \\ \vdots & \cdots\ddots & \vdots \\ P_{m,1}I & \cdots\cdots & P_{m,m}I \end{pmatrix}.$$

These new definitions are quite useful. For example, how likely is the particle to be at $(X_{j,j}, X_{i,i})$? The probability for this is just $|v_{i,j}|^2$ (see exercises below). After all, the state v is now interpreted as a grid function, with two indices.

12.14.2 Commutativity

Thus, $X \otimes I$ is completely different from $I \otimes X$: the former acts on the horizontal x-rows, whereas the latter acts on the vertical y-columns in the grid. Because they act in different directions, these matrices commute with each other.

Furthermore, although X and P don't commute, $X \otimes I$ and $I \otimes P$ do:

$$(X \otimes I)(I \otimes P) = \begin{pmatrix} P_{1,1}X & \cdots\cdots & P_{1,m}X \\ \vdots & \ddots\cdots & \vdots \\ \vdots & \cdots\ddots & \vdots \\ P_{m,1}X & \cdots\cdots & P_{m,m}X \end{pmatrix} = (I \otimes P)(X \otimes I).$$

Likewise, $P \otimes I$ and $I \otimes X$ do commute with each other:

$$(P \otimes I)(I \otimes X) = (I \otimes X)(P \otimes I).$$

Let's extend this to a yet higher dimension.

12.14.3 Three-Dimensional Grid

So far, we've only considered scalar position and momentum: x and p. Let's move on to a three-dimensional position:

$$r \equiv \begin{pmatrix} x \\ y \\ z \end{pmatrix}$$

(Chapter 2, Section 2.3.1). This way, r encapsulates three degrees of freedom: the x-, y-, and z-positions. Likewise, p is now a three-dimensional vector: the linear momentum, containing the x-, y-, and z-scalar momenta. In quantum mechanics, each component should be mirrored by a matrix.

For this purpose, assume now that $n = m^3$, for some integer m. This way, our original state v can now be interpreted not only as a vector but also as a grid function:

$$v \in \mathbb{C}^n \cong \mathbb{C}^{m \times m \times m}.$$

Let's take our original $m \times m$ matrices X and P, and place them in suitable tensor products. This way, we obtain extended $n \times n$ matrices, which may help observe position and momentum in the x-, y-, and z-coordinates:

$$R_x \equiv X \otimes I \otimes I$$
$$R_y \equiv I \otimes X \otimes I$$
$$R_z \equiv I \otimes I \otimes X$$
$$P_x \equiv P \otimes I \otimes I$$
$$P_y \equiv I \otimes P \otimes I$$
$$P_z \equiv I \otimes I \otimes P.$$

These new definitions are quite useful. For example, how likely is the particle to be at $(X_{k,k}, X_{j,j}, X_{i,i})$? The probability for this is just $|v_{i,j,k}|^2$ (see exercises below). After all, the state v is now interpreted as a grid function, with three indices.

Do these new matrices commute with each other? Well, it depends: if they act in different directions, then they do (Section 12.14.2). For example,

$$[R_x, P_y] = (0).$$

If, on the other hand, they act in the same direction, then they don't. For example,

$$[R_x, P_x] = i\bar{h} I \otimes I \otimes I.$$

This is used next.

12.15 Angular Momentum

12.15.1 Angular Momentum Component

Thanks to these new matrices, we can now define a new kind of observable: angular-momentum component. This mirrors the original (deterministic) angular momentum:

$$L_x \equiv R_y P_z - R_z P_y$$
$$L_y \equiv R_z P_x - R_x P_z$$
$$L_z \equiv R_x P_y - R_y P_x.$$

Indeed, as in the classical case, these definitions are cyclic: the coordinates x, y, and z are shifted cyclically.

Are these legitimate observables? Well, are they Hermitian? Thanks to commutativity (Sections 12.14.2–12.14.3), they indeed are. For instance,

$$L_x^h = (R_y P_z - R_z P_y)^h = P_z^h R_y^h - P_y^h R_z^h = P_z R_y - P_y R_z = R_y P_z - R_z P_y = L_x.$$

This mirrors the x-component in the vector product $r \times p$ (Chapter 2, Section 2.3.2). Here, however, we have an algebraic advantage: two matrices can combine to form the third one. This shows once again how clever it was to use matrices to model physical observables.

12.15.2 Using the Commutator

The above matrices don't commute with each other. On the contrary: thanks to Sections 12.6.4 and 12.14.2–12.14.3, they have a nonzero commutator:

$$\begin{aligned}
[L_x, L_y] &= [R_y P_z - R_z P_y, R_z P_x - R_x P_z] \\
&= [R_y P_z, R_z P_x] - [R_y P_z, R_x P_z] - [R_z P_y, R_z P_x] + [R_z P_y, R_x P_z] \\
&= R_y P_x [P_z, R_z] - R_y R_x [P_z, P_z] - P_y P_x [R_z, R_z] + P_y R_x [R_z, P_z] \\
&= -i\bar{h} R_y P_x - (0) - (0) + i\bar{h} P_y R_x \\
&= i\bar{h} L_z.
\end{aligned}$$

The same works in the other components as well:

$$[L_x, L_y] = i\bar{h} L_z$$
$$[L_y, L_z] = i\bar{h} L_x$$
$$[L_z, L_x] = i\bar{h} L_y.$$

After all, in the angular momentum, the coordinates x, y, and z are shifted cyclically. In summary, L_x, L_y, and L_z don't commute. On the contrary: each two have a nonzero commutator – the third one (times $i\bar{h}$).

12.15.3 Ladder Operator: Raising an Eigenvalue

Let's use the above to raise an eigenvalue. For instance, let u be an eigenvector of L_z, with the eigenvalue λ:

$$L_z u = \lambda u.$$

Since L_z is Hermitian, λ must be real. How to raise it? For this purpose, define a new matrix:

$$T \equiv L_x + i L_y.$$

It doesn't commute with L_z. On the contrary: they have a nonzero commutator:

$$[L_z, T] = [L_z, L_x + iL_y]$$
$$= [L_z, L_x] + i\,[L_z, L_y]$$
$$= i\bar{h}L_y + \bar{h}L_x$$
$$= \bar{h}T.$$

Thanks to Section 12.11.1, we can now raise λ: if Tu is a nonzero vector, then it is an eigenvector of L_z as well, with the new eigenvalue $\lambda + \bar{h}$. This way, T serves as a ladder operator, to help "climb" up. Furthermore, this procedure can now repeat time and again, producing bigger and bigger eigenvalues, until hitting the zero vector, and reaching the maximal eigenvalue of L_z.

These eigenvalues are the only values that the z-angular momentum may take. This is indeed quantum mechanics: angular momentum is no longer continuous. On the contrary: it may take some discrete values only.

12.15.4 Lowering an Eigenvalue

This also works the other way around: not only to raise but also to lower λ. For this purpose, redefine T as

$$T \equiv L_x - iL_y.$$

It doesn't commute with L_z. On the contrary: they have a nonzero commutator:

$$[L_z, T] = [L_z, L_x - iL_y]$$
$$= [L_z, L_x] - i\,[L_z, L_y]$$
$$= i\bar{h}L_y - \bar{h}L_x$$
$$= -\bar{h}T.$$

Thanks to Section 12.11.1, we can now lower λ: if Tu is a nonzero vector, then it is an eigenvector of L_z as well, with a new eigenvalue: $\lambda - \bar{h}$. This way, our new T serves as a new ladder operator, to help go down the ladder. Furthermore, this procedure can now repeat time and again, until hitting the zero vector, and reaching the minimal eigenvalue of L_z.

So far, we've studied the eigenvalues and eigenvectors of L_z. The same can now be done for L_x and L_y as well. Let's combine these matrices to form the entire angular momentum.

12.15.5 Angular Momentum

Let's place the above matrices as blocks in a new rectangular $3n \times n$ matrix:

$$L \equiv \begin{pmatrix} L_x \\ L_y \\ L_z \end{pmatrix}.$$

This is the nondeterministic angular momentum. It mirrors the deterministic angular momentum $r \times p$ (Chapter 2, Section 2.3.2). In the following exercises, we'll see an interesting example: spin.

12.16 Exercises: Spin

12.16.1 Potential: Divergence of Flux

1. Consider a ball, centered at the origin: $(0, 0, 0)$. Hint: see Figure 12.1.
2. Assume that the ball has charge -1.
3. Outside the ball, consider a point r. What is the electrostatic field at r? Hint: it is radial and outgoing, and its magnitude is $1/\|r\|^2$. Thus, it must be $r/\|r\|^3$.
4. Around the ball, draw a sphere of radius $\|r\|$.
5. What is its surface area? Hint: $4\pi\|r\|^2/3$.
6. On the sphere, draw a little circle around r.
7. What is its surface area? Hint: $\|r\|^2$ (times a constant, which we disregard).
8. What is the flux of the electrostatic field through the circle? Hint: up to a constant,

$$\text{flux} = \text{field} \cdot \text{surface area} \sim \frac{r}{\|r\|^3}\|r\|^2 = \frac{r}{\|r\|}.$$

9. What is the gradient of a scalar function? Hint: vector of partial derivatives.
10. How is it denoted? Hint: by ∇, followed by the scalar function.
11. For example, what is the gradient of $1/\|r\|$? Hint:

$$\nabla\left(\frac{1}{\|r\|}\right) = \nabla\left((x^2 + y^2 + z^2)^{-1/2}\right)$$
$$= -\frac{1}{2}(x^2 + y^2 + z^2)^{-3/2}(2x, 2y, 2z)^t$$
$$= -\frac{r}{\|r\|^3}.$$

12. Is this familiar? Hint: this is the good old field (with a minus sign).
13. What is the physical meaning of this? Hint: the field is minus the gradient of the potential, as illustrated in Figure 12.1.
14. What is the divergence of a vector function? Hint: sum of partial derivatives.
15. How is it denoted? Hint: by ∇, followed by the vector function.
16. For example, what is the divergence of r? Hint:

$$\nabla r = \nabla\left((x, y, z)^t\right) = \frac{\partial x}{\partial x} + \frac{\partial y}{\partial y} + \frac{\partial z}{\partial z} = 1 + 1 + 1 = 3.$$

17. What is the divergence of the flux of the electrostatic field through the above circle? Hint: up to a constant,

$$\nabla\,(\text{flux}) = \nabla\,(\text{field} \cdot \text{surface area})$$
$$\sim \nabla\left(\frac{r}{\|r\|}\right)$$
$$= \frac{\nabla r}{\|r\|} + r^t\nabla\left(\frac{1}{\|r\|}\right)$$
$$= \frac{3}{\|r\|} - r^t\frac{r}{\|r\|^3}$$
$$= \frac{3}{\|r\|} - \frac{1}{\|r\|}$$
$$= \frac{2}{\|r\|}.$$

18. Is this familiar?
19. What is this physically? Hint: the electrostatic potential (times 2).

12.16.2   Eigenvalues and Eigenvectors

1. What is an observable? Hint: an $n \times n$ Hermitian matrix (or operator).
2. What can you say about its eigenvalues? Hint: they are real (Chapter 26, Section 26.4.4).
3. Consider two eigenvectors, corresponding to two distinct eigenvalues. What can you say about them? Hint: they are orthogonal to each other (Chapter 26, Section 26.4.5).
4. Can you make them orthonormal? Hint: just normalize them to have norm 1.
5. Consider now two (linearly independent) eigenvectors that share the same eigenvalue. Can you make them orthogonal to each other? Hint: take one of them, calculate its projection onto the other one (using their inner product), and subtract this projection from the original vector. This is called the Gram-Schmidt process.
6. Normalize them to have norm 1.
7. How many (linearly independent) eigenvectors does an observable have? Hint: n.
8. How many (distinct) eigenvalues may an observable have? Hint: at most n.
9. A degenerate observable has less than n eigenvalues. What can you say about its eigenvectors? Hint: there are two (linearly independent) eigenvectors that share the same eigenvalue.
10. Could the position matrix X be degenerate? Hint: X must have distinct elements on its main diagonal, to stand for distinct positions that the particle may take.

12.16.3   Hamiltonian and Energy Levels

1. Consider the Hamiltonian of the harmonic oscillator (Section 12.12.1). Is it Hermitian? Hint: X and P are Hermitian, and so are their squares.
2. Is it a legitimate observable?
3. May it have an eigenvalue with a nonzero imaginary part? Hint: a Hermitian matrix can have real eigenvalues only.
4. May it have a negative eigenvalue? Hint: the number operator must have a nonnegative expectation (Sections 12.11.3–12.11.4).
5. What is the minimal eigenvalue of the Hamiltonian?
6. May it be zero? Hint: it must be bigger than the minimal eigenvalue of the number operator, which is zero.
7. What is the physical meaning of this eigenvalue? Hint: this is the minimal energy of the harmonic oscillator.
8. May the harmonic oscillator have no energy at all? Hint: no! Its minimal energy is positive.

12.16.4   The Ground State and Its Conservation

1. In the Hamiltonian of the harmonic oscillator, look again at the minimal eigenvalue. What is the corresponding eigenvector? Hint: the ground state: the state of minimal energy.

2. How does it look like? Hint: a Gaussian distribution (Figure 12.6).
3. At the ground state, what is the probability to have a particular amount of energy? Hint: at probability 1, it has the minimal energy. It can't have any other energy level. This is deterministic.
4. Can the ground state change dynamically in time? Hint: no! Energy must remain at its minimum.
5. Consider now yet another state: some other eigenvector of the Hamiltonian. Can it change dynamically in time? Hint: no! Energy must remain the same eigenvalue of the Hamiltonian.
6. How much energy may the harmonic oscillator have? Hint: the allowed energy levels are the eigenvalues of the Hamiltonian: $\bar{h}\omega(k + 1/2)$ ($k \geq 0$).
7. To model the Hamiltonian well, what must the dimension n be? Hint: n should better be infinite. This way, one can start from the ground state, and raise eigenvalues time and again, designing infinitely many new states, with more and more energy.
8. This way, is there a maximal energy? Hint: the above process is unstoppable. Indeed, the zero vector is never reached (Section 12.11.7).

12.16.5 Coherent State and Its Dynamics

1. In a coherent state, what is the probability to have a particular amount of energy? Hint: see Figure 12.8.
2. Is this deterministic? Hint: no! Each energy level has its own (nonzero) probability.
3. Is this constant in time? Hint: yes! Each energy level only precesses, with no effect on the probability.
4. Is the coherent state constant in time? Hint: no! Each energy level precesses dynamically (at its own frequency).
5. Does this affect the probability to have a particular amount of energy? Hint: no! Each energy level keeps the same absolute value, and the same probability.
6. Could this affect any observable that commutes with the Hamiltonian? Hint: no! It has the same eigenvectors, which only precess, with no effect on the probabilities.
7. What is the physical meaning of this? Hint: this observable behaves like the total energy: it obeys a conservation law as well.
8. Still, could this kind of precession affect an observable that doesn't commute with the Hamiltonian? Hint: yes, due to interference (either constructive or destructive). (See also Chapter 23, Section 23.12.)

12.16.6 Entanglement

1. Consider now a particle in a two-dimensional $m \times m$ grid. How does the state v look like? Hint: v is now a grid function: a complex function $v_{i,j}$, defined for every $1 \leq i, j \leq m$.
2. What does it mean that v has norm 1? Hint: its sum of squares is 1:

$$\sum_{i=1}^{m} \sum_{j=1}^{m} |v_{i,j}|^2 = 1.$$

3. How likely is the particle to be at $(X_{j,j}, X_{i,i})$? Hint: the probability for this is $|v_{i,j}|^2$.

4. How likely is the y-position to be $X_{i,i}$? Hint: sum the above probability over the ith row in the grid:

$$\sum_{k=1}^{m} |v_{i,k}|^2.$$

5. Assuming that $y = X_{i,i}$, how likely is the x-position to be $X_{j,j}$? Hint: the probability for this is

$$\frac{|v_{i,j}|^2}{\sum_{k=1}^{m} |v_{i,k}|^2}.$$

Indeed, look at the ith row on its own.

6. From the above formulas, how likely is the particle to be at $(X_{j,j}, X_{i,i})$? Hint: multiply the above probabilities by each other:

$$\frac{|v_{i,j}|^2}{\sum_{k=1}^{m} |v_{i,k}|^2} \sum_{k=1}^{m} |v_{i,k}|^2 = |v_{i,j}|^2.$$

7. Is this as expected?

8. Could $v_{i,j}$ be written as a product of the form $u_i w_j$, where u is defined in one column only, and w is defined in one row only? Hint: no! Look at two different columns: each may have a different distribution of probabilities for the y-position. Likewise, look at two different rows: each may have a different distribution of probabilities for the x-position.

9. This is called entanglement: the y-position depends on the x-position, or interacts with it, or is *entangled* to it.

12.16.7 Angular Momentum and Its Eigenvalues

1. Consider now a particle in a three-dimensional $m \times m \times m$ grid. How does the state v look like? Hint: v is now a grid function: a complex function $v_{i,j,k}$, defined for every $1 \le i, j, k \le m$.

2. What does it mean that v has norm 1? Hint: its sum of squares is 1:

$$\sum_{i=1}^{m} \sum_{j=1}^{m} \sum_{k=1}^{m} |v_{i,j,k}|^2 = 1.$$

3. How likely is the particle to be at $(X_{k,k}, X_{j,j}, X_{i,i})$? Hint: the probability for this is $|v_{i,j,k}|^2$.

4. How likely is the z-position to be $X_{i,i}$? Hint: sum the above probability over $j, k = 1, 2, \ldots, m$:

$$\sum_{j=1}^{m} \sum_{k=1}^{m} |v_{i,j,k}|^2.$$

5. Consider now an angular-momentum component like L_z (Section 12.15.1). Is it Hermitian?

6. Is it a legitimate observable?

7. Are its eigenvalues real? Hint: see Chapter 26, Section 26.4.4.

8. Are its eigenvectors orthogonal to each other? Hint: see Chapter 26, Section 26.4.5.

9. Make them orthonormal. Hint: see exercises above.

10. Consider some eigenvalue of L_z. Consider an n-dimensional state $v \in \mathbb{C}^n$. How likely is the z-angular momentum to be the same as this eigenvalue? Hint: normalize v to have norm 1. Then, take its inner product with the relevant (orthonormal) eigenvector. Finally, take the absolute value of this inner product, and square it up.

11. Must zero be an eigenvalue of L_z? Hint: yes – look at the constant eigenvector, or any other grid function that is insensitive to interchanging the x- and y-coordinates: $x \leftrightarrow y$ (Section 12.14.3).

12. Conclude that L_z must have a nontrivial null space.

13. Given a positive eigenvalue of L_z, show that its negative counterpart must be an eigenvalue as well. Hint: interpret the eigenvector as a grid function. Interchange the x- and y-coordinates: $x \leftrightarrow y$. This makes a new eigenvector, with the negative eigenvalue.

14. Show that L_z has a few eigenvalues of the form

$$0, \; \pm\bar{h}, \; \pm 2\bar{h}, \; \pm 3\bar{h}, \; \cdots$$

(a finite list). Hint: see Sections 12.15.3–12.15.4.

15. Show that L_z may in theory have a few more eigenvalues of the form

$$\pm\frac{\bar{h}}{2}, \; \pm\frac{3\bar{h}}{2}, \; \pm\frac{5\bar{h}}{2}, \; \pm\frac{7\bar{h}}{2}, \; \cdots.$$

Hint: use symmetry considerations to make sure that, in this (finite) list, the minimal and maximal eigenvalues have the same absolute value.

12.16.8 Spin-One

1. Define new 3×3 matrices:

$$S_x \equiv \bar{h} \begin{pmatrix} 0 & 0 & 0 \\ 0 & 0 & -i \\ 0 & i & 0 \end{pmatrix}$$

$$S_y \equiv \bar{h} \begin{pmatrix} 0 & 0 & i \\ 0 & 0 & 0 \\ -i & 0 & 0 \end{pmatrix}$$

$$S_z \equiv \bar{h} \begin{pmatrix} 0 & -i & 0 \\ i & 0 & 0 \\ 0 & 0 & 0 \end{pmatrix},$$

where $i \equiv \sqrt{-1}$ is the imaginary number.

2. These are called spin matrices.

3. Show that these definitions are cyclic under the shift $x \to y \to z \to x$.

4. Are these matrices Hermitian?

5. Are they legitimate observables?

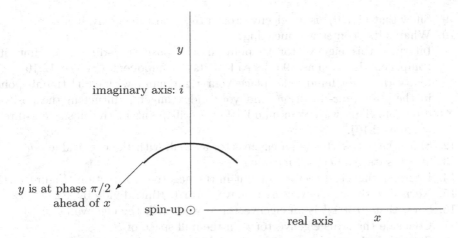

Fig. 12.10. How likely is the boson to have spin-up? Look at the eigenvector $(1, i, 0)^t/\sqrt{2}$. Thanks to the right-hand rule, it points from the page towards your eye, as indicated by the '$\odot$' at the origin. Now, calculate its inner product with the (normalized) state v. Finally, take the absolute value of this inner product, and square it up.

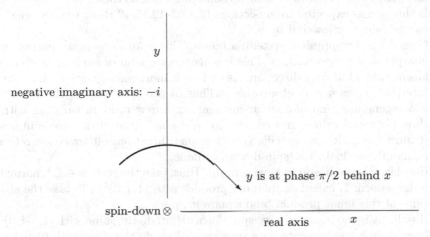

Fig. 12.11. How likely is the boson to have spin-down? Look at the eigenvector $(1, -i, 0)^t/\sqrt{2}$. Thanks to the right-hand rule, it points deep into the page, as indicated by the '$\otimes$' at the origin. Now, calculate its inner product with the (normalized) state v. Finally, take the absolute value of this inner product, and square it up.

6. Calculate their commutator, and show that it is cyclic:

$$[S_x, S_y] = i\bar{h}S_z$$
$$[S_y, S_z] = i\bar{h}S_x$$
$$[S_z, S_x] = i\bar{h}S_y.$$

7. Conclude that these matrices mirror the angular-momentum components in Section 12.15.2.

8. Focus, for instance, on S_z. What are its eigenvectors and eigenvalues?

9. Show that $(1, i, 0)^t$ is an eigenvector of S_z, with the eigenvalue $\bar{h}$.
10. What is its geometrical meaning?
11. Interpret this eigenvector to point in the positive z-direction. Hint: its y-component is at phase 90° ahead of its x-component (Figure 12.10). Now, follow the right-hand rule: place your right hand with your thumb pointing in the positive x-direction, and your index finger pointing in the positive y-direction. This way, your middle finger will point in the positive z-direction (Figure 12.10).
12. Show that $(1, -i, 0)^t$ is an eigenvector as well, with the eigenvalue $-\bar{h}$.
13. What is its geometrical meaning?
14. Interpret this eigenvector to point in the negative z-direction (Figure 12.11).
15. Normalize these eigenvectors to have norm 1. Hint: divide by $\sqrt{2}$.
16. Show that $(0, 0, 1)^t$ is an eigenvector as well, with the eigenvalue 0.
17. Conclude that this eigenvector is in the null space of S_z.
18. Show that these eigenvectors are orthogonal to each other. Hint: calculate their inner product, and don't forget the complex conjugate.
19. Conclude that they are not only orthogonal but also orthonormal.
20. Is this as expected from a Hermitian matrix? Hint: yes – a Hermitian matrix must have real eigenvalues and orthonormal eigenvectors.
21. Is this also as expected from Sections 12.15.3–12.15.4? Hint: yes – an eigenvalue can be raised or lowered by $\bar{h}$.
22. Consider a new physical system: a boson. This is an elementary particle with a new physical property: spin. This is a degenerate kind of angular momentum: it has no value, but only direction. As such, it is more mathematical than physical.
23. Interpret S_z as a new observable, telling us the spin around the z-axis. This is a degenerate kind of angular momentum: a new random variable, with only three potential values. Indeed, the boson "spins" around the z-axis in a rather strange way: with no specific rate, but just direction: either counterclockwise (spin-up), or clockwise (spin-down), or none.
24. How likely is the boson to have spin-up? Hint: take the state $v \in \mathbb{C}^3$, normalize it to have norm 1, calculate its inner product with $(1, i, 0)^t/\sqrt{2}$, take the absolute value of this inner product, and square it up.
25. How likely is the boson to have spin-down? Hint: do the same with $(1, -i, 0)^t/\sqrt{2}$.
26. How likely is the boson to have spin-zero? Hint: do the same with $(0, 0, 1)^t$. The result is $|v_3|^2$.
27. Repeat the above exercises for S_x as well. This makes a new observable: spin-right, in the positive x-direction.
28. Repeat the above exercises for S_y as well. This makes a new observable: spin-in, in the positive y-direction.

12.16.9 Pauli Matrices and Spin-One-Half

1. The above spin is also called spin-one, because the maximal eigenvalue is 1 (times $\bar{h}$).
2. Consider now a new physical system: a fermion. (For example, an electron or a proton or a neutron.)
3. This is a simpler kind of particle. It has a simpler kind of spin, called spin-one-half, because its maximal observation is 1/2 (times $\bar{h}$).

4. For this purpose, consider a new state of a lower dimension: $v \in \mathbb{C}^2$.
5. Define the 2×2 Pauli matrices:

$$\sigma_z \equiv \begin{pmatrix} 1 & 0 \\ 0 & -1 \end{pmatrix}$$

$$\sigma_x \equiv \begin{pmatrix} 0 & 1 \\ 1 & 0 \end{pmatrix}$$

$$\sigma_y \equiv \begin{pmatrix} 0 & -i \\ i & 0 \end{pmatrix}.$$

6. Are these matrices Hermitian?
7. Are they legitimate observables?
8. Multiply the Pauli matrices by $\bar{h}/2$:

$$\tilde{S}_z \equiv \frac{\bar{h}}{2}\sigma_z$$

$$\tilde{S}_x \equiv \frac{\bar{h}}{2}\sigma_x$$

$$\tilde{S}_y \equiv \frac{\bar{h}}{2}\sigma_y.$$

9. These are also called spin matrices.
10. Calculate the commutator of these matrices, and show that it is cyclic:

$$\left[\tilde{S}_z, \tilde{S}_x\right] = i\bar{h}\tilde{S}_y$$

$$\left[\tilde{S}_x, \tilde{S}_y\right] = i\bar{h}\tilde{S}_z$$

$$\left[\tilde{S}_y, \tilde{S}_z\right] = i\bar{h}\tilde{S}_x.$$

11. Conclude that these matrices mirror the angular-momentum components in Section 12.15.2.
12. What are the eigenvectors of $\tilde{S}_z$? Hint: the standard unit vectors $(1,0)^t$ and $(0,1)^t$.
13. Are they orthonormal?
14. What are their eigenvalues?
15. What are the eigenvectors of $\tilde{S}_x$? Hint: $(1,1)^t$ and $(1,-1)^t$.
16. Are they orthogonal to each other?
17. Normalize them to have norm 1.
18. What are their eigenvalues? Hint: $\pm\bar{h}/2$.
19. Is this as expected? Hint: yes – a Hermitian matrix must have real eigenvalues and orthonormal eigenvectors.
20. Is this as in Sections 12.15.3–12.15.4? Hint: yes – an eigenvalue may be raised or lowered by $\bar{h}$.
21. What are the eigenvectors of $\tilde{S}_y$? Hint: $(1,i)^t$ and $(1,-i)^t$.
22. Are they orthogonal to each other? Hint: calculate their inner product, and don't forget the complex conjugate.
23. Normalize them to have norm 1. Hint: divide by $\sqrt{2}$.

24. What are their eigenvalues?
25. As in spin-one above, interpret these eigenvectors to indicate spin. Hint: see Figures 12.10–12.11.
26. Show that these matrices have the same determinant:

$$\det\left(\tilde{S}_z\right) = \det\left(\tilde{S}_x\right) = \det\left(\tilde{S}_y\right) = -\frac{\bar{h}^2}{4}.$$

Hint: for the definition of the determinant, see Figure 2.11.
27. Show that these matrices have the same square:

$$\tilde{S}_z^2 = \tilde{S}_x^2 = \tilde{S}_y^2 = \frac{\bar{h}^2}{4}I,$$

where I is the 2×2 identity matrix.
28. Does this agree with the eigenvalues calculated above? Hint: take an eigenvector, and apply the matrix to it twice.
29. Consider now both spin and position at the same time. To tell us about both, how should the state look like? Hint: it should be a $2n$-dimensional vector:

$$v \in \mathbb{C}^{2n} :$$

a complex grid function, defined on a new $2 \times n$ grid.
30. Consider now a new physical system: two fermions (say, two electrons). Could they have exactly the same state in $\mathbb{C}^{2n}$? Hint: no! This is Pauli's exclusion principle.

12.16.10   Polarization

1. The Pauli matrices could be used to observe not only spin-one-half but also a completely different physical property, in a completely different physical system.
2. For this purpose, consider now a new physical system: a photon.
3. The photon is a boson. As such, it has a state in $\mathbb{C}^3$, to help model its spin-one.
4. The spin could also be presented as a state in $\mathbb{C}^2$, to help model a new physical property: polarization.
5. Indeed, the photon is not only a particle but also a light ray, or an electromagnetic wave. As such, it travels in some direction: say upwards, in the positive z-direction. At the same time, it also oscillates in the horizontal x-y plane.
6. To help observe this, the Pauli matrices could also serve as new observables. For example, $\tilde{S}_z$ tells us how likely the photon is to oscillate in the x- or y-direction: in a (normalized) state $v \equiv (v_1, v_2)^t$, the probability to oscillate in the x-direction is $|v_1|^2$, whereas the probability to oscillate in the y-direction is $|v_2|^2$. How is this related to the eigenvectors of $\tilde{S}_z$? Hint: v_1 is the inner product of v with $(1,0)^t$, and v_2 is the inner product of v with $(0,1)^t$.
7. At the same time, $\tilde{S}_x$ tells us how likely the photon is to oscillate obliquely, at angle $45°$ in between the x- and y-axes. To calculate the probability for this, just take the eigenvector $(1,1)^t/\sqrt{2}$, calculate its inner product with v, take the absolute value, and square it up.
8. Write the result as a formula in terms of v_1 and v_2.

9. Finally, $\tilde{S}_y$ tells us how likely the photon is to make circles in the x-y plane (Figures 12.10–12.11). To calculate the probability to make circles counterclockwise, take the eigenvector $(1, i)^t/\sqrt{2}$, calculate its inner product with v (don't forget the complex conjugate), take the absolute value, and square it up.

10. Write the result as a formula in terms of v_1 and v_2.

11. To calculate the probability to make circles clockwise, on the other hand, do the same with the orthogonal eigenvector $(1, -i)^t/\sqrt{2}$.

12. Later on, we'll study this in the deeper context of Lie algebras, and introduce a complete geometrical model to explain both spin and polarization at the same time.

13
Quantum Chemistry: Electronic Structure

Look at an individual electron, placed at some point in space. It has an electric charge: minus one. Why doesn't it repel itself? This paradox is answered in quantum mechanics: the electron is not really at one individual point, but at many points at the same time, each with a certain probability.

This makes a distribution: a positive function, whose graph is above the real axis. What is the probability to find the electron in between two points on the real axis? To find out, integrate the area above these points, just below the graph of the function. On the other hand, what is the probability to find the electron at a specific point on the real axis? Zero! After all, the vertical issuing from the point has no width or area at all. Fortunately, in practice, we only need finite precision: it is good enough to locate the electron in a tiny interval, within a tiny error.

Let's use quantum mechanics for a practical purpose: to uncover the structure of the electrons in the atom. Indeed, in quantum mechanics, the position of each electron is a random variable: we can't tell it for sure, but only at some probability. Likewise, energy and momentum are nondeterministic as well: we don't know what they are precisely, but only with some uncertainty. This is not because we are ignorant, but because nature is stochastic.

Fortunately, we can still tell where each electron *might* be, and how likely this is. For this purpose, we need its wave function: its orbital.

In quantum mechanics, electrons are often indistinguishable from each other. To model this, the wave function should take the form of a determinant. This clever algebraic trick leads to a simple formula for the (expected) energy, kinetic and potential alike. This indeed models indistinguishable electrons of the same spin.

How to find the concrete wave function (or orbital)? For this purpose, we need the total energy of the electron, including the entire electrostatic energy due to attraction to the nucleus and repulsion from the other electrons. Again, this is a random variable: it can never be calculated explicitly, but only in terms of expectation. Still, this is good enough for our purpose: solving for the orbital as an eigenvector, with a physical eigenvalue: its energy level.

13.1 Wave Function

13.1.1 Particle and Its Wave Function

Consider a particle in the three-dimensional Cartesian space. In classical mechanics, it has a deterministic position: (x, y, z). In quantum mechanics, on the other hand, its position is nondeterministic: a random variable, known at some probability only.

In Chapter 12, Section 12.14.3, we've already met the state v: a grid function, defined on a uniform $m \times m \times m$ grid. This tells us the (nondeterministic) position of the particle in 3-D. How likely is it to be at $(X_{k,k}, X_{j,j}, X_{i,i})$? The probability for this is $|v_{i,j,k}|^2$.

Now, let's extend this, and make it not only discrete but also continuous. Instead of v, let's talk about a wave function $w(x, y, z)$, defined on the entire three-dimensional Cartesian space. This way, the particle could now be everywhere: not only on a discrete grid but also in just any point in 3-D. How likely is it to be at $(x, y, z) \in \mathbb{R}^3$? The probability for this is just $|w(x, y, z)|^2$.

This makes sense: after all, the position should better be a continuous random variable, which may take just any value, not necessarily in a discrete grid. For this purpose, however, the sums and inner products used in Chapter 12 should be replaced by integrals. In particular, to make a legitimate probability, w must be normalized to satisfy

$$\int \int \int |w(x, y, z)|^2 dx\, dy\, dz = 1,$$

where each integral sign integrates over one spatial coordinate, from $-\infty$ to ∞. This can be viewed as an extension of the vector norm defined in Chapter 26, Section 26.2.4. In this sense, w has norm 1. Later on, we'll make sure that this normalization condition indeed holds.

So far, we've mainly talked about one observable: position. Still, there is yet another important observable, which can never be continuous: energy. Indeed, only some energy levels are allowed, and the rest remain nonphysical. This is indeed quantum mechanics: energy comes in discrete quantities.

How to uncover the wave function w? This will be discussed below. For simplicity, we use atomic units, in which the particle has mass 1, and Planck constant is 1 as well.

13.1.2 Entangled Particles

Consider now two particles that may interact and even collide with each other. (In the language of quantum mechanics, they are *entangled* to each other.) In this case, the particles don't have independent wave functions, but just one joint wave function: $w(r_1, r_2)$, where $r_1 = (x_1, y_1, z_1)$ and $r_2 = (x_2, y_2, z_2)$ are their possible positions. Once $w(r_1, r_2)$ is solved for, it must also be normalized to satisfy

$$\int \int \int \int \int \int |w(r_1, r_2)|^2 dx_1\, dy_1\, dz_1\, dx_2\, dy_2\, dz_2 = 1.$$

This way, $|w(r_1, r_2)|^2$ may indeed serve as a legitimate probability function, to tell us how likely the particles are to be at r_1 and r_2 at the same time (respectively).

13.1.3 Disentangled Particles

Unfortunately, there are still a few problems with this model. First, what about three or four or more particles? The dimension soon gets too high to handle! Besides, even with just two particles, the joint wave function is not very informative: it doesn't give us any information about each individual particle on its own. Therefore, it may make sense to assume that the particles don't interact with each other, so their wave function can be factored as a product of the form

$$w(r_1, r_2) = v^{(1)}(r_1)v^{(2)}(r_2),$$

where $v^{(1)}(r_1)$ and $v^{(2)}(r_2)$ are the wave functions of the individual particles. (In the language of quantum mechanics, they are now *disentangled* from each other.) Later on, we'll improve on this model yet more, to handle not only distinguishable but also indistinguishable particles.

13.2 Electrons in Their Orbitals

13.2.1 Atom: Electrons in Orbitals

Consider now a special kind of particle: an electron. More precisely, consider an atom with M (disentangled) electrons. In particular, look at the nth electron ($1 \leq n \leq M$). It has a (nondeterministic) position $r_n \equiv (x_n, y_n, z_n)$ in the three-dimensional Cartesian space.

Where is the electron? We'll never know for sure! After all, measuring the position is not a good idea – it can change the original wave function forever, with no return. Without doing this, the best we can tell is that the electron *could* be at r_n. The probability for this is $|v^{(n)}(r_n)|^2$, where $v^{(n)}$ is the wave function of the nth electron: its orbital.

In general, $v^{(n)}$ is a complex function, defined in the entire three-dimensional Cartesian space. Like every complex number, $v^{(n)}$ has a polar decomposition. In it, what matters is the absolute value. The phase, on the other hand, has no effect on the probability $|v^{(n)}|^2$. Still, it does play an important role in the dynamics of the system: it tells us the (linear and angular) momentum of the electron, at least nondeterministically. Later on, we'll study this in the deeper context of Lie algebras.

The function $v^{(n)}(r_n)$ is also known as the nth orbital: it tells us where the nth electron could be found in the atom. Unfortunately, $v^{(n)}$ is not yet known. To uncover it, we must solve a generalized eigenvalue problem.

13.2.2 Potential Energy and Its Expectation

What is the potential energy in the atom? This is a random variable, so we can never tell it for sure. Fortunately, we can still tell its expectation. For this purpose, assume again that the electrons are entangled to each other, so

$$w(r_1, r_2, \ldots, r_M)$$

is their joint wave function. This way, $|w(r_1, r_2, \ldots, r_M)|^2$ is the probability to find them at $r_1, r_2, \ldots, r_n$ at the same time (respectively). Later on, we'll write w more explicitly.

The potential energy has a few terms, coming from electrostatics. The first term comes from attraction to the nucleus, placed at the origin (Figure 12.1). To have its expectation, take the probability $|w|^2$ to find the electrons at a particular position, multiply by the potential $1/\|r\|$, sum over all the electrons, and integrate:

$$-\int\int\int \cdots \int\int\int \sum_{n=1}^{M} \frac{|w|^2}{\|r_n\|} dx_1 dy_1 dz_1 dx_2 dy_2 dz_2 \cdots dx_M dy_M dz_M.$$

This is a $3M$-dimensional integral: each integral sign integrates over one individual coordinate, from $-\infty$ to ∞. Later on, we'll simplify this considerably.

On top of this, there is yet more potential energy. This adds more terms, coming from the electrostatic repulsion of every two electrons from each other:

$$\int\int\int \cdots \int\int\int \sum_{i=1}^{M} \sum_{n=i+1}^{M} \frac{|w|^2}{\|r_i - r_n\|} dx_1 dy_1 dz_1 dx_2 dy_2 dz_2 \cdots dx_M dy_M dz_M.$$

This sums all pairs of electrons, indexed by $1 \le i < n \le M$. This way, each pair appears only once, not twice.

These are the Coulomb integrals. Let's go ahead and simplify them.

13.3 Distinguishable Electrons

13.3.1 Hartree Product

Unfortunately, w is not informative enough: it mixes different orbitals with each other. To avoid this, let's separate variables. For this purpose, assume again that the electrons are disentangled from each other. so w can again be written as a product of orbitals:

$$w(r_1, r_2, \ldots, r_M) = v^{(1)}(r_1) v^{(2)}(r_2) \cdots v^{(M)}(r_M).$$

This is the Hartree product. Thanks to it, we'll have more information about the nth individual electron, and how likely it is to be at r_n. In fact, the probability for this will be $|v^{(n)}(r_n)|^2$.

To have this kind of factorization, the electrons must be not only disentangled but also distinguishable from each other. For example, two electrons can be distinguished by spin: one has spin-up, and the other has spin-down (Figure 12.2). This leaves no room for confusion: each has its own identity. This, however, is not always the case. Electrons may have the same spin, and be completely indistinguishable from one another. Such electrons must have a more complicated wave function. Still, for the time being, let's assume that the above factorization is valid.

13.3.2 Potential Energy of a Hartree Product

In the Hartree product, each individual orbital should better make a legitimate probability function:

$$\int \int \int |v^{(i)}(r)|^2 dxdydz = 1, \quad 1 \le i \le M.$$

In this case, the expectation of the potential energy simplifies to read

$$-\sum_{n=1}^{M} \int \int \int \frac{|v^{(n)}|^2}{\|r\|} dxdydz$$

$$+\sum_{i=1}^{M} \sum_{n=i+1}^{M} \int \int \int \int \int \int |v^{(i)}(r)|^2 \frac{1}{\|r - \tilde{r}\|} |v^{(n)}(\tilde{r})|^2 dxdydzd\tilde{x}d\tilde{y}d\tilde{z}.$$

Here, both $r \equiv (x, y, z)$ and $\tilde{r} \equiv (\tilde{x}, \tilde{y}, \tilde{z})$ are dummy variables, integrated upon in this six-dimensional integral. This is why there is no need to index them by i or n any more.

13.4 Indistinguishable Electrons

13.4.1 Indistinguishable Electrons

Unfortunately, two electrons can be distinguished from each other only if they have a different spin: one has spin-up, and the other has spin-down (Figure 12.2). If, on the other hand, they have the same spin, then they can never be distinguished from each other.

Thus, it makes sense to split our electrons into two disjoint subsets. For this purpose, let $0 \le L \le M$ be a new integer number. Now, assume that the L former electrons have spin-up, and the $M - L$ latter electrons have spin-down.

Let's focus on the L former electrons. What is their joint wave function? Well, it can no longer be a simple Hartree product, which tells us nothing about their indistinguishability.

13.4.2 Pauli's Exclusion Principle: Slater Determinant

What is the wave function of the L former electrons? This is a Slater determinant: the determinant of a new $L \times L$ matrix:

$$\frac{1}{\sqrt{L!}} \det \left(\left(v^{(n)}(r_i) \right)_{1 \le i, n \le L} \right).$$

This way, the electrons satisfy Pauli's exclusion principle: two electrons can never have the same state (same spin and also same orbital) at the same time, or the above matrix would have two identical columns, and its determinant would vanish.

To form the Slater determinant, we must first redefine the determinant from scratch, and see its algebraic properties. For this purpose, we need some more background in discrete math.

13.5 Permutations and Their Group

13.5.1 Permutation

To define the Slater determinant, we must introduce yet another important tool: permutation. For this purpose, consider the set of n natural numbers:

$$\{1,\ 2,\ 3,\ \ldots, n\}$$

(for some natural number n). A permutation is a one-to-one mapping from this set onto itself. This means that each natural number $1 \le i \le n$ is mapped to a distinct natural number $1 \le p(i) \le n$. This is denoted by

$$p\left(\{1,\ 2,\ 3,\ \ldots, n\}\right).$$

13.5.2 Switch

For example, the switch

$$(1 \to 3)$$

switches 1 with 3: at the same time, 1 maps to 3, 3 maps back to 1, and the rest of the numbers remain unchanged: 2 maps to 2, 4 maps to 4, and so on. For this reason, the switch is symmetric: it can also be written as

$$(1 \to 3) = (3 \to 1).$$

After the switch, the list of n natural numbers takes a new order:

$$\{3,\ 2,\ 1,\ 4,\ 5,\ \ldots, n\}.$$

We also say that the switch is odd: it picks a minus sign. This is denoted by

$$e((1 \to 3)) = -1.$$

13.5.3 Cycle

A cycle, on the other hand, can be more complicated. For example,

$$(1 \to 3 \to 2)$$

maps 1 to 3, 3 to 2, and 2 back to 1 (at the same time). This is why the cycle is indeed cyclic: it can also be written as

$$(1 \to 3 \to 2) = (2 \to 1 \to 3) = (3 \to 2 \to 1).$$

After the cycle, the new order is

$$\{2,\ 3,\ 1,\ 4,\ 5,\ \ldots, n\}.$$

Indeed, what does this cycle do? It maps 1 to 3. To make room, both 3 and 2 must shift one space leftwards. As a matter of fact, this can be written as the composition (or product) of two switches:

$$(3 \to 2 \to 1) = (3 \to 2)(2 \to 1).$$

This is read leftwards: first, 1 switches with 2, producing

$$\{2, \ 1, \ 3, \ 4, \ 5, \ \ldots, n\}.$$

Then, it switches with 3 as well, producing

$$\{2, \ 3, \ 1, \ 4, \ 5, \ \ldots, n\},$$

as required.

Why is this decomposition useful? Because it tells us that the cycle is even, not odd. Indeed, each switch picks a minus sign, which cancel each other:

$$e\left((3 \to 2 \to 1)\right) = e\left((3 \to 2)\right) e\left((2 \to 1)\right) = (-1)(-1) = 1.$$

Let's introduce a short notation for this cycle:

$$[3 \to 1] \equiv (3 \to 2 \to 1).$$

As discussed above, this cycle is even:

$$e([3 \to 1]) = 1.$$

This is also called a 3-cycle. The switch, on the other hand, is also called a 2-cycle.

Likewise, we can also write a yet longer cycle – a 4-cycle:

$$[4 \to 1] \equiv (4 \to 3 \to 2 \to 1).$$

This cycle can be decomposed as the composition of three switches:

$$[4 \to 1] = (4 \to 3)(3 \to 2)(2 \to 1).$$

Again, this is read leftwards: 1 switches with 2, then with 3, then with 4. The result is

$$\{2, \ 3, \ 4, \ 1, \ 5, \ 6, \ \ldots, \ n\},$$

as required. This is why this cycle is odd, not even:

$$e([4 \to 1]) = -1,$$

and so on.

13.5.4 Decomposition of a Permutation

How does a general permutation look like? Well, let's design one. For this purpose, let's look at 1. Where to map it? Well, assume that 1 maps to some $1 \le k \le n$. This occupies $k = p(1)$. Now, where to map other numbers? Well, they can't map to k any more. As a matter of fact, the rest of the numbers (from 2 to n) could be mapped in two stages:

- First, mix them among themselves (using a smaller permutation q).
- Then, shift those numbers that lie from 2 to k one space leftwards.

This way, k is not used, as required. In summary, the original permutation has been decomposed as

$$p\left(\{1, \ 2, \ 3, \ \ldots, n\}\right) = [k \to 1] q\left(\{2, \ 3, \ 4, \ \ldots, n\}\right),$$

for a unique (smaller) permutation q that mirrors p.

13.5.5 Group of Permutations

Let's place all permutations on $\{1, 2, 3, \ldots, n\}$ in a new group:

$$P\left(\{1,\ 2,\ 3,\ \ldots, n\}\right).$$

What is a group? It is a set of mathematical objects, closed under some algebraic operation, which is associative, but not necessarily commutative. Furthermore, the group must also contain an identity object. Finally, each object must have its own inverse.

In our case, the objects are the permutations, and the operation is composition. Indeed, the composition of two permutations is a legitimate permutation in its own right. Furthermore, we have the identity permutation that changes nothing. Finally, each individual permutation has its (unique) inverse.

Thanks to Section 13.5.4, the entire permutation group can be written as the union of smaller groups:

$$P\left(\{1,\ 2,\ 3,\ \ldots, n\}\right) = \cup_{k=1}^{n} [k \to 1] P\left(\{2,\ 3,\ 4,\ \ldots, n\}\right).$$

Note that, in this union, the k-cycles $[k \to 1]$ have alternating signs: even, odd, even, odd, and so on. In other words,

$$e([k \to 1]) = (-1)^{k-1}.$$

This will be useful later.

13.5.6 Number of Permutations

To use the above group more easily, let's denote it by

$$P \equiv P\left(\{1, 2, 3, \ldots, n\}\right)$$

for short. How big is P? In other words, how many permutations are there in P? Well, in the above union, k could take n possible values: from 1 to n. Thanks to mathematical induction, P must contain $n!$ different permutations:

$$|P| = n!.$$

Half of them are odd, and half are even. To see this, pick some odd permutation $q \in P$, say

$$q \equiv (1 \to 2).$$

This way, for every permutation $p \in P$,

$$e(qp) = e(q)e(p) = -e(p).$$

Thus, the invertible mapping

$$p \to qp$$

maps odd to even permutation, and even to odd.

13.6 Matrix and Its Determinant

13.6.1 The Matrix

Let's use our permutations in linear algebra. For this purpose, consider an $n \times n$ matrix:

$$A \equiv (a_{i,j})_{1 \leq i,j \leq n}.$$

More explicitly,

$$A \equiv \begin{pmatrix} a_{1,1} & a_{1,2} & a_{1,3} & \cdots & a_{1,n} \\ a_{2,1} & a_{2,2} & a_{2,3} & \cdots & a_{2,n} \\ & & \cdots & & \\ & & \cdots & & \\ a_{n,1} & a_{n,2} & a_{n,3} & \cdots & a_{n,n} \end{pmatrix}.$$

In this picture, i indexes the rows, and j indexes the columns. This may give some geometrical intuition about the matrix.

13.6.2 The Determinant

What is the determinant of A? This is a new number: sum of products. Each product multiplies n distinct elements: one from each row, and one from each column:

$$\det(A) \equiv \sum_{p \in P} e(p) a_{1,p(1)} a_{2,p(2)} a_{3,p(3)} \cdots a_{n,p(n)}.$$

Why does this definition make sense? Because it has some attractive properties.

13.6.3 Interchanging Columns

What happens if we interchange two columns? Not much: the determinant only picks a minus sign. For example, let's interchange the first and second column. For this purpose, define again the switch

$$q \equiv (1 \to 2).$$

As discussed in Section 13.5.6, each individual permutation $p \in P$ is mirrored by a unique permutation $qp \in P$. Therefore, to sum over all p's, we can equally well sum over all qp's:

$$\det(A) \equiv \sum_{p \in P} e(p) a_{1,p(1)} a_{2,p(2)} a_{3,p(3)} \cdots a_{n,p(n)}$$

$$= \sum_{p \in P} e(qp) a_{1,qp(1)} a_{2,qp(2)} a_{3,qp(3)} \cdots a_{n,qp(n)}$$

$$= -\sum_{p \in P} e(p) a_{1,qp(1)} a_{2,qp(2)} a_{3,qp(3)} \cdots a_{n,qp(n)},$$

which is minus the determinant of a matrix in which the first and second columns have been interchanged.

13.6.4 Duplicate Columns

What about a matrix with two identical columns? From the above, its determinant is the minus of itself. In other words, its determinant must vanish. This will be useful later.

13.6.5 The Transpose Matrix and Its Determinant

Now, let's interchange the roles of rows and columns. This produces the transpose matrix, denoted by A^t:

$$A^t_{i,j} \equiv a_{j,i}, \quad 1 \leq i, j \leq n.$$

What is its determinant? The same as that of A. Indeed, each permutation $p \in P$ is mirrored by its unique inverse, denoted by p^{-1}. Therefore, to sum over all p's, we can equally well sum over all p^{-1}'s. Furthermore, in each product, we can reorder:

$$\det(A) \equiv \sum_{p \in P} e(p) a_{1,p(1)} a_{2,p(2)} a_{3,p(3)} \cdots a_{n,p(n)}$$

$$= \sum_{p^{-1} \in P} e(p) a_{1,p(1)} a_{2,p(2)} a_{3,p(3)} \cdots a_{n,p(n)}$$

$$= \sum_{p^{-1} \in P} e\left(p^{-1}\right) a_{p^{-1}(1),1} a_{p^{-1}(2),2} a_{p^{-1}(3),3} \cdots a_{p^{-1}(n),n}$$

$$= \sum_{p \in P} e\left(p\right) a_{p(1),1} a_{p(2),2} a_{p(3),3} \cdots a_{p(n),n}$$

$$= \det\left(A^t\right).$$

13.6.6 Interchanging Rows

As discussed above, we can safely interchange the roles of rows and columns. Therefore, interchanging rows has the same effect as interchanging columns: the determinant only picks a minus sign. As a result, a matrix with two identical rows has a zero determinant as well. This will be useful later.

13.6.7 Determinant of a Product

Consider now two matrices of order n:

$$A \equiv (a_{i,j})_{1 \leq i,j \leq n} \quad \text{and} \quad B \equiv (b_{i,j})_{1 \leq i,j \leq n}.$$

What is the determinant of AB? This is just

$$\det(AB) \equiv \sum_{p \in P} e(p)(AB)_{1,p(1)}(AB)_{2,p(2)} \cdots (AB)_{n,p(n)}.$$

In this sum, we have products of n factors, each of the form

$$(AB)_{i,p(i)} \equiv \left(\sum_{j=1}^{n} a_{i,j} b_{j,p(i)} \right).$$

How to multiply these factors from $i = 1, 2, 3, \ldots, n$? Better open parentheses, and pick one particular j from each such factor. Which j to pick? Well, there is no point to pick the same j from two different factors, say the ith and kth factors. After all, the resulting product will be soon canceled with a similar product, obtained from a permutation of the form

$$(i \to k)p,$$

which mirrors p: it is nearly the same as p, but also switches i and k on top, picking a minus sign.

So, we better focus on a more relevant option: pick a different j from each factor, say

$$j = q(i),$$

for some permutation $q \in P$. This way,

$$\det(AB)$$
$$\equiv \sum_{p,q \in P} e(p) a_{1,q(1)} b_{q(1),p(1)} a_{2,q(2)} b_{q(2),p(2)} \cdots a_{n,q(n)} b_{q(n),p(n)}$$
$$= \sum_{p,q^{-1} \in P} e(p) a_{q^{-1}(1),1} b_{1,pq^{-1}(1)} a_{q^{-1}(2),2} b_{2,pq^{-1}(2)} \cdots a_{q^{-1}(n),n} b_{n,pq^{-1}(n)}$$
$$= \sum_{p,q \in P} e(p) a_{q(1),1} b_{1,pq(1)} a_{q(2),2} b_{2,pq(2)} \cdots a_{q(n),n} b_{n,pq(n)}$$
$$= \sum_{r,q \in P} e(r) e(q) a_{q(1),1} b_{1,r(1)} a_{q(2),2} b_{2,r(2)} \cdots a_{q(n),n} b_{n,r(n)}$$
$$= \det\left(A^t\right) \det(B)$$
$$= \det(A) \det(B).$$

In summary, the determinant of the product is indeed the product of the determinants:

$$\det(AB) = \det(A) \det(B).$$

Let's use this result further.

13.6.8 Orthogonal and Unitary Matrix

Thanks to the above discussion, if O is an orthogonal matrix, then its determinant is either 1 or -1:

$$1 = \det\left(O^t O\right) = \det\left(O^t\right) \det(O) = (\det(O))^2.$$

Furthermore, the determinant of the Hermitian adjoint is the complex conjugate of the original determinant:

$$\det\left(A^h\right) = \det\left(\bar{A}^t\right) = \det\left(\overline{A^t}\right) = \overline{\det}(A).$$

As a result, if U is a unitary matrix, then its determinant is a complex number of absolute value 1:

$$1 = \det\left(U^h U\right) = \det\left(U^h\right) \det(U) = |\det(U)|^2.$$

This will be useful below.

13.7 Determinant: Recursive Definition

13.7.1 Minors

Thanks to our permutations, we were able to define the determinant explicitly, with a few interesting properties. Still, to see more, better define the determinant in a new way: recursively. To do this, we must first define a slightly smaller matrix: the minor.

Let $1 \leq i, j \leq n$ be two fixed indices. Define a new $(n-1) \times (n-1)$ matrix: take A, and drop its ith row and jth column. The result is indeed smaller: just $n-1$ rows and $n-1$ columns. This is the (i, j)th minor of A, denoted by $A^{(i,j)}$.

13.7.2 Recursive Definition

Thanks to the minors, we can now go ahead and define the determinant recursively. If A is very small, and contains one element only, then its determinant is just this element. If, on the other hand, A is bigger, then its determinant is a "sum" of the determinants of those minors obtained by dropping the first row:

$$\det(A) \equiv \begin{cases} a_{1,1} & \text{if } n = 1 \\ \sum_{j=1}^{n}(-1)^{j+1}a_{1,j}\det\left(A^{(1,j)}\right) & \text{if } n > 1. \end{cases}$$

This kind of recursion could also be viewed as mathematical induction on $n = 1, 2, 3, \ldots$. Indeed, for $n = 1$, $\det(A)$ is just the only element in A: $\det(A) \equiv a_{1,1}$. For $n > 1$, on the other hand, look at the minors, and calculate their determinants recursively, using the induction hypothesis. This completes the induction step, as required.

Why is this the same as the original definition in Section 13.6.2? To see this, use again mathematical induction on n, and use the union in Section 13.5.5 to mirror the minors.

13.7.3 Working With Columns

As we've seen in Section 13.6.5, the roles of rows and columns could interchange, with no effect on the determinant. Thus, instead of working with rows, we could equally well work with columns:

$$\det(A) = \begin{cases} a_{1,1} & \text{if } n = 1 \\ \sum_{i=1}^{n}(-1)^{i+1}a_{i,1}\det\left(A^{(i,1)}\right) & \text{if } n > 1. \end{cases}$$

This will be useful below.

13.7.4 Multilinearity

Let α be some scalar (number). Let's multiply the first column by α. What is the effect on the determinant? Well, thanks to the above formula, the determinant is multiplied by α as well. In particular, by picking $\alpha = 0$, one can see that a zero column produces a zero determinant.

Furthermore, let's look at the first column, and write it as the sum of two vectors: $v + w$. Let B be a new matrix that is nearly the same as A: only its first column is different – it is the same as v. Likewise, let C be yet another new matrix that is nearly the same as A: only its first column is different – it is the same as w. Thanks to the above formula, we now have

$$det(A) = det(B) + det(C).$$

This means that the determinant is linear in terms of the first column. For example, one could add to the first column any multiple of any other column, with no effect on the determinant.

Thanks to Section 13.6.3, linearity holds not only in terms of the first column but also in terms of any other column as well. Furthermore, thanks to Section 13.6.5, linearity holds not only for columns but also for rows. In summary, the determinant is multilinear.

13.8 Cramer's Formula

13.8.1 Matrix as a Mapping

Let's view A as a mapping of an n-dimensional vector v:

$$v \to u \equiv Av,$$

where $u \equiv Av$ is a new n-dimensional vector, whose ith component is defined as the following sum:

$$u_i \equiv \sum_{j=1}^{n} a_{i,j} v_j.$$

13.8.2 Nonsingular Matrix: Nonzero Determinant

Assume also that A has a nonzero determinant:

$$\det(A) \neq 0.$$

As we'll see below, A is then nonsingular (invertible). In other words, there is a (unique) inverse matrix A^{-1} that stands for the inverse mapping:

$$v = A^{-1}u.$$

Assume now that u is already given, and we seek v. Still, A^{-1} is too expensive to calculate. Could v be uncovered without calculating A^{-1} explicitly? In other words, how to solve the equation

$$Av = u?$$

This is a more realistic task: A is "inverted" not explicitly but only implicitly: find just the unknown vector v, not the entire inverse matrix A^{-1}.

13.8.3 How to Invert Implicitly?

Thanks to the determinant, we can now do just this: solve for v only. For this purpose, let's focus on its ith component: v_i.

Let B be a new matrix that is nearly the same as A. Only its ith column is different: it is the same as the known vector u. This way, we can now define the ith component of v:

$$v_i \equiv \frac{det(B)}{det(A)}.$$

This is called Cramer's formula (or rule).

Why does this make sense? To see this, let's start from a simple case, in which u is a standard unit vector – its first component is 1, and the others vanish:

$$u_k \equiv \begin{cases} 1 & \text{if} \quad k = 1 \\ 0 & \text{if} \quad k \neq 1. \end{cases}$$

Now, let's focus again on some $1 \leq i \leq n$. How does B look like? Well, its ith column is as simple as u: it has only one nonzero element: $B_{1,i} = 1$. Therefore,

$$det(B) = (-1)^{i+1} det\left(A^{(1,i)}\right)$$

(because, in our recursive definition, only one term survives).

To prove Cramer's formula, let C be yet another matrix that is nearly the same as A: only its first row may be different: it is the same as the ith row. This has no effect on those minors that drop the first row:

$$C^{(1,j)} = A^{(1,j)}, \quad 1 \leq j \leq n.$$

What is the determinant of C? Well, it depends: if $i = 1$, then $C = A$, so $det(C) = det(A)$. If, on the other hand, $i \neq 1$, then C has duplicate rows, so $det(C) = 0$.

We are now ready to verify Cramer's formula:

$$(Av)_i = \sum_{j=1}^{n} a_{i,j} v_j$$

$$= \frac{1}{det(A)} \sum_{j=1}^{n} (-1)^{j+1} a_{i,j} det\left(A^{(1,j)}\right)$$

$$= \frac{1}{det(A)} \sum_{j=1}^{n} (-1)^{j+1} C_{1,j} det\left(C^{(1,j)}\right)$$

$$= \frac{det(C)}{det(A)}$$

$$= \begin{cases} 1 & \text{if} \quad i = 1 \\ 0 & \text{if} \quad i \neq 1 \end{cases}$$

$$= u_i,$$

as required.

Likewise, u could be any other standard unit vector, with 1 at the second (or any other) component, and 0 elsewhere. Finally, thanks to multilinearity, u could even be a general vector, as required.

This concludes our algebraic tools. We are now ready to benefit from them geometrically.

13.9 Orbitals and Their Canonical Form

13.9.1 The Overlap Matrix and Its Diagonal Form

Let's use the above in the context of functions, defined in the three-dimensional Cartesian space. The overlap of two functions is obtained by integrating them against one another. This can be viewed as an extension of the inner product, defined in Chapter 26, Section 26.2.1. This way, we can talk about orthogonality, and even orthonormality.

In particular, once this is calculated for every two orbitals, we obtain the $L \times L$ overlap matrix:

$$O \equiv (O_{i,n})_{1 \leq i,n \leq L} \equiv \left(\int \int \int \bar{v}^{(i)}(r) v^{(n)}(r) dx dy dz \right)_{1 \leq i,n \leq L}.$$

(Although we call it 'O', this is not necessarily an orthogonal matrix.)

A proper orbital should have norm 1: overlap 1 with itself. This way, it makes a legitimate probability function. Still, by now, we don't have this property as yet: the main-diagonal elements $O_{i,i}$ may still be different from 1.

Fortunately, O is Hermitian and positive semidefinite (Chapter 26, Section 26.5.1). As such, it can be diagonalized by a unitary matrix U, independent of r:

$$O = U^h D U,$$

where

$$D \equiv \mathrm{diag}\,(D_{1,1}, D_{2,2}, \ldots, D_{L,L})$$

is a diagonal matrix, with the eigenvalues of O on its main diagonal:

$$D_{n,n} \geq 0, \quad 1 \leq n \leq L.$$

Often, the orbitals are linearly independent of each other: they have no linear combination that vanishes (almost) everywhere. In this case, O is not only positive semidefinite but also positive definite – its eigenvalues are strictly positive:

$$D_{n,n} > 0, \quad 1 \leq n \leq L.$$

In this case, the overlap matrix has a positive determinant:

$$\det(O) = \det(D) = D_{1,1} D_{2,2} \cdots D_{L,L} > 0.$$

13.9.2 Unitary Transformation

Let's use the unitary matrix $\bar{U}$ to transform the original orbitals to the new orbitals

$$u^{(i)} \equiv \sum_{j=1}^{L} \bar{U}_{i,j} v^{(j)}.$$

What is the overlap matrix of these new orbitals? Let's look at its (i, n)th element. Since U is independent of r, this element takes the form

$$\int \int \int \bar{u}^{(i)}(r) u^{(n)}(r) dx dy dz = \int \int \int \sum_{j=1}^{L} U_{i,j} \bar{v}^{(j)}(r) \sum_{k=1}^{L} \bar{U}_{n,k} v^{(k)}(r) dx dy dz$$

$$= \sum_{j=1}^{L} \sum_{k=1}^{L} U_{i,j} \left(\int \int \int \bar{v}^{(j)}(r) v^{(k)}(r) dx dy dz \right) \bar{U}_{n,k}$$

$$= \sum_{j=1}^{L} \sum_{k=1}^{L} U_{i,j} O_{j,k} \bar{U}^t_{k,n}$$

$$= \left(U O \bar{U}^t \right)_{i,n}$$

$$= D_{i,n}.$$

Since D is diagonal, the new orbitals are orthogonal to each other: they have zero overlap with each other. Let's go ahead and use this property.

13.9.3 Slater Determinant and Its Overlap

What is the Slater determinant? Well, like every determinant, it is the sum of $L!$ different products, using $L!$ different permutations of $1, 2, \ldots, L$. This determinant is insensitive to the above unitary transformation. Indeed, once written in terms of the new orbitals, it just picks a complex factor of absolute value 1: $\det(U)$. This has no effect on its overlap with itself:

$$\frac{1}{L!} \int \int \int \cdots \int \int \int \left| \det \left(\left(v^{(n)}(r_i) \right)_{1 \leq i, n \leq L} \right) \right|^2 dx_1 dy_1 dz_1 \cdots dx_L dy_L dz_L$$

$$= \frac{1}{L!} \int \int \int \cdots \int \int \int \left| \det \left(\left(v^{(n)}(r_i) \right)_{1 \leq i, n \leq L} \bar{U}^t \right) \right|^2 dx_1 dy_1 dz_1 \cdots dx_L dy_L dz_L$$

$$= \frac{1}{L!} \int \int \int \cdots \int \int \int \left| \det \left(\left(u^{(n)}(r_i) \right)_{1 \leq i, n \leq L} \right) \right|^2 dx_1 dy_1 dz_1 \cdots dx_L dy_L dz_L$$

$$= D_{1,1} D_{2,2} \cdots D_{L,L}.$$

Why? Because the latter integrand is just the complex conjugate of the determinant times the determinant itself. In both, better pick the same permutation, or there would be no contribution at all (thanks to the orthogonality of the transformed orbitals). But the same permutation has no effect: it just interchanges the dummy variables integrated upon. Thanks to the coefficient $1/L!$, the above formula indeed holds.

13.9.4 The Canonical Form

So, it is convenient to work with the transformed orbitals, which have a diagonal overlap matrix. In other words, they are orthogonal to each other. If they are also linearly independent of each other, then they can also be normalized:

$$u^{(i)} \leftarrow D_{i,i}^{-1/2} u^{(i)}.$$

This is indeed their canonical form. In it, they are also orthonormal. In other words, their overlap matrix is just the identity matrix.

In summary, we've got a simple algorithm to normalize the original Slater determinant. First, calculate the original overlap matrix. Then, diagonalize it. Then, use the unitary matrix to transform the orbitals. Finally, normalize the transformed orbitals, to obtain their canonical form. In terms of these final orbitals, the new Slater determinant indeed has overlap 1 with itself, as required. This is why the canonical form is so useful.

Later on, we'll make sure to have the canonical form automatically for free, with no need to use this algorithm at all. Therefore, we can assume that the orbitals are already in their canonical form. (Later on, we'll see that this assumption is indeed plausible.) This will help simplify the expected energy considerably.

13.10 Expected Energy

13.10.1 Coulomb and Exchange Integrals

Yet another Slater determinant can also be defined for the $M - L$ latter orbitals of the remaining spin-down electrons. In summary, our up-to-date wave function takes the form of a product of two Slater determinants:

$$w(r_1, r_2, \ldots, r_M)$$
$$\equiv \frac{1}{\sqrt{L!}} \det\left(\left(v^{(n)}(r_i)\right)_{1 \le i,n \le L}\right) \frac{1}{\sqrt{(M-L)!}} \det\left(\left(v^{(n)}(r_i)\right)_{L < i,n \le M}\right).$$

With this new wave function, what is the expected potential energy? Fortunately, it is much simpler than the general form in Section 13.2.2. In fact, it is nearly as simple as in Section 13.3.2: it contains just one more double sum of new integrals – the exchange integrals.

To see this, as discussed above, assume that $v^{(1)}, v^{(2)}, \ldots, v^{(L)}$ are already in their canonical form: orthonormal in terms of overlap. Likewise, assume that $v^{(L+1)}, v^{(L+2)}, \ldots, v^{(M)}$ are orthonormal as well. Later on, we'll make sure that this is indeed the case.

In the expected energy, we often integrate on the probability function $|w|^2$. It is sometimes more convenient to write it as

$$|w|^2 = \bar{w} \cdot w.$$

This way, the potential due to attraction to the nucleus simplifies to read

$$-\int\int\int\cdots\int\int\int\sum_{n=1}^{M}\frac{|w|^2}{\|r_n\|}dx_1dy_1dz_1dx_2dy_2dz_2\cdots dx_Mdy_Mdz_M$$

$$=-\int\int\int\cdots\int\int\int\bar{w}\sum_{n=1}^{M}\frac{1}{\|r_n\|}wdx_1dy_1dz_1dx_2dy_2dz_2\cdots dx_Mdy_Mdz_M$$

$$=-\sum_{n=1}^{M}\int\int\int\bar{v}^{(n)}\frac{1}{\|r\|}v^{(n)}dxdydz.$$

Why is this correct? Well, let's focus on the former Slater determinant in w. Like every determinant, it is just the sum of products, each uses a different permutation of $1, 2, \ldots, L$. Now, to have a nonzero integral, better pick the same permutation in $\bar{w}$ as in w. (Otherwise, thanks to orthogonality, there would be no contribution at all.) But the same permutation has no effect: it just interchanges the dummy variables integrated upon. Thanks to the coefficient $1/\sqrt{L!}$, the above formula indeed holds.

In the Coulomb integrals, on the other hand, things are not so simple. To have a nonzero contribution, one has two options. In the main option, pick the same permutation in $\bar{w}$ as in w. This will produce new Coulomb integrals of the form

$$\sum_{i=1}^{M}\sum_{n=i+1}^{M}\int\int\int\int\int\int|v^{(i)}(r)|^2\frac{1}{\|r-\tilde{r}\|}|v^{(n)}(\tilde{r})|^2dxdydzd\tilde{x}d\tilde{y}d\tilde{z}.$$

Still, this is not the only option: in $\bar{w}$, one could also pick a slightly different permutation, in which i and n switch on top ($1 \le i, n \le L$, or $L < i, n \le M$). For example, if p is picked in w, then pick

$$(n \to i)p$$

in $\bar{w}$. If p is even (odd), then $(n \to i)p$ is odd (even):

$$e\left((n \to i)p\right) = e\left((n \to i)\right)e(p) = -e(p).$$

This will produce the so-called exchange integrals, with a minus sign:

$$-\sum_{i=1}^{M}\sum_{\substack{n>i \text{ same spin}}}\int\int\int\int\int\int\bar{v}^{(i)}(r)v^{(i)}(\tilde{r})\frac{1}{\|r-\tilde{r}\|}\bar{v}^{(n)}(\tilde{r})v^{(n)}(r)dxdydzd\tilde{x}d\tilde{y}d\tilde{z}$$

Let's see what this means for each individual orbital.

13.10.2 Effective Potential Energy

Now, let's focus on the nth electron only. What is the effective potential that it feels? Well, it feels attraction to the nucleus:

$$-\int\int\int\bar{v}^{(n)}\frac{1}{\|r\|}v^{(n)}dxdydz.$$

On top of this, it also feels repulsion from all other electrons:

$$+\sum_{i=1}^{M}\int\int\int\int\int\int |v^{(i)}(r)|^2 \frac{1}{\|r-\tilde{r}\|}|v^{(n)}(\tilde{r})|^2 dxdydzd\tilde{x}d\tilde{y}d\tilde{z}.$$

Here one may ask: does it feel any repulsion from itself? No, it doesn't. There is one fictitious term in this sum: the term for which $i = n$. Don't worry: it will drop soon.

On top of this, it also feels the exchange force from all other electrons of the same spin:

$$-\sum_{i,\text{ same spin as }n}\int\int\int\int\int\int \bar{v}^{(i)}(r)v^{(i)}(\tilde{r})\frac{1}{\|r-\tilde{r}\|}\bar{v}^{(n)}(\tilde{r})v^{(n)}(r)dxdydzd\tilde{x}d\tilde{y}d\tilde{z}.$$

Here one may ask: does it feel any exchange force from itself? No, it doesn't. There is one fictitious term in the above sum: the term for which $i = n$. Fortunately, it cancels the fictitious term introduced above.

Thanks to these fictitious terms, we have uniformity: the sums go over $i = 1, 2, \ldots, M$, including $i - n$. This way, the nth orbital will solve the same equation as the other orbitals, as discussed below.

13.10.3 Kinetic Energy

On top of this, the nth electron also has its own kinetic energy:

$$\frac{1}{2}\int\int\int \nabla^t \bar{v}^{(n)} \cdot \nabla v^{(n)} dxdydz,$$

where '∇' gives the vector of partial derivatives (and '∇^t' gives the row vector).

13.10.4 Schrodinger Equation – Integral Form

Together, all these terms must sum to the expected energy of the nth electron:

$$E\int\int\int |v^{(n)}|^2 dxdydz,$$

where E is a constant energy level: an eigenvalue of the Hamiltonian. After all, in quantum mechanics, energy comes in discrete quantities. Only these energy levels are allowed.

This is indeed Schrodinger equation for the nth electron: its kinetic and effective potential energy (including the exchange terms) sum up to its entire expected energy. Later on, once Lie algebras are introduced, we'll also see the time-dependent Schrodinger equation.

13.11 The Hartree-Fock System

13.11.1 Basis Functions – The Coefficient Matrix

So far, our orbital has been a function in the three-dimensional Cartesian space. This is too general. To help uncover the orbital, we better approximate it by a piecewise-polynomial function. For this purpose, let's write it as a linear combination of basis functions:

$$v^{(n)} \doteq \sum_{j=1}^{K} c_j \psi_j,$$

where the ψ_j's are the basis functions, and the c_j's are their (unknown) complex coefficients. (Actually, c_j depends on n as well, but we can disregard this, since we focus on one particular n.) Let's plug this in the effective energy in Section 13.10.2, term by term. For this purpose, in each term, replace $\bar{v}^{(n)}$ by $\bar{\psi}_l$, and $v^{(n)}$ by ψ_j. (Although ψ_l is often real, we still take its complex conjugate, to be on the safe side.) This will assemble the (l, j)th element in the coefficient matrix A:

$a_{l,j}$

$$\equiv -\int\int\int \bar{\psi}_l \frac{1}{\|r\|} \psi_j \, dx \, dy \, dz$$

$$+ \sum_{i=1}^{M} \int\int\int\int\int\int |v^{(i)}(r)|^2 \frac{1}{\|r - \tilde{r}\|} \bar{\psi}_l(\tilde{r})\psi_j(\tilde{r}) \, dx \, dy \, dz \, d\tilde{x} \, d\tilde{y} \, d\tilde{z}$$

$$- \sum_{i, \text{ same spin as } n} \int\int\int\int\int\int \bar{v}^{(i)}(r)v^{(i)}(\tilde{r}) \frac{1}{\|r - \tilde{r}\|} \bar{\psi}_l(\tilde{r})\psi_j(r) \, dx \, dy \, dz \, d\tilde{x} \, d\tilde{y} \, d\tilde{z}$$

$$+ \frac{1}{2} \int\int\int \nabla^t \bar{\psi}_l \cdot \nabla \psi_j \, dx \, dy \, dz$$

$(1 \leq l, j \leq K)$. This way, A is no longer a constant matrix. On the contrary: it depends on the orbitals – the unknown $v^{(i)}$'s in the above sums. Still, for fixed orbitals, A is Hermitian, as required.

13.11.2 The Mass Matrix

Now, define also the mass matrix B. Its (l, j)th element is just the overlap of ψ_l with ψ_j:

$$b_{l,j} \equiv \int\int\int \bar{\psi}_l \psi_j \, dx \, dy \, dz$$

$(1 \leq l, j \leq K)$.

How to solve for the unknown c_j's? For this purpose, place them in a new K-dimensional vector:

$$\mathbf{c} \equiv (c_1, \ c_2, \ c_3, \ \ldots, \ c_K)^t.$$

This way, we can now plug our discrete approximation in. The effective energy in Sections 13.10.2–13.10.3 takes now the discrete form

$$\mathbf{c}^h A \mathbf{c} = E \mathbf{c}^h B \mathbf{c},$$

where E is the (unknown) energy level. Thus, this is a nonlinear equation, with two types of unknowns: the vector $\mathbf{c}$, and the scalar E.

13.11.3 Pseudo-Eigenvalue Problem

How to make sure that the orbitals are indeed in their canonical form? In other words, how to make sure that same-spin orbitals are indeed orthonormal in terms of overlap? Fortunately, same-spin orbitals solve the same generalized eigenvalue problem, with the same (Hermitian) coefficient matrix A. Assume that they have distinct generalized eigenvalues (energy levels). (Otherwise, assume that they have already passed a Gram-Schmidt orthogonalization.) Thanks to the exercises at the end of Chapter 26, they are indeed orthogonal to each other: have zero overlap with one another. Once normalized properly, they are indeed in their canonical form, as assumed all along.

The mass matrix B is the same for all orbitals. The coefficient matrix A, on the other hand, is not. It depends on the orbitals, or, more precisely, on their spin. For $1 \le n \le L$ (spin-up), A comes in one form. For $L < n \le M$ (spin-down), on the other hand, A has a different form. Still, only same-spin orbitals should be orthogonal to each other. Fortunately, they share the same A, and solve the same pseudo-eigenvalue problem.

What should the energy level E be? Well, it should be minimal. For this purpose, we need to solve a pseudo-eigenvalue problem:

$$Ac = EBc.$$

The term "pseudo" reminds us that this is actually a nonlinear system: A depends on the unknown orbitals. Still, for fixed orbitals (say the solutions), A is Hermitian, as required.

Furthermore, on the right-hand side, we have yet another symmetric matrix: B. This is a generalized eigenvalue problem (as in the exercises at the end of Chapter 26), whose eigenvectors indeed produce orthogonal orbitals of zero overlap with one another. Once normalized properly, they are indeed in their canonical form, as discussed above.

13.12 Exercises: Orbitals and Their Canonical Form

13.12.1 Permutation – Product of Switches

1. For some $1 \le i < k \le n$, consider a switch of the form $(i \to k)$. What does it do? Hint: at the same time, i maps to k, and k maps back to i.
2. Show that it is symmetric:

$$(i \to k) = (k \to i).$$

3. What is its inverse? Hint: itself.
4. Consider a cycle of the form

$$[k \to i] \equiv (k \to k-1 \to k-2 \to \cdots \to i+1 \to i).$$

What does it do? Hint: at the same time, k maps to $k-1$, $k-1$ maps to $k-2$, ..., $i+1$ maps to i, and i maps back to k.
5. Is it symmetric as well? Hint: only if $k = i+1$.
6. Write it as a product (composition) of switches:

$$[k \to i] = (k \to k-1)(k-1 \to k-2) \cdots (i+1 \to i).$$

7. What is its inverse? Hint:

$$[k \to i]^{-1} = (i \to i+1)(i+1 \to i+2)\cdots(k-1 \to k) = [i \to k].$$

8. Write the original switch $(i \to k)$ as a composition of two such cycles. Hint:

$$(i \to k) = [i \to k-1] \circ [k \to i].$$

9. Conclude once again that the original switch is odd. Hint:

$$
\begin{aligned}
e((i \to k)) &= e\left([i \to k-1] \circ [k \to i]\right) \\
&= e\left([i \to k-1]\right) e\left([k \to i]\right) \\
&= (-1)^{k-1-i+k-i} \\
&= -1.
\end{aligned}
$$

10. Consider now a more general 3-cycle:

$$(l \to k \to i).$$

What does it do? Hint: at the same time, l maps to k, k maps to i, and i maps back to l.

11. Write it as a product of two switches. Hint:

$$(l \to k \to i) = (l \to k)(k \to i).$$

12. Conclude that it is even.

13. Consider now a general permutation

$$p \in P.$$

Write it as a product of general cycles. Hint: start from 1. It must map to some number, which must map to some other number, and so on, until returning back to 1. This completes one general cycle. The rest is a disjoint (smaller) permutation, which can benefit from an induction hypothesis.

14. Conclude that every permutation can be written as a product of general cycles, each written as a product of switches, each written as a product of two more elementary cycles, each written as a product of most elementary switches, as above.

15. In the language of group theory, the elementary switches (that switch two neighbors with each other) *generate* the entire group of permutations.

16. What can you say about the determinant of the transpose?

17. What can you say about the determinant of the Hermitian adjoint?

18. What can you say about the determinant of a product of two matrices?

19. What can you say about the determinant of an orthogonal matrix?

20. What can you say about the determinant of a unitary matrix?

13.12.2 How to Have The Canonical Form?

1. Consider the unitary transformation in Section 13.9.2. How does it affect the Slater determinant? Hint: it multiplies it by $\det(U^h)$.

2. Does this affect the absolute value of the Slater determinant? Hint: from Section 13.6.8,

$$|\det(U)| = |\det(U^h)| = 1.$$

3. Consider the original overlap matrix O (Section 13.9.1). Is it Hermitian?
4. Is it positive semidefinite?
5. Is it positive definite? Hint: only if the original orbitals are linearly independent of each other: they have no linear combination that vanishes (almost) everywhere.
6. What does "almost" mean? Hint: see Chapter 4, Section 4.6.5.
7. Show that

$$\det(O) = \det(D) = D_{1,1} D_{2,2} \cdots D_{n,n}.$$

Hint: O and D share the same determinant.

8. Is this positive? Hint: only if the original orbitals are linearly independent of each other: they have no linear combination that vanishes (almost) everywhere.
9. How to normalize the original Slater determinant without calculating the unitary transformation explicitly? Hint: just divide by $\sqrt{\det(O)}$.
10. Is this necessary? Hint: no! Next, we'll obtain the canonical form automatically for free.
11. Why is the canonical form attractive? Hint: it helps simplify the integrals in the expected (potential) energy.
12. Show that same-spin orbitals are eigenfunctions of the same generalized eigenvalue problem, with the same (Hermitian) coefficient matrix, and the same mass matrix on the right-hand side. This way, if they have different generalized eigenvalues (energy levels), then they are indeed orthogonal to each other: have zero overlap with one another. Once normalized properly, they are indeed in their canonical form, as assumed all along. Hint: see exercises below.
13. Show that the coefficient matrix A in Section 13.11.1 is indeed Hermitian, provided that the orbitals in the integrand are fixed.
14. Show that the mass matrix in Section 13.11.2 is indeed symmetric.
15. Look at the pseudo-eigenvalue problem in Section 13.11.3. In what sense is it "pseudo?" Hint: nonlinearity.
16. In what sense is it "generalized?" Hint: the mass matrix can be a general Hermitian matrix, not only the identity matrix.
17. If two (pseudo) eigenvectors share the same generalized eigenvalue E (energy level), could we still assume that they are orthogonal to each other? Hint: yes, assume that they have already passed a (generalized) Gram-Schmidt orthogonalization (with respect to the mass matrix).
18. If, on the other hand, they have different E's, are they still orthogonal to each other? Hint: yes, see exercises at the end of Chapter 26.
19. Conclude that the orbitals formed from them have zero overlap with each other.
20. Conclude that, once normalized properly, these orbitals are indeed in their canonical form.
21. Conclude that, in retrospect, it was indeed plausible to simplify the Coulomb and exchange integrals in Section 13.10.1.

Part VI

Introduction
to Lie Algebras
and Their Applications

Introduction to Lie Algebras and their Applications

So far, we've discussed a few branches in physics. What do they have in common? Well, they have a common algebraic structure: Lie algebras.

In this part, we introduce Lie algebras from scratch. Furthermore, we uncover their inner structure, using a very simple tool: mathematical induction. In particular, we decompose the Lie algebra in terms of simple ideals, useful in quantum and Hamiltonian mechanics.

What are these simple ideals? Well, they are Lie algebras in their own right, which may take a familiar face: they contain matrices that could stand for observables in quantum mechanics. Furthermore, once exponentiated, these matrices make a new Lie group, containing a few kinds of useful matrices: rotation matrices in geometry, and Lorentz matrices in special relativity.

Thanks to Lie algebras, we can also introduce Hamiltonian mechanics, including a few conservation and symmetry laws. Finally, at the end of the book, a treat is waiting: a complete geometrical model to explain both spin and polarization at the same time. This shows quite vividly how useful Lie algebras are in modern physics as well.

The Jordan Form
of a Matrix
and Its Extension
to Algebras

To introduce Lie algebras, we need some material from linear and modern algebra. For a start, we need a fundamental theorem in number theory: the Chinese remainder theorem. This will help in linear algebra: design the Jordan decomposition of a matrix. This will help in modern algebra as well: design a subalgebra, in which a given derivation has a Jordan form as well. This is indeed a good preparation work for Lie algebras and their applications.

14.1 Nilpotent Matrix and Generalized Eigenvectors

14.1.1 Nilpotent Matrix

Let B be a square (real or complex) matrix. Assume that, if raised to a sufficiently high power, then it would vanish. In other words, for some natural number $m \geq 1$,

$$B^m = (0)$$

(the zero matrix). Then we say that B is nilpotent. Clearly, m is not unique: it could increase, and still satisfy the above equation. Later on, we'll see how to pick a good m.

What does this mean for an individual vector? To see this, let u be a nonzero vector. Clearly,

$$B^m u = (0)u = \mathbf{0}$$

(the zero vector). Still, there is no need to go to a power as high as m: there may be a (smaller) number

$$1 \leq k \leq m$$

such that

$$B^k u = \mathbf{0}.$$

Unlike m, which depends on B only, k depends on u too. Still, like m, k is not unique: it could increase, and still satisfy this equation. Let's pick a proper k.

14.1.2 Cycle – Invariant Subspace

Assume that k is the minimal number that satisfies the above equation. This way,

$$B^{k-1}u \neq 0$$

is an eigenvector of B, with the eigenvalue 0:

$$B\left(B^{k-1}u\right) = B^k u = 0.$$

In summary, we can start from u, and apply B time and again, producing a new cycle of k new vectors:

$$u, \; Bu, \; B^2 u, \; \ldots, \; B^{k-1}u.$$

These are called generalized eigenvectors. Only the latter is a legitimate eigenvector: the former, on the other hand, are not.

Clearly, the cycle is invariant under B: each vector is mapped to the next one. (The only exception is the latter vector, which is mapped to zero.) For this reason, the cycle spans an invariant subspace.

In the simple case in which $k = 1$, the cycle is very short: it contains u only. In this special case, u is a legitimate eigenvector in its own right.

14.1.3 Generalized Eigenvectors – Linear Independence

Let's order the cycles in terms of decreasing length. For example, assume that the longest cycle contains four vectors:

$$u, \; Bu, \; B^2 u, \; B^3 u.$$

Furthermore, assume that the second longest cycle contains three vectors:

$$v, \; Bv, \; B^2 v.$$

These seven vectors are generalized eigenvectors. Are they linearly independent? To make sure, we need one more assumption.

Assume that the eigenvectors $B^3 u$ and $B^2 v$ are linearly independent of one another. Are the above seven vectors linearly independent as well?

To check on this, assume that

$$c_0 u + c_1 Bu + c_2 B^2 u + c_3 B^3 u + d_0 v + d_1 Bv + d_2 B^2 v = 0.$$

Must all these coefficients vanish:

$$c_0 = c_1 = c_2 = c_3 = d_0 = d_1 = d_2 = 0?$$

To check on this, place them in the top diagram in Figure 14.1: the longer cycle at the top row, and the shorter at the bottom row. Now, apply B to the entire linear combination:

$$B\left(c_0 u + c_1 Bu + c_2 B^2 u + c_3 B^3 u + d_0 v + d_1 Bv + d_2 B^2 v\right) = 0.$$

Since $B^4 u = B^3 v = 0$, this simplifies to read

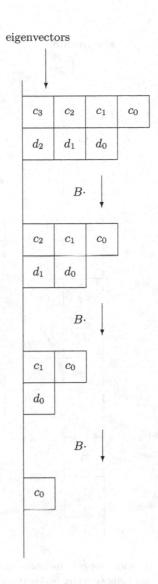

Fig. 14.1. If the eigenvectors $B^3 u$ and $B^2 v$ are linearly independent, then so are all their generalized eigenvectors. Indeed, apply B to the entire linear combination time and again, shifting the diagram leftwards. Those cells that pass the left edge drop, leaving the upper-right cell only, which must therefore vanish: $c_0 = 0$.

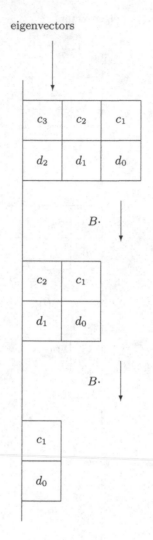

Fig. 14.2. Once c_0 has dropped, repeat the above process once again, shifting leftwards as before. This time, only two cells survive. Since the eigenvectors are linearly independent, we must have $d_0 = c_1 = 0$. Drop them, and repeat the same process again, and so on.

$$c_0 Bu + c_1 B^2 u + c_2 B^3 u + d_0 Bv + d_1 B^2 v = \mathbf{0}.$$

This is illustrated geometrically in Figure 14.1: the entire diagram shifts leftwards, and the leftmost cells drop. Next, apply B once again. This shifts the diagram once again, dropping yet another column from the left. Finally, apply B for the last time. This shifts the diagram once again, leaving one last cell on the right: c_0.

We can therefore conclude that

$$c_0 = 0,$$

as required. This "shaves" the rightmost cell from the original diagram. The process can now restart all over again, as in Figure 14.2, to conclude that

$$c_1 = d_0 = 0.$$

This "shaves" the second-right column as well, and so on. In the end, we'll conclude that all the coefficients must indeed vanish, as asserted.

This is not limited to cycles of length three or four only: it could be extended to longer cycles as well, using a nested mathematical induction.

Furthermore, this is not limited to two cycles: it is true for one cycle as well. This gives us our first bonus: the vectors in a cycle are linearly independent of each other. Therefore, the length of the cycle cannot possibly exceed the dimension of our linear space (the order of B).

Moreover, the above process is not limited to two cycles only: it could be extended to many more cycles as well. We therefore get our second bonus: if the eigenvectors are linearly independent, then so are also all the generalized eigenvectors in their cycles.

14.1.4 Linear Dependence

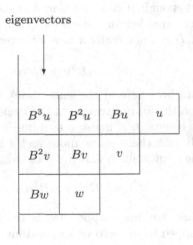

Fig. 14.3. If there is no new eigenvector any more, then there is no new generalized eigenvector either. Indeed, if Bw depends linearly on B^3u and B^2v, then w depends linearly on their cycles (the vectors in the first and second rows).

Next, let's work the other way around: assume that there is no new eigenvector any more. Is it also true that there is no new *generalized* eigenvector any more?

To check on this, let w be some vector, with a rather short cycle of length 2 only:

eigenvectors

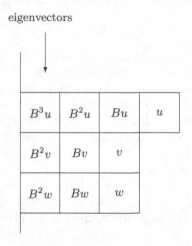

Fig. 14.4. Furthermore, if B^2w depends linearly on B^3u and B^2v, then both w and Bw depend linearly on the two former cycles (the vectors in the two upper rows).

$$w, \ Bw.$$

This new cycle is short enough: it could be placed at the bottom row in our diagram, and still obey our decreasing-length convention (Figure 14.3).

Now, assume that Bw is not really a new eigenvector, but depends linearly on the former ones:

$$Bw = \alpha B^3u + \beta B^2v$$

(where α and β are not both zero at the same time). Now, what about w? Does it also depend linearly on the former vectors in the cycles above it?

Well, in the above equation, B maps w to a linear combination of eigenvectors. Still, w is not the only one: there is one more vector mapped in the same way! To design it, just shift the coefficients α and β rightwards, to the column just above w:

$$B\left(\alpha B^2u + \beta Bv\right) = \alpha B^3u + \beta B^2v.$$

So, we got two vectors that are mapped by B to the same vector. Thus, their difference must be mapped to the zero vector, and must therefore be an eigenvector in its own right:

$$w - \alpha B^2u - \beta Bv = \gamma B^3u + \delta B^2v.$$

So, w is not really new, as asserted.

Next, assume that w makes a slightly longer cycle:

$$w, \ Bw, \ B^2w.$$

This is still short enough: it could be placed at the bottom row, without violating our decreasing-length rule (Figure 14.4).

Assume now that the "new" eigenvector B^2w is not really new: it depends linearly on the former eigenvectors:

$$B^2w = \alpha B^3u + \beta B^2v.$$

As discussed above, we can then write Bw as a linear combination as well:

$$Bw = \alpha B^2u + \beta Bv + \gamma B^3u + \delta B^2v.$$

So, B maps w to this linear combination. Still, w is not the only one: there is yet another vector mapped in the same way. To design it, just shift the coefficients α, β, γ, and δ rightwards:

$$B\left(\alpha Bu + \beta v + \gamma B^2u + \delta Bv\right) = \alpha B^2u + \beta Bv + \gamma B^3u + \delta B^2v.$$

So, we got two vectors that are mapped by B to the same vector. Thus, their difference must be mapped to zero, and is therefore an eigenvector in its own right:

$$w - \alpha Bu - \beta v - \gamma B^2u - \delta Bv = \zeta B^3u + \eta B^2v.$$

So, w is not really new: it depends linearly on the previous generalized eigenvectors.

This is not limited to this small example. In fact, it could be extended to many more rows and columns, using a nested mathematical induction.

In summary, if there is a really new w, then there is also a really new eigenvector. The cycle leading to it can't be too long, or it would have been used earlier (as in our convention). Thus, the entire cycle must be really new. After all, we already know that all generalized eigenvectors in all these cycles must be linearly independent of each other (Section 14.1.3). As such, they could help form a new basis: Jordan basis.

14.1.5 Jordan Basis

Recall that B is nilpotent:

$$B^m = (0).$$

Recall also that m is not unique: if m increases, then the above equation is still valid. How to pick a proper m?

Assume that m is as big as the length of the longest cycle. This way, we can indeed start from the longest cycle, and design shorter and shorter cycles, until having a complete basis for our linear space.

For example, m could be as big as the order of B. In this case, no cycle could possibly exceed m:

$$k \leq m.$$

(Otherwise, the vectors in the cycle would depend linearly on each other, which is impossible, as discussed at the end of Section 14.1.3.) Thus, nothing could stop us: we could go on and on, and design shorter and shorter cycles, until the entire basis of m vectors is complete.

14.1.6 Jordan Blocks

Let's order our cycles in a reverse order. In the above example, this gives

$$\{\ldots,\; B^2v,\; Bv,\; v,\; B^3u,\; B^2u,\; Bu,\; u\}.$$

How does B act on this list? Well, each vector is mapped to the previous one. (The only exceptions are the eigenvectors, which are mapped to zero.)

Now, in the above list, represent each vector by a standard unit vector (one component is 1, and the others are 0). This way, B takes a particularly simple form:

$$B = \begin{pmatrix} \ddots & & & & & & & \\ & \ddots & & & & & & \\ & & 0 & 1 & & & & \\ & & & 0 & 1 & & & \\ & & & & 0 & & & \\ & & & & & 0 & 1 & \\ & & & & & & 0 & 1 \\ & & & & & & & 0 & 1 \\ & & & & & & & & 0 \end{pmatrix}.$$

This is the Jordan form of B. Most elements in it are zero. (This is why they are missing – we don't even bother to write them.) Only one diagonal is nonzero: the superdiagonal, just above the main diagonal. In our example, there are two Jordan blocks: at the lower-right corner, there is a 4×4 block, with three 1's in the superdiagonal. Before it, there is a 3×3 block, with just two 1's in the superdiagonal.

In general, what is the order of B? It is the total number of vectors in all cycles.

Next, look at higher powers of B: B^2, B^3, ... In these powers, what happens to the Jordan blocks? Well, the 1's shift rightwards, to the next higher superdiagonal, until they exceed the Jordan block at the upper-right corner, and drop. This is why, in our example,

$$B^4 = (0).$$

What happens in a more general matrix, not necessarily nilpotent? Well, there is only one change: in the Jordan form, the main diagonal could be nonzero as well. Let's see this in detail.

14.2 Matrix and Its Jordan Form

14.2.1 Characteristic Polynomial, Eigenvalue, and Multiplicity

So far, we've studied the nilpotent matrix B. Next, let's discuss a more general matrix A (of order n), nilpotent or not. Its Jordan form is slightly different: the main-diagonal elements are not necessarily zero any more.

The characteristic polynomial of A could be written like this:

$$\det(A - \lambda I) = (\lambda_1 - \lambda)(\lambda_2 - \lambda)(\lambda_3 - \lambda) \cdots (\lambda_n - \lambda),$$

where the λ_i's (the eigenvalues of A) are not necessarily distinct: some of them could be the same. To avoid this, better rewrite this as

$$\det(A - \lambda I) = (\lambda_1 - \lambda)^{n_1}(\lambda_2 - \lambda)^{n_2}(\lambda_3 - \lambda)^{n_3}\cdots(\lambda_k - \lambda)^{n_k}$$

for some new $k \geq 1$, and new λ_i's that are now *distinct*, each with its own multiplicity: n_i. Clearly, the multiplicities sum to n:

$$n_1 + n_2 + n_3 + \cdots + n_k = n.$$

14.2.2 Matrix Block – Invariant Subspace

Let's focus on the first eigenvalue: λ_1. Let's subtract λ_1 from the main diagonal in A, and write the result in a block form:

$$A - \lambda_1 I = \begin{pmatrix} C - \lambda_1 I & \\ E & J - \lambda_1 I \end{pmatrix},$$

where I is an identity matrix of a suitable order.

In the above, what is J? To see this, let

$$V \equiv \{v \mid (A - \lambda_1 I)^{n_1} v = 0\}$$

be the null space of $(A - \lambda_1 I)^{n_1}$. Clearly, this subspace is invariant under $A - \lambda_1 I$. This is why the upper-right block is missing in the above decomposition. Indeed, define J as A, restricted to this subspace only:

$$J \equiv A \mid_V.$$

This is why J appears in the lower-right corner: it acts on those standard unit vectors that span V.

14.2.3 Matrix Block and Its Jordan Form

As a matter of fact, we already know these vectors. Indeed, thanks to the above definition, $J - \lambda_1 I$ is nilpotent:

$$(J - \lambda_1 I)^{n_1} = (A - \lambda_1 I)^{n_1} \mid_V = (0).$$

Thus, $J - \lambda_1 I$ (and indeed J itself as well) has a Jordan form, as in Section 14.1.6:

$$J = \begin{pmatrix} \ddots & & & & & & & \\ & \ddots & & & & & & \\ & & \lambda_1 & 1 & & & & \\ & & & \lambda_1 & 1 & & & \\ & & & & \lambda_1 & & & \\ & & & & & \lambda_1 & 1 & \\ & & & & & & \lambda_1 & 1 \\ & & & & & & & \lambda_1 \end{pmatrix}.$$

This is a more general Jordan form: there are nonzero elements not only on the superdiagonal but also on the main diagonal.

14.2.4 Matrix Block and Its Characteristic Polynomial

How big is J? To see this, decompose the characteristic polynomial of A:

$$\det(A - \lambda I) = \det(C - \lambda I)\det(J - \lambda I),$$

where I is an identity matrix of a suitable order.

So, how could the characteristic polynomial of J look like? Well, it could only contain factors from the characteristic polynomial of A. Could it contain a factor like $\lambda_2 - \lambda$? No! After all, no eigenvector corresponding to λ_2 could ever belong to V.

Thus, the characteristic polynomial of J must look like

$$\det(J - \lambda I) = (\lambda_1 - \lambda)^l$$

for some natural number l. This also follows from the Jordan form of J, illustrated above.

What could l be? Could

$$l > n_1?$$

No! After all, the characteristic polynomial of J could only contain a factor from the characteristic polynomial of A. On the other hand, could

$$l < n_1?$$

Well, if this were true, then C would have the factor $\lambda_1 - \lambda$ in its own characteristic polynomial as well. This way, C would also have an eigenvector $w \neq 0$, corresponding to the eigenvalue λ_1. But then we could design a new eigenvector for A as well: just add a few dummy components to w:

$$(A - \lambda_1 I)\begin{pmatrix} w \\ 0 \end{pmatrix} = \begin{pmatrix} 0 \\ Ew \end{pmatrix},$$

so

$$(A - \lambda_1 I)^{l+1}\begin{pmatrix} w \\ 0 \end{pmatrix} = (A - \lambda_1)^l\begin{pmatrix} 0 \\ Ew \end{pmatrix} = \begin{pmatrix} 0 \\ (J - \lambda_1 I)^l\, Ew \end{pmatrix} = \begin{pmatrix} 0 \\ (0)Ew \end{pmatrix} = \begin{pmatrix} 0 \\ 0 \end{pmatrix},$$

so

$$(A - \lambda_1)^{n_1}\begin{pmatrix} w \\ 0 \end{pmatrix} = 0$$

as well. In summary, we've managed to design a new vector that can't possibly exist: on one hand, it lies outside V (because $w \neq 0$). On the other hand, it lies in the null space of $(A - \lambda_1 I)^{n_1}$, which is nothing but V!

This is a contradiction. We must therefore have

$$l = n_1.$$

In summary, J is an $n_1 \times n_1$ matrix block. Its Jordan form is already illustrated above.

Now, there is nothing special about λ_1: the same could be done for any other eigenvalue as well.

Thus, in its final Jordan form, A will be block-diagonal, with k such Jordan blocks along its main diagonal. To design this explicitly, we need some background in number theory.

14.3 Greatest Common Divisor

14.3.1 Integer Division With Remainder

Let's write the Jordan form in a more elegant way. Later on, this will help prove Cartan's criterion in Lie algebras (Chapter 20).

For this purpose, let $n > m > 0$ be two natural numbers. What is the ratio n/m? Here, we don't mean the usual fraction n/m, but the maximal integer number that doesn't exceed it:

$$\left\lfloor \frac{n}{m} \right\rfloor = \max \left\{ j \in \mathbb{Z} \mid j \le \frac{n}{m} \right\},$$

where $\mathbb{Z}$ is the set of integer numbers. This way, we can now divide n by m with remainder: represent n as

$$n = km + l,$$

where

$$k = \left\lfloor \frac{n}{m} \right\rfloor,$$

and l is the remainder (or residual) that is too small to be divided by m:

$$0 \le l < m.$$

This l is also called n modulus m:

$$l = (n \bmod m).$$

14.3.2 Congruence: Same Remainder

The above equation could also be written as

$$n \equiv l \bmod m.$$

This means that n and l are related to each other: they are *congruent* modulus m:

$$n = km + l$$
$$l = 0 \cdot m + l.$$

This means that both n and l are indistinguishable modulus m, and have the same remainder: l:

$$(n \bmod m) = l = (l \bmod m).$$

In other words, their difference $n - l$ could be divided by m (evenly, with no remainder at all):

$$m \mid n - l.$$

This is indeed a mathematical equivalence relation. (Check!)

14.3.3 Common Divisor

Consider now the special case in which there is no remainder at all:

$$(n \bmod m) = 0.$$

In this case, we say that m divides n:

$$m \mid n.$$

In general, however, m doesn't divide n. Still, m may share a common divisor with n: a third number that divides both n and m.

For example, if both n and m are even, then 2 is a common divisor. Still, there could be other common divisors as well. What is the maximal one?

14.3.4 The Euclidean Algorithm

How to calculate the greatest common divisor of n and m? Well, there are two possibilities: if m divides n, then m itself is the greatest common divisor. If not, then

$$\begin{aligned}
\mathrm{GCD}(n, m) &= \mathrm{GCD}(km + l, m) \\
&= \mathrm{GCD}(l, m) \\
&= \mathrm{GCD}(m, l) \\
&= \mathrm{GCD}(m, n \bmod m).
\end{aligned}$$

So, instead of the original numbers

$$n > m,$$

we can now work with two smaller numbers:

$$m > l = (n \bmod m).$$

In summary, we have the recursive formula:

$$\mathrm{GCD}(n, m) = \begin{cases} m & \text{if } (n \bmod m) = 0 \\ \mathrm{GCD}(m, n \bmod m) & \text{if } (n \bmod m) > 0. \end{cases}$$

How to prove this? Use mathematical induction on $n = 2, 3, 4, \ldots$ (see below). This is the Euclidean algorithm.

14.3.5 The Extended Euclidean Algorithm

Thanks to this kind of mathematical induction, one could also write the greatest common divisor as a linear combination:

$$\mathrm{GCD}(n, m) = an + bm,$$

for some integer numbers a and b (positive or negative or zero). Indeed, if m divides n, then this is easy:

$$GCD(n, m) = m = 0 \cdot n + 1 \cdot m,$$

as required. If not, then the induction hypothesis tells us that

$$GCD(n, m) = GCD(m, l) = \tilde{a} \cdot m + \tilde{b} \cdot l$$

for some new integer numbers $\tilde{a}$ and $\tilde{b}$. Now, in this formula, substitute

$$l = n - km.$$

This gives

$$
\begin{aligned}
GCD(n, m) &= GCD(m, l) \\
&= \tilde{a} \cdot m + \tilde{b} \cdot l \\
&= \tilde{a} \cdot m + \tilde{b} (n - km) \\
&= \tilde{b} \cdot n + \left(\tilde{a} - k \cdot \tilde{b} \right) m.
\end{aligned}
$$

To have the desired linear combination, just define the new coefficients

$$
\begin{aligned}
a &= \tilde{b} \\
b &= \tilde{a} - k \cdot \tilde{b} = \tilde{a} - \left\lfloor \frac{n}{m} \right\rfloor \tilde{b}.
\end{aligned}
$$

This is the extended Euclidean algorithm.

14.3.6 The Extended Euclidean Algorithm With Modulus

Unfortunately, the coefficients a and b could be negative. Worse, they could be too big, and even exceed n. How to avoid this? Well, throughout the above algorithm (including the recursion), introduce just one change: take modulus n. This way, we can extend the induction hypothesis yet more, and assume that both $\tilde{a}$ and $\tilde{b}$ are between 0 and $n - 1$:

$$0 \leq \tilde{a}, \tilde{b} < n.$$

Now, rewrite the induction step as

$$
\begin{aligned}
a &= \tilde{b} \\
s &= \left\lfloor \frac{n}{m} \right\rfloor \tilde{b} \\
b &= \tilde{a} - s.
\end{aligned}
$$

To make sure that a and b are between 0 and $n - 1$ as well, make a slight change: take modulus n:

$$
\begin{aligned}
a &= \tilde{b} \\
s &= \left(\left\lfloor \frac{n}{m} \right\rfloor \tilde{b} \right) \bmod n \\
b &= \begin{cases} \tilde{a} - s & \text{if } \tilde{a} \geq s \\ \tilde{a} + n - s & \text{if } \tilde{a} < s. \end{cases}
\end{aligned}
$$

This way, thanks to the induction step, a and b are kept between 0 and $n - 1$ as well:

$$0 \le a, s, b < n,$$

as required. Furthermore, modulus n, nothing has changed. Thus, we have a new linear combination modulus n:

$$\mathrm{GCD}(n, m) \equiv (an + bm) \bmod n.$$

This will be useful later.

14.4 Operations With Modulus

14.4.1 Coprime

Let's use the greatest common divisor to define the coprime. For this purpose, consider a natural number n. What is a coprime of n? This is any other number p (prime or not) that shares no common divisor with n:

$$\mathrm{GCD}(n, p) = 1.$$

There may be a few legitimate coprimes. Which one to pick? Better pick a moderate one. For this purpose, start from an initial guess, say the prime number

$$p = 5.$$

Does it divide n? If not, then pick it as our desired coprime. If, on the other hand, p divides n, then we must keep looking:

$$p \leftarrow \text{ the next prime number,}$$

and so on. In at most $(\log_2 n)$ guesses, we'll eventually find a new prime number p that doesn't divide n any more, and can therefore serve as its legitimate coprime.

14.4.2 Multiplication With Modulus

Let's consider yet another task. Let n and j be two natural numbers. How to calculate the product

$$nj \bmod m?$$

Unfortunately, both n and j could be very long, and contain many digits. Their product nj could be even longer, and not easy to store on the computer or work with. How to avoid this?

Fortunately, both n and j could be written in the form in Section 14.3.2:

$$n = \left\lfloor \frac{n}{m} \right\rfloor m + (n \bmod m)$$

$$j = \left\lfloor \frac{j}{m} \right\rfloor m + (j \bmod m).$$

In the right-hand side, only the latter term is interesting. The former term, on the other hand, is a multiple of m, which is going to drop anyway. Therefore, instead of nj, better calculate

$$(n \bmod m)(j \bmod m) \bmod m.$$

In other words, the modulus operation could be carried out not only after but also before starting to multiply. This way, we only multiply moderate numbers, which never exceed $m - 1$.

14.4.3 Power With Modulus

The same idea could be used time and again to calculate a power like

$$n^k \bmod m.$$

Here, even for a moderate k, n^k may be too long to store or use. Thanks to the above idea, we can now avoid this: in the algorithm in Chapter 25, Section 25.5.2, before using n^2, take modulus m:

$$\text{power}(n, k, m) = \begin{cases} 1 & \text{if } k = 0 \\ n \bmod m & \text{if } k = 1 \\ n \cdot \text{power}(n^2 \bmod m, (k-1)/2, m) \bmod m & \text{if } k > 1 \text{ and } k \text{ is odd} \\ \text{power}(n^2 \bmod m, k/2, m) & \text{if } k > 1 \text{ and } k \text{ is even.} \end{cases}$$

Better yet, before starting the calculation, substitute

$$n \leftarrow (n \bmod m).$$

This avoids the large number n^2 in the first place. Instead, the recursion is applied to the moderate number

$$(n \bmod m)^2 \le (m-1)^2.$$

This way, throughout the entire calculation, all intermediate products are kept moderate, and never exceed $(m-1)^2$. To prove this, use mathematical induction on k (see exercises below).

Still, this seems a bit too pedantic. After all, the modulus operation could be costly as well, so why apply it so often? Better apply it only when absolutely necessary: when detecting an inner (temporary) variable too long to store or use efficiently.

14.4.4 Inverse With Modulus

Let p and q be some given coprime numbers (prime or not):

$$\text{GCD}(p, q) = 1.$$

Another important task is to find the inverse of q modulus p, denoted by

$$q^{-1} \bmod p.$$

What is this inverse? It is the unique solution x of the equation

$$qx \equiv 1 \bmod p.$$

Does x always exist? Moreover, is it unique? To prove this, consider the set S, containing the integer numbers from 0 to $p - 1$:

$$S = \{0, 1, 2, \ldots, p - 1\}.$$

Define the new mapping

$$M : S \to S$$

by
$$M(s) = (qs \bmod p), \quad s \in S.$$

Our aim is to show that M maps some number to 1. This number will then serve as the desired solution x. To show this, let's look at M, and study its properties.

Is M one-to-one? To check on this, consider two numbers $a, b \in S$ (say, $a \geq b$). Assume that
$$M(a) = M(b).$$

This means that
$$q(a - b) \equiv 0 \bmod p.$$

In other words, p divides $q(a - b)$:
$$p \mid q(a - b).$$

Since p shares no common divisor with q, it must divide $a - b$:
$$a \equiv b \bmod p.$$

Thus, M is indeed one-to-one. As such, it preserves the total number of elements in S:
$$|M(S)| = |S| = p.$$

Thus, M is a one-to-one mapping from S onto S. Thus, it has an inverse mapping:
$$M^{-1} : S \to S.$$

In particular, we can now define
$$x = M^{-1}(1).$$

In summary, the inverse of q modulus p exists uniquely:
$$\left(q^{-1} \bmod p\right) = M^{-1}(1).$$

14.4.5 Using The Extended Euclidean Algorithm

How to solve for x in practice? For this purpose, assume that $q < p$. Otherwise, just substitute
$$q \leftarrow q - p$$

time and again, until q becomes small enough. After all, this makes no difference to x.

Now, use the extended Euclidean algorithm with modulus (Section 14.3.6). This way, we can now write
$$1 = \mathrm{GCD}(p, q) \equiv (ap + bq) \equiv bq \bmod p,$$

for some integer numbers a and b that lie between 0 and $p-1$. Obviously, the desired solution is just
$$x = b.$$

In summary, the inverse of q modulus p is now available explicitly:
$$\left(q^{-1} \bmod p\right) = b.$$

This will be useful below.

14.5 The Chinese Remainder Theorem

14.5.1 Equation With Modulus

To design the Jordan decomposition, we need some background in number theory. In particular, we are going to prove the Chinese remainder theorem. The proof is not only theoretical but also constructive: it actually designs a practical algorithm.

Consider k natural numbers (prime or not), greater than 1:

$$p_1, \ p_2, \ p_3, \ \ldots, \ p_k > 1.$$

Assume that they are mutually coprime: if you pick two of them as you like, then they share no common divisor. In other words, their greatest common divisor is 1:

$$\mathrm{GCD}(p_i, p_j) = 1 \quad (1 \leq i \neq j \leq k).$$

Consider now the first equation:

$$x \equiv d_1 \bmod p_1,$$

where $0 \leq d_1 < p_1$ is a given (integer) number, and x is the desired (integer) solution.

What could x be? Well, this seems too easy: just define

$$x = d_1.$$

Still, this is not what we want. Here, we want x to have a special form.

14.5.2 Using The Coprime

Let P be the product of all the p_i's:

$$P = p_1 p_2 p_3 \cdots p_k.$$

Likewise, let P_1 be the product of all the p_i's but the first one:

$$P_1 = p_2 p_3 p_4 \cdots p_k = \frac{P}{p_1}.$$

We want x to have the special form

$$x = P_1 x_1.$$

Could we indeed have such an x? Well, to have x, we must first have x_1.

What could x_1 be? Well, we already know what it is. After all, p_2, p_3, p_4, $\ldots$, p_k are coprime to p_1. Therefore, their product P_1 must be coprime to p_1 as well:

$$\mathrm{GCD}(P_1, p_1) = 1.$$

Let $\tilde{x}_1$ be the inverse of P_1 modulus p_1:

$$P_1 \tilde{x}_1 \equiv 1 \bmod p_1.$$

In fact, $\tilde{x}_1$ could be calculated by the extended Euclidean algorithm (Sections 14.4.4–14.4.5). Now, define

$$x_1 = d_1 \tilde{x}_1.$$

This way,

$$P_1 x_1 \equiv P_1 d_1 \tilde{x}_1 \equiv d_1 \bmod p_1,$$

as required. So, $P_1 x_1$ is indeed a legitimate solution to our original equation.

This is not the end of it. To $P_1 x_1$, we could now add a multiple of p_1. For example, let P_2 be the product of all the p_i's but p_2:

$$P_2 = p_1 p_3 p_4 p_5 \cdots p_k = \frac{P}{p_2}.$$

This way, to $P_1 x_1$, we could also add P_2, and design a new solution of the form

$$P_1 x_1 + P_2.$$

Indeed, this makes no harm: since P_2 is a multiple of p_1, our equation is still satisfied.

What is all this good for? After all, we've already solved the original equation long ago! Well, let's introduce more equations.

14.5.3 System of Equations With Modulus

The above equation is easy enough to solve. Still, in the above, we managed to design a special kind of solution. Why is this good? To see this, consider now a system of k equations with modulus:

$$x \equiv d_1 \bmod p_1$$
$$x \equiv d_2 \bmod p_2$$
$$x \equiv d_3 \bmod p_3$$
$$\vdots$$
$$x \equiv d_k \bmod p_k,$$

where

$$0 \le d_1 < p_1$$
$$0 \le d_2 < p_2$$
$$0 \le d_3 < p_3$$
$$\vdots$$
$$0 \le d_k < p_k,$$

are given numbers.

All these equations must be solved simultaneously by the same x. This is not so easy any more. How to find a new x that satisfies all these equations at the same time?

Like before, let's solve each individual equation, and obtain its own solution. This way, we'll have not only x_1 but also $x_2, x_3, x_4, \ldots, x_k$. From these, the new x will be obtained as a linear combination.

For this purpose, define more products of the form

$$P_1 = \frac{P}{p_1}$$

$$P_2 = \frac{P}{p_2}$$

$$P_3 = \frac{P}{p_3}$$

$$\vdots$$

$$P_k = \frac{P}{p_k}.$$

Each of these products contains all of the p_i's but one. Let's use them to define our new x:

$$x = P_1 x_1 + P_2 x_2 + P_3 x_3 + \cdots + P_k x_k,$$

where

$$P_1 x_1 \equiv d_1 \bmod p_1$$
$$P_2 x_2 \equiv d_2 \bmod p_2$$
$$P_3 x_3 \equiv d_3 \bmod p_3$$
$$\vdots$$
$$P_k x_k \equiv d_k \bmod p_k.$$

This is the desired x that solves all the above equations. For example, it solves the first equation. Indeed, it is not much different from $P_1 x_1$: it just contains a few more multiples of p_1 that make no harm. The same is true for the other equations as well, as required.

14.5.4 Uniqueness

In the above process, x could become too large. To make it look better, let's make it moderate again: subtract P from it. After all, as discussed above, this makes no harm: since P is a product of all the p_i's, all equations are still satisfied.

By subtracting P time and again, we can eventually make x as small as

$$0 \le x < P.$$

In this range, x is also unique. Indeed, assume that

$$0 \le y < P$$

is also a solution to the same system. Without loss of generality, assume also that

$$x \ge y.$$

Must x and y be the same? Well, look at their difference:

$$0 \le x - y < P.$$

It solves the homogeneous system

$$x - y \equiv 0 \bmod p_1$$
$$x - y \equiv 0 \bmod p_2$$
$$x - y \equiv 0 \bmod p_3$$
$$\vdots$$
$$x - y \equiv 0 \bmod p_k.$$

In other words, $x - y$ is divided by all the p_i's:

$$p_1, \ p_2, \ p_3, \ldots, \ p_k \mid x - y.$$

Thus, their product must divide $x - y$ as well:

$$P \mid x - y.$$

As a result, we must have

$$x - y = 0,$$

or

$$x = y.$$

Thus, in the above range, x is indeed unique. This is the Chinese remainder theorem.

14.6 How to Use The Remainder?

14.6.1 Remainder in Integer Division

What is a remainder? Well, we've already used it in a few places. In Section 14.3.1, we've considered two natural numbers $n > m > 0$, and divided them with remainder. In other words, we wrote n in the new form

$$n = lm + r,$$

where l is some nonnegative integer number, and the remainder r is smaller than m:

$$0 \leq r < m.$$

This was indeed the key to the (extended) Euclidean algorithm (Sections 14.3.4–14.3.5).

14.6.2 Remainder in a Binary Number

The remainder is also useful for other purposes, not explicitly but implicitly. Indeed, in Chapter 25, Section 25.4.2, we'll write a natural number k in its binary form:

$$k = b_m b_{m-1} b_{m-2} \cdots b_1 b_0,$$

where $b_m, b_{m-1}, \ldots, b_1, b_0$ are the binary digits (0 or 1), ordered rightwards: from the most significant to the least significant one. This is actually a binary polynomial:

$$k = b_0 + b_1 \cdot 2 + b_2 \cdot 2^2 + \cdots + b_m \cdot 2^m = \sum_{j=0}^{m} b_j 2^j = q(2),$$

where q is the polynomial

$$q(x) = \sum_{i=0}^{m} b_i x^i,$$

and x is a (formal and abstract) variable.

Given a natural number k, how to uncover its binary form? For this purpose, just divide by 2 with remainder:

$$k = 2l + b_0.$$

What is this remainder? We already know it: the least significant digit b_0. So, we are indeed on the right track to forming the binary representation, digit by digit.

What are the other digits? To find out, use recursion on the new number l, used in the above formula. Its own least significant digit will then serve as b_1, and so on. Let's extend this to a more general polynomial as well.

14.6.3 Remainder in a Polynomial

In the above, we've considered the binary representation of a natural number. Thanks to the remainder, we managed to obtain the binary digits explicitly, digit by digit. Indeed, by dividing the original number by 2, we've uncovered its least meaningful digit: b_0. This could go on and on recursively, uncovering the rest of the digits as well.

These digits are just coefficients in a binary polynomial. This motivates the following idea: why only binary polynomial? Why not do the same in just any polynomial of any kind? This is actually Horner's algorithm.

14.6.4 Remainder in Horner's Algorithm

In Chapter 25, Section 25.2.4, we'll consider a polynomial of degree n:

$$p(x) = \sum_{i=0}^{n} a_i x^i,$$

where x is the (independent) variable, and the a_i's are given coefficients: not necessarily binary any more, but real or even complex.

Now, the monomial x is a polynomial in its own right, although quite simple. How to divide by it with remainder? Just like we've done before with numbers: write p in the new form

$$p(x) = a_0 + x\hat{p}(x),$$

where $\hat{p}(x)$ is a new polynomial:

$$\hat{p}(x) = a_1 + a_2 x + a_3 x^2 + \cdots + a_n x^{n-1} = \sum_{i=0}^{n-1} a_{i+1} x^i.$$

In Chapter 25, Section 25.2.4, we'll use this to help calculate p at a given argument x. Here, on the other hand, we want more: to actually divide p with remainder.

Indeed, we've already got our remainder: the constant a_0. After all, this is a "polynomial" in its own right, albeit degenerate and boring. Furthermore, it is also "smaller" than x. In what sense? In sense of degree: x is of degree 1, whereas a_0 is of degree 0 only.

Thus, most of what we've done in number theory could now be extended to polynomials as well. In particular, this applies to the Chinese remainder theorem as well. This will also have a nice application in linear algebra.

14.7 Using The Chinese Remainder Theorem

14.7.1 Characteristic Polynomial, Root, and Multiplicity

Let's go back to our $n \times n$ matrix A. Denote its characteristic polynomial in a slightly different way:

$$P(x) = (x - \lambda_1)^{n_1} (x - \lambda_2)^{n_2} (x - \lambda_3)^{n_3} \cdots (x - \lambda_k)^{n_k},$$

where the λ_i's are the *distinct* eigenvalues of A. This way, each eigenvalue is still a root of P. Indeed, at the argument $x = \lambda_i$, P vanishes:

$$P(\lambda_i) = 0, \quad 1 \le i \le k.$$

Each λ_i comes with its own multiplicity: n_i. Together, the multiplicities sum to n:

$$n_1 + n_2 + n_3 + \cdots + n_k = n.$$

14.7.2 Multiplicity and Jordan Subspace

The multiplicity n_i has yet another interpretation: it is also the dimension of the null space of $(A - \lambda_i I)^{n_i}$, denoted by V_i:

$$V_i = \{v \mid (A - \lambda_i I)^{n_i} v = \mathbf{0}\}.$$

(See Section 14.2.4.)

The new subspace V_i is the ith Jordan subspace. This will be useful in the Jordan decomposition below.

We are now ready to interpret the Chinese remainder theorem, with polynomials rather than numbers. In particular, instead of the p_i's used in the original version, we'll have new polynomials: factors taken from the characteristic polynomial P.

14.7.3 The Chinese Remainder Theorem With Polynomials

In the Chinese remainder theorem, we have to solve a system of equations modulus p_i: mutually coprime numbers. Here, on the other hand, we replace numbers with polynomials. After all, we've already established that numbers are just a special kind of polynomials. So, consider now the new system of equations:

$$Q \equiv \lambda_1 \bmod (x - \lambda_1)^{n_1}$$
$$Q \equiv \lambda_2 \bmod (x - \lambda_2)^{n_2}$$
$$Q \equiv \lambda_3 \bmod (x - \lambda_3)^{n_3}$$

$$\vdots$$

$$Q \equiv \lambda_k \bmod (x - \lambda_k)^{n_k}.$$

To these equations, better add one more equation:

$$Q \equiv 0 \bmod x.$$

The reason for this will become clear later. Still, this extra equation is not always necessary. For example, if $\lambda_1 = 0$, then we already have a stronger equation of the form

$$Q \equiv 0 \bmod x^{n_1}.$$

So, there is no need to add any new equation. Instead, just raise the power, to obtain a yet stronger equation:

$$Q \equiv 0 \bmod x^{n_1+1}.$$

This is done for the sake of uniformity. Our system is now complete.

In this system, what is Q? This is the unknown polynomial: $Q(x)$. In each equation, Q is congruent to a degenerate "polynomial" of degree zero: the constant λ_i. This congruence is modulus a factor from the characteristic polynomial: $(x - \lambda_i)^{n_i}$.

14.8 Matrix and Its Jordan Decomposition

14.8.1 The Solution Polynomial

Thanks to the Chinese remainder theorem, we can now assume that we've already solved the above system, and discovered the desired polynomial $Q(x)$.

How does Q look like? Well, thanks to the extra equation in Section 14.7.3, Q has the factor x in it. Thus, once written as a sum, Q has no constant term at all. On the contrary: it starts from the linear term x^1 onwards. This is useful in the context of Lie algebras, to help prove Cartan's criterion (Chapter 20).

Furthermore, Q is quite special: it is unique modulus $xP(x)$. In other words, Q is the only polynomial of degree n or less that solves the equations: no other polynomial of degree n or less could be added to it.

Let's go ahead and use $Q(x)$ in practice. By now, what does x stand for? Well, x is still unspecified: it could be just anything. Let's go ahead and give it a concrete meaning.

14.8.2 The Diagonal Part

For this purpose, instead of the argument x, plug the original matrix A. This way, $Q(A)$ is indeed a legitimate $n \times n$ matrix: a linear combination of A and its powers.

What's so good about $Q(A)$? Well, it is diagonal. (More precisely, it is diagonalizable or semisimple, but we're not pedant.) In what sense? Well, thanks to the ith equation, it could be written as a constant, plus an irrelevant term: $(A - \lambda_i I)^{n_i}$.

Why is this term irrelevant? Because, in the null space of $(A - \lambda_i I)^{n_i}$, it is always zero. (This is our Jordan subspace V_i.) For this reason, it is completely invisible in V_i, and has no effect whatsoever. The only relevant term in V_i is the constant $\lambda_i I$. This is true for every $1 \le i \le k$.

14.8.3 The Nilpotent Error

Finally, $Q(A)$ has another desirable property: it approximates A well. In other words, the error $A - Q(A)$ is "small." In what sense? Well, thanks to the ith equation, it satisfies

$$A - Q(A) \equiv A - \lambda_i I \mod (A - \lambda_i I)^{n_i}.$$

Once restricted to V_i, this becomes even simpler:

$$(A - Q(A))\,|_{V_i} = (A - \lambda_i I)\,|_{V_i}.$$

What is the n_ith power of this? Well, since V_i is invariant under $A - \lambda_i I$, we simply have

$$(A - Q(A))^{n_i}\,|_{V_i} = (A - \lambda_i I)^{n_i}\,|_{V_i} = (0).$$

Furthermore, every greater power (such as the nth power) is zero as well:

$$(A - Q(A))^n\,|_{V_i} = (0).$$

This is true for all $1 \le i \le k$. In summary, the error has a vanishing power:

$$(A - Q(A))^n = (0).$$

For this reason, the error is nilpotent: in some sense, it is "small" and negligible. In this sense, the diagonal part $Q(A)$ approximates A pretty well.

14.8.4 The Jordan Decomposition

In summary, thanks to the Chinese remainder theorem, A has been written as the sum of two parts: its diagonal part, plus a less important part – a nilpotent error:

$$A = Q(A) + (A - Q(A)).$$

Let's use this in a few examples.

14.9 Example: Space of Polynomials

14.9.1 Polynomial and Its Differentiation

Consider a polynomial of degree n or less:

$$p(x) = a_0 + a_1 x + a_2 x^2 + \cdots + a_n x^n = \sum_{i=0}^{n} a_i x^i.$$

This could be viewed as an $(n+1)$-dimensional vector:

$$(a_0, a_1, a_2, \ldots, a_n).$$

On polynomials like p, define the differentiation operator:

$$Dp = p'.$$

Fortunately, this is a linear operator. As such, it indeed has a Jordan form. As a matter of fact, the Jordan decomposition is just

$$D = (0) + D.$$

Why? Because the "error" D is nilpotent! Indeed, its nth power is the nth derivative of p, which is just a constant:

$$D^n p = p^{(n)} = n! a_n$$

(Chapter 25, Section 25.6.3). For this reason, the next derivative vanishes:

$$D^{n+1} p = p^{(n+1)} = 0,$$

which means nilpotency.

14.9.2 The Jordan Block

In fact, D has just one Jordan block. How to design it? Pick a degenerate "polynomial:" the constant 1. Is this an eigenvector of D? To check on this, just apply D to it:

$$D1 = 1' = 0.$$

So, the eigenvalue is just zero.

By now, our basis contains just one member: the constant 1. How to design more members? Use the following process: take the antiderivative – the linear monomial x:

$$Dx = x' = 1.$$

Next, pick the antiderivative of x: $x^2/2$. This process could continue on and on, introducing more and more monomials (of higher and higher power) into the same Jordan basis. For this purpose, just take the antiderivative time and again, producing the new basis

$$\left\{ 1, x, \frac{x^2}{2}, \frac{x^3}{3!}, \ldots, \frac{x^n}{n!} \right\}.$$

This basis spans the entire space: all polynomials of degree n or less.

How does D act on this basis? Well, D maps each member to the previous one. The only exception is the first member: D maps it to zero. Thus, in this basis, D indeed takes its Jordan form:

$$D = \begin{pmatrix} 0 & 1 & & & & \\ & \ddots & \ddots & & & \\ & & \ddots & \ddots & & \\ & & & 0 & 1 & \\ & & & & 0 & 1 \\ & & & & & 0 \end{pmatrix}.$$

This is indeed an $(n+1) \times (n+1)$ matrix, as required. It has just n nonzero elements: the 1's in its superdiagonal, just above the main diagonal. Its powers, on the other hand, have fewer and fewer nonzero elements:

$$D^2 = \begin{pmatrix} 0 & & 1 & & & \\ & \ddots & & \ddots & & \\ & & \ddots & & \ddots & \\ & & & 0 & & 1 \\ & & & & 0 & & 1 \\ & & & & & 0 & \\ & & & & & & 0 \end{pmatrix}$$

$$D^3 = \begin{pmatrix} 0 & & 1 & & & \\ & \ddots & & \ddots & & \\ & & \ddots & & \ddots & \\ & & & 0 & & 1 \\ & & & & 0 & \\ & & & & & 0 \\ & & & & & & 0 \end{pmatrix}$$

$$\vdots$$

$$D^n = \begin{pmatrix} 0 & & & 1 \\ & \ddots & & \\ & & \ddots & \\ & & & 0 \\ & & & & 0 \\ & & & & & 0 \\ & & & & & & 0 \end{pmatrix}.$$

Finally, the next power vanishes:

$$D^{n+1} = (0),$$

which shows nilpotency once again.

14.9.3 Polynomials of Any Degree

In the above, n was never specified. Therefore, one could ask: why stop at n at all? Why not go on and on indefinitely, and consider the infinite-dimensional space that contains all polynomials of just any degree? This way, we'll still have the same Jordan form

$$D = (0) + D,$$

where the "error" D is now "nilpotent" in a rather nonstandard sense: a given polynomial p, if differentiated sufficiently many times, then the result is zero. In fact, if p is of degree n, then

$$D^{n+1}p = 0.$$

Still, $n + 1$ is no longer global: it depends on p. This is no longer nilpotency in its standard sense.

Again, there is just one Jordan subspace, which is actually very big: the entire space, including all polynomials. Unlike before, however, this space is no longer finite-dimensional, but infinite-dimensional. This way, D is now an infinite matrix, or an operator.

14.10 Eigenfunctions

14.10.1 Exponent: An Eigenfunction

Consider now the exponent function $\exp(\lambda x)$, where λ is a given constant (real or even complex). How to differentiate it? Like a composite function:

$$D \exp(\lambda x) = \lambda \exp'(\lambda x) = \lambda \exp(\lambda x).$$

So, we've got an eigenfunction of D, with a new eigenvalue: λ.

As an eigenfunction, this exponent lies in the null space of $D - \lambda$:

$$(D - \lambda) \exp(\lambda x) = 0.$$

Thus, this exponent belongs to a new Jordan subspace. How to design the entire subspace?

14.10.2 Polynomial times Exponent: Leibniz Rule

To design the entire Jordan subspace, let's consider not only an individual exponent but also a product: an exponent times a polynomial. How to differentiate it? As in Leibniz rule (or the product rule):

$$D(p \exp(\lambda x)) = (Dp) \exp(\lambda x) + pD \exp(\lambda x).$$

14.10.3 Virtual Binary Tree

This could also be written as

$$(D - \lambda)(p \exp(\lambda x)) = (Dp) \exp(\lambda x) + p(D - \lambda) \exp(\lambda x).$$

This could be viewed as a new (virtual) binary tree (Figure 14.5). At the head, there is the original operator $D - \lambda$. Then, it splits into two branches. At the end of the left branch, attach the operator D. At the end of the right branch, on the other hand, attach the operator $D - \lambda$.

This is only a two-level tree. In the next higher power, on the other hand, we get a three-level tree.

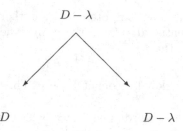

Fig. 14.5. The virtual two-level tree. At the head, there is the original operator $D - \lambda$. At the bottom, on the other hand, it splits into two operators: D at the bottom-left, and $D - \lambda$ at the bottom-right.

14.10.4 Three-Level Tree: Pascal's Triangle

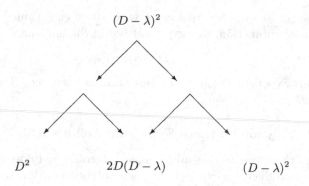

Fig. 14.6. The virtual three-level tree. At the head, there is the original operator $(D-\lambda)^2$. At the bottom, on the other hand, it splits: D^2 at the bottom-left, $(D-\lambda)^2$ at the bottom-right, and $2D(D - \lambda)$ at the bottom-middle, as in Pascal's triangle.

Likewise, to help split $(D - \lambda)^2$ as well, use the above trick not only once but twice:

$$(D - \lambda)^2(p\exp(\lambda x))$$
$$= (D - \lambda)(D - \lambda)(p\exp(\lambda x))$$
$$= (D - \lambda)\left((Dp)\exp(\lambda x) + p(D - \lambda)\exp(\lambda x)\right)$$
$$= (D^2 p)\exp(\lambda x) + 2(Dp)(D - \lambda)\exp(\lambda x) + p(D - \lambda)^2\exp(\lambda x).$$

This is illustrated in Figure 14.6. Still, this is a rather nonstandard "tree:" at the bottom-middle, both inner branches unite, producing the new coefficient 2. This is as in the second level in Pascal's triangle:

$$1, \ 2, \ 1.$$

Let's go ahead and extend this to yet higher powers as well. This will mirror the binomial formula.

14.10.5 How to Mirror The Binomial Formula?

Next, let's move on to a yet more difficult task: to split $(D - \lambda)^k$:

$$(D - \lambda)^k (p \exp(\lambda x)) = \sum_{i=0}^{k} \binom{k}{i} D^i p (D - \lambda)^{k-i} \exp(\lambda x)$$
$$= \exp(\lambda x) D^k p$$
$$= \exp(\lambda x) p^{(k)}.$$

Indeed, in the above sum, only one term survives: the term in which $i = k$. All the rest, on the other hand, drop. Now, p must have some finite degree. If k is bigger than the degree of p, then the above vanishes.

In summary, we finally got our (infinite-dimensional) Jordan subspace: all polynomials times $\exp(\lambda x)$. This is spanned by its Jordan basis: all monomials times $\exp(\lambda x)$. This could be written as

$$\mathrm{span}_{k=0}^{\infty} \left\{ x^k \exp(\lambda x) \right\}.$$

This is one Jordan subspace. Now, the same could be done for each individual λ (real or even complex). In summary, we have infinitely many Jordan subspaces. This way, D acts on a rather big function space, containing all polynomials times all exponents. This is spanned by its basis: all monomials times all exponents. This could be written as

$$\mathrm{span}_{k \geq 0, \lambda \in \mathbb{C}} \left\{ x^k \exp(\lambda x) \right\}.$$

Let's go ahead and extend this to a yet more formal and abstract setting.

14.11 Algebra and Its Derivation

14.11.1 Leibniz Rule

What is Leibniz rule? So far, it was the same as the product rule, telling us how to differentiate a product of two functions:

$$(fg)' = f'g + fg'.$$

Now, on the other hand, we extend this to a more formal and abstract setting. Indeed, here f and g are not necessarily functions, but just members of an algebra: a vector space with one more operation: product (multiplication).

14.11.2 Product – Multiplication

In a linear space, vectors may have a product or not. In an algebra, on the other hand, they must: if f and g are in the algebra, then their product fg must belong to the algebra as well. In other words, the algebra is closed under multiplication.

This is a vector-times-vector multiplication. Don't confuse it with the usual scalar-times-vector multiplication. So far, we didn't specify yet how it words. Later on, we'll see a few examples.

Still, this kind of multiplication could be rather nonstandard: it is distributive (bilinear), but not necessarily commutative or even associative. (See exercises below.)

14.11.3 Derivation

In an algebra, consider a linear operator D: a linear mapping that acts on the algebra, and maps each member to some member. Here, D doesn't have to mean differentiation any more: it is much more general.

Is D a derivation? Well, to deserve to be called a derivation, D must behave like one. In other words, it must obey a Leibniz rule, in its general version:

$$D(fg) = (Df)g + fDg.$$

This way, in algebraic terms, D indeed behaves like a legitimate derivation: to act upon a product, first act on the first member, then on the second member, then sum up.

This is indeed a most general setting. Thanks to it, the differentiation operator discussed so far can now be viewed as just a special case of the present operator D.

14.12 Product and Its Derivation

14.12.1 Two-Level Tree

In its new version, the Leibniz rule still makes a two-level binary tree. This is illustrated in Figure 14.7, which mirrors Figure 14.5.

Here, however, the picture is more symmetric. For this purpose, use two parameters: τ and λ. In our algebra, consider the operator $D - \tau - \lambda$. Thanks to Leibniz rule, here is how it acts on a product:

$$(D - \tau - \lambda)(fg) = ((D - \tau)f)g + f(D - \lambda)g.$$

This way, the original operator $D - \tau - \lambda$ splits into two parts: $D - \tau$ acts on f, and $D - \lambda$ acts on g.

What's so good about this formula? Well, if f and g are eigenvectors of D with the eigenvalues τ and λ (respectively), then fg is an eigenvector as well, with the new eigenvalue $\tau + \lambda$ (provided that $fg \neq 0$). Next, let's extend this formula to a yet higher power as well.

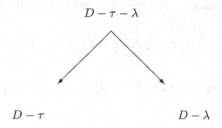

Fig. 14.7. The virtual two-level tree. At the head, there is the original operator $D - \tau - \lambda$. At the bottom, on the other hand, it splits into two operators: $D - \tau$ at the bottom-left, and $D - \lambda$ at the bottom-right.

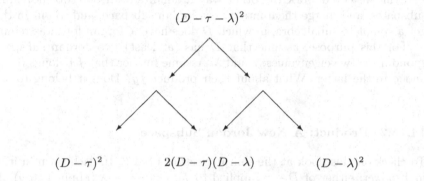

Fig. 14.8. The virtual three-level tree. At the head, there is the original operator $(D - \tau - \lambda)^2$. At the bottom, on the other hand, it splits: $(D - \tau)^2$ at the bottom-left, $(D - \lambda)^2$ at the bottom-right, and $2(D - \tau)(D - \lambda)$ at the bottom-middle, as in Pascal's triangle.

14.12.2   Multilevel Tree

How does $(D - \tau - \lambda)^2$ split? This already makes a three-level tree (Figure 14.8, which mirrors Figure 14.6). At the bottom-middle, the two inner branches unite, as in the second level in Pascal's triangle:

$$(D - \tau - \lambda)^2(fg) = (D - \tau - \lambda)(D - \tau - \lambda)(fg)$$
$$= (D - \tau - \lambda)\left(((D - \tau)f)\,g + f(D - \lambda)g\right)$$
$$= \left((D - \tau)^2 f\right)g + 2(D - \tau)f(D - \lambda)g + f(D - \lambda)^2 g.$$

14.12.3 Pascal's Triangle – Binomial Formula

This is how the first and second powers split. What about the kth power? It splits as in the binomial formula – as in the kth level in Pascal's triangle:

$$(D - \tau - \lambda)^k(fg) = \sum_{i=0}^{k} \binom{k}{i} (D - \tau)^i f (D - \lambda)^{k-i} g.$$

We can now see that Section 14.10.5 is nothing but a special case, with $\tau = 0$.

14.13 Product and Its Jordan Subspace

14.13.1 Two Members of Two Jordan Subspaces

Does D have a Jordan decomposition? We don't know! After all, the Jordan decomposition was designed only in finite dimension (Section 14.8.4). Our algebra, on the other hand, could be bigger than that: infinite-dimensional.

Still, we could work the other way around: start from two individual Jordan subspaces, and merge them into a new Jordan subspace, and so on. In the end, we get a complete subalgebra, in which D does have a Jordan form, as required.

For this purpose, assume that D has (at least) two Jordan subspaces, corresponding to two eigenvalues: τ and λ. Assume further that f belongs to the former, and g to the latter. What about their product fg? Does it belong to any Jordan subspace?

14.13.2 Product: A New Jordan Subspace

To check on this, look at the sum in Section 14.12.3. In each term in it, there is a high power: either of $D - \tau$ (applied to f), or of $D - \lambda$ (applied to g). Therefore, one could pick k so large that all these terms vanish. Thus, the product fg indeed belongs to a new Jordan subspace, corresponding to a new eigenvalue: the sum $\tau + \lambda$.

14.13.3 Example: Polynomials times Exponents

As a matter of fact, we've already seen an example of this. Indeed, in Section 14.10.5, we looked at a product: a polynomial times an exponent. Each belongs to its own Jordan subspace: the polynomial is associated with eigenvalue 0, and the exponent with eigenvalue λ.

What about their product? Does it belong to any Jordan subspace? It sure does: to the Jordan subspace corresponding to the sum: the "new" eigenvalue $0 + \lambda = \lambda$.

Thanks to our new multilevel tree (Figure 14.8), we can now extend this to $\tau \neq 0$ as well. For this purpose, consider the product

$$q(x) \exp(\tau x) \cdot p(x) \exp(\lambda x).$$

In it, look at the first function: the polynomial q times the exponent $\exp(\tau x)$. This belongs to the Jordan subspace corresponding to eigenvalue τ. This is multiplied by yet another function: the polynomial p times the exponent $\exp(\lambda x)$. This belongs to yet another Jordan subspace, corresponding to eigenvalue λ. What about their product? It is just

$$q \ \exp(\tau x) \cdot p \ \exp(\lambda x) = (qp) \exp((\tau + \lambda)x),$$

which indeed belongs to a new Jordan subspace, corresponding to a new eigenvalue: the sum $\tau + \lambda$.

14.14   Derivation on a Subalgebra

14.14.1   Restriction to a Subalgebra

Unfortunately, our original algebra may be too big: it may contain members that belong to no Jordan subspace at all. To fix this, let's restrict our discussion to a subalgebra, spanned by the generalized eigenvectors of D: complete Jordan subspaces.

Why is this a legitimate subalgebra? Because it is closed under multiplication. Indeed, as discussed above, if you pick two members of two Jordan subspaces, then their product lies in some Jordan subspace in the subalgebra. Thus, the subalgebra is indeed a legitimate algebra in its own right: closed under both addition and multiplication.

As a matter of fact, we've just seen an example of a subalgebra.

14.14.2   Example: Polynomials times Exponents

In the beginning, we've started with a special case of algebra: C^∞, which contains all smooth functions. In this algebra, you could differentiate as many times as you like. Still, we never considered all of it. Instead, we focused on a small subalgebra only, spanned by polynomials times exponents. Why? Because, in this subalgebra, differentiation has a Jordan form. Indeed, this subalgebra splits into disjoint Jordan subspaces, each of a very special kind: a unique exponent, times just any polynomial.

The entire algebra, C^∞, is too big: in it, differentiation has no Jordan form. Only in the above subalgebra does differentiation have a Jordan form. Let's see yet another example.

14.14.3   Example: Finite-Dimensional Algebra

For yet another example, consider just any finite-dimensional algebra. (In the exercises below, we'll see a few examples in physics.) In Section 14.8.4, we've already seen that every linear mapping (including the present D) has a Jordan form.

Now, the present D is a special kind of linear operator: a derivation. Indeed, although it is not a standard differentiation as in calculus, it still obeys Leibniz rule (in its formal version).

Once f and g are picked from two Jordan subspaces, their product fg must belong to a third Jordan subspace, corresponding to the sum of their eigenvalues. Let's go back to the general case (finite-dimensional or not), and see an interesting property.

14.14.4 Is The Diagonal Part a Derivation?

In our subalgebra, we can safely decompose D in its Jordan decomposition:

$$D = \Lambda + (D - \Lambda),$$

where Λ is diagonal, and the error $D - \Lambda$ is nilpotent. If our subalgebra is infinite-dimensional, then this kind of "nilpotency" is rather nonstandard (Section 14.9.3).

We already know that D is a derivation: it obeys a Leibniz rule. What about its diagonal part, Λ? Is it a derivation in its own right? In other words: does Λ obey a Leibniz rule as well? To check on this, pick two members of our subalgebra: f and g. Assume that f belongs to the Jordan subspace corresponding to τ. Because Λ is diagonal,

$$\Lambda f = \tau f.$$

Likewise, assume that g belongs to the Jordan subspace corresponding to λ. Because Λ is diagonal,

$$\Lambda g = \lambda g.$$

Now, what about their product? As discussed above, it must belong to the Jordan subspace corresponding to the sum $\tau + \lambda$:

$$\begin{aligned}
\Lambda(fg) &= (\tau + \lambda)(fg) \\
&= \tau(fg) + \lambda(fg) \\
&= (\tau f)g + f\lambda g \\
&= (\Lambda f)g + f\Lambda g.
\end{aligned}$$

This is nothing but Leibniz rule for Λ. In summary, we got what we wanted: Λ satisfies its own Leibniz rule, and is therefore a legitimate derivation in its own right.

14.15 Exercises: Hermitian Matrix and Its Eigenbasis

14.15.1 Natural Numbers and Their Binary Representation

1. Consider a natural number n. Consider its binary representation. What is the least significant (rightmost) binary digit?
2. Assume that n is even. What is its least significant binary digit? Hint: 0.
3. Assume that n is odd. What is its least significant binary digit? Hint: 1.
4. No matter whether n is odd or even, what is its least significant binary digit? Hint: $n \bmod 2$.
5. Divide n by 2 with remainder. What is the remainder? Hint: $n \bmod 2$.
6. How is the remainder related to the least significant binary digit? Hint: they are the same.
7. What is the next (more significant) binary digit? Hint: divide n by 2 with remainder. Disregard the remainder. Look at the result: $\lfloor n/2 \rfloor$. Find its least significant binary digit.
8. What are the rest of the binary digits in n? Hint: repeat the above recursively: drop the rightmost digit time and again (Section 14.6.2).

9. How is this related to Horner's algorithm? Hint: both use the same recursion: divide by x (or 2) with remainder, to shift a list of numbers time and again.

10. What is the difference? Hint: in the above, we shift the (unknown) binary digits, to uncover them one by one. In Horner's algorithm, on the other hand, we work the other way around: shift the (well-known) coefficients, and use them one by one, to uncover the (unknown) value.

11. Repeat the above, but this time with base 3 rather than 2.

12. Pick an arbitrarily long natural number at random. Look at its representation in base 3. How likely is it to contain only the digits 0 and 2, but never the digit 1? Hint: the probability for this is as low as zero (Chapter 5).

13. How to add two (arbitrarily long) natural numbers to each other? Hint: write the bigger number in the upper line. Write the other number just below it. Start from the rightmost (least significant) digit. Add digit by digit, all the way leftwards. As your teacher always said, don't forget the carry!

14. How to add two polynomials to each other? Hint: likewise: monomial by monomial.

15. How to subtract two natural numbers from each other? Hint: assume that the former is bigger. (Otherwise, interchange their roles, and attach a minus sign to the result.) Write it in the upper line. Write the other number just below it. Start from the rightmost digit. Work vertically, digit by digit, all the way leftwards.

16. How to subtract two polynomials from each other? Hint: likewise, monomial by monomial.

17. How to multiply two natural numbers by each other? Hint: vertically, leftwards, digit by digit.

18. Is there a better way? Hint: try FFT (Chapter 5 in [44]).

19. How to multiply two polynomials by each other? Hint: likewise, monomial by monomial.

20. Is there a better way?

21. How to divide two natural numbers by each other (with remainder)? Hint: vertically, rightwards, digit by digit (assuming that the latter is nonzero).

22. How to divide two polynomials by each other (with remainder)? Hint: likewise, monomial by monomial.

23. After the division, look at the remainder. What is its meaning? Hint: the remainder is just the result of the modulus operation: the former modulus the latter.

24. Use the above to write the Euclidean algorithm for polynomials as well, and find their greatest common divisor: a new polynomial that divides them both (evenly).

25. Write the extended Euclidean algorithm (Section 14.3.5) for polynomials as well: write their greatest common divisor as their "sum," with some "coefficients," which could be polynomials in their own right.

26. Apply the Chinese remainder theorem, with polynomials instead of numbers.

27. Why is this legitimate?

28. If $\lambda_1 \neq \lambda_2$, are the polynomials $(x - \lambda_1)^{n_1}$ and $(x - \lambda_2)^{n_2}$ coprime?

29. In what sense is the solution $Q(x)$ unique? Hint: no polynomial of degree n or less could ever be added to it.

30. Show that, if A is an $n \times n$ matrix, then $Q(A)$ is an $n \times n$ matrix as well.

31. Show that $Q(A)$ is diagonal (semisimple).
32. Show that the error $A - Q(A)$ is nilpotent.

14.15.2 Nilpotent Hermitian Matrix

1. Let B be a nilpotent matrix of order l. Could B have a nonzero eigenvalue? Hint: assume that B had a nonzero eigenvalue $\lambda \neq 0$, with an eigenvector $v \neq \mathbf{0}$. Then we'd have
$$B^i v = \lambda^i v \neq \mathbf{0}, \quad i = 1, 2, 3, \ldots,$$
and B could never be nilpotent.
2. Conclude that B has one eigenvalue only: 0.
3. Look at a short cycle of B, of length 2 only:
$$w, \ Bw.$$
What can you say about Bw? Hint: it is an eigenvector of B, corresponding to the eigenvalue 0:
$$B(Bw) = \mathbf{0}.$$
4. Assume now that B is also Hermitian. Could such a cycle exist? Hint: no! Indeed, this would lead to a contradiction:
$$0 = (\mathbf{0}, w) = (B(Bw), w) = (Bw, Bw) > 0.$$
5. So, B has no cycle of length 2. Still, could B have a slightly longer cycle (of length 3), like
$$w, \ Bw, \ B^2 w?$$
Hint: no! Indeed, since B is Hermitian, this would lead to a contradiction:
$$0 = (\mathbf{0}, Bw) = (B(B^2 w), Bw) = (B^2 w, B^2 w) > 0.$$
6. Could B have a longer cycle (of length 4 or more)? Hint: no! To prove this, use the same technique.
7. Conclude that B has only very short "cycles," of length 1 each.
8. Conclude that B has no generalized eigenvectors, but only genuine eigenvectors.
9. Conclude that B has l linearly-independent eigenvectors.
10. Orthonormalize them. Hint: use a Gram-Schmidt process.
11. What is their joint eigenvalue? Hint: 0 (see above).
12. Conclude that B is the zero matrix.

14.15.3 Hermitian Matrix: Orthonormal Eigenbasis

1. Let A be a Hermitian matrix (nilpotent or not).
2. Let λ_1 be an eigenvalue of A (of multiplicity n_1).
3. What can you say about λ_1? Hint: it is real (because A is Hermitian).
4. Pick two different eigenvectors from two different Jordan blocks of A. What can you say about them? Hint: they are orthogonal to one another (because they have different eigenvalues).
5. Define the subspace V, as in Section 14.2.2.

6. What is V? Hint: V is the null space of $(A - \lambda_1 I)^{n_1}$.
7. Define the matrix J, as in Section 14.2.2.
8. What is J? Hint: J is the restriction of A to V.
9. What is the order of J? Hint: n_1.
10. Is J Hermitian? Hint: yes. Indeed, for every two vectors $p, q \in V$, J behaves just like A:

$$(Jp, q) = (Ap, q) = (p, Aq) = (p, Jq).$$

11. Let I be the identity matrix of order n_1. Look at the matrix $J - \lambda_1 I$.
12. Is it Hermitian? Hint: yes (because λ_1 is real).
13. Is it nilpotent? Hint: yes. Indeed, look at its n_1st power. This is just the zero matrix. Indeed, for every vector $q \in V$,

$$(J - \lambda_1 I)^{n_1} q = (A - \lambda_1 I)^{n_1} q = \mathbf{0}.$$

14. Conclude that $J - \lambda_1 I$ is nilpotent and Hermitian at the same time.
15. Conclude that

$$J - \lambda_1 I = (0)$$

(the zero matrix).
16. Conclude that

$$J = \lambda_1 I.$$

17. Do the same for the other Jordan blocks of A as well.
18. Conclude that A has a diagonal Jordan form, corresponding to an orthonormal eigenbasis.

15

Design Your Lie Algebra

In this chapter, we'll get a taste of Lie algebras. In particular, we'll study the adjoint representation and the exponent of a mapping, and demonstrate their attractive property: preserving the Lie brackets. Then, we'll design a few specific Lie algebras, used later in quantum mechanics and special relativity.

This chapter has a special structure: it is organized in terms of exercises, followed by hints or even complete solutions. This serves a pedagogical purpose: to get you to design your Lie algebra on your own, and see quite vividly how it develops from scratch.

15.1 Derivation and Its Exponent

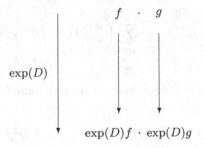

Fig. 15.1. Thanks to the binomial formula, $\exp(D)$ preserves multiplication. In other words, order doesn't matter: either multiply in the upper level and then apply $\exp(D)$ to fg, or apply $\exp(D)$ to f and g separately, and then multiply in the lower level.

1. What is an algebra? Hint: a vector space with a new operation: product or multiplication (Chapter 14, Section 14.11.1). This is a vector-times-vector multiplication: don't confuse it with the usual scalar-times-vector multiplication.

2. Is it closed under multiplication? In what sense?
3. What are the properties of this product? Hint: distributive (bilinear), but not necessarily commutative or even associative (see below).
4. Let D be a derivation in the algebra (Chapter 14, Section 14.11.3). Let f and g be some members of the algebra. Write a Leibniz rule for D. Hint:

$$D(fg) = Df \cdot g + fDg.$$

5. Assume that D is nilpotent. What does this mean? Hint: for some (sufficiently large) m, D^m is the zero mapping: it maps every member to the zero member.
6. Show that, for every $l \geq m$, D^l is the zero mapping as well. Hint: as a linear mapping, D maps zero to zero.
7. Define its exponent:

$$\exp(D) = \sum_{i=0}^{\infty} \frac{D^i}{i!}.$$

8. Is this really infinite? Hint: only the m leading terms survive.
9. Show that this is already in the Jordan form. Hint: the first term is just the identity mapping. This is the diagonal part. The rest is the "tail"

$$D + \frac{D^2}{2} + \frac{D^3}{3!} + \cdots + \frac{D^{m-1}}{(m-1)!} = \sum_{i=1}^{m-1} \frac{D^i}{i!}.$$

This is nilpotent. To see this, look at its mth power: it contains only high powers of D that vanish.
10. What is the inverse of $\exp(D)$? Hint: $\exp(-D)$.
11. Write it in terms of the above tail. Hint: take the tail, and look at its (finite) power series, with alternating signs.
12. How does the derivation D and its powers act on a product like fg? Hint: mirror the binomial formula:

$$D^n(fg) = \sum_{i=0}^{n} \binom{n}{i} D^i f \cdot D^{n-i} g.$$

(Figures 14.7–14.8).
13. From this formula, is D^2 a derivation in its own right? Hint: no! It obeys no Leibniz rule:

$$D^2(fg) = D^2 f \cdot g + 2Df \cdot Dg + fD^2 g \neq D^2 f \cdot g + fD^2 g$$

(in general).
14. Likewise, for any $1 < l < m$, is D^l a derivation?
15. Consider the real exponent function: $\exp(x)$. Recall that

$$\exp(x + y) = \exp(x) \exp(y).$$

16. So far, we've seen how D and its powers act on a product like fg. How does $\exp(D)$? Hint: mirror the above formulas, and obtain

$$\exp(D)(fg) = \exp(D)f \cdot \exp(D)g.$$

17. In summary, thanks to the above exercises, what can you say about $\exp(D)$? Hint: it is an invertible mapping that preserves multiplication (Figure 15.1).
18. Is it a derivation? Hint: no! It obeys no Leibniz rule:

$$\exp(D)(fg) = \exp(D)f \cdot \exp(D)g \neq \exp(D)f \cdot g + f \cdot \exp(D)g$$

(in general).

19. Let E be yet another mapping of the same kind: linear, invertible, product preserving, but not necessarily a derivation.
20. Denote its inverse by E^{-1}.
21. Look at the composition of E with its inverse:

$$EE^{-1} = E^{-1}E.$$

Show that this is just the identity mapping that changes nothing.

22. Show that the identity mapping is of the same kind: linear, invertible, product preserving, but not a derivation.
23. Look at a triple composition:

$$EDE^{-1}.$$

How does it act? Hint: leftwards, in three stages: first, apply E^{-1}. Then, apply D. Finally, apply E.

24. Is it a derivation in its own right? Hint: apply it to a product of the form Ef times Eg:

$$
\begin{aligned}
EDE^{-1}(Ef \cdot Eg) &= ED(fg) \\
&= E(Df \cdot g + fDg) \\
&= EDfEg + EfEDg \\
&= EDE^{-1}EfEg + EfEDE^{-1}Eg.
\end{aligned}
$$

This is indeed a legitimate Leibniz rule.

25. What are its powers? Hint:

$$\left(EDE^{-1}\right)^{l} = ED^{l}E^{-1}$$

($l \geq 0$). Indeed, the inner E and E^{-1} cancel out.

26. Is EDE^{-1} nilpotent?
27. Conclude that it has an exponent.
28. How does its exponent look like? Hint:

$$\exp\left(EDE^{-1}\right) = E\exp(D)E^{-1}.$$

29. Use the original algebra to design a new algebra. Hint: see below.
30. Look at the set of linear mappings from the original algebra into itself. Is this a legitimate linear space?
31. To multiply two such mappings F and G, use composition: $F \circ G$.
32. How does $F \circ G$ act on a member of the original algebra? Hint: leftwards. First, apply G. On the result, apply F.
33. Is the new algebra indeed closed under this kind of multiplication? Hint: $F \circ G$ is indeed a linear mapping as well. To see this, open parentheses in $F \circ G(\alpha f + \beta g)$, where α and β are some scalars.

34. Is this multiplication distributive (bilinear)? Hint: let H be yet another linear mapping. Open parentheses in $(\alpha F + \beta G) \circ H$ and in $H \circ (\alpha F + \beta G)$.
35. Is this multiplication associative? Hint: yes, composition is.
36. Is it commutative? Hint: no, composition is not (in general).
37. Give a few examples of such linear mappings. Hint: D and its powers and $\exp(D)$.
38. In general, linear mappings may not commute. Still, some special mappings may commute with each other. For example, does D commute with $\exp(D)$? Hint: yes, both are polynomials in D.
39. On our new algebra, define the new operator '$F\circ$' that converts the old mapping G into the new mapping $F \circ G$.
40. Is it linear? Hint: Open parentheses in $F \circ (\alpha G + \beta H)$.
41. Is it a derivation? Hint: no! It obeys no Leibniz rule:

$$F \circ (G \circ H) = (F \circ G) \circ H \neq (F \circ G) \circ H + G \circ (F \circ H)$$

(in general).

15.2 Lie Algebra and Its Adjoint Representation

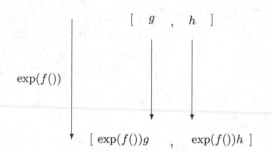

Fig. 15.2. Thanks to the binomial formula, $\exp(f())$ preserves the Lie brackets. In other words, order doesn't matter: either calculate the Lie brackets in the upper level and then apply $\exp(f())$ to $[g, h]$, or apply $\exp(f())$ to g and h separately, and then calculate the Lie brackets in the lower level.

1. Consider now a special kind of algebra: a Lie algebra [16, 20, 25, 30, 52]. In it, the product is replaced by a new notation – the Lie brackets:

$$fg \leftarrow [f, g].$$

2. Furthermore, in a Lie algebra, it is also assumed that this product is alternative:

$$[f, f] = \mathbf{0}$$

(the zero vector).

3. Use alternativity and bilinearity to prove antisymmetry as well:

$$[f, g] = -[g, f].$$

Hint:

$$\mathbf{0} = [f + g, f + g] = [f, f] + [f, g] + [g, f] + [g, g] = [f, g] + [g, f].$$

4. Finally, a Lie algebra must also satisfy the Jacobi identity (see below).
5. Let h be a fixed member of our Lie algebra. (In physics, h is known as the Hamiltonian.) Define a special kind of "differentiation:"

$$f' = [f, h].$$

6. To be a legitimate derivation, it must satisfy a Leibniz rule. How should it look like? Hint: our product is denoted now by Lie brackets. Therefore, Leibniz rule takes the form

$$[f, g]' = [f', g] + [f, g'].$$

7. Write this more explicitly. Hint: thanks to the definition of f', g', and $[f, g]'$,

$$[[f, g], h] = [[f, h], g] + [f, [g, h]].$$

8. Use antisymmetry to obtain the well-known Jacobi identity:

$$[[f, g], h] + [[h, f], g] + [[g, h], f] = \mathbf{0}$$

(the zero vector).

9. Is zero an eigenvalue of this derivation? What is its eigenvector? Hint: $[h, h] = \mathbf{0}$.
10. Assume that the Lie algebra is finite-dimensional. Does this derivation have a Jordan form?
11. In this Jordan form, look at the diagonal part. Is it a derivation in its own right? In other words, does it obey its own Leibniz rule? Hint: see Chapter 14, Section 14.14.4.
12. Consider again a general Lie algebra, finite-dimensional or not. Interpret the above derivation the other way around – as a mapping of h:

$$f(h) = [f, h].$$

13. This way, f is now interpreted not only as an individual member but also as a mapping on every member like h.
14. This new mapping is also called the adjoint of the original member f. It is often denoted by

$$\text{ad}(f) \equiv [f, \cdot].$$

This way, it could be applied to h:

$$\text{ad}(f)(h) = [f, \cdot]h = [f, h] = f(h).$$

15. Is this a linear mapping? Hint: our Lie brackets are distributive (bilinear).
16. Likewise, interpret the member $[f, g]$ as a mapping as well. Hint: on each given h, the member $[f, g]$ acts like this:

$$[f, g](h) = [[f, g], h].$$

17. This is called the adjoint of the original member $[f, g]$.
18. Use this to rewrite the Jacobi identity as

$$[f, g](h) = -g(f(h)) + f(g(h)).$$

19. Better yet, drop the argument h, and rewrite this as a whole:

$$[f, g] = f \circ g - g \circ f.$$

20. What is the left-hand side? Hint: it is a mapping: the adjoint of the original member $[f, g]$.
21. Use this formula to explain why the Lie brackets are often called the commutator.
22. How is this formula different from the original Jacobi identity? Hint: it talks about mappings of the entire Lie algebra, not just about individual members.
23. Observe that this formula actually *defines* new brackets, acting on two adjoint mappings, f and g, and producing their commutator, denoted also by $[f, g]$ for short.
24. Are these new brackets still antisymmetric? Hint: they must be, because they are inherited from the old ones. In fact, they are just a new interpretation (or representation): each member is represented by its own adjoint.
25. Prove this yet more directly, using the explicit definition on the right-hand side. Hint: let f_1 and f_2 be two members (interpreted as their own adjoints). Thanks to the above formula,

$$\begin{aligned}
[f_1, f_2] &= f_1 \circ f_2 - f_2 \circ f_1 \\
&= -(f_2 \circ f_1 - f_1 \circ f_2) \\
&= -[f_2 f_1].
\end{aligned}$$

26. Do these new brackets still satisfy their own Jacobi identity? Hint: they must, because they are just a new interpretation (or representation): each member is represented by its own adjoint.
27. Prove this yet more directly, using the explicit definition on the right-hand side. Hint: let f_1, f_2, and f_3 be three members (interpreted as their own adjoints). Apply the above definition twice:

$$\begin{aligned}
[[f_1, f_2], f_3] &= [f_1, f_2] \circ f_3 - f_3 \circ [f_1, f_2] \\
&= (f_1 \circ f_2 - f_2 \circ f_1) \circ f_3 - f_3 \circ (f_1 \circ f_2 - f_2 \circ f_1) \\
&= f_1 \circ f_2 \circ f_3 - f_2 \circ f_1 \circ f_3 - f_3 \circ f_1 \circ f_2 + f_3 \circ f_2 \circ f_1.
\end{aligned}$$

Here, the first term is mirrored by the third one: they differ only in sign and in a cyclic shift of indices:

$$(1, 2, 3) \to (3, 1, 2) \to (2, 3, 1).$$

Likewise, the second term is mirrored by the fourth one. Therefore, if you carry out these shifts one by one, and sum up all 12 terms, then you must get the zero vector, as required.

28. Conclude that these new brackets (on mappings, not members) may indeed be viewed as legitimate Lie brackets.

29. Conclude that these mappings indeed make a legitimate Lie algebra as well.
30. Go one step further, and write

$$f(g) = [f, g] = f \circ g - g \circ f.$$

31. Better yet, drop the g, and obtain

$$f() = f \circ - \circ f.$$

32. Here, what is $f\circ$? Hint: an operator: it takes a mapping g, and composes f with it, to produce a new mapping: $f \circ g$.
33. Is it linear? Hint: see the end of the previous section.
34. Is it a derivation? Hint: no! See the end of the previous section.
35. What is $\circ f$? Hint: an operator: it takes a mapping g, and composes it with f, to produce a new mapping: $g \circ f$.
36. Is it linear?
37. Is it a derivation?
38. In the above formula, what is the right-hand side? Hint: the difference between these two.
39. The above formula tells us that this is the same as the left-hand side. What is the original meaning of the left-hand side? Hint: an operator: it takes a special kind of mapping – the adjoint of the member g. To this member, it applies f, to produce the new member $f(g) = [f, g]$. Finally, it looks at the adjoint again, and interprets $[f, g]$ as a mapping rather than an individual member.
40. What if f was no adjoint, but just some linear mapping on some vector space? Could you still define $f()$ as an operator acting on a general linear mapping g? Hint: use the right-hand side as a new definition.
41. Better yet, define the new operator $f()$ even for an f that is a member of some abstract algebra, with some associative product '$\circ$' defined in it. Hint: use the right-hand side as a new definition.
42. Why must this product be associative? Hint: to make sure that $f()$ is a derivation:

$$\begin{aligned} f(g \circ h) &= f \circ g \circ h - g \circ h \circ f \\ &= f \circ g \circ h + (-g \circ f \circ h + g \circ f \circ h) - g \circ h \circ f \\ &= (f \circ g \circ h - g \circ f \circ h) + (g \circ f \circ h - g \circ h \circ f) \\ &= f(g) \circ h + g \circ f(h). \end{aligned}$$

43. Repeat this with the roles of g and h interchanged, subtract, and obtain the Jacobi identity once again.
44. The new operator $f()$ is also called the adjoint of the original mapping f.
45. Is it linear?
46. Assume that $f\circ$ is nilpotent. What does this mean? Hint: for some (sufficiently large) k, $(f\circ)^k$ is the zero operator: it converts every mapping to the zero mapping.
47. Define its exponent:

$$\exp(f\circ) = \sum_{i=0}^{\infty} \frac{(f\circ)^i}{i!}.$$

48. Is this really infinite? Hint: only the k leading terms survive.
49. Show that this is already in the Jordan form. Hint: the first term is just the identity operator. This is the diagonal part. The rest, on the other hand, is nilpotent. To see this, take its kth power: it contains only high powers of $f\circ$ that vanish.
50. What is the inverse operator? Hint: $\exp(-f\circ)$.
51. Write it in terms of the above nilpotent part. Hint: take the above nilpotent part, and look at its (finite) power series, with alternating signs.
52. Likewise, look at $\circ f$. Assume that it is nilpotent as well (with $\tilde{k}$ rather than k).
53. Define its exponent as well.
54. What is its Jordan form?
55. What is its inverse?
56. Now, look at $f()$. What is its nth power? Hint: thanks to the binomial formula,

$$(f())^n = (f\circ - \circ f)^n = \sum_{i=0}^{n}(-1)^{n-i}\binom{n}{i}(f\circ)^i(\circ f)^{n-i}$$

$(n \geq 0)$.
57. How does this operator act on a given mapping g? Hint: in the right-hand side, plug g in the middle: in between $(f\circ)^i$ and $(\circ f)^{n-i}$.
58. Is $f()$ nilpotent? Hint: in the above formula, pick n as large as $k + \tilde{k}$. This way, in the right-hand side, all terms vanish.
59. The original f is said to be adjoint-nilpotent. Why? Hint: because its adjoint $f()$ is nilpotent.
60. Define the exponent of $f()$. In the "infinite" series, how many terms survive? Hint: at most $k + \tilde{k}$.
61. Show that this exponent is already in the Jordan form as well.
62. What is its inverse? Hint: $\exp(-f())$.
63. Show that

$$\exp(f()) = \exp(f\circ - \circ f) = \exp(f\circ)\exp(-\circ f).$$

64. How does this act on a given mapping g? Hint: in the right-hand side, plug g in the middle.
65. Use this to have the Jordan form of $\exp(f())$ in yet another way.
66. Use this to have the inverse of $\exp(f())$ in yet another way.
67. Show that $\exp(f())$ preserves the new Lie brackets (Figure 15.2): for every two mappings g and h as above,

$$\exp(f())([g, h]) = [\exp(f())g, \exp(f())h],$$

or, more explicitly,

$$\exp(f())(g \circ h - h \circ g) = \exp(f())g \circ \exp(f())h \; - \; \exp(f())h \circ \exp(f())g.$$

Hint: in the previous section, this is proved even for a general derivation in a general algebra. Our new Lie algebra is just a special case: its product uses the new Lie brackets.

68. In view of the above hint, must $f()$ have the form

$$f() = f \circ - \circ f?$$

Could $f()$ be a more general derivation of the form

$$f() = [f, \cdot],$$

in a Lie algebra with no 'o' at all, as in the beginning of this section? Hint: yes! Indeed, the special form $f() = f \circ - \circ f$ was only used to prove nilpotency. Therefore, it could be dropped, provided that we already know from another source that the derivation $f() = [f, \cdot]$ is nilpotent.

69. In summary, thanks to the above exercises, what can you say about $\exp(f())$? Hint: it is an invertible operator that preserves the Lie brackets.

70. Is it a derivation? Hint: no! See the previous section.

71. Let E be yet another operator of the same kind: linear, invertible, Lie-bracket preserving, but not necessarily a derivation. Look at a triple composition of three operators:

$$E \circ f() \circ E^{-1}.$$

How does it act on a given mapping g? Hint: leftwards, in three stages: first, apply E^{-1} (the inverse of E). Then, apply the operator $f()$. Finally, apply E.

72. Is it a derivation in its own right? Hint: yes – see the previous section.

73. Is it an adjoint of anything? Hint: see below.

74. Design a new mapping: take the old mapping f, and apply E to it. Hint: the new mapping is Ef.

75. What is its adjoint? Hint: the new operator $(Ef)()$.

76. Write it as a triple composition of three operators:

$$(Ef)() = E \circ f() \circ E^{-1}.$$

Hint: look at the right-hand side. Apply it to Eg:

$$E \circ f() \circ E^{-1} Eg = E \circ f() g$$
$$= Ef(g)$$
$$= E[f, g]$$
$$= [Ef, Eg]$$
$$= (Ef)(Eg).$$

77. Is it nilpotent?

78. Conclude that it has an exponent.

79. Write its exponent as a new triple composition:

$$\exp((Ef)()) = \exp\left(E \circ f() \circ E^{-1}\right) = E \exp(f()) E^{-1}.$$

Hint: see the previous section.

80. Finally, look at things the other way around: conclude that every triple composition of the form $E \exp(f()) E^{-1}$ can be written more compactly as the exponent of some adjoint. Hint:

$$E \exp(f()) E^{-1} = \exp\left(E \circ f() \circ E^{-1}\right) = \exp\left((Ef)()\right).$$

81. Illustrate this geometrically in a commutative diagram. Hint: see Figure 15.3.

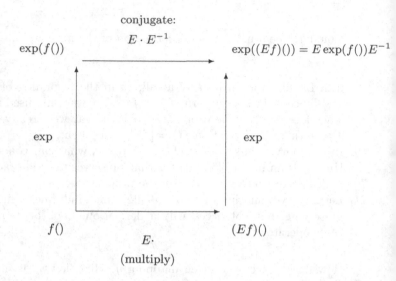

conjugate:

$$E \cdot E^{-1}$$

$\exp(f())$ ———————→ $\exp((Ef)()) = E \exp(f()) E^{-1}$

exp   exp

$f()$ $(Ef)()$

$E \cdot$

(multiply)

Fig. 15.3. A commutative diagram: the operator $f()$ is at the lower-left corner. How to get to the upper-right corner? There are two equivalent ways: either multiply by E (bottom edge) and then take the exponent (right edge), or take the exponent first (left edge) and then conjugate: multiply by E from the left, and by E^{-1} from the right (top edge). In either way, you obtain the same result at the upper-right corner.

15.3 Examples: Linear Lie Algebras

1. Design a Lie algebra with a geometrical meaning.
2. Look at the three-dimensional Cartesian space. Is it a Lie algebra?
3. look at the vector product in Chapter 2. Use it to introduce new Lie brackets:

$$[u, v] \equiv u \times v,$$

where u and v are any three-dimensional vectors.

4. Is it alternative? Hint: it sure is:

$$[u, u] = u \times u = \mathbf{0}.$$

5. Conclude that it is antisymmetric as well.
6. Does it satisfy the Jacobi identity? Hint: first, check a triplet of standard unit vectors: $\mathbf{i}$, $\mathbf{j}$, and $\mathbf{k}$. Thanks to alternativity,

$$[[\mathbf{i}, \mathbf{j}], \mathbf{k}] = (\mathbf{i} \times \mathbf{j}) \times \mathbf{k} = \mathbf{k} \times \mathbf{k} = \mathbf{0}.$$

Next, check yet another typical triplet: $\mathbf{i}$, $\mathbf{i}$, and $\mathbf{j}$. Thanks to alternativity and antisymmetry,

$$[[\mathbf{i}, \mathbf{i}], \mathbf{j}] + [[\mathbf{j}, \mathbf{i}], \mathbf{i}] + [[\mathbf{i}, \mathbf{j}], \mathbf{i}]$$
$$= \mathbf{0} + [[\mathbf{j}, \mathbf{i}], \mathbf{i}] - [[\mathbf{j}, \mathbf{i}], \mathbf{i}]$$
$$= \mathbf{0}.$$

Now, there is nothing special about $\mathbf{i}$ and $\mathbf{j}$: the same could be done for any two standard unit vectors. Finally, use bilinearity to extend this to any triplet of any three vectors as well.

7. Design yet another Lie algebra with a physical meaning.
8. For this purpose, look at the $n \times n$ (complex) matrices.
9. Do they make a Lie algebra? Hint: the commutator indeed satisfies the Jacobi identity (Section 15.2).
10. This is the general linear Lie algebra, denoted by gl_n.
11. More precisely, its full name is

$$gl_n \equiv gl(n, \mathbb{C}).$$

Here, however, we always deal with complex matrices, so there is no need to write '$\mathbb{C}$.'

12. Every subalgebra of gl_n is called a linear Lie algebra. Give an example.
13. Look at the 3×3 matrices that model the spin of a boson. (See exercises at the end of Chapter 12.) Do they span a Lie algebra? Hint: they are closed under the commutator. (Span with imaginary coefficients only.)
14. Is this a linear Lie algebra? Hint: it is a subalgebra of gl_3.
15. How is it related to the three-dimensional Cartesian space discussed above? Hint: they are isomorphic to each other. Indeed, in both, the Lie brackets are cyclic.
16. Conclude once again that the vector product obeys the Jacobi identity.
17. Look at the 2×2 Pauli matrices that model both the spin of a fermion and the polarization of an electromagnetic wave. (See exercises at the end of Chapter 12.) Do they span a Lie algebra? Hint: they are closed under the commutator. (Span with imaginary coefficients only.)
18. Is this a linear Lie algebra? Hint: it is a subalgebra of gl_2.
19. How is it related to the previous two examples? Hint: all three are isomorphic to each other. Indeed, in all three, the Lie brackets are cyclic.
20. Look at gl_n: the general linear Lie algebra. In particular, this is a vector space. Design a basis for it. Hint: let $1 \leq p, q \leq n$ be two fixed natural numbers. Define a new $n \times n$ matrix $E^{(p,q)}$ with only one nonzero element – at the pth row and qth column:

$$\left(E^{(p,q)}\right)_{i,j} = \begin{cases} 1 & \text{if } i = p \text{ and } j = q \\ 0 & \text{otherwise.} \end{cases}$$

21. Use the tensor

$$\delta_i^p \equiv \begin{cases} 1 & \text{if } i = p \\ 0 & \text{if } i \neq p \end{cases}$$

to write this more compactly. Hint:

$$\left(E^{(p,q)}\right)_{i,j} = \delta_i^p \delta_j^q.$$

22. How many such matrices are there? Hint: n^2.

23. Are they linearly independent?

24. What is the dimension of gl_n? Hint: n^2.

25. What is the product of two such matrices? Hint: let $1 \le p, q, r, s \le n$ be four fixed natural numbers. To have a nonzero element in the product, we must have $q = r$. Indeed, on the right-hand side, only the term with $j = q = r$ survives:

$$\sum_{j=1}^{n} \left(E^{(p,q)} \right)_{i,j} \left(E^{(r,s)} \right)_{j,k} = \sum_{j=1}^{n} \delta_i^p \delta_j^q \delta_j^r \delta_k^s$$

$$= \delta_q^r \delta_i^p \delta_k^s$$

$$= \delta_q^r \left(E^{(p,s)} \right)_{i,k}.$$

26. Write this yet more compactly:

$$E^{(p,q)} E^{(r,s)} = \delta_q^r E^{(p,s)}.$$

27. Interchange (p, q) with (r, s):

$$E^{(r,s)} E^{(p,q)} = \delta_p^s E^{(r,q)}.$$

28. Subtract these formulas, and obtain the commutator:

$$\left[E^{(p,q)}, E^{(r,s)} \right] = E^{(p,q)} E^{(r,s)} - E^{(r,s)} E^{(p,q)}$$

$$= \delta_q^r E^{(p,s)} - \delta_p^s E^{(r,q)}$$

$$= \begin{cases} E^{(p,s)} & \text{if } q = r \text{ and } p \ne s \\ -E^{(r,q)} & \text{if } p = s \text{ and } q \ne r \\ E^{(p,p)} - E^{(q,q)} & \text{if } q = r \ne p = s \\ (0) & \text{otherwise.} \end{cases}$$

29. Conclude that the commutator is trace-free: it must have a zero trace.

30. Extend this to any two matrices: their a commutator must be trace-free. Hint: write each matrix as a linear combination of the E's, open parentheses, and use bilinearity.

15.4 Trace-Free Matrices

1. Look at the trace-free matrices of order n. Do they make a Lie algebra? Hint: the commutator is trace-free as well.

2. This is the special linear Lie algebra, denoted by sl_n.

3. Is it indeed linear? Hint: $sl_n \subset gl_n$.

4. Design a basis for it. Hint: take $E^{(p,q)}$ with $p \ne q$, and also $E^{(r,r)} - E^{(r+1,r+1)}$ $(1 \le r < n)$.

5. How many matrices are there in this basis? Hint: $n^2 - 1$.

6. What is the dimension of sl_n? Hint: $n^2 - 1$.

7. Let L be yet another Lie algebra that lies in between:

$$sl_n \subset L \subset gl_n.$$

8. Is L linear? Hint: $L \subset gl_n$.
9. What is its dimension? Hint: either n^2 or $n^2 - 1$.
10. What is L? Hint: if L has dimension n^2, then $L = gl_n$. If, on the other hand, L has dimension $n^2 - 1$, then $L = sl_n$.
11. Conclude that there is no Lie algebra strictly between sl_n and gl_n.
12. sl_n is also called A_l, where $l = n - 1$.
13. In terms of l, what is the dimension of A_l? Hint: $l^2 + 2l$.
14. For example, look at $sl_2 = A_1$.
15. Design a basis for it. Hint: the basis could contain the following matrices:

$$e \equiv \begin{pmatrix} 0 & 1 \\ 0 & 0 \end{pmatrix}, \quad f \equiv \begin{pmatrix} 0 & 0 \\ 1 & 0 \end{pmatrix}, \quad \text{and} \quad h \equiv \begin{pmatrix} 1 & 0 \\ 0 & -1 \end{pmatrix}.$$

16. What is the dimension? Hint: 3.
17. Does this agree with the general rule in the previous exercises? Hint: here, $n = 2$ and $l = 1$, so $n^2 - 1 = l^2 + 2l = 3$, as required.
18. Show that

$$[e, f] = h.$$

19. Does this agree with the general formula for the E's in Section 15.3?
20. Show also that

$$[h, e] = 2e, \quad \text{and} \quad [h, f] = -2f.$$

21. Does this agree with the general formula for the E's in Section 15.3?
22. Look at the new mapping

$$[h, \cdot] : sl_2 \to sl_2,$$

defined by

$$v \to [h, v], \quad v \in sl_2.$$

Is this familiar? Hint: this is just the adjoint of h, denoted by $h(v)$ in Section 15.2.
23. What are its eigenvectors? Hint: e, f, and h.
24. What are their eigenvalues? Hint: 2, -2, and 0, respectively.

15.5 Triangular Matrices

1. Next, look at upper-triangular matrices only. What basis should be used to span them? Hint: take those $E^{(p,q)}$'s with $p \le q$.
2. How does their commutator look like? Hint: recall that

$$\left[E^{(p,q)}, E^{(r,s)} \right] = E^{(p,q)} E^{(r,s)} - E^{(r,s)} E^{(p,q)}$$

$$= \delta_q^r E^{(p,s)} - \delta_p^s E^{(r,q)}$$

$$= \begin{cases} E^{(p,s)} & \text{if } q = r \text{ and } p \ne s \\ -E^{(r,q)} & \text{if } p = s \text{ and } q \ne r \\ E^{(p,p)} - E^{(q,q)} & \text{if } q = r \ne p = s \\ (0) & \text{otherwise.} \end{cases}$$

In the context of upper-triangular matrices, assume that $p \le q$ and $r \le s$. Then, to have a nonzero commutator, we can never use the third option, but only the first or second option: either $p \leq q = r \leq s$, or $r \leq s = p \leq q$. Furthermore, to avoid $p = q = r = s$, we must also have a strict inequality: either $p < s$ or $r < q$ (respectively). In either case, the commutator is *strictly* upper triangular.

3. Extend this to every linear combination of $E^{(p,q)}$'s with $p \leq q$: their commutator must be strictly upper triangular. Hint: open parentheses, and use bilinearity.

4. Look at the upper-triangular matrices of order n. Do they make a Lie algebra? Hint: they are closed under the commutator.

5. Denote this Lie algebra by

$$T_0 \equiv \text{span} \left\{ E^{(p,q)} \mid q - p \geq 0 \right\}.$$

6. Is T_0 linear? Hint: $T_0 \subset gl_n$.

7. What is its dimension? Hint: $n(n+1)/2$.

8. Look at the strictly upper-triangular matrices of order n. Do they make a Lie algebra? Hint: they are closed under the commutator.

9. Denote this Lie algebra by

$$T_1 \equiv \text{span} \left\{ E^{(p,q)} \mid q - p \geq 1 \right\}.$$

10. Is T_1 linear? Hint: $T_1 \subset T_0 \subset gl_n$. Furthermore, $T_1 \subset sl_n \subset gl_n$.

11. What is its dimension? Hint: $n(n-1)/2$.

12. Likewise, for any $k \geq 0$, look at yet stricter matrices (with yet fewer nonzero diagonals), and define yet smaller Lie algebras:

$$T_k \equiv \text{span} \left\{ E^{(p,q)} \mid q - p \geq k \right\}.$$

13. Pick an E from T_k, and another from T_j (for some $j, k \geq 0$). Where is their commutator? Hint: in T_{j+k}.

14. Use bilinearity to extend this: pick one matrix from T_k, and another from T_j. Then, their commutator is in $T_{j+k} \subset T_j \cap T_k$.

15. In particular, when $j = k$, the commutator is in $T_{2k} \subset T_k$.

16. Conclude that T_k is closed under the commutator.

17. Conclude that

$$T_k \subset T_{k-1} \subset T_{k-2} \subset \cdots \subset T_1 \subset T_0$$

are subalgebras of each other.

18. So far, we've seen that a member from T_j and a member from T_k have their commutator in T_{j+k}. Next, work the other way around: pick an $E^{(p,q)} \in T_{j+k}$. Could it be written as a commutator? (Assume that $j + k > 0$.) Hint: since $q - p \geq j + k > 0$,

$$E^{(p,q)} = E^{(p,p+j)} E^{(p+j,q)} = \left[E^{(p,p+j)}, E^{(p+j,q)} \right].$$

19. In the above exercise, why must $j+k > 0$? Hint: if $j = k = 0$, then the above not always works. For example, $E^{(1,1)} \in T_{j+k} = T_0$ could never be a commutator. Why? Because its trace is 1, not 0.

20. Give a new example of a strictly upper-triangular matrix. Hint: a nilpotent Jordan block as in Chapter 14, Section 14.1.6 has 1's on the superdiagonal only, and 0 elsewhere.

21. Write it as a sum of the above E's.
22. Use this to calculate its powers.
23. Does this agree with Chapter 14, Section 14.1.6?
24. Show that its kth power is in T_k ($k \geq 0$).
25. Conclude that its nth power is the zero matrix.
26. Conclude that it is indeed nilpotent.
27. Look at two matrices: $E^{(r,r)}$ and $E^{(p,p)}$ ($1 \leq r, p \leq n$). Are they strictly upper triangular?
28. Are they upper triangular?
29. Are they diagonal?
30. Do they commute with each other?
31. What is their commutator? Hint: the zero matrix.
32. Is this also true for every two diagonal matrices? Hint: they commute, and are commutator-free.
33. Look at all the diagonal matrices of order n. Do they make a Lie algebra? Hint: their commutator is the zero matrix, which is diagonal as well.
34. Do they have a physical meaning? Hint: a real diagonal matrix could serve as a position matrix in quantum mechanics (Chapter 12, Section 12.5.2).
35. Is this a linear Lie algebra? Hint: it is a subalgebra of $T_0 \subset gl_n$.
36. What is its dimension? Hint: n.

15.6 Orthogonal Lie Algebra: D_l

1. Next, let S be an $n \times n$ symmetric matrix:

$$S = S^t.$$

2. Look at a new $n \times n$ (complex) matrix X that is S-antisymmetric:

$$SX = -X^t S.$$

3. Look at all such X's. Do they make a Lie algebra? Hint: look at yet another matrix $\tilde{X}$ that is S-antisymmetric as well:

$$S\tilde{X} = -\tilde{X}^t S.$$

This way, their product satisfies

$$SX\tilde{X} = -X^t S\tilde{X} = X^t \tilde{X}^t S = \left(\tilde{X}X\right)^t S.$$

Thus, their commutator is S-antisymmetric as well:

$$S[X, \tilde{X}] = S\left(X\tilde{X} - \tilde{X}X\right) = \left(\tilde{X}X - X\tilde{X}\right)^t S = -[X, \tilde{X}]^t S,$$

as required.
4. This is the orthogonal Lie algebra, denoted by o_n.
5. Is o_n linear? Hint: $o_n \subset gl_n$.
6. If S is also nonsingular, then o_n is said to be nondegenerate.

7. For example, assume that n is even:

$$n = 2l.$$

Assume also that S can be written in a block form:

$$S = \begin{pmatrix} & I \\ I & \end{pmatrix},$$

where I is the $l \times l$ identity matrix.

8. Write X in a block form as well:

$$X = \begin{pmatrix} M & N \\ P & Q \end{pmatrix},$$

where M, N, P, and Q are some $l \times l$ blocks.

9. Write the above condition in a block form as well:

$$\begin{aligned} SX &= \begin{pmatrix} & I \\ I & \end{pmatrix} \begin{pmatrix} M & N \\ P & Q \end{pmatrix} \\ &= \begin{pmatrix} P & Q \\ M & N \end{pmatrix} \\ &= -X^t S \\ &= -\begin{pmatrix} M^t & P^t \\ N^t & Q^t \end{pmatrix} \begin{pmatrix} & I \\ I & \end{pmatrix} \\ &= -\begin{pmatrix} P^t & M^t \\ Q^t & N^t \end{pmatrix}. \end{aligned}$$

10. In this form, each block gives us a new condition. The upper-left block tells us that P must be antisymmetric. How many degrees of freedom are there in P? Hint: $l(l-1)/2$.

11. Likewise, the lower-right block tells us that N must be antisymmetric as well. How many degrees of freedom are there in N? Hint: $l(l-1)/2$.

12. Finally, the upper-right and lower-left blocks tell us that

$$Q = -M^t.$$

Thus, Q and M can't be picked freely. Only one of them can, but the other cannot. How many degrees of freedom are there in them? Hint: l^2.

13. What is the total number of degrees of freedom in X? Hint: $2l^2 - l$.

14. What is the trace of X? Hint: thanks to the latter condition, X must be trace-free.

15. In the above setting, o_n has a special name: D_l.

16. What is its dimension? Hint: $2l^2 - l$.

17. Conclude that $D_l \subset sl_{2l}$.

15.7 Simplectic Lie Algebra: C_l

1. Next, introduce a little change. Assume now that S is not symmetric but *antisymmetric*:

$$S = -S^t.$$

2. As before, look at an $n \times n$ (complex) matrix X that is S-antisymmetric:

$$SX = -X^t S.$$

3. Look at all such X's. Do they still make a Lie algebra? Hint: look at yet another matrix $\tilde{X}$ that is S-antisymmetric as well:

$$S\tilde{X} = -\tilde{X}^t S.$$

This way, their product still satisfies

$$SX\tilde{X} = -X^t S\tilde{X} = X^t \tilde{X}^t S = \left(\tilde{X}X\right)^t S.$$

Thus, as before, their commutator is S-antisymmetric as well:

$$S[X, \tilde{X}] = S\left(X\tilde{X} - \tilde{X}X\right) = \left(\tilde{X}X - X\tilde{X}\right)^t S = -[X, \tilde{X}]^t S.$$

4. This is the simplectic Lie algebra, denoted by sp_n.
5. Is sp_n linear? Hint: $sp_n \subset gl_n$.
6. If S is also nonsingular, then sp_n is said to be nondegenerate.
7. For example, assume again that n is even:

$$n = 2l.$$

Assume also that S can be written in a block form:

$$S = \begin{pmatrix} & I \\ -I & \end{pmatrix}.$$

8. Write X in this block form as well:

$$X = \begin{pmatrix} M & N \\ P & Q \end{pmatrix},$$

where M, N, P, and Q are some $l \times l$ new blocks.
9. Write the above condition in a block form as well:

$$SX = \begin{pmatrix} & I \\ -I & \end{pmatrix}\begin{pmatrix} M & N \\ P & Q \end{pmatrix}$$
$$= \begin{pmatrix} P & Q \\ -M & -N \end{pmatrix}$$
$$= -X^t S$$
$$= -\begin{pmatrix} M^t & P^t \\ N^t & Q^t \end{pmatrix}\begin{pmatrix} & I \\ -I & \end{pmatrix}$$
$$= -\begin{pmatrix} -P^t & M^t \\ -Q^t & N^t \end{pmatrix}.$$

10. In this form, each block gives us a new condition. The upper-left block tells us that P must be symmetric. How many degrees of freedom are there in P? Hint: $l(l+1)/2$.

11. Likewise, the lower-right block tells us that N must be symmetric as well. How many degrees of freedom are there in N? Hint: $l(l+1)/2$.

12. Finally, the upper-right and lower-left blocks tell us that

$$Q = -M^t.$$

Thus, Q and M can't be picked freely. Only one of them can, but the other cannot. How many degrees of freedom are there in them? Hint: l^2.

13. What is the total number of degrees of freedom in X? Hint: $2l^2 + l$.

14. What is the trace of X? Hint: thanks to the latter condition, X must be trace-free.

15. In the above setting, sp_n has a special name: C_l.

16. What is its dimension? Hint: $2l^2 + l$.

17. Conclude that $C_l \subset sl_{2l}$.

15.8 Orthogonal Lie Algebra: B_l

1. Let's return to o_n: assume that S is not antisymmetric but *symmetric* again. Let's make a new change. Assume now that n is not even but odd:

$$n = 2l + 1.$$

2. Assume that S can now be written in a new block form, with one more row and one more column:

$$S = \begin{pmatrix} 1 & & \\ & & I \\ & I & \end{pmatrix}.$$

3. Write X in a new block form as well:

$$X = \begin{pmatrix} d & y & z \\ u & M & N \\ v & P & Q \end{pmatrix},$$

where M, N, P, and Q are new $l \times l$ blocks, y and z are some row vectors, u and v are some column vectors, and d is some scalar.

4. Write the above condition in this block form as well:

$$\begin{aligned}
SX &= \begin{pmatrix} 1 & & \\ & & I \\ & I & \end{pmatrix} \begin{pmatrix} d & y & z \\ u & M & N \\ v & P & Q \end{pmatrix} \\
&= \begin{pmatrix} d & y & z \\ v & P & Q \\ u & M & N \end{pmatrix} \\
&= -X^t S \\
&= -\begin{pmatrix} d & u^t & v^t \\ y^t & M^t & P^t \\ z^t & N^t & Q^t \end{pmatrix} \begin{pmatrix} 1 & & \\ & & I \\ & I & \end{pmatrix} \\
&= -\begin{pmatrix} d & v^t & u^t \\ y^t & P^t & M^t \\ z^t & Q^t & N^t \end{pmatrix}.
\end{aligned}$$

5. In this form, the conditions are now

$$P = -P^t$$
$$N = -N^t$$
$$Q = -M^t$$
$$d = 0$$
$$y = -v^t$$
$$z = -u^t.$$

6. Compare this to the former orthogonal Lie algebra: D_l. How many new degrees of freedom are there here? Hint: we have now $2l$ new degrees of freedom to pick u and v freely.
7. What is the total number of degrees of freedom in X? Hint: $2l^2 + l$.
8. What is the trace of X? Hint: thanks to the above conditions, X is trace-free.
9. In the above case, o_n has a special name: B_l.
10. What is its dimension? Hint: $2l^2 + l$.
11. Is there a simplectic Lie algebra of the same dimension? Hint: C_l.
12. Show that $B_l \subset sl_{2l+1}$.

16

Ideals
and Isomorphism Theorems

Later on, we'll also introduce a few useful Lie groups in quantum mechanics and special relativity. The group may have a normal subgroup, useful to define the factor (quotient) group [27, 45]. In Lie algebras, this structure is mirrored: the original Lie algebra may include an ideal: a subalgebra in its own right. This may help design the factor Lie algebra, and prove all sorts of isomorphism theorems, to mirror group theory.

16.1 Ideals

16.1.1 Two Subalgebras

Our original Lie algebra can now be used to produce new Lie algebras. We've already seen a few examples of subalgebras: Lie algebras in their own right. Here, we'll use subalgebras to produce yet more interesting Lie algebras.

Let L be a Lie algebra. Let

$$K, M \subset L$$

be two subalgebras. Let's introduce a new notation:

$$[K, M] \equiv \operatorname{span}\{[k, m] \mid k \in K, \ m \in M\}.$$

What is this? Well, this is a new subspace of L, spanned by vectors of the form $[k, m]$, where k is in K, and m is in M. Thanks to antisymmetry, we immediately have

$$[K, M] = [M, K].$$

Still, is this a subalgebra in its own right? In other words, is it closed under the Lie brackets? In other words, if we picked two vectors of the form $[k, m]$ and $[\tilde{k}, \tilde{m}]$ and placed them in Lie brackets, would the result satisfy

$$[[k, m], [\tilde{k}, \tilde{m}]] \in [K, M]?$$

Not necessarily! This is why $[K, M]$ is only a subspace, but not necessarily a subalgebra.

16.1.2 Sum of Subalgebras

Let's continue to "play" with our subalgebras. Let's sum them up:

$$K + M \equiv \{k + m \mid k \in K, \ m \in M\}.$$

Clearly, this is a subspace. Indeed, it is linear: closed under addition and scalar multiplication. Still, is it a subalgebra as well? To check on this, pick two such vectors, and place them in Lie brackets:

$$[k + m, \tilde{k} + \tilde{m}] = [k, \tilde{k}] + [k, \tilde{m}] + [m, \tilde{k}] + [m, \tilde{m}].$$

In general, we know nothing about $[k, \tilde{m}]$ or $[m, \tilde{k}]$: they could lie outside $K + M$. Thus, $K + M$ is only a subspace, but not necessarily a subalgebra.

Nevertheless, we can still look at a special case, in which both $[K, M]$ and $K + M$ are legitimate subalgebras.

16.1.3 Ideal

Let's look at a special subspace

$$I \subset L.$$

(Don't confuse this with the identity matrix I, used often in linear algebra.) To be a legitimate subalgebra, I must be closed under the Lie brackets:

$$[I, I] \subset I.$$

Let's require even more:

$$[I, L] \subset I.$$

This way, I is called an ideal of L.

16.1.4 Two Ideals

Let I and J be two ideals of L. What about their sum? Is it a legitimate ideal as well? Well, look at three vectors: $i \in I$, $j \in J$, and $l \in L$. Thanks to bilinearity,

$$[i + j, l] = [i, l] + [j, l] \in [I, L] + [J, L] \subset I + J,$$

as required. Furthermore, this is also true for every linear combination, so

$$[I + J, L] \subset I + J,$$

as required.

Next, what about $[I, J]$? Is it a legitimate ideal as well? Well, look again at three vectors: $i \in I$, $j \in J$, and $l \in L$. Thanks to the Jacobi identity,

$$[[i, j], l] = -[[l, i], j] - [[j, l], i] \in [[I, L], J] + [[J, L], I] \subset [I, J] + [J, I] = [I, J],$$

as required. Thanks to bilinearity, this also extends to any linear combination of $[i, j]$'s. Thus,

$$[[I, J], L] \subset [I, J],$$

as required.

16.2 Homomorphism and Its Kernel

16.2.1 Homomorphism and Isomorphism

What is a homomorphism between groups? This is a mapping that preserves the group operation: multiplication. If the homomorphism is invertible, then it is also called isomorphism.

How is this relevant to Lie algebras? Well, a Lie algebra is first of all a linear space, which is a group in its own right. There are just a few technical changes:

- The group operation is now called addition, not multiplication.
- This is denoted by the plus sign: '+'.
- The unit element is now the zero vector: $\mathbf{0}$.
- The "inverse" is just the negative vector.
- Since addition is commutative, every subgroup is normal.

Still, this is not the whole story. After all, a vector space is more than just a group. Thus, we have a new requirement: our homomorphism must preserve not only addition but also scalar multiplication. In other words, it must be linear.

Still, this is not the whole story. In the context of algebras, we have yet another requirement: our homomorphism must also preserve the product operation. In a Lie algebra, for example, it must preserve the Lie brackets. On top of this, if the homomorphism is also invertible, then it is also called isomorphism (Chapter 15, Section 15.1).

16.2.2 Kernel – Null Space

As discussed above, a Lie algebra is in particular a linear space. Therefore, a linear mapping must map $\mathbf{0}$ to $\mathbf{0}$. Still, if this mapping is only a homomorphism (not an isomorphism), then it maps a lot of vectors to $\mathbf{0}$. These vectors make a new subspace: the null space. In group theory, this is called kernel.

In the context of Lie algebras, we use both terms: kernel or null space. In an isomorphism, in particular, the kernel is quite small: it contains just one vector: $\mathbf{0}$.

16.2.3 The Kernel: A New Ideal

Let ϕ be a homomorphism on a Lie algebra L. How does ϕ behave? First of all, ϕ is linear. What does this mean? Pick two vectors $k, l \in L$, and a scalar α. Then, ϕ preserves both addition and scalar multiplication:

$$\phi(k + l) = \phi(k) + \phi(l)$$
$$\phi(\alpha k) = \alpha \phi(k).$$

Still, this is not the whole story. After all, L is not only a vector space but also a Lie algebra. Thus, to be a legitimate homomorphism, ϕ must also preserve the Lie brackets:

$$\phi([k, l]) = [\phi(k), \phi(l)]$$

(Chapter 15, Section 15.2). Thanks to this, ϕ is indeed a legitimate homomorphism. We then say that our original Lie algebra L is homomorphic to the image of ϕ.

Assume now that k is in the kernel of ϕ:

$$\phi(k) = \mathbf{0}.$$

This way,

$$\phi([k, l]) = [\phi(k), \phi(l)] = [\mathbf{0}, \phi(l)] = \mathbf{0}.$$

Thus, $[k, l]$ is in the kernel as well. In summary, the kernel is not only a subspace but also an ideal in its own right. This will be useful below.

16.2.4 The Factor (Quotient) Lie Algebra

So far, we started from a given homomorphism, and used it to design a new ideal: its kernel. Next, let's work the other way around: start from a given ideal, and use it to design a new homomorphism.

Let $I \subset L$ be an ideal. In particular, I is a subalgebra. In particular, I is a subspace. In particular, I is a subgroup with respect to the addition operation '+'. Since addition is commutative, I is actually a normal subgroup. Thus, we can go ahead and design the factor group.

For this purpose, however, we must use the plus sign '+'. This produces a new Lie algebra – the factor Lie algebra:

$$L/I \equiv \{x + I \mid x \in L\}.$$

Here, each vector takes the form

$$x + I \equiv \{x + i \mid i \in I\}.$$

(Here, i could be just any vector in I.) Instead of x, one could also pick any other representative of the form $x + \tilde{i}$ (for a fixed $\tilde{i} \in I$). Indeed,

$$x + I \equiv \{x + i \mid i \in I\} = \{x + \tilde{i} + i \mid i \in I\} = x + \tilde{i} + I.$$

Now, pick two such members:

$$x + I, \; y + I \in L/I.$$

We already know how to add them to each other:

$$(x + I) + (y + I) \equiv (x + y) + I.$$

By now, L/I is just a group. Indeed, it has just one algebraic operation: addition. Still, this is not the whole story. It is time to go ahead and introduce yet another algebraic operation. How to multiply by a scalar? Like this:

$$\alpha(x + I) \equiv \alpha x + I.$$

As before, this is independent of the particular choice of representative. Indeed, pick some $\tilde{i} \in I$, and use it to design a new representative. Fortunately, this has no effect on scalar multiplication:

$$\alpha(x + \tilde{i} + I) \equiv \alpha(x + \tilde{i}) + I = \alpha x + \alpha \tilde{i} + I = \alpha x + I = \alpha(x + I).$$

By now, L/I is not only a group but also a linear space. Indeed, the above operations are distributive in two ways:

$$
\begin{aligned}
\alpha((x+I)+(y+I)) &= \alpha(x+y+I) \\
&= \alpha(x+y)+I \\
&= \alpha x + \alpha y + I \\
&= \alpha x + I + \alpha y + I \\
&= \alpha(x+I) + \alpha(y+I),
\end{aligned}
$$

and

$$
\begin{aligned}
(\alpha+\beta)(x+I) &= (\alpha+\beta)x + I \\
&= \alpha x + \beta x + I \\
&= \alpha x + I + \beta x + I \\
&= \alpha(x+I) + \beta(x+I).
\end{aligned}
$$

In the exercises below, we'll prove this in yet another way.

By now, L/I is a legitimate linear space. To prove this, we've never used the fact that I was an ideal. It is time to use this, and define new Lie brackets in L/I:

$$[x+I, y+I] \equiv [x,y] + I.$$

Again, this is independent of the particular choice of representative. To see this, pick two vectors $i, \tilde{i} \in I$, and use them to design two new representatives. Fortunately, the Lie brackets still give the same result:

$$
\begin{aligned}
[x+i+I, y+\tilde{i}+I] &\equiv [x+i, y+\tilde{i}] + I \\
&= [x,y] + [i,y] + [x,\tilde{i}] + [i,\tilde{i}] + I \\
&= [x,y] + I
\end{aligned}
$$

(because I is an ideal). Fortunately, these new Lie brackets are also bilinear nd alternative, and also satisfy the Jacobi identity. (See exercises below.) Thus, L/I is indeed a legitimate Lie algebra.

16.2.5 The Canonical Homomorphism

By now, we've never defined any concrete homomorphism on our original Lie algebra L. It is time to do this. This way, we'll get to see its algebraic structure more vividly.

Define a new mapping

$$\pi : L \to L/I$$

by picking the relevant member from the factor Lie algebra:

$$\pi(x) \equiv x + I, \quad x \in L.$$

Let's study this new mapping. Is π linear? To check on this, pick two vectors $x, y \in L$, and a scalar α. Then, π preserves addition:

$$\pi(x+y) = (x+y) + I = (x+I) + (y+I) = \pi(x) + \pi(y).$$

Furthermore, π also preserves scalar multiplication:

$$\pi(\alpha x) = \alpha x + I = \alpha(x + I) = \alpha\pi(x).$$

Thus, π is indeed linear. Does it also preserve the Lie brackets? Let's check:

$$\pi([x, y]) \equiv [x, y] + I = [x + I, y + I] = [\pi(x), \pi(y)].$$

Thus, π is indeed a legitimate homomorphism from L *onto* L/I. In summary, we indeed worked the other way around: thanks to the given ideal I, we managed to design a new homomorphism: π. Thanks to π, we also get a bonus for free: since π preserves the Lie brackets, we can deduce once again that L/I is indeed a legitimate Lie algebra: its Lie brackets are indeed bilinear and alternative, and also satisfy the Jacobi identity. (See exercises below.)

16.2.6 The Canonical Homomorphism and Its Kernel

Furthermore, let's study π a little more: what is its kernel? To answer this, let's look at L/I: what is its zero vector? This is just I. Indeed, pick some member of the form $x + I \in L/I$. Now, add I to it:

$$(x + I) + I = (x + I) + (\mathbf{0} + I) = (x + \mathbf{0}) + I = x + I.$$

Thus, adding I changes nothing. Thus, I is indeed the zero "vector" in L/I.

What is the kernel of π? To find out, pick some vector $x \in L$. Now, to be in the kernel, x must be mapped to I:

$$\pi(x) = x + I = I.$$

To satisfy this, x must belong to I:

$$x \in I.$$

Thus, the kernel of π is exactly I, as asserted. In summary, we indeed worked the other way around: we started from a given ideal I, and managed to design a new homomorphism, whose kernel is I. This leads to a few interesting isomorphism theorems.

16.3 Isomorphism Theorems

16.3.1 The First Isomorphism Theorem

As in group theory, here too we have a few important isomorphism theorems. The first one deals with the following question. Let ϕ be a homomorphism on a Lie algebra L. How to write ϕ as a composition of two homomorphisms?

For this purpose, let I be an ideal of L, included in the kernel of ϕ:

$$x \in I \Rightarrow \phi(x) = \mathbf{0}.$$

Let

$$\pi : L \to L/I$$

be the canonical homomorphism. Then, the original homomorphism ϕ could split into two stages:

- First, apply π, to map to some member of L/I.
- To L/I, apply a new homomorphism, named ψ.

In short, there is a unique homomorphism ψ on L/I, such that

$$\phi = \psi \circ \pi.$$

How to define ψ? Easy:

$$\psi(x + I) \equiv \phi(x), \quad x \in L.$$

This way, ψ doesn't depend on any particular representative. Indeed, for any fixed $\tilde{i} \in I$,

$$\psi(x + \tilde{i} + I) = \phi(x + \tilde{i}) = \phi(x) + \phi(\tilde{i}) = \phi(x) + \mathbf{0} = \phi(x).$$

Furthermore, ψ is linear on L/I. To see this, pick some $x, y \in L$, and some scalar α. Then,

$$\psi(x + y + I) = \phi(x + y) = \phi(x) + \phi(y) = \psi(x + I) + \psi(y + I),$$

and

$$\psi(\alpha x + I) = \phi(\alpha x) = \alpha \phi(x) = \alpha \psi(x + I).$$

Finally, ψ also preserves the Lie brackets in L/I:

$$\psi([x + I, y + I]) = \psi([x, y] + I) = \phi([x, y]) = [\phi(x), \phi(y)] = [\psi(x + I), \psi(y + I)]].$$

In summary, ψ is indeed a legitimate homomorphism on L/I, as required. In other words, L/I is homomorphic to the image of ϕ.

What happens in the special case in which I is the same as the kernel of the original homomorphism ϕ? In this case, ψ is also one-to-one. To see this, pick $x, y \in L$ such that $x + I \neq y + I$. This way,

$$\psi(x + I) = \phi(x) \neq \phi(y) = \psi(y + I),$$

as asserted. Thus, ψ is an invertible isomorphism. This way, L/I is not only homomorphic but also *isomorphic* to the image of ϕ. This mirrors the fundamental theorem of homomorphism in group theory [27, 45].

16.3.2 The Second Isomorphism Theorem

The above theorem mirrors the fundamental theorem of homomorphism in group theory. The next theorem, on the other hand, mirrors the second isomorphism theorem in group theory.

Let L be a Lie algebra. Let

$$I, J \subset L$$

be two ideals. What about their intersection $I \cap J$? Clearly, it is a subspace: closed under addition and scalar multiplication. Still, is it an ideal as well? To check on this, pick two vectors: $x \in I \cap J$, and $l \in L$. Place them in Lie brackets:

$$[x, l] \in I, \quad \text{and} \quad [x, l] \in J$$

(because both I and J are ideals). In short,

$$[x, l] \in I \cap J.$$

This extends to every linear combination as well. Thus,

$$[I \cap J, L] \subset I \cap J,$$

as required.

We can now go ahead and define a new homomorphism

$$\phi : J \to \frac{I + J}{I}$$

by

$$\phi(j) \equiv j + I, \quad j \in J.$$

This mirrors an analogous theorem in group theory. (In Chapter 5 in [45], I plays the role of S, J plays the role of T, and ϕ is the same as ξ.) Here, however, the group operation is addition, denoted by the plus sign: '$+$'. Since addition is commutative, every subgroup is indeed normal.

Here, however, we know more: I, J, $I + J$, and $I \cap J$ are not only subgroups (with respect to addition) but also ideals. This way, ϕ is linear: it preserves not only addition but also scalar multiplication. To see this, pick some vector $j \in J$, and some scalar α. Then,

$$\phi(\alpha j) \equiv \alpha j + I = \alpha(j + I) = \alpha \phi(j).$$

Furthermore, ϕ has yet another advantage: it also preserves Lie brackets. To see this, pick two vectors $j, \tilde{j} \in J$. Then,

$$\phi([j, \tilde{j}]) \equiv [j, \tilde{j}] + I = [j + I, \tilde{j} + I] = [\phi(j), \phi(\tilde{j})] .$$

So, ϕ is indeed a legitimate homomorphism. Finally, what is the kernel of ϕ? Thanks to group theory, we already know the answer: $I \cap J$. Moreover, as in group theory, ϕ is not only *into* but also *onto*. Thus,

$$\frac{J}{I \cap J} \simeq \frac{I + J}{I}.$$

(See the end of Section 16.3.1.) Unlike in group theory, this isomorphism tells us much more. It preserves the entire structure of a Lie algebra: not only addition but also scalar multiplication and Lie brackets.

16.3.3 The Third Isomorphism Theorem

The next theorem also mirrors an analogous theorem in group theory. To see this, let L be a Lie algebra. Let I and J be two ideals of L, such that $I \subset J$. In what follows, I and J will play the role of S and T in Chapter 5 in [45]. Likewise, let

$$\pi : L \to L/I$$

and

$$\pi' : L/I \to \frac{L/I}{J/I}$$

play the role of ξ and ξ' there: canonical homomorphisms with the following properties: they are both *onto*, linear, and Lie-bracket preserving. Thus, their composition

$$\pi'\pi : L \to \frac{L/I}{J/I}$$

has the same properties: *onto*, linear, and Lie-bracket preserving. Furthermore, as in group theory, its kernel is exactly J. Therefore,

$$L/J \simeq \frac{L/I}{J/I}.$$

(See the end of Section 16.3.1.) This is indeed the third isomorphism theorem. Unlike in group theory, however, here the isomorphism is much more powerful: it preserves not only addition but also scalar multiplication and Lie brackets.

Exercises:
Solvability and Nilpotency

We are now ready to introduce some properties that a Lie algebra may have. On one hand, it may be solvable, and even nilpotent. Nilpotency is stronger: if the Lie algebra is nilpotent, then it must also be solvable. (Don't confuse nilpotency of a Lie algebra with nilpotency of an individual matrix: these are different things.)

On the other hand, a Lie algebra may be semisimple, and even simple. Simplicity is stronger: if the Lie algebra is simple, then it must also be semisimple.

The two latter properties disagree with the two former. Indeed, if a Lie algebra is simple (or even only semisimple), then it can never be solvable, let alone nilpotent.

To help introduce these properties better, this chapter is arranged in exercises (followed by hints or even complete solutions). This way, you get to see the theory more vividly, and develop it from scratch, step by step, exercise by exercise.

17.1 The Canonical Homomorphism

1. Let L be a Lie algebra. Let $I \subset L$ be an ideal. What does this mean? Hint: $[I, L] \subset I$.
2. Look at L/I. Is this a legitimate group?
3. What is the group operation? Hint: addition, denoted by '+'.
4. In this group, what is the unit element? Hint: $\mathbf{0} + I = I$.
5. In this group, what is the inverse of a given vector of the form $x + I$? Hint: $-x + I$.
6. Is L/I a linear space as well? Hint: see Chapter 16, Section 16.2.4.
7. How many distributive laws does it satisfy? Hint: see Chapter 16, Section 16.2.4.
8. Look at the canonical mapping π. Where is it defined? Hint: on L.
9. What is its image? Hint: $\pi(L) = L/I$.
10. Is π linear? Hint: see Chapter 16, Section 16.2.5.
11. Use this to prove once again that L/I satisfies two distributive laws. Hint: pick two vectors $x, y \in L$, and two scalars: α and β. Write the distributive laws in L:

$$\alpha(x + y) = \alpha x + \alpha y$$
$$(\alpha + \beta)x = \alpha x + \beta x.$$

Now, since π is linear, it can be safely applied to both sides:

$$\alpha(x + y + I) = \alpha(x + I) + \alpha(y + I)$$
$$(\alpha + \beta)(x + I) = \alpha(x + I) + \beta(x + I).$$

12. Do the same for the associative law for addition. Hint: first, write it in L: for every $x, y, z \in L$,
$$(x + y) + z = x + (y + z).$$
Then, apply π to both sides:
$$(x + y + I) + z + I = x + I + (y + z + I).$$

13. Conclude once again that L/I is indeed a legitimate linear space.
14. Look again at $\pi : L \to L/I$. Does it preserve the Lie brackets? Hint: see Chapter 16, Section 16.2.5.
15. Use this to prove that L/I is indeed a legitimate Lie algebra: its Lie brackets are indeed bilinear and alternative, and also satisfy the Jacobi identity. Hint: pick three vectors $x, y, z \in L$, and a scalar α. Write the properties in L – bilinearity, alternativity, and Jacobi identity:

$$[x + y, z] = [x, z] + [y, z]$$
$$[\alpha x, z] = \alpha[x, z]$$
$$[x, x] = 0$$
$$[[x, y], z] + [[z, x], y] + [[y, z], x] = 0.$$

Now, since π is linear and preserves the Lie brackets, it can be safely applied to both sides:

$$[x + y + I, z + I] = [x + I, z + I] + [y + I, z + I]$$
$$[\alpha x + I, z + I] = \alpha[x + I, z + I]$$
$$[x + I, x + I] = I$$

$$[[x + I, y + I], z + I] + [[z + I, x + I], y + I] + [[y + I, z + I], x + I] = I.$$

16. This is true not only for π but also for any homomorphism ϕ: its image $\phi(L)$ is a legitimate Lie algebra in its own right. Explain why.

17.2 The Derived Algebra

1. Let L be a Lie algebra. Look at $\{0\}$: the degenerate subspace that contains one vector only: the zero vector. Is this a legitimate subspace? Hint: is it closed under addition and scalar multiplication? Well, $0 + 0 = 0$, and $\alpha 0 = 0$.
2. Is it a subalgebra? Hint: is it closed under the Lie brackets? Well, thanks to bilinearity, $[0, 0] = 0$.
3. Furthermore, is it an ideal? Hint: thanks to bilinearity, $[0, l] = 0$ (for every $l \in L$).
4. Conclude that this is the smallest ideal.
5. What is the biggest ideal? Hint: see below.
6. Look at L itself. Is this an ideal?

7. Look at $[L, L]$. What is this? Hint: the subspace spanned by all vectors of the form $[l, m]$, for every $l, m \in L$.
8. Is $[L, L] \subset L$? Hint: L is closed under the Lie brackets.
9. $[L, L]$ is called the derived algebra.
10. Is $[L, L]$ an ideal? Hint: in Chapter 16, Section 16.1.4, set $I = J = L$.
11. On the other hand, is $L \subset [L, L]$? Hint: not always!
12. Give an example of $L = [L, L]$. Hint: let L be the three-dimensional Cartesian space (Chapter 15, Section 15.3). Write each standard unit vector as the vector product of the other two.
13. Give yet another example of $L = [L, L]$. Hint: look at the Pauli matrices, with imaginary coefficients (Chapter 15, Section 15.3). Write each matrix as the commutator of the other two.
14. Next, look at yet another special case: $L = sl_2$. In this case, is $L = [L, L]$? Hint: check on a basis: $e, f, h \in sl_2$ (Chapter 15, Section 15.4). Could these be written as a commutator? They sure can:

$$e = \frac{[h, e]}{2}, \quad f = \frac{[h, f]}{-2}, \quad \text{and} \quad h = [e, f].$$

15. Extend this to sl_n as well. Hint: scan $i = 1, 2, \ldots, n - 1$. For each fixed i, start from $j = i+1$. As before, write $E^{(i,j)}$, $E^{(j,i)}$, and $E^{(i,i)} - E^{(j,j)}$ as a commutator. Next, fill the ith row (and column): for $j > i + 1$, write as a commutator:

$$E^{(i,j)} = E^{(i,i)} E^{(i,j)} = \left[E^{(i,i)} - E^{(i+1,i+1)}, E^{(i,j)}\right]$$

$$E^{(j,i)} = E^{(j,i)} E^{(i,i)} = \left[E^{(j,i)}, E^{(i,i)} - E^{(i+1,i+1)}\right].$$

16. Conclude that

$$sl_n = [sl_n, sl_n].$$

17. Conclude also that

$$sl_n = [sl_n, sl_n] = [gl_n, gl_n].$$

Hint: since $sl_n \subset gl_n$,

$$sl_n = [sl_n, sl_n] \subset [gl_n, gl_n].$$

On the other hand, since the commutator is trace-free (Chapter 15, Section 15.3),

$$[gl_n, gl_n] \subset sl_n.$$

18. Conclude that both gl_n and sl_n, although different, have the same derived algebra.

17.3 Automorphisms and Their Lie Group: Cartan Subalgebra

1. What is a homomorphism?
2. What is an isomorphism?
3. An isomorphism is an invertible homomorphism. Give an example.

4. An isomorphism from the Lie algebra onto itself is called an automorphism. Give an example.

5. For example, look at the identity operator. Is this an automorphism? Why?

6. For yet another example, look at the operator E in Chapter 15, Section 15.2. Is this an automorphism? Why?

7. Look at all automorphisms. Is this a group?

8. What is the group operation? Hint: composition of automorphisms.

9. What is the unit element? Hint: the identity automorphism.

10. This group is denoted by $aut(L)$.

11. For yet another example, look at the operator $f() = [f, \cdot]$ (Chapter 15, Section 15.2), and take its exponent. Is this an automorphism? Why?

12. This is called an inner automorphism.

13. Look at many exponents: not only of $f()$ but also of all its scalar multiples. Do they make a new subgroup of inner automorphisms? Hint: since all scalar multiples of $f()$ commute with each other, exponentiation converts addition to multiplication.

14. This is called a one-parameter subgroup of the Lie group.

15. Is it Abelian (commutative)?

16. We also say that this subgroup is *generated* by f.

17. Multiply $f()$ by a real coefficient t, and take the exponent.

18. Recall that this preserves the Lie brackets:

$$\exp(tf())[g, h] = [\exp(tf())g, \exp(tf())h],$$

where g and h are in the Lie algebra.

19. Use Leibniz rule to differentiate this with respect to t at $t = 0$. Hint: differentiating $\exp(tf())$ at $t = 0$ gives just the original operator $f()$. This gives

$$f([g, h]) = [f(g), h] + [g, f(h)],$$

or

$$[f, [g, h]] = [[f, g], h] + [g, [f, h]].$$

20. Is this familiar? Hint: this is just the Leibniz rule (or the Jacobi identity) in our Lie algebra.

21. In this sense, differentiation is the inverse of exponentiation.

22. Furthermore, look not only at $f()$ but also at many other nilpotent derivations that commute with each other (zero commutator). Do they span a new subalgebra? Hint: since zero is in it, it is indeed closed under the Lie brackets, as required.

23. Look at such a subalgebra of a maximal dimension.

24. This is called a Cartan subalgebra.

25. Look at the members of a Cartan subalgebra. Now, look at their exponents. Do they make a new (Abelian) subgroup of inner automorphisms? Hint: since the members of a Cartan subalgebra commute with each other, exponentiation converts addition to multiplication.

26. Look again at an inner automorphism: the exponent of a nilpotent derivation of the form $f() = [f, \cdot]$ (Chapter 15, Section 15.2).

27. Such inner automorphisms *generate* yet more automorphisms: compositions of inner automorphisms.

28. Is this a subgroup of $aut(L)$? Hint: the identity automorphism is inner. Indeed, it is the exponent of zero.
29. This subgroup is denoted by

$$int(L) \subset aut(L).$$

30. Is it normal? Hint: first, consider a generator (an inner automorphism). Conjugate it with just any $E \in aut(L)$: plug it in between E and E^{-1}. Do you obtain a new inner automorphism? (See Figure 15.3.) Finally, conjugate not only an individual generator but also a composition of generators.
31. This could be viewed as a representation of the Lie group.

17.4 The Adjoint Representation and Its Trace

1. Let L be a Lie algebra. What is a homomorphism on L? Hint: a linear mapping (into some other Lie algebra) that preserves Lie brackets.
2. What is its image?
3. Is the image a legitimate Lie algebra as well? Hint: is it closed under addition, scalar multiplication, and Lie brackets?
4. A representation is a special kind of homomorphism. In what way? Hint: its image is in gl_n (for some natural number n.).
5. Let ϕ be a representation. How does it act on the derived algebra $[L, L]$? Hint:

$$\phi([l, m]) = [\phi(l), \phi(m)], \quad l, m \in L.$$

(Extend this linearly to every linear combination of $[l, m]$'s.)
6. What is the meaning of the right-hand side? Hint: commutator of two matrices of order n.
7. What is its trace? Hint: zero (Chapter 15, Section 15.3).
8. Conclude that the image of the derived algebra is trace-free:

$$\phi([L, L]) \subset sl_n.$$

9. Furthermore, assume that L is the same as its derived algebra: $L = [L, L]$.
10. Conclude that $\phi(L)$ is trace-free:

$$\phi(L) = \phi([L, L]) \subset sl_n.$$

11. Verify this for a concrete example: $L = sl_2$. Hint: recall that

$$sl_2 = [sl_2, sl_2]$$

(Section 17.2).
12. Recall also that it has a basis of three members. Hint: see Chapter 15, Section 15.4.
13. Write them as standard unit vectors:

$$e \sim \begin{pmatrix} 1 \\ 0 \\ 0 \end{pmatrix}$$

$$h \sim \begin{pmatrix} 0 \\ 1 \\ 0 \end{pmatrix}$$

$$f \sim \begin{pmatrix} 0 \\ 0 \\ 1 \end{pmatrix}.$$

14. Define ϕ concretely: the adjoint representation:

$$\phi(e) = [e, \cdot]$$
$$\phi(h) = [h, \cdot]$$
$$\phi(f) = [f, \cdot]$$

(Chapter 15, Sections 15.2 and 15.4).

15. For instance, look at the mapping $\phi(h) = [h, \cdot]$. What are its eigenvectors? Hint:

$$\phi(h)(e) = [h, e] = 2e$$
$$\phi(h)(h) = [h, h] = 0h$$
$$\phi(h)(f) = [h, f] = -2f.$$

16. In the above basis, how does $\phi(h)$ look like? Hint: it is diagonal:

$$\phi(h) \sim \begin{pmatrix} 2 & & \\ & 0 & \\ & & -2 \end{pmatrix}.$$

Indeed, this matrix has three eigenvectors: e, h, and f (written as standard unit vectors).

17. In the above basis, how does $\phi(e)$ look like? Hint: it has nonzero elements on one superdiagonal only:

$$\phi(e) \sim \begin{pmatrix} & -2 & \\ & & 1 \\ & & \end{pmatrix}.$$

18. In the above basis, how does $\phi(f)$ look like? Hint: it has nonzero elements on one subdiagonal only:

$$\phi(f) \sim \begin{pmatrix} & & \\ -1 & & \\ & 2 & \end{pmatrix}.$$

19. What is the trace of these matrices? Hint: zero.
20. Why? Hint: because they are in the image of ϕ, which is trace-free.
21. What about any other representation, not necessarily the adjoint? Is its image still trace-free? Hint: the same algebraic properties still hold.

22. Extend this to sl_k, for any natural number k:

$$\text{trace}\left(\phi\left(sl_k\right)\right) = \text{trace}\left(\phi\left([sl_k, sl_k]\right)\right) = \text{trace}\left([\phi\left(sl_k\right), \phi\left(sl_k\right)]\right) = 0,$$

where ϕ is now a representation of sl_k, say the adjoint:

$$\phi(l) = [l, \cdot], \quad l \in sl_k.$$

Hint: see Section 17.2.

17.5 The Derived Series

1. Define a series of (smaller and smaller) Lie algebras: L, its derived algebra, the derived algebra of the derived algebra, and so on.
2. More precisely, use mathematical induction:

$$L^{(0)} \equiv L$$
$$L^{(i+1)} \equiv \left[L^{(i)}, L^{(i)}\right] \quad (i \geq 0).$$

3. This is called the derived series.
4. Show that, in this series, each member is an ideal of L. Hint: use mathematical induction, and Chapter 16, Section 16.1.4.
5. Show that, in this series, each member is included in the previous ones. Hint: use mathematical induction.
6. Let i be a fixed natural number. Look at $L^{(i)}$. What is its own derived series? Hint:

$$\left(L^{(i)}\right)^{(j)} = L^{(i+j)} \quad (j \geq 0).$$

7. For example, set $L = sl_n$ (for some fixed n). What is its derived series? Hint: a constant series: $L^{(i)} = sl_n$ $(i \geq 0)$.
8. In this case, the derived series never shrinks to the zero Lie algebra. In such a case, we say that L is not solvable.
9. Conclude that sl_n is not solvable.
10. For yet another example, set $L = gl_n$. What is its derived series? Hint: $L^{(i)} = sl_n$ $(i > 0)$.
11. Conclude that gl_n is not solvable.

17.6 Solvable vs. Simple Lie Algebra

1. Recall the notations in Chapter 15, Section 15.5: T_0 contains upper-triangular matrices, T_1 contains strictly upper-triangular matrices, and so on: as k increases, the matrices in T_k have fewer and fewer nonzero diagonals.
2. What is the derived algebra of T_0? Hint: $[T_0, T_0] = T_1$. Indeed, on one hand, $[T_0, T_0] \subset T_1$ (because the commutator is strictly upper triangular). On the other hand, $T_1 = [T_0, T_1] \subset [T_0, T_0]$ (see below).
3. If $j + k > 0$, what is T_{j+k}? Hint: $T_{j+k} = [T_j, T_k]$ (Chapter 15, Section 15.5).
4. In particular, what is T_{2k}? Hint: $T_{2k} = [T_k, T_k]$ $(k > 0)$.

5. Set $L \equiv T_0$. What is its derived series? Hint:

$$L^{(0)} \equiv T_0$$
$$L^{(i)} = T_{2^{i-1}} \quad (i > 0).$$

6. Let i increase until 2^{i-1} is as big as the order of the matrices: $2^{i-1} \geq n$. Show that, for such a big i, $L^{(i)}$ is the zero Lie algebra.
7. In such a case, we say that L is solvable.
8. Conclude that T_0 is solvable.
9. Likewise, calculate the derived series of T_k (for some fixed $k \geq 0$).
10. Conclude that T_k is solvable as well.
11. Could you prove this without calculating its derived series explicitly? Hint: compare the derived series of T_k to that of T_0: each member in the former is included in the corresponding member in the latter (by mathematical induction).
12. Let L be a Lie algebra. What is its largest ideal? Hint: L itself (Section 17.2).
13. What is its smallest ideal? Hint: the ideal that contains one vector only: the zero vector.
14. What happens if the derived algebra is as small as this ideal:

$$[L, L] = \{\mathbf{0}\}?$$

Hint: in this case, all vectors commute with each other, or have a zero commutator. We then say that L is commutative or Abelian.

15. In this case, is there a nontrivial ideal, different from both L and $\{\mathbf{0}\}$? Hint: in this case, every subalgebra is an ideal.
16. If L isn't Abelian, then it could still have a nontrivial ideal. If it doesn't, then it is called simple.
17. If L is simple, what is its derived algebra? Hint: since L is not Abelian, $[L, L] \neq \{\mathbf{0}\}$. Since L has no nontrivial ideal, we must have $[L, L] = L$.
18. What is its derived series? Hint: it must be constant: $L^{(i)} = L$ $(i \geq 0)$.
19. Conclude that a simple Lie algebra can never be solvable.
20. Conclude that a solvable Lie algebra can never be simple. Hint: if it were, then it could never be solvable in the first place.
21. Give an example of a simple Lie algebra. Hint: sl_2. Indeed, let $I \subset sl_2$ be an ideal. We need to show that $I = sl_2$ or $I = \{\mathbf{0}\}$. Indeed, what vector could be in I? If $h \in I$, then $[h, e] = 2e \in I$ and $[h, f] = -2f \in I$ as well. Likewise, if $e \in I$, then $[e, f] = h \in I$ as well. In either case, $I = sl_2$. Now, could I contain a more complicated vector like

$$v \equiv c_e e + c_h h + c_f f,$$

for some coefficients c_e, c_h, and c_f? Well, as an ideal, I must then contain more vectors:

$$[v, e], \quad [[v, e], e], \quad [v, f], \quad [[v, f], f] \quad \in I.$$

Again, either $c_e = c_h = c_f = 0$ or $e, h, f \in I$.

22. Could sl_2 be solvable? Hint: never! Its derived series is constant, and never shrinks to the zero Lie algebra.
23. Likewise, look at the three-dimensional Cartesian space, with vector product. Is it a simple Lie algebra? Hint: use a similar proof.

24. Could it be solvable? Hint: no!
25. Could a Lie algebra be neither simple nor solvable? Hint: for example, gl_n is not simple. Indeed, it has a nontrivial ideal: sl_n. Still, gl_n is not solvable either (Section 17.5).

17.7 Solvability and a Homomorphism

1. Let L be a Lie algebra. Let $K \subset L$ be a subalgebra. What can you say about its derived series? Hint: $K^{(i)} \subset L^{(i)}$ (by mathematical induction on $i = 0, 1, 2, \ldots$).
2. Conclude that, if L is solvable, then so is K.
3. Furthermore, let ϕ be a homomorphism on L. Look at the homomorphic image: $\phi(L)$. What is its derived algebra? Hint:

$$(\phi(L))^{(1)} = [\phi(L), \phi(L)] = \phi([L, L]) = \phi\left(L^{(1)}\right).$$

4. Assume that we already have the derived series of L. How to obtain the derived algebra of $\phi(L)$ for free, with no need to calculate any Lie brackets? Hint: just take $L^{(1)}$, and apply ϕ to it.
5. Moreover, how to obtain the entire derived series of $\phi(L)$ for free? Hint: see below.
6. In the above formula, replace L by $L^{(i)}$ (for some fixed $i \geq 0$). Hint:

$$\left(\phi\left(L^{(i)}\right)\right)^{(1)} = \phi\left(L^{(i+1)}\right).$$

7. To simplify this formula, use an induction hypothesis:

$$\phi\left(L^{(i)}\right) = (\phi(L))^{(i)}.$$

What do you get? Hint:

$$(\phi(L))^{(i+1)} = \phi\left(L^{(i+1)}\right).$$

8. Use this as an induction step to prove that

$$(\phi(L))^{(i)} = \phi\left(L^{(i)}\right) \quad (i \geq 0).$$

9. Obtain the derived series of $\phi(L)$ for free, with no calculation at all. Hint: scan the derived series of L. To each Lie algebra in it, just apply ϕ.
10. Conclude that, if L is solvable, then so is $\phi(L)$.
11. Let L be a Lie algebra, and let $I \subset L$ be an ideal. Let $\pi : L \to L/I$ be the canonical homomorphism (Section 17.1). In the above formula, replace ϕ by π. Hint:

$$(L/I)^{(i)} = (\pi(L))^{(i)} = \pi\left(L^{(i)}\right) = L^{(i)}/I \quad (i \geq 0).$$

12. Assume that we already have the derived series of L. How could we obtain the derived series of L/I for free? Hint: scan the derived series of L. Each Lie algebra should be divided by I.

13. Use the above formula to show that, if L is solvable, then so is L/I. Hint: in this case, for a sufficiently large i, $L^{(i)} = \{0\}$, so the right-hand side is as small as $L^{(i)}/I = \{0 + I\} = \{I\}$, where I is the zero vector in L/I.

14. Could you prove this in yet another way? Hint: L/I is the homomorphic image of L under π.

15. Assume now the other way around: we already know that L/I is solvable, but don't know about L. For a sufficiently large i, how does the above formula look like? Hint: for a sufficiently large i, the left-hand side is as small as $\{I\}$ (because I is the zero vector in L/I). Thus, the left-hand side could be replaced by $\{I\}$:

$$\{I\} = L^{(i)}/I.$$

16. Conclude that, for such a big i,

$$L^{(i)} \subset I.$$

17. Assume now that not only L/I but also I is solvable (but we still don't know about L). Conclude that, for the above big i, $L^{(i)}$ is solvable as well. Hint: it is included in I.

18. Conclude that, if both I and L/I are solvable, then L itself is solvable as well.

19. Let L be a Lie algebra. Let $I, J \subset L$ be two ideals. Assume that J is solvable. What about $I \cap J$? Is it solvable as well? Hint: $I \cap J \subset J$.

20. What about $J/(I \cap J)$? Is it solvable as well? Hint: let $\pi : J \to J/(I \cap J)$ be the canonical homomorphism. We already proved that the homomorphic image $\pi(J) = J/(I \cap J)$ is solvable as well.

21. What about $(I + J)/I$? Is it solvable as well? Hint: use the second isomorphism theorem (Chapter 16, Section 16.3.2).

22. Finally, assume now that not only J but also I is solvable. What about $I + J$? Is it solvable as well? Hint: since both I and $(I + J)/I$ are solvable, $I + J$ is solvable as well.

17.8 Radical and Semisimplicity

1. Let L be a Lie algebra. Let $R \subset L$ be the maximal solvable ideal. Then, R is called the radical (if exists).

2. If L is solvable, what is its radical? Hint: in this case, $R = L$.

3. Is the radical unique? Hint: let $I \subset L$ be another solvable ideal. Then, $R + I$ is yet another solvable ideal. Since R is maximal, we must have $R + I \subset R$, so $I \subset R$ as well. In summary, R includes all solvable ideals.

4. If $R = \{0\}$, then we say that L is semisimple.

5. If there is no radical at all, how to define semisimplicity? Hint: L is semisimple if it has no solvable ideal but $\{0\}$.

6. If L is simple, must it also be semisimple? Hint: if L is simple, then its only ideals are L (which is not solvable) and $\{0\}$.

7. If L is solvable, could it still be semisimple? Hint: if L is solvable, then $R = L$, so L is not semisimple.

8. If L is semisimple, could it still be solvable? Hint: if it were, then it could never be semisimple in the first place.

9. Is L/R semisimple? Hint: pick a solvable ideal in L/R. How could it look like? Well, it could be written as I/R, where $I \subset L$ is an ideal. (As a matter of fact, I is the origin of the above ideal under the canonical homomorphism.) Since both R and I/R are solvable, I is solvable as well. Since R is maximal, $I \subset R$. Thus, $I/R = \{R\}$ contains just one "vector:" R, which is the zero vector in L/R.

17.9 The Lower Central Series

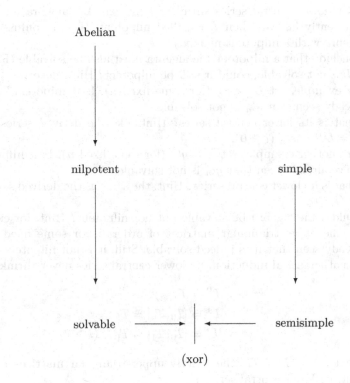

Fig. 17.1. On one hand, a simple Lie algebra is also semisimple, so it can never be solvable, let alone nilpotent, let alone Abelian. On the other hand, a nilpotent Lie algebra is also solvable, so it can never be semisimple, let alone simple.

1. Let L be a Lie algebra. Define a new series of (smaller and smaller) Lie algebras. Again, use mathematical induction:

$$L^0 \equiv L$$
$$L^{i+1} \equiv \left[L, L^i\right] \quad (i \geq 0).$$

2. This is called the lower central series.

3. In this series, look at the first and second members. Are they familiar? Hint: they are the same as in the derived series: first L, followed by its derived algebra.
4. Show that, in this series, each member is an ideal of L. Hint: use mathematical induction.
5. Show that, in this series, each member is included in the previous ones. Hint: use mathematical induction.
6. Scan the lower central series. Show that each member includes the corresponding member in the derived series:

$$L^{(i)} \subset L^i, \quad i \geq 0.$$

Hint: use mathematical induction.

7. If the lower central series shrinks to the zero Lie algebra, or $L^i = \{0\}$ for a sufficiently large i, then L is called nilpotent. (Don't confuse a nilpotent Lie algebra with a nilpotent matrix.)
8. Conclude that a nilpotent Lie algebra must also be solvable (Figure 17.1).
9. If L is not solvable, could it still be nilpotent? Hint: never!
10. For example, set $L = sl_n$ (for some fixed n). Is it nilpotent? Hint: no! We've already seen that sl_n is not solvable.
11. What is its lower central series? Hint: like the derived series, it is constant: $L^i = L^{(i)} = sl_n$ $(i \geq 0)$.
12. For another example, set $L = gl_n$ (for some fixed n). Is it nilpotent? Hint: no! We've already seen that gl_n is not solvable.
13. What is its lower central series? Hint: the same as the derived series: $L^i = L^{(i)} = sl_n$ $(i > 0)$.
14. Could a Lie algebra be solvable but not nilpotent? Hint: for example, set $L = T_0$ (the upper-triangular matrices of order n, for some fixed $n > 1$). We've already seen that it is indeed solvable. Still, it is not nilpotent. Indeed, thanks to mathematical induction, its lower central series never shrinks:

$$L^0 = T_0$$
$$L^1 = [T_0, T_0] = T_1$$
$$L^i = [T_0, T_1] = T_1 \quad (i \geq 2).$$

15. What about $L = T_1$ (the strictly upper-triangular matrices)? Is it nilpotent? Hint: its lower central series is

$$L^i = T_{i+1} \quad (i \geq 0).$$

Indeed, thanks to mathematical induction,

$$L^0 = L = T_1$$
$$L^{i+1} = [L, L^i] = [T_1, T_{i+1}] = T_{i+2} \quad (i \geq 0).$$

For a sufficiently large i, this indeed shrinks to the zero Lie algebra.

16. Likewise, show that T_k is nilpotent not only for $k = 1$ but also for every $k \geq 1$.

17.10   Nilpotency and a Homomorphism

1. Let L be a Lie algebra. Let $K \subset L$ be a subalgebra. What can you say about its lower central series? Hint: $K^i \subset L^i$ (by mathematical induction).
2. Conclude that, if L is nilpotent, then so is K.
3. Furthermore, let $i \geq 0$ be fixed, and assume now that $K \subset L^i$. What can you say now about the lower central series of K? Hint: for all $j \geq 0$, $K^j \subset L^{i+j}$ (by mathematical induction).
4. For example, set $K = L^i$ (for some fixed i). What do you get? Hint: $(L^i)^j \subset L^{i+j}$ $(j \geq 0)$.
5. Moreover, let ϕ be a homomorphism on L. Look at the homomorphic image: $\phi(L)$. What is its derived algebra? Hint:

$$(\phi(L))^1 = [\phi(L), \phi(L^0)] = \phi([L, L^0]) = \phi(L^1).$$

6. Assume that we already have the lower central series of L. How to obtain the derived algebra of $\phi(L)$ for free, with no need to calculate any Lie brackets? Hint: just take L^1, and apply ϕ to it.
7. Moreover, how to obtain the entire lower central series of $\phi(L)$ for free? Hint: see below.
8. In the above formula, replace 0 by i, and 1 by $i+1$ (for a fixed $i \geq 0$), to obtain

$$(\phi(L))^{i+1} = \phi(L^{i+1}) \quad (i \geq 0).$$

Hint: use the induction hypothesis $\phi(L)^i = \phi(L^i)$ to prove the induction step:

$$
\begin{aligned}
(\phi(L))^{i+1} &= [\phi(L), (\phi(L))^i] \\
&= [\phi(L), \phi(L^i)] \\
&= \phi([L, L^i]) \\
&= \phi(L^{i+1}).
\end{aligned}
$$

9. In summary, thanks to mathematical induction, we've just proved that

$$(\phi(L))^i = \phi(L^i) \quad (i \geq 0).$$

10. Obtain the lower central series of $\phi(L)$ for free, with no calculation at all. Hint: scan the lower central series of L. To each Lie algebra in it, just apply ϕ.
11. Conclude that, if L is nilpotent, then so is $\phi(L)$.
12. Let L be a Lie algebra, and let $I \subset L$ be an ideal. Let $\pi : L \to L/I$ be the canonical homomorphism (Section 17.1). In the above formula, replace ϕ by π. Hint:

$$(L/I))^i = (\pi(L))^i = \pi(L^i) = L^i/I \quad (i \geq 0).$$

13. Assume that we already have the lower central series of L. How could we obtain the lower central series of L/I for free? Hint: scan the lower central series of L. Each Lie algebra should be divided by I.
14. Use the above formula to show that, if L is nilpotent, then so is L/I. Hint: in this case, for a sufficiently large i, $L^i = \{0\}$, so the right-hand side is as small as $L^i/I = \{0 + I\} = \{I\}$, where I is the zero vector in L/I.

15. Could you prove this in yet another way? Hint: L/I is the homomorphic image of L under π.

16. Assume now the other way around: we already know that L/I is nilpotent, but don't know about L. For a sufficiently large i, how does the above formula look like? Hint: for a sufficiently large i, the left-hand side is as small as $\{I\}$ (because I is the zero vector in L/I). Thus, the left-hand side could be replaced by $\{I\}$:

$$\{I\} = L^i/I.$$

17. Conclude that, for such a big i,

$$L^i \subset I.$$

18. Assume now that not only L/I but also I is nilpotent. Must L be nilpotent as well? Hint: not necessarily. Indeed, $L^i \subset I$ only tells us that $(L^i)^j \subset I^j$ $(j \geq 0)$, but L^{i+j} could be bigger than that.

19. Give one more condition to make sure that L is nilpotent as well. Hint: assume that I is a special kind of ideal: the center of L, containing those vectors that commute with all other vectors:

$$I = \{c \in L \mid [c, l] = \mathbf{0}, \; l \in L\}.$$

This way,

$$[L, I] = \{\mathbf{0}\}.$$

For the above big i, we therefore have

$$L^{i+1} = [L, L^i] \subset [L, I] = \{\mathbf{0}\}.$$

Thus, in this case, L is indeed nilpotent.

20. Let L be a nilpotent Lie algebra. Assume that L is nontrivial: $L \neq \{\mathbf{0}\}$. What about the center of L? Could it be trivial? Hint: let k be the minimal integer for which $L^k = \{\mathbf{0}\}$. Since L is nontrivial and nilpotent, $k > 0$. Now, look at L^{k-1}. It must be nontrivial, or k wouldn't be minimal any more. Moreover, L^{k-1} is included in the center of L. Indeed, it commutes with L:

$$[L, L^{k-1}] = L^k = \{\mathbf{0}\}.$$

Thus, the center of L is indeed nontrivial, as asserted.

21. Moreover, let ϕ be the adjoint homomorphism on L (Section 17.4). Show that, for each individual $x \in L$, $\phi(x)$ is a nilpotent matrix. More precisely, its kth power vanishes:

$$(\phi(x))^k(y) = \mathbf{0}, \quad x, y \in L.$$

Hint: use mathematical induction to show that

$$(\phi(x))^i(y) \in L^i, \quad x, y \in L, \; i = 0, 1, 2, 3, \ldots.$$

22. Conclude that $\phi(x)$ is a nilpotent matrix, in the sense in Chapter 14, Section 14.1.1.

23. Conclude that x is adjoint-nilpotent: its adjoint is a nilpotent matrix (Chapter 15, Section 15.2).

24. Note that this is true for every $x \in L$. In summary, if L is a nilpotent Lie algebra, then all vectors in it are adjoint-nilpotent.

25. What about the other way around? If all vectors in L are adjoint-nilpotent, must L be a nilpotent Lie algebra? Hint: this will follow from Engel's theorems below.

Nilpotency
and Engel's Theorems

A linear Lie algebra contains matrices. Are they nilpotent? Here we continue to study this.

A general Lie algebra, on the other hand, may contain more abstract members, with no notion of nilpotency at all. Still, we can now ask: are they adjoint-nilpotent? After all, the adjoint is a linear mapping, or a matrix, which could be nilpotent or not. If it is, then the original member is said to be adjoint-nilpotent: its adjoint is nilpotent.

How to make sure that all members are adjoint-nilpotent? In the above exercises, we've seen a sufficient condition for this: if the Lie algebra is nilpotent (its lower central series shrinks to zero), then each member in it is indeed adjoint-nilpotent.

This, however, is the easy direction: a global property (the nilpotency of the entire Lie algebra) leads to a local property: each individual member is adjoint-nilpotent. What about the other way around? Suppose that we didn't know whether our Lie algebra is nilpotent or not. Instead, we knew something else: all its members are adjoint-nilpotent. Could we then safely deduce that our Lie algebra is nilpotent?

This is indeed the difficult direction. After all, what we know is only local and particular: each individual member is adjoint-nilpotent. What we want to deduce, on the other hand, is far more global and general: nilpotency of the entire Lie algebra as a whole. Before proving this, let's ask more questions of the same kind.

18.1 Common Eigenvector

18.1.1 Nilpotent Matrices: Local vs. Global Properties

In Chapter 14, we've studied a nilpotent matrix. We've seen that it has (at least one) eigenvector with eigenvalue zero. Furthermore, in a suitable basis, it has a (strictly upper-triangular) Jordan form: 1's on the superdiagonal only, and 0 elsewhere.

Still, two matrices could disagree: each could have a different eigenvector, and a different basis in which it takes its Jordan form. Here, on the other hand, we'll learn better: in a linear Lie algebra, if all matrices are nilpotent, then they share not only their common eigenvector but also their common basis, in which they are strictly upper triangular. In this sense, these properties become not only local but also global.

18.1.2 Nilpotent Matrix and Its Adjoint

To prove this, we'll use a nice property that we've already seen in Chapter 15, Section 15.2: if x is a nilpotent matrix (or mapping), then it also makes two nilpotent operators: $x\circ$ (left multiplication) and $\circ x$ (right multiplication). The difference between these two is the adjoint operator: $x() \equiv x \circ - \circ x$. Thanks to the binomial formula, this adjoint is nilpotent as well. This way, in the following proof, we can move from the original Lie algebra to its adjoint representation, and use the induction hypothesis there.

18.1.3 Induction on The Dimension of L

Consider a linear Lie algebra:

$$L \subset gl_n$$

(for some natural number n). This way, what are the members of L? They are $n \times n$ matrices. Assume that all of them are nilpotent. As discussed above, they are then also adjoint-nilpotent. Furthermore, each of them has its own eigenvector of eigenvalue zero. Still, is this also true globally? In other words, do they also share a *common* eigenvector of this kind?

To prove this, let's use mathematical induction on the dimension of L. Note that the dimension of L could be different from n. For example, if $L = sl_n$, then its dimension is $n^2 - 1 \neq n$ (Chapter 15, Section 15.4).

First, let's assume that L is one-dimensional. This doesn't mean that $n = 1$: n may be bigger. What it means is that L contains one matrix only (up to scalar multiplication). More precisely, for some nonzero matrix $y \neq (0)$,

$$L = \{\alpha y \mid \alpha \text{ a scalar}\}.$$

How to find an eigenvector of y? We already know how: since y is nilpotent, there is some natural number $l \geq 1$ for which $y^l = (0)$. Pick some nonzero vector $v_0 \neq \mathbf{0}$. Then we have $y^l v_0 = (0)v_0 = \mathbf{0}$. Now, l is not necessarily minimal: let k be the *minimal* natural number for which $y^k v_0 = \mathbf{0}$. This way, the required eigenvector is $y^{k-1}v_0$. Indeed,

$$y\left(y^{k-1}v_0\right) = y^k v_0 = \mathbf{0}.$$

18.1.4 The Induction Step

So far, L was one-dimensional. Practically, it contained one matrix only. Assuming that this matrix was nilpotent, it was easy enough to design an eigenvector for it. Still, could we extend this to a few (nilpotent) matrices, and design a common eigenvector?

Assume now that L is at least two-dimensional. Assume also that we already know how to design a common eigenvector (of eigenvalue zero) for every lower-dimensional Lie algebra of nilpotent matrices. (This is the induction hypothesis.) How to extend this to L as well?

For this purpose, let $K \subset L$ be a maximal proper subalgebra: a maximal subalgebra that is still different from L, and can't be extended any more, not even by one vector. In other words, there is no other subalgebra that lies strictly in between

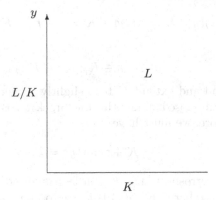

Fig. 18.1. The original Lie algebra L splits into two parts: a maximal proper subalgebra K at the horizontal, and L/K at the vertical, spanned by the new vector y. The induction hypothesis is used on K and $\phi(K)$, which are both lower-dimensional.

K and L. Clearly, K is at least one-dimensional. After all, K could be spanned by just one nonzero matrix, like y above.

Fortunately, K is lower-dimensional. Thanks to the induction hypothesis, all x's in K share a common eigenvector $v_0 \neq 0$, for which $xv_0 = \mathbf{0}$. Still, this is not good enough: what about those x's that are left outside K? They may not share v_0 as an eigenvector! We need to fix this.

Unfortunately, we don't know yet that K is an ideal. Still, we can go ahead and divide L by K, not as Lie algebras, but only as linear spaces, as in group theory. After all, to divide a vector space, we don't have to have an ideal. Indeed, only at the very end of Chapter 16, Section 16.2.4, did we use the ideal property, to make sure that the Lie brackets are well-defined. Still, so long as we don't need any Lie brackets in it, L/K may serve as a suitable vector space, with legitimate linear mappings (or matrices) defined on it (Figure 18.1).

Look at some matrix $x \in L$. What is its adjoint? This is a mapping $[x, \cdot]$ from L into L. Here, however, we use the factor adjoint representation, denoted by ϕ. This way, $\phi(x)$ is a linear mapping from L/K into L/K.

Why are we doing this? Because we'd like to use the induction hypothesis to design an eigenvector of the form $y + K \in L/K$. As an eigenvector, it will be nonzero: $y + K \neq \mathbf{0} + K = K$, or $y \notin K$. This way, we'll be able to extend K by a new vector: y.

Still, because we don't know yet that K is an ideal, ϕ can't be defined on the whole of L. It must be restricted, and defined in K only:

$$\phi(k)(l + K) \equiv [k, l] + K, \quad k \in K, \, l \in L.$$

This way, $\phi(k)$ is indeed a legitimate mapping, defined on L/K. Why? Well, if we replaced l by $l + \tilde{k}$ (for just any $\tilde{k} \in K$), then nothing would change:

$$\phi(k)(l + \tilde{k} + K) \equiv [k, l + \tilde{k}] + K = [k, l] + [k\tilde{k}] + K = [k, l] + K.$$

Both K and $\phi(K)$ are lower-dimensional, and can benefit from the induction hypothesis. Thus, the $\phi(x)$'s share a common eigenvector $y + K \in L/K$, for which

$$\phi(x)(y + K) = [x, y] + K = \mathbf{0} + K = K, \quad x \in K.$$

In other words,

$$[x, y] \in K, \quad x \in K.$$

So, we can go ahead and extend K to a slightly bigger subalgebra: $K + \text{span}\{y\}$. After all, this is still closed under the Lie brackets. But K is a maximal proper subalgebra. Therefore, we must have

$$K + \text{span}\{y\} = L.$$

We can now see in retrospect that K is a legitimate ideal after all. Therefore, ϕ is actually trivial: for each $x \in K$, $\phi(x)$ is the zero mapping: it maps L/K to its own zero vector: K. Still, we don't care about these things any more.

So far, we used the induction hypothesis on $\phi(K)$. Let's use it on K itself as well, to design a common eigenvector $v_0 \neq \mathbf{0}$:

$$x v_0 = \mathbf{0}, \quad x \in K.$$

Now, is $y v_0 = \mathbf{0}$ as well? If so, then we're done: v_0 is indeed our desired eigenvector. Otherwise, let's move on to $y v_0$, and check whether it could serve as our common eigenvector. After all, for every $x \in K$, we already know that $[x, y] \in K$, so

$$x y v_0 = x y v_0 - \mathbf{0} = x y v_0 - y x v_0 = (xy - yx) v_0 = [x, y] v_0 = \mathbf{0}.$$

So, $y v_0$ is indeed an eigenvector of every $x \in K$, with eigenvalue zero. If $y v_0$ is an eigenvector of y as well, then we're done. Otherwise, then more iterations are needed: continue to $y^2 v_0$, and so on, until $y^{k-1} v_0$, which is our desired eigenvector: not only of every $x \in K$ but also of every $x \in L$, as required.

18.2 Strictly Upper-Triangular Form

18.2.1 First Coordinate: The Common Eigenvector

The common eigenvector designed above could now mark the direction of a new coordinate. This way, this vector is denoted by a standard unit vector: its first component is 1, and the rest are zero. Our nilpotent matrix must then have a zero column on the left:

$$\begin{pmatrix} 0 & u^t \\ \mathbf{0} & U \end{pmatrix} \begin{pmatrix} 1 \\ 0 \\ \vdots \\ 0 \end{pmatrix} = \mathbf{0}.$$

Here, u^t is some $(n-1)$-dimensional row vector, to make the first row. Furthermore, U is some $(n-1) \times (n-1)$ block, corresponding to the $n-1$ other coordinates (to be specified recursively, using the induction hypothesis). These coordinates may be not orthogonal to the first one. Still, we are not afraid of this. After all, we've already seen oblique coordinates in Figure 11.10.

18.2.2 Invariant Block

How to specify the other $n - 1$ coordinates? Recursively and implicitly, of course: in a mathematical induction on n. After all, blocks like U make a new subalgebra of gl_{n-1}. Indeed, they are invariant under multiplication: upon multiplying two matrices of the above form, their lower-right blocks multiply each other, to produce a new lower-right block in the product. For example,

$$\begin{pmatrix} 0 & u^t \\ \mathbf{0} & U \end{pmatrix} \begin{pmatrix} 0 & v^t \\ \mathbf{0} & V \end{pmatrix} = \begin{pmatrix} 0 & v^t V \\ \mathbf{0} & UV \end{pmatrix}.$$

For this reason, the lower-right blocks are invariant under the commutator as well:

$$\left[\begin{pmatrix} 0 & u^t \\ \mathbf{0} & U \end{pmatrix}, \begin{pmatrix} 0 & v^t \\ \mathbf{0} & V \end{pmatrix} \right] = \begin{pmatrix} 0 & u^t \\ \mathbf{0} & U \end{pmatrix} \begin{pmatrix} 0 & v^t \\ \mathbf{0} & V \end{pmatrix} - \begin{pmatrix} 0 & v^t \\ \mathbf{0} & V \end{pmatrix} \begin{pmatrix} 0 & u^t \\ \mathbf{0} & U \end{pmatrix}$$

$$= \begin{pmatrix} 0 & u^t V \\ \mathbf{0} & UV \end{pmatrix} - \begin{pmatrix} 0 & v^t U \\ \mathbf{0} & VU \end{pmatrix}$$

$$= \begin{pmatrix} 0 & u^t V - v^t U \\ \mathbf{0} & UV - VU \end{pmatrix}$$

$$= \begin{pmatrix} 0 & u^t V - v^t U \\ \mathbf{0} & [U,V] \end{pmatrix}.$$

Moreover, the lower-right block is also invariant under power:

$$\begin{pmatrix} 0 & u^t \\ \mathbf{0} & U \end{pmatrix}^l = \begin{pmatrix} 0 & v^t U^{l-1} \\ \mathbf{0} & U^l \end{pmatrix}, \quad l \geq 1.$$

For this reason, if the original matrix is nilpotent, then the above is zero (for a sufficiently large l), so U is nilpotent as well.

18.2.3 Mathematical Induction on n

Thanks to the induction hypothesis, we can now design $n - 1$ new coordinates in such a way that blocks like U and V become strictly upper triangular as well. This way, the original matrix is indeed strictly upper triangular, as asserted.

Or did we forget something? What about the case $n = 1$? This is easy enough: a nilpotent "matrix" in gl_1 has just one entry: (0). This is indeed strictly upper triangular, as required.

18.2.4 Nilpotent Linear Lie Algebra

In summary, our original Lie algebra is quite special: in the above basis, it contains strictly upper triangular matrices only:

$$L \subset T_1$$

(Chapter 15, Section 15.5). So, we can also deduce a global property: L is a nilpotent Lie algebra.

Still, by now, we proved this for a linear Lie algebra only. What about a more abstract Lie algebra that contains no matrices at all? In this case, there is no multiplication at all, let alone any notion of power or nilpotency. At best, the members could be adjoint-nilpotent (because the adjoint is a linear mapping, or a matrix). If all members are adjoint-nilpotent, could we safely deduce a global property: that the entire Lie algebra is nilpotent as a whole? This is indeed Engel's theorem.

18.3 Engel's Theorem

18.3.1 Finite-Dimensional Lie Algebra

We arrive now at our main question. Let L be a (finite-dimensional) Lie algebra (not necessarily linear). Assume that all its members are adjoint-nilpotent. In other words, all matrices in $\phi(L)$ are nilpotent, where ϕ is the adjoint representation:

$$\phi(l) = [l, \cdot], \quad l \in L.$$

In this case, we already know that $\phi(L)$ is nilpotent. Still, is L nilpotent as well?

18.3.2 The Adjoint Mappings and Their Common Eigenvector

Thanks to Section 18.1.4, we already know that the $\phi(l)$'s have a common eigenvector $v_0 \in L$:

$$\phi(l)(v_0) = [l, v_0] = \mathbf{0}, \quad l \in L.$$

This means that v_0 commutes with all members of L. In other words, v_0 is in Z – the center of L.

18.3.3 Dividing by The Center

To prove that L is nilpotent, let's use mathematical induction on its dimension. If L is one-dimensional, then it is Abelian, and therefore nilpotent, as asserted.

Assume now that L is at least two-dimensional. We've already seen that it has a nontrivial center:

$$Z \neq \{0\}.$$

Let's look at L/Z (Figure 18.2). What is its adjoint? This is defined by

$$\phi(l + Z)(m + Z) = [l + Z, m + Z] = [l, m] + Z, \quad l, m \in L.$$

This is the factor adjoint representation (Section 18.1.4).

18.3.4 Induction on The Dimension of L

Clearly, for each individual $l \in L$, this is a nilpotent mapping from L/Z into itself. Thanks to the induction hypothesis, we can now deduce a global property: L/Z is nilpotent. Thanks to Chapter 17, Section 17.10, L is nilpotent as well, as asserted. This completes the induction step to prove Engel's theorem.

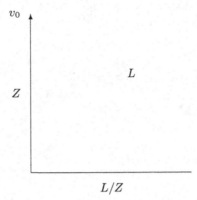

Fig. 18.2. The original Lie algebra L splits into two parts: its center Z (containing the vector v_0) at the vertical, and L/Z at the horizontal. The induction hypothesis is used on L/Z, which is lower-dimensional.

Weight Space
and Lie's Lemma and Theorem

So far, we studied a linear Lie algebra of nilpotent matrices, and designed a common eigenvector of eigenvalue zero. This eigenvector spans a new subspace: the weight space. As a matter of fact, the weight space doesn't have to be one-dimensional: it could be spanned by a few common eigenvectors.

Below, we extend this further: the eigenvalue doesn't have to be zero any more. Furthermore, the matrices don't have to be nilpotent any more. Instead, we focus on some matrices of special interest. Each matrix of this kind may have its own eigenvalue for the common eigenvectors in the weight space.

Moreover, Lie's lemma tells us that the weight space remains invariant under our Lie algebra: if you pick a common eigenvector, then you could go ahead and apply just any matrix to it, and still obtain

- either a new common eigenvector of the same eigenvalue,
- or the zero vector.

This powerful property may help understand better the structure of a (finite-dimensional) solvable Lie algebra:

- It could be decomposed in a chain of nested ideals.
- It has a nilpotent derived algebra.
- If it is also linear, then all matrices in it are upper triangular (in some basis).

The latter property tells us that a linear solvable Lie algebra must be a subalgebra of T_0. This mirrors a well-known property: a linear *nilpotent* Lie algebra must be a subalgebra of T_1 (in some basis, as in Chapter 18, Section 18.2.4).

19.1 Weight Space

19.1.1 Linear Functional – Eigenvalue

Let L be a linear Lie algebra, containing matrices of the same order. Let $I \subset L$ be an ideal. On I, define a new linear functional:

$$\lambda : I \to \mathbb{C}.$$

This is a linear mapping that assigns a scalar to each matrix in I.

Let's look at a simple example: I is the same as L, λ is the zero functional, and all the matrices in L are nilpotent. In this case, we already know that there is (at least one) common eigenvector v_0 of eigenvalue 0 (Chapter 18, Section 18.1.4). Let v_0 span a new subspace:

$$V \equiv \operatorname{span}\{v_0\}.$$

For simplicity, we assume that V is one-dimensional. In practice, however, there could be a few (linearly independent) common eigenvectors. In such a case, V is two-dimensional or more, but this doesn't matter. After all, these eigenvectors are indistinguishable in terms of eigenvalue, and could be treated as one and the same.

19.1.2 More General Common Eigenvector

the above is just an example. In general, I could be smaller than L, and the weight space V contains common eigenvectors, shared by every matrix $i \in I$:

$$V \equiv V_\lambda \equiv \{v \mid iv = \lambda(i)v, \ i \in I\}.$$

This way, for each individual matrix $i \in I$, v is an eigenvector, with the eigenvalue $\lambda(i)$ (depending on i, but not on v). In the above example, λ is identically zero, so V contains the common eigenvector v_0 of eigenvalue zero, as required. In a more general case, on the other hand, to each individual matrix in I, λ assigns a suitable eigenvalue. This way, V still contains common eigenvectors: this time, however, they may have nonzero eigenvalues that depend on the matrix under consideration, and may change from matrix to matrix in I.

$V \equiv V_\lambda$ is called a weight space. All vectors in V have the same eigenvalue: it depends on the matrix $i \in I$ only. For this reason, all vectors in V could be regarded as one and the same. This is why we can safely assume that V is one-dimensional.

Given an individual matrix $i \in I$, how to uncover the eigenvalue assigned to it? Just apply λ to i, and you'd find the eigenvalue: $\lambda(i)$. This depends on the matrix only, but not on the vector: $\lambda(i)$ is the same for all vectors in V. (This is why we could assume that V was one-dimensional.) Still, the eigenvalue may change (linearly) from some $i \in I$ to some other $i' \in I$. This way, we do better than in linear algebra: in one go, we have the eigenvector and eigenvalue of not only one but many matrices at the same time.

19.2 Lie's Lemma

19.2.1 Lie's Lemma: Invariant Weight Space

Lie's lemma says that the weight space is invariant (remains unchanged) under our (linear) Lie algebra:

$$LV \subset V.$$

What does this mean? This means that, for each individual matrix $l \in L$, for each individual vector $v \in V$, lv is still in the weight space:

$$lv \in V.$$

What does this mean? This means that, for each individual matrix $i \in I$, not only v but also lv is

- an eigenvector (of the same eigenvalue: $\lambda(i)$),
- or the zero vector.

In other words,

$$ilv = \lambda(i)lv.$$

This means that

$$\mathbf{0} = ilv - \lambda(i)lv = ilv - liv = (il - li)v = [i,l]v = i'v = \lambda(i')v$$

(for some other i' in the ideal I). So, it is sufficient to prove is that

$$\lambda(i') = 0, \quad i' \in [I, L].$$

19.2.2 Invariant Subspace and Its Basis

To prove this, let's design a new basis (for a new subspace): start from $v \neq \mathbf{0}$, and apply l to it, time and again:

$$v, \; lv, \; l^2v, \; l^3v, \; \ldots, \; l^{k-1}v,$$

where k is the first natural number for which l^kv depends linearly on the previous vectors.

In this new basis, how does v look like? Well, v is represented by a standard unit vector, with the first component 1, and the rest 0:

$$v \sim \begin{pmatrix} 1 \\ 0 \\ 0 \\ \vdots \\ 0 \end{pmatrix}.$$

Likewise, how does lv look like? Well, in our new basis, lv is represented by the second standard unit vector:

$$lv \sim \begin{pmatrix} 0 \\ 1 \\ 0 \\ \vdots \\ 0 \end{pmatrix},$$

and so on.

Next, let's restrict i, i', and l to the new subspace spanned by this new basis. As a result, we obtain three new $k \times k$ matrices. Let's denote them by B, B', and A, respectively.

How does A look like? Well, in our new basis, A is a Lanczos matrix: it has 1's on the subdiagonal, some entries on the rightmost column, and 0 elsewhere:

$$A = \begin{pmatrix} & & & & * \\ 1 & & & & * \\ & 1 & & & * \\ & & 1 & & * \\ & & & \ddots & * \\ & & & 1 & * \end{pmatrix}.$$

On the other hand, how does B look like? Well, we already know one thing about it: it has eigenvector v of eigenvalue $\lambda(i)$:

$$Bv = iv = \lambda(i)v.$$

To fit this, the first (leftmost) column in B must look like this: the upper-left entry must be $\lambda(i)$, and the other entries below it must vanish.

How about the second column? Well, to obtain it, we must apply B to lv:

$$Blv = ilv = [i,l]v + liv = i'v + \lambda(i)lv = \lambda(i')v + \lambda(i)lv.$$

Thus, the second column in B looks like this: $\lambda(i)$ on the main diagonal, $\lambda(i')$ above it, and zeroes below it.

We can now use this formula to design the next column as well:

$$Bl^2v = il(lv) = [i,l]lv + lilv = i'lv + l(ilv) = i'lv + l\left(\lambda(i')v + \lambda(i)lv\right).$$

What is the highest power of l here? This is in the latter term: $\lambda(i)l^2v$, which contributes the entry $\lambda(i)$ on the main diagonal, with nonzeroes only above it, and zeroes below it.

19.2.3 Mathematical Induction in The Block

We can already see the pattern that emerges: B (and similarly B') is upper triangular, with a constant value on its main diagonal: $\lambda(i)$. This could be proved by mathematical induction on $j = 1, 2, 3, \ldots, k-1$:

$$Bl^jv = il(l^{j-1}v) = [i,l]l^{j-1}v + lil^{j-1}v = i'l^{j-1}v + l(il^{j-1}v).$$

Indeed, what do we have here? There are two terms: the former is the previous column in B', which is too short to contribute: it has zeroes at the jth component and below it (thanks to the induction hypothesis about B'). The latter, on the other hand, is obtained from the previous column in B, pushed downwards, contributing $\lambda(i)$ to the (j,j)th main-diagonal entry.

19.2.4 Trace-Free Commutator

This kind of mathematical induction could be carried out on B and B' at the same time. Thus, B' has the same structure: upper triangular, with a constant value on its main diagonal: $\lambda(i')$. But B' is just a restriction (or representation) of i' to our new (invariant) k-dimensional subspace, spanned by the above basis. This kind of representation preserves the Lie brackets (as in Chapter 18, Section 18.2.2). Thus, like i', B' is a trace-free commutator (Chapter 15, Section 15.3), so

$$0 = \text{trace}(i') = \text{trace}(B') = k\lambda(i').$$

In summary, we've just proved that

$$\lambda(i') = 0,$$

as required. This completes the proof of Lie's lemma.

19.3 Lie's Theorem

19.3.1 Linear Solvable Lie Algebra

Let's use Lie's lemma to prove Lie's theorem as well. This will help design a weight space for a solvable Lie algebra.

Let L be a linear solvable Lie algebra:

$$L \subset gl_n.$$

Let's use Lie's lemma to design a weight space for L. To do this, let's use mathematical induction on the dimension of L.

19.3.2 Induction on The Dimension of L

If L has dimension zero, then it contains just one matrix: the zero matrix. Define λ to be the zero functional. This way, the weight space $V \equiv V_\lambda$ could be really big – it could contain just any n-dimensional vector:

$$(0)v = \mathbf{0} = 0v = \lambda v, \quad v \in V.$$

19.3.3 Dividing by The Derived Algebra

Assume now that L is at least one-dimensional. Since L is solvable, it is different from its derived algebra:

$$[L, L] \neq L.$$

Thus, $[L, L]$ is lower-dimensional. Therefore, the factor Lie algebra

$$L/[L, L]$$

is at least one-dimensional. Furthermore, it is Abelian. Indeed, for every two vectors $x, y \in L$,

$$[x + [L, L], y + [L, L]] = [x, y] + [L, L] = [L, L],$$

which is the zero "vector" in $L/[L, L]$.

19.3.4 Abelian Factor Lie Algebra and Its Subspaces

Since $L/[L, L]$ is Abelian, every subspace of it is a legitimate ideal. Let T be a maximal proper subspace: smaller than $L/[L, L]$, and cannot be extended any more. Clearly, the dimension of T is smaller by 1 than the dimension of $L/[L, L]$:

$$\dim(T) = \dim(L/[L, L]) - 1.$$

19.3.5 Canonical Homomorphism and Its Kernel

Now, let

$$\pi : L \to L/[L, L]$$

be the canonical homomorphism that maps L onto $L/[L, L]$ (Chapter 16, Section 16.2.6). What is its kernel? This is just $[L, L]$. Let's use it to calculate dimension.

19.3.6 Dimension and Kernel

What is the difference between the dimension of L and the dimension of $L/[L,L]$? This is the same as the dimension of the kernel of $\pi - [L,L]$:

$$\dim([L,L]) = \dim(L) - \dim(L/[L,L]).$$

Now, look at $T \subset L/[L,L]$. We've already seen that the dimension of T is smaller by 1 than the dimension of $L/[L,L]$. Let $K \subset L$ contain those vectors that are mapped by π into T:

$$K \equiv \{l \in L \mid \pi(l) \in T\}.$$

Clearly, K is an ideal of L. As such, K is solvable as well. If we only knew that K was lower-dimensional, then we could apply the induction hypothesis to it.

19.3.7 What Is The Dimension of K?

Now, if you restricted π to K only, then the kernel would remain the same: $[L,L]$. Thus, the dimension of $[L,L]$ is also the difference between the dimension of K and the dimension of T:

$$\dim([L,L]) = \dim(K) - \dim(T).$$

As a result, the dimension of K is smaller by 1 than the dimension of L:

$$\dim(K) = \dim(L) - 1.$$

This is the key for the induction step.

19.3.8 Decomposition of L

For this reason, we can now pick a nonzero vector $y \in L \setminus K$, and obtain

$$L = K + \mathrm{span}(y).$$

(Compare with the decomposition in Chapter 18, Section 18.1.4, where λ was much more boring: the zero functional.)

19.3.9 The Induction Step

We can now apply the induction hypothesis to K, and design a new linear functional λ on K, with a new weight space $V \equiv V_\lambda$. Our only job is to define λ on y as well. Later on, we'll extend λ linearly to the entire L, as required.

19.3.10 Using Lie's Lemma

Fortunately, we can now apply Lie's lemma to L, with its ideal $K \subset L$. This means that V is indeed invariant under y. Thus, y could be restricted to V. Once restricted, it has a Jordan form there, including (at least one) eigenvector $v_0 \in V$, with a new eigenvalue, which we denote by $\lambda(y)$:

$$yv_0 = \lambda(y)v_0.$$

19.3.11   Extend The Functional Linearly

We can now go ahead and extend λ linearly to the entire L, as required. Our up-to-date weight space $V \equiv V_\lambda$ will then contain at least one vector: v_0, a common eigenvector of all matrices of the form $l \in L$, each with its own eigenvalue $\lambda(l)$ (depending on l only, not on v_0, as required). This completes the proof of Lie's theorem.

19.4 Upper-Triangular Form

19.4.1 Designing a Basis

Recall that L is a linear solvable Lie algebra. How does it look like? Well, thanks to Lie's theorem, all matrices in it have a common eigenvector: v_0. Thus, one could just repeat the proof in Chapter 18, Sections 18.2.1–18.2.4, with just one change: the upper-left entry is no longer zero, but $\lambda(l)$ (depending on the relevant matrix $l \in L$ only). This designs a new basis, in which all matrices in L are upper triangular.

19.4.2 Upper-Triangular Form

Thus, in the above basis,
$$L \subset T_0$$
(Chapter 15, Section 15.5).

This also works the other way around: if
$$L \subset T_0,$$
then L is linear and solvable (Chapter 17, Section 17.6). In summary, a Lie algebra is linear and solvable if and only if it is a subalgebra of T_0 (in some basis).

Next, let L be solvable and finite-dimensional (but not necessarily linear any more). How does it look like?

19.5 Solvable m-Dimensional Lie Algebra

19.5.1 The Adjoint Representation

Let L be a solvable Lie algebra (not necessarily linear) of dimension m. Let ϕ be the adjoint representation:

$$\phi(l) \equiv [l, \cdot] : L \to L, \quad l \in L.$$

This way, $\phi(L)$ contains linear mappings from L into itself, which could also be written as $m \times m$ matrices. Thus, $\phi(L)$ is a solvable subalgebra of gl_m, for which Lie's theorem works. With the basis in Section 19.4.2, these are $m \times m$ upper-triangular matrices:

$$\phi(L) \subset T_0 \subset gl_m.$$

19.5.2 Basis for a Solvable Lie Algebra

So far, we applied Lie's theorem to a linear solvable Lie algebra: $\phi(L)$. This way, the basis in Section 19.4.2 contains m (linearly independent) vectors from L. In this basis, each adjoint like $[l, \cdot]$ could be written as an upper-triangular $m \times m$ matrix.

Let's list the above basis, vector by vector:

$$v_0, v_1, v_2, \ldots, v_{m-1}.$$

In other words, L is spanned by these vectors:

$$L = \text{span}\,(v_0, v_1, v_2, v_3, \ldots, v_{m-1}).$$

Let's group these vectors in nested ideals.

19.5.3 Decomposition: Chain of Nested Ideals

Next, use this list to define new ideals of L:

$$L_0 \equiv \{0\}$$
$$L_1 \equiv \text{span}\,\{v_0\}$$
$$L_2 \equiv \text{span}\,\{v_0, v_1\}$$
$$L_3 \equiv \text{span}\,\{v_0, v_1, v_2\}$$
$$\cdots \quad \cdots$$
$$L_m \equiv \text{span}\,\{v_0, v_1, v_2, \ldots, v_{m-1}\} = L.$$

Are these really ideals? Well, each L_i is indeed invariant under the (upper-triangular) $[l, \cdot]$'s, and is therefore a legitimate ideal of L. In summary, we have a new decomposition in terms of a new chain of nested ideals:

$$L_0 \subset L_1 \subset L_2 \subset \cdots \subset L_m = L,$$

and the dimension of L_i is exactly i:

$$\dim\,(L_i) = i, \quad 0 \leq i \leq m.$$

19.6 Nilpotent Derived Algebra

19.6.1 Using The Adjoint Representation

Next, look at the derived algebra: $[L, L]$. Apply ϕ to it:

$$\phi([L, L]) = [\phi(L), \phi(L)].$$

As a matter of fact, this is true for just any representation ϕ. Here, however, we are particularly interested in the special case in which ϕ is the adjoint representation. In this setting, let's write this formula more explicitly.

19.6.2 Member of The Derived Algebra

Pick some x in the derived algebra:

$$x \in [L, L].$$

Let's start from a simple case, in which x could be written as

$$x = [x', x'']$$

(for some $x', x'' \in L$).

Now, the adjoint representation preserves the Lie brackets (Chapter 15, Section 15.2). Therefore, the adjoint of x could be written as a commutator of two (upper-triangular) matrices:

$$[x, \cdot] = [[x', x''], \cdot] = [[x' \cdot], [x'', \cdot]] \in T_1$$

(because both $[x', \cdot]$ and $[x'', \cdot]$ are upper triangular in the above basis, so their commutator is strictly upper triangular, as proved in Chapter 15, Section 15.3).

This is true not only for this simple x but also for a more complicated x: any linear combination of such $[x', x'']$'s.

19.6.3 Adjoint-Nilpotent Members of The Derived Algebra

In summary, in the above basis,

$$[x, \cdot] \in T_1.$$

Thus, x is adjoint-nilpotent. But this could be done for just any $x \in [L, L]$. Thanks to Engel's theorem (Chapter 18, Section 18.3.4), we also have a global property: $[L, L]$ is a nilpotent Lie algebra.

19.6.4 Nilpotency of The Derived Algebra

This also works the other way around: if $[L, L]$ is nilpotent, then L is solvable. In summary, a finite-dimensional Lie algebra is solvable if and only if its derived algebra is nilpotent.

20

Cartan's Criterion
for Solvability

In this chapter, we introduce Cartan's criterion to check whether a given Lie algebra is solvable. This will be useful later. We start from a linear Lie algebra. Then, we turn to a more general Lie algebra.

Let L be a linear Lie algebra, containing matrices of the same order. How could we tell whether it is solvable? Cartan's criterion offers a simple computational test: scan all matrices $y \in L$, and all matrices $x \in [L, L]$ (the derived algebra). If the product xy is always trace-free, then all x's are nilpotent. As a result, L is indeed solvable.

This is good enough so long as L is linear, so x and y are matrices, with a well-defined trace. But what if L wasn't linear? In this case, the above test doesn't make sense any more. Instead, we must talk about the adjoint. This way, the test takes the following form. Instead of x, look at its adjoint: $[x, \cdot]$. Likewise, instead of y, look at its adjoint: $[y, \cdot]$. Both are linear mappings, or matrices. As such, they have a well-defined trace. If their product is always trace-free, then x is adjoint-nilpotent. Thanks to Engel's theorem, $[L, L]$ is nilpotent, so L is indeed solvable. This is Cartan's criterion in its most general form.

20.1 Criterion for Nilpotency of a Matrix

20.1.1 Two Subspaces of gl_n

Consider a familiar Lie algebra: gl_n, containing all $n \times n$ (complex) matrices. In particular, this is also a vector space. As such, we can talk about its subspaces.

Consider two subspaces that include one another:

$$A \subset B \subset gl_n.$$

Now, in gl_n, each matrix l has an adjoint $[l, \cdot]$ that maps gl_n into gl_n as follows:

$$m \to [l, m], \quad m \in gl_n.$$

We are particularly interested in those adjoints that map B into A:

$$B \to A.$$

In other words, we focus on those l's for which

$$[l, B] \equiv \{[l, b] \mid b \in B\} \subset A.$$

Later on, we'll specify both A and B.

20.1.2 Well-Behaved Mapping of Matrices

Let's pick a matrix x whose adjoint is of the above kind:

$$[x, \cdot] : B \to A,$$

or

$$[x, B] \subset A.$$

What about the square of this adjoint? Does it behave in the same way? Well,

$$[x, [x, B]] \subset [x, A] \subset [x, B] \subset A.$$

Thus, $[x, \cdot]^2$ is well-behaved too: it maps B into A as well. Likewise, every power of $[x, \cdot]$ is well-behaved too. Moreover, every linear combination of powers of $[x, \cdot]$ is well-behaved too: it maps B into A, as required. In other words,

$$P([x, \cdot]) : B \to A,$$

where P is a polynomial (with complex coefficients) with no constant term.

20.1.3 Criterion for Nilpotency

Assume that the above matrix x has not only a well-behaved adjoint but also another attractive property: for every matrix y with a well-behaved adjoint, the product xy is trace-free:

$$\text{trace}(xy) = 0.$$

Then, x is nilpotent.

This is a criterion for the nilpotency of an individual matrix: x. Let's go ahead and prove this.

20.1.4 The Jordan Form

How does x look like? Well, let's write it in its Jordan form (Chapter 14):

$$x = \begin{pmatrix} \lambda_1 & 1 & & & & & \\ & \lambda_2 & \ddots & & & & \\ & & \lambda_3 & 1 & & & \\ & & & \ddots & 0 & & \\ & & & & \ddots & 1 & \\ & & & & & \lambda_{n-1} & \ddots \\ & & & & & & \lambda_n \end{pmatrix} = \Lambda + (x - \Lambda),$$

where Λ is the diagonal part that contains the λ_i's, and $x - \Lambda$ is the nilpotent part that has 1's on the superdiagonal, and 0 elsewhere. Here, we allow duplication: some λ_i's could be the same – we don't mind.

Recall the notation in Chapter 15, Section 15.3: $E^{(p,q)}$ is the matrix that has one nonzero entry only: 1, in the pth row and qth column, and 0 elsewhere. This way, we could write Λ as

$$\Lambda = \sum_{k=1}^{n} \lambda_k E^{(k,k)}.$$

20.1.5 The Adjoint and Its Jordan Form

Furthermore, to write the adjoint of Λ, we could simply write how it acts on a basis: the $E^{(i,j)}$'s. This way, in the above sum, only two terms survive: either $k = i$, or $k = j$:

$$
\begin{aligned}
\left[\Lambda, , E^{(i,j)}\right] &= \left[\sum_{k=1}^{n} \lambda_k E^{(k,k)}, E^{(i,j)}\right] \\
&= \left[\lambda_i E^{(i,i)} + \lambda_j E^{(j,j)}, E^{(i,j)}\right] \\
&= \left[\lambda_i E^{(i,i)}, E^{(i,j)}\right] + \left[\lambda_j E^{(j,j)}, E^{(i,j)}\right] \\
&= \lambda_i \left[E^{(i,i)}, E^{(i,j)}\right] + \lambda_j \left[E^{(j,j)}, E^{(i,j)}\right] \\
&= \lambda_i E^{(i,j)} - \lambda_j E^{(i,j)} \\
&= (\lambda_i - \lambda_j) E^{(i,j)}.
\end{aligned}
$$

What does this mean? Well, $E^{(i,j)}$ serves here as an eigenvector of the adjoint of Λ, with a new eigenvalue: $\lambda_i - \lambda_j$. To see this more clearly, let's parse $E^{(i,j)}$ as a long n^2-dimensional vector – a standard unit vector, with just one nonzero component, and 0 elsewhere. This way, $[\Lambda, \cdot]$ takes the form of a big $n^2 \times n^2$ diagonal matrix:

$$
\left[\Lambda, E^{(i,j)}\right] =
\begin{pmatrix} \ddots & & & \\ & \ddots & & \\ & & \lambda_i - \lambda_j & \\ & & & \ddots \\ & & & & \ddots \end{pmatrix}
\begin{pmatrix} 0 \\ \vdots \\ 0 \\ 1 \\ 0 \\ \vdots \\ 0 \end{pmatrix} \leftarrow (i,j).
$$

Thus, not only Λ but also $[\Lambda, \cdot]$ is diagonal. As a bonus, we obtain for free the Jordan decomposition, not only of x but also of $[x, \cdot]$:

$$[x, \cdot] = [\Lambda, \cdot] + [x - \Lambda, \cdot].$$

After all, since $x - \Lambda$ is nilpotent, its adjoint $[x - \Lambda, \cdot]$ is nilpotent as well (thanks to the binomial formula, as in Chapter 15, Section 15.2). Therefore, we could also write

$$[\Lambda, \cdot] = Q([x, \cdot]),$$

where Q is a polynomial (with complex coefficients) with no constant term (Chapter 14, Sections 14.8.1–14.8.4). Thus, $[\Lambda, \cdot]$ is well-behaved too: it maps B into A (Section 20.1.2). This will be useful below.

20.1.6 Rational Linear Functional

Still, Λ could be complex: the λ_i's could be complex, with an imaginary part, containing $\sqrt{-1}$. This is not good for us. How to avoid this?

Well, the above could be done not only for Λ but also for any other diagonal matrix. To design this, look at a new linear space:

$$\left\{ \sum_{k=1}^{n} q_k \lambda_k \mid q_k \in \mathbb{Q}, \ 1 \le k \le n \right\},$$

where $\mathbb{Q}$ is the set of rational numbers. This way, the λ_k's take the role of vectors, and their coefficients are rational. This makes a new linear space, closed under addition and scalar multiplication. Indeed, such a linear combination could be multiplied by a rational scalar, yielding a new linear combination. Furthermore, two such linear combinations could be added to one another, yielding a new linear combination.

We may not need all the λ_k's: some of them might happen to share a common unit: a (real or complex) number that divides them evenly. In such a case, some λ_k's could drop, leaving only a few independent λ_k's. In this case, the dimension of our new linear space could be smaller than n. For example, we might end up with only three independent λ_k's, say

$$\lambda_1 = 1, \ \lambda_2 = \sqrt{2} + \sqrt{-3}, \ \lambda_3 = \sqrt{5} + \sqrt{-7}$$

(or the square root of any other prime number). In this case, our new linear space is only three-dimensional.

Our aim is to get rid of the $\sqrt{-1}$ that may lie in the complex λ_k's. For this, define a new linear functional on our new linear space:

$$f : \left\{ \sum_{k=1}^{n} q_k \lambda_k \mid q_k \in \mathbb{Q}, \ 1 \le k \le n \right\} \to \mathbb{Q}$$

that maps each linear combination to some rational number. For this purpose, define f at the independent λ_k's only. For example, pick some $1 \le j \le n$, and define $f(\lambda_j) = 1$, and $f(\lambda_k) = 0$ at all the other independent λ_k's. Then, extend this linearly, for just any rational coefficient.

20.1.7 Rational Diagonal Matrix

Now, for $1 \le k \le n$, list the new rational $f(\lambda_k)$'s along the main diagonal of a new diagonal matrix:

$$y \equiv \begin{pmatrix} f(\lambda_1) & & & & \\ & f(\lambda_2) & & & \\ & & f(\lambda_3) & & \\ & & & \ddots & \\ & & & & f(\lambda_n) \end{pmatrix} = \sum_{k=1}^{n} f(\lambda_k) E^{(k,k)}.$$

Note that, thanks to linearity,

$$f(\lambda_i) - f(\lambda_j) = f(\lambda_i - \lambda_j), \quad 1 \le i, j \le n.$$

Thus, what we did for Λ in Section 20.1.5, we can also do for the new diagonal matrix y, and write its adjoint as a big $n^2 \times n^2$ diagonal matrix:

$$\left[y, E^{(i,j)} \right] = \begin{pmatrix} \ddots & & & \\ & \ddots & & \\ & & f(\lambda_i - \lambda_j) & \\ & & & \ddots \\ & & & & \ddots \end{pmatrix} \begin{pmatrix} 0 \\ \vdots \\ 0 \\ 1 \\ 0 \\ \vdots \\ 0 \end{pmatrix} \leftarrow (i,j).$$

20.1.8 Polynomials to Relate Diagonal Matrices

What's the difference between $[\Lambda, \cdot]$ and $[y, \cdot]$? Well, both are diagonal $n^2 \times n^2$ matrices, but the former is complex, whereas the latter is rational. This is what we wanted.

Next, using Lagrangian interpolation, we can now design a new polynomial R that agrees with f:

$$R(\lambda_i - \lambda_j) = f(\lambda_i - \lambda_j), \quad 1 \le i, j \le n.$$

Furthermore, by setting $i = j$, we immediately see that

$$R(0) = f(0) = 0,$$

so R has no constant term.

We can now use our polynomials to relate our diagonal matrices to each other:

$$[y, \cdot] = R([\Lambda, \cdot]) = R(Q([x, \cdot])) = R \circ Q([x, \cdot]).$$

Moreover,

$$R \circ Q(0) = R(Q(0)) = R(0) = 0,$$

so $R \circ Q$ has no constant term either. Therefore, $[y, \cdot]$ is as well-behaved as $[x, \cdot]$: it maps B into A, as required (Section 20.1.2).

20.1.9 Using The Trace-Free Property

Thanks to our original assumption in Section 20.1.3, we now have

$$0 = \text{trace}(xy) = \sum_{k=1}^{n} \lambda_k f(\lambda_k).$$

Now, let's apply f to both sides of this equation. On the left, thanks to linearity, $f(0) = 0$. On the right, f is applied to the λ_k's (which play the role of vectors), not to the $f(\lambda_k)$'s (which are their rational coefficients). Therefore, we obtain

$$0 = \sum_{k=1}^{n} f^2(\lambda_k).$$

Fortunately, f was designed cleverly: rational, with no $\sqrt{-1}$ at all. Thus, we must have

$$f(\lambda_1) = f(\lambda_2) = f(\lambda_3) = \cdots = f(\lambda_n) = 0.$$

But this could be done for just any linear functional f, no matter how it was defined. Thus, we must also have

$$\lambda_1 = \lambda_2 = \lambda_3 = \cdots = \lambda_n = 0,$$

so x is indeed nilpotent, as asserted. This proves our criterion for nilpotency.

20.2 Nilpotency of The Derived Algebra

20.2.1 Trace of a Product

To use the above criterion, we must first see some properties of the trace. Let x, y, and z be matrices of the same order. We already know that the commutator is trace-free:

$$\text{trace}([x, y]) = 0$$

(Chapter 15, Section 15.3). In other words,

$$\text{trace}(xy - yx) = 0,$$

or

$$\text{trace}(xy) = \text{trace}(yx).$$

Thus, the trace is insensitive to the order in which the matrices are multiplied: the order could reverse, and the trace remains the same.

Likewise, with three matrices, we still have

$$\text{trace}([y, xz]) = 0,$$

or

$$\text{trace}(yxz) = \text{trace}(xzy).$$

Thus, the trace is insensitive to a cyclic permutation, in which y jumps to the end:

$$yxz \rightarrow xzy$$

has no effect on the trace.

Let's move on to a more complicated "product." This is a mixed triple "product:" x and y in Lie brackets, times z:

$$\text{trace}([x,y]z) = \text{trace}(xyz) - \text{trace}(yxz)$$
$$= \text{trace}(xyz) - \text{trace}(xzy)$$
$$= \text{trace}(x[y,z]).$$

Thus, the trace is insensitive to a change like

$$[x,y]z \rightarrow x[y,z],$$

in which the Lie brackets shift one space rightwards, to capture y and z. Let's go ahead and use this.

20.2.2 Nilpotency of The Derived Algebra

Let

$$L \subset gl_n$$

be a linear Lie algebra. We can now specify the subspaces $A \subset B \subset gl_n$ in Section 20.1.2:

$$B \equiv L \quad \text{and} \quad A \equiv [L,L].$$

To be well-behaved, an adjoint like $[y, \cdot]$ must satisfy

$$[y,L] \subset [L,L].$$

This is clearly true for every $y \in L$, but not necessarily for $y \notin L$.

Assume that, for every $x \in [L,L]$ and $y \in L$,

$$\text{trace}(xy) = 0.$$

Pick some $x \in [L,L]$. Consider first a simple case, in which x could be written as

$$x = [x', x''],$$

for some $x', x'' \in L$. (Later on, we'll consider linear combinations as well.) Thanks to our assumption, for every $y \in L$,

$$\text{trace}(xy) = 0.$$

Still, is this good enough? What about $y \notin L$? How can we be sure that xy is still trace-free? Fortunately, we can. Indeed, for every well-behaved $[y, \cdot]$ (no matter whether $y \in L$ or not),

$$[y,L] = [y,B] \subset A = [L,L].$$

Therefore, from Section 20.2.1,

$$\text{trace}(xy) = \text{trace}([x',x'']y) = \text{trace}(x'[x'',y]) = \text{trace}([x'',y]x') = 0,$$

as follows from our assumption, applied to the product of $[x'', y] \in [L, L]$ times $x' \in L$.

This can now be extended linearly to every linear combination of $[x', x'']$'s as well. Thus, xy is trace-free not only for $y \in L$ but also for $y \notin L$, provided that $[y, \cdot]$ is well-behaved.

This is quite what we need. Thanks to the criterion in Section 20.1.3, x is nilpotent.

20.3 Cartan's Criterion for Solvability

20.3.1 Linear Lie Algebra and Its Solvability

Thus, our assumption is not too strong:

$$\text{trace}(xy) = 0, \quad x \in [L, L], \; y \in L.$$

From this, we managed to deduce more:

$$\text{trace}(xy) = 0, \quad x \in [L, L], \; [y, L] \subset [L, L]$$

(no matter whether $y \in L$ or not). This is what we need to show that these x's are nilpotent. This way, $[L, L]$ contains nilpotent matrices only. Thanks to this, we also have a global property: $[L, L]$ is a nilpotent Lie algebra (Chapter 18, Section 18.2.4). As a result, L is solvable. This is Cartan's little criterion, limited to a *linear* Lie algebra like L. But what about a more general Lie algebra, not necessarily linear?

20.3.2 Finite-Dimensional Lie Algebra

Let L be a more general (finite-dimensional) Lie algebra (linear or not). How to write our assumption? After all, L may contain most abstract members, with no trace at all!

Let ϕ denote the adjoint representation of L:

$$\phi(l) \equiv [l, \cdot] : L \to L, \quad l \in L.$$

This way, $\phi(L)$ is indeed a linear Lie algebra. We can then write our assumption for it: assume that

$$\text{trace}(xy) = 0, \quad x \in \phi([L, L]), \; y \in \phi(L).$$

After all,

$$\phi([L, L]) = [\phi(L), \phi(L)],$$

so the above assumption indeed implies that all x's in $[\phi(L), \phi(L)]$ are nilpotent. In other words, all members in $[L, L]$ are adjoint-nilpotent. Thanks to Engel's theorem, we also have a global property: $[L, L]$ is a nilpotent Lie algebra (Chapter 18, Section 18.3.4). As a result, L is solvable. This is Cartan's criterion in its general terms: it applies to L, linear or not.

Killing Form
and Simple Ideal Decomposition

We can now define the Killing form, which has a lot of applications in quantum mechanics and relativity. In short, the Killing form ia the trace of the adjoint. Thanks to the Killing form, we have a computational algorithm to test whether a given Lie algebra is semisimple or not. If it is, then it can also be decomposed in a standard way: as the direct sum of its own simple ideals.

Actually, this mirrors a familiar vector space, decomposed as the direct sum of its own one-dimensional axes. Thus, to study a semisimple Lie algebra, it is sufficient to study its components: simple Lie algebras like A_l, B_l, C_l, and D_l (Chapter 15), with their applications in quantum mechanics and relativity.

For pedagogical reasons, this chapter is arranged in exercises, followed by their solutions or at least guidelines. This way, you get to study new material actively, as if you developed it on your own.

21.1 Killing Form on an m-Dimensional Lie Algebra

1. Let L be a Lie algebra. Let $I \subset L$ be an ideal. Define the derived series of I. Hint: use mathematical induction, as in Chapter 17, Section 17.5:

$$I^{(0)} \equiv I$$
$$I^{(i+1)} \equiv \left[I^{(i)}, I^{(i)} \right] \quad (i \geq 0).$$

2. Show that, in this series, each member is an ideal (not only of I but also) of L. Hint: use mathematical induction, and Chapter 16, Section 16.1.4.

3. Define also the lower central series of I. Hint: use mathematical induction, as in Chapter 17, Section 17.9:

$$I^0 \equiv I$$
$$I^{i+1} \equiv [I, I^i] \quad (i \geq 0).$$

4. Show that, in this series, each member is an ideal (not only of I but also) of L. Hint: use mathematical induction, and Chapter 16, Section 16.1.4.

5. Assume that L is finite-dimensional. For each two vectors $x, y \in L$, define the Killing form: take the adjoint of x, multiply it by the adjoint of y, and calculate the trace:

$$\kappa(x, y) \equiv \text{trace}([x, \cdot] \cdot [y, \cdot]).$$

6. Show that κ is bilinear: linear in x, and also in y.
7. Show that κ is symmetric:

$$\kappa(x, y) = \kappa(y, x).$$

Hint: as in Chapter 20, Section 20.2.1,

$$\kappa(x, y) \equiv \text{trace}([x, \cdot] \cdot [y, \cdot]) = \text{trace}([y, \cdot] \cdot [x, \cdot]) = \kappa(y, x).$$

8. Pick three vectors: $x, y, z \in L$. Show that κ is insensitive to shifting the Lie brackets one space rightwards:

$$\kappa([x, y], z) = \kappa(x, [y, z]).$$

Hint: thanks to Chapter 20, Section 20.2.1,

$$
\begin{aligned}
\kappa([x, y], z) &\equiv \text{trace}\left([[x, y], \cdot] \cdot [z, \cdot]\right) \\
&= \text{trace}\left([[x, \cdot], [y, \cdot]] \cdot [z, \cdot]\right) \\
&= \text{trace}\left([x, \cdot] \cdot [[y, \cdot], [z, \cdot]]\right) \\
&= \text{trace}\left([x, \cdot] \cdot [[y, z], \cdot]\right) \\
&= \kappa(x, [y, z]).
\end{aligned}
$$

9. Let S contain those x's that are "orthogonal" to all y's:

$$S \equiv \{x \in L \mid \kappa(x, L) = 0\} \equiv \{x \in L \mid \kappa(x, y) = 0, \; y \in L\}.$$

10. Is S a subspace? Hint: κ is linear.
11. Conclude that S is the maximal subspace for which

$$\kappa(S, L) = 0.$$

12. Is S an ideal of L? Hint: pick three vector: $s \in S$, and $y, \tilde{y} \in L$. Is $[s, y] \in S$ as well? It sure is:

$$\kappa([s, y], \tilde{y}) = \kappa(s, [y, \tilde{y}]) = 0.$$

13. Conclude that S is the maximal ideal for which

$$\kappa(S, L) = 0.$$

21.2 Ideal and Its Basis

1. Assume that L is m-dimensional ($m \geq 1$). Let $0 \leq r \leq m$ be a fixed integer number. Let $I \subset L$ be an ideal, spanned by the following basis vectors:

$$I = \text{span}(x_1, x_2, x_3, \ldots, x_r).$$

2. Introduce $m - r$ more basis vectors, to span the entire Lie algebra:

$$L = \text{span}(x_1, x_2, x_3, \ldots, x_m).$$

3. Pick two vectors: $x \in I$, and $y \in L$.
4. The adjoint of y is
$$[y, \cdot] : L \to L$$
(a linear mapping, or an $m \times m$ matrix).
5. The adjoint of x, on the other hand, is
$$[x, \cdot] : L \to I$$
(a linear mapping, or an $r \times m$ matrix). Why?
6. Write both adjoints in a block form:
$$[x, \cdot] = \begin{pmatrix} [x, \cdot]_{rr} & * \\ 0 & 0 \end{pmatrix}$$
and
$$[y, \cdot] = \begin{pmatrix} [y, \cdot]_{rr} & * \\ 0 & * \end{pmatrix}$$
(where the subscript 'rr' indicates an $r \times r$ block, '$*$' stands for some block, and '0' stands for a zero block.
7. Use this block form to multiply:
$$[x, \cdot] \cdot [y, \cdot] = \begin{pmatrix} [x, \cdot]_{rr}[y, \cdot]_{rr} & * \\ 0 & 0 \end{pmatrix}.$$

8. Observe that $[x, \cdot]$ "wins:" the product takes the same structure as $[x, \cdot]$.
9. Use this to calculate the trace:
$$\text{trace}([x, \cdot] \cdot [y, \cdot]) = \text{trace}\left([x, \cdot]_{rr}[y, \cdot]_{rr}\right).$$

10. Conclude that dropping $L \setminus I$ doesn't change κ. Indeed, let κ_I be the Killing form of I rather than L. If $x, y \in I$, then
$$\kappa_I(x, y) = \text{trace}\left([x, \cdot]_{rr}[y, \cdot]_{rr}\right) = \text{trace}([x, \cdot] \cdot [y, \cdot]) = \kappa(x, y).$$

11. Conclude that, for the ideal S in Section 21.1,
$$\kappa_S(S, S) = \kappa(S, S) = 0.$$

12. In particular,
$$\kappa_S([S, S], S) = \kappa([S, S], S) = 0.$$

13. Is this familiar? Hint: this is Cartan's criterion that tells us that S is solvable (Chapter 20, Section 20.3.2).
14. Conclude that
$$S \subset R,$$
where R is the radical: the maximal solvable ideal of L (Chapter 17, Section 17.8).
15. Conclude that, if $S \neq \{0\}$, then $R \neq \{0\}$ as well, so L is not semisimple.

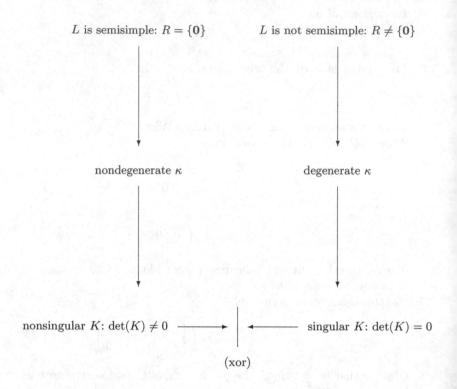

L is semisimple: $R = \{\mathbf{0}\}$

L is not semisimple: $R \neq \{\mathbf{0}\}$

nondegenerate κ

degenerate κ

nonsingular K: $\det(K) \neq 0$ $\longrightarrow$ $\longleftarrow$ singular K: $\det(K) = 0$

(xor)

Fig. 21.1. A semisimple Lie algebra (of a zero radical, and no solvable ideal but zero) has a nondegenerate Killing form (with a nonsingular matrix of a nonzero determinant). A not-semisimple Lie algebra (of a nonzero radical), on the other hand, has a degenerate Killing form (with a singular matrix of zero determinant).

21.3 Nilpotent Ideal and Its Basis

1. Assume that the above ideal I is nilpotent. This way, its lower central series must shrink to zero. Let $n \geq 0$ be the first integer number for which $I^n = \{\mathbf{0}\}$.
2. Design the basis of I cleverly, to span the nested ideals in the lower central series. Say,

$$I^{n-1} = \text{span}\,(x_1, x_2)\,,$$
$$I^{n-2} = \text{span}\,(x_1, x_2, x_3, x_4)\,,$$
$$I^{n-3} = \text{span}\,(x_1, x_2, x_3, x_4, x_5, x_6, x_7)\,,$$
$$I^{n-4} = \text{span}\,(x_1, x_2, x_3, x_4, x_5, x_6, x_7, x_8)\,,$$

and so on.

3. Show that, for each $x \in I$, each member of the lower central series is mapped like this:

$$[x, \cdot] : I^i \to I^{i+1}, \qquad 0 \leq i < n.$$

4. Conclude that $[x, \cdot]_{rr}$ has a strictly upper staircase form. Say,

$$[x, \cdot]_{rr} = \begin{pmatrix} * & * & * & * & * & * & * & & * \\ & * & * & * & * & * & * & & * \\ & & * & * & * & * & & & * \\ & & & * & * & * & * & & * \\ & & & & & * & * & & * \\ & & & & & * & * & & * \\ & & & & & & * & * & * \\ & & & & & & & * & * \\ & & & & & & & & \ddots & * \\ & & & & & & & & & \ddots \end{pmatrix}.$$

5. Conclude that $[x, \cdot]_{rr}$ is strictly upper triangular, with zero trace:

$$[x, \cdot]_{rr} \in T_1 \subset sl_r.$$

6. Show that, for each $y \in L$, each I^i is invariant under the adjoint of y:

$$[y, \cdot] : I^i \to I^i, \quad i \geq 0.$$

Hint: each I^i is a legitimate ideal (not only of I but also) of L (Section 21.1).

7. Conclude that $[y, \cdot]_{rr}$ has an upper staircase form that mirrors $[x, \cdot]_{rr}$:

$$[y, \cdot]_{rr} = \begin{pmatrix} * & * & * & * & * & * & * & * & & * \\ * & * & * & * & * & * & * & * & & * \\ & & * & * & * & * & * & * & & * \\ & & * & * & * & * & * & * & & * \\ & & & & * & * & * & * & & * \\ & & & & * & * & * & * & & * \\ & & & & * & * & * & * & & * \\ & & & & & & * & * & & * \\ & & & & & & & & \ddots & * \\ & & & & & & & & & \ddots \end{pmatrix}.$$

8. Show that $[x, \cdot]_{rr}$ "wins:" the product $[x, \cdot]_{rr}[y, \cdot]_{rr}$ has the same strictly upper staircase form, just like $[x, \cdot]_{rr}$.

9. Conclude that $[x, \cdot]_{rr}[y, \cdot]_{rr}$ is strictly upper triangular, with zero trace:

$$[x, \cdot]_{rr}[y, \cdot]_{rr} \in T_1 \subset sl_r.$$

10. Conclude that

$$\kappa(x, y) = \text{trace}([x, \cdot] \cdot [y, \cdot]) = \text{trace}\,([x, \cdot]_{rr}[y, \cdot]_{rr}) = 0.$$

11. Conclude that
$$\kappa(x, L) = 0.$$

Hint: the above is true for every $y \in L$.

12. Conclude that
$$\kappa(I, L) = 0.$$

Hint: the above is true for every $x \in I$.

13. Conclude that the nilpotent ideal I is included in the ideal S of Section 21.1:

$$I \subset S.$$

14. Must S be nilpotent? Hint: no.

15. Must L include a nonzero nilpotent ideal? Hint: no.

16. Give a condition to make sure that it does.

17. Show that, if L has a nonzero radical $R \neq \{0\}$, then R includes a nonzero nilpotent ideal $I \neq \{0\}$. Hint: since R is solvable, $[R, R]$ is nilpotent (Chapter 19, Section 19.6.4)). If $[R, R] \neq \{0\}$, then define $I \equiv [R, R]$. If, on the other hand, $[R, R] = \{0\}$, then define $I \equiv R$.

18. Conclude that, if $R \neq \{0\}$, then $S \neq \{0\}$ as well. Hint: the above nilpotent nonzero ideal I is included in S.

19. Show that, if $S \neq \{0\}$, then $R \neq \{0\}$ as well. Hint: $S \subset R$ (Section 21.2).

20. Conclude that $R = \{0\}$ if and only if $S = \{0\}$.

21. Recall that, if $R = \{0\}$, then we say that L is semisimple (has no solvable ideal but zero).

22. S is also called the radical of κ. If it is zero, then we say that κ is nondegenerate.

23. Conclude that L is semisimple if and only if κ is nondegenerate (Figure 21.1).

24. Use the above basis to define a new $m \times m$ matrix K:

$$K_{i,j} \equiv \kappa(x_i, xj), \quad 1 \leq i, j \leq m.$$

25. Is K the same as its own transpose? Hint:

$$K_{i,j} = \kappa(x_i, x_j) = \kappa(x_j, x_i) = K_{j,i},$$

or

$$K = K^t.$$

26. Is K Hermitian? Hint: not necessarily: if $K_{i,j}$ contains an imaginary part, then

$$K_{i,j} = K_{j,i} \neq \bar{K}_{j,i}.$$

27. Show that S is the same as the null space of $K = K^t$.

28. Conclude that κ is nondegenerate if and only if K is nonsingular, or

$$\det(K) \neq 0$$

(as illustrated in Figure 21.1).

29. Use the above formula as a computational criterion to make sure that L is semisimple.

30. Prove the above (strictly) upper staircase structure in yet another way. Hint: see below.

31. Recall that $I \subset L$ is a nilpotent ideal, with a lower central series that shrinks to zero.

32. Pick two vectors: $x \in I$, and $y \in L$.

33. Look again at the product of two matrices:

$$[x, \cdot] \cdot [y, \cdot].$$

34. How does it act on a given vector $l \in L$? Hint: in two stages: first, apply $[y, \cdot]$ to l, to produce $[y, l] \in L$. Then, apply $[x, \cdot]$ to $[y, l]$, to produce $[x, [y, l]] \in I$. Indeed, because x is in the ideal I, the result is in I as well.

35. Conclude that

$$[x, \cdot] \cdot [y, \cdot] : L \to I.$$

36. For short, denote the new vector $[x, [y, l]]$ by $i \in I$. Apply $[x, \cdot] \cdot [y, \cdot]$ to it. How does it act? Hint: in two stages: forst, apply $[y, \cdot]$ to i, to produce $[y, i] \in I$. Then, apply $[x, \cdot]$ to $[y, i]$, to produce $[x, [y, i]] \in [I, I]$.

37. Continue this process. Can you recognize a pattern?

38. Can you write a law about the matrix $[x, \cdot] \cdot [y, \cdot]$ and its powers, and how they act on a given vector? Hint: see below.

39. Use mathematical induction to prove that

$$([x, \cdot] \cdot [y, \cdot])^{i+1} : L \to I^i \quad (i \geq 0).$$

Hint: for $i = 0$, we've already proved this. For $i = 1, 2, 3, \ldots$, on the other hand, thanks to the induction hypothesis,

$$
\begin{aligned}
([x, \cdot] \cdot [y, \cdot])^{i+1} (L) &= [x, \cdot] \cdot [y, \cdot] ([x, \cdot] \cdot [y, \cdot])^i (L) \\
&\subset [x, \cdot] \cdot [y, \cdot] (I^{i-1}) \\
&= [x, \cdot] \cdot [y, I^{i-1}] \\
&\subset [x, \cdot] \cdot I^{i-1} \\
&= [x, I^{i-1}] \\
&\subset [I, I^{i-1}] \\
&= I^i.
\end{aligned}
$$

40. Conclude that $[x, \cdot] \cdot [y, \cdot]$ is a nilpotent matrix. Hint: since I is a nilpotent ideal,

$$([x, \cdot] \cdot [y, \cdot])^{n+1} : L \to I^n = \{0\}.$$

41. Conclude that

$$\kappa(x, y) = \text{trace}([x, \cdot] \cdot [y, \cdot]) = 0.$$

42. Conclude that

$$\kappa(x, L) = 0.$$

Hint: the above is true for every $y \in L$.

43. Conclude that $x \in S$.

44. Conclude that $I \subset S$. Hint: the above is true for every $x \in I$.

45. Conclude once again that, if $R \neq \{0\}$, then $S \neq \{0\}$ as well. Hint: so far, we proved that S includes all nilpotent ideals of L. Now, since R is solvable, $[R, R]$ is nilpotent (Chapter 19, Section 19.6.4)), and included in S. If $[R, R] = \{0\}$, then R is Abelian (and nilpotent), and included in S. In either case, $S \neq \{0\}$. This is illustrated in Figure 21.1 on the right.

21.4 Structure Constants

1. To calculate κ, let $1 \leq i, j \leq m$ be two fixed natural numbers. Now, look at $[x_i, x_j]$. This is a new vector in L. As such, it could be written as a linear combination:

$$[x_i, x_j] = \sum_{l=1}^{m} c_{i,j}^l x_l,$$

for some (complex) coefficients $c_{i,j}^l$. Use Einstein's summation convention to write this better. Hint: as in Chapter 11, Section 11.13.3, drop the '$\sum$' symbol:

$$[x_i, x_j] = c_{i,j}^l x_l.$$

2. These $c_{i,j}^l$'s are called the structure constants. They depend not only on L but also on the chosen basis.

3. What law must the structure constants obey, no matter what basis was chosen? Hint: thanks to alternativity, for each fixed i,

$$\mathbf{0} = [x_i, x_i] = c_{i,i}^l x_l.$$

(Here, i appears as a lower index only, so it is not summed over.) Now, since the x_l's are linearly independent, their coefficients must vanish:

$$c_{i,i}^l = 0, \qquad 1 \leq i, l \leq m.$$

4. Could you write yet another law? Hint: thanks to antisymmetry,

$$\mathbf{0} = [x_i, x_j] + [x_j, x_i] = c_{i,j}^l x_l + c_{j,i}^l x_l = \left(c_{i,j}^l + c_{j,i}^l \right) x_l.$$

Now, since the x_l's are linearly independent, their coefficients must vanish:

$$c_{i,j}^l + c_{j,i}^l = 0, \qquad 1 \leq i, j, l \leq m.$$

5. Could you write all these laws in just one line? Hint: for each fixed $1 \leq l \leq m$, the $c_{i,j}^l$'s make a new $m \times m$ antisymmetric matrix:

$$\left(c_{i,j}^l \right)_{1 \leq i, j \leq m}$$

(with two lower indices: i and j).

6. Could you write yet another law? Hint: let $1 \leq i, j, k \leq m$ be fixed. Then,

$$[[x_i, x_j], x_k] = [c_{i,j}^l x_l, x_k] = c_{i,j}^l [x_l, x_k] = c_{i,j}^l c_{l,k}^n x_n.$$

Now, look at the triplet (i, j, k). Carry out a cyclic permutation on it, and sum up three expressions: with (i, j, k), (k, i, j), and (j, k, i). Thanks to the Jacobi identity,

$$\begin{aligned}\mathbf{0} &= [[x_i, x_j], x_k] + [[x_k, x_i], x_j] + [[x_j, x_k], x_i] \\ &= \left(c_{i,j}^l c_{l,k}^n + c_{k,i}^l c_{l,j}^n + c_{j,k}^l c_{l,i}^n \right) x_n.\end{aligned}$$

Now, since the x_n's are linearly independent, their coefficients must vanish:

$$c_{i,j}^l c_{l,k}^n + c_{k,i}^l c_{l,j}^n + c_{j,k}^l c_{l,i}^n = 0 \quad (1 \leq i < j < k \leq m, \quad \text{and} \quad 1 \leq n \leq m).$$

7. Use the above antisymmetry to rewrite this as

$$c_{i,j}^l c_{k,l}^n + c_{k,i}^l c_{j,l}^n + c_{j,k}^l c_{i,l}^n = 0 \quad (1 \le i < j < k \le m, \text{ and } 1 \le n \le m).$$

8. In these laws, we only picked $i < j < k$. Why? Hint: any other pick would give no new law, but only repeat an old law.

9. Conclude that, to be good candidates to serve as legitimate structure constants for any basis, the $c_{i,j}^n$'s must obey the above laws.

10. So far, we fixed l, and looked at the $m \times m$ antisymmetric matrix

$$\left(c_{i,j}^l\right)_{1 \le i,j \le m},$$

with two lower indices: i and j. But what if we fixed i, not l? What is the meaning of the new $m \times m$ matrix

$$C^{(i)} \equiv \left(c_{i,j}^l\right)_{1 \le l,j \le m},$$

with the upper index l and the lower index j? Hint: see below.

11. Represent our basis in terms of m-dimensional standard unit vectors:

$$x_1 \sim \begin{pmatrix} 1 \\ 0 \\ 0 \\ 0 \\ \vdots \\ 0 \end{pmatrix}, \quad x_2 \sim \begin{pmatrix} 0 \\ 1 \\ 0 \\ 0 \\ \vdots \\ 0 \end{pmatrix}, \quad x_3 \sim \begin{pmatrix} 0 \\ 0 \\ 1 \\ 0 \\ \vdots \\ 0 \end{pmatrix}, \quad x_j \sim \begin{pmatrix} 0 \\ \vdots \\ 0 \\ 1 \\ 0 \\ \vdots \\ 0 \end{pmatrix} \leftarrow j,$$

and so on.

12. To these vectors, apply the matrix $[x_i, \cdot]$:

$$[x_i, x_j] = [x_i, \cdot] \, x_j \sim [x_i, \cdot] \begin{pmatrix} 0 \\ \vdots \\ 0 \\ 1 \\ 0 \\ \vdots \\ 0 \end{pmatrix} = \begin{pmatrix} c_{i,j}^1 \\ c_{i,j}^2 \\ c_{i,j}^3 \\ \vdots \\ c_{i,j}^m \end{pmatrix}.$$

13. Conclude that, for each fixed i,

$$[x_i, \cdot] = \left(c_{i,j}^l\right)_{1 \le l,j \le m},$$

where l stands for the row index, and j for the column index.

14. Write this more compactly:

$$[x_i, \cdot] = C^{(i)}, \quad 1 \le i \le m.$$

15. Multiply two such matrices:

$$[x_i, \cdot] \cdot [x_k, \cdot] = C^{(i)} C^{(k)}.$$

16. In this product, focus on the (l, n)th element:

$$\left([x_i, \cdot] \cdot [x_k, \cdot]\right)_{l,n} = \left(C^{(i)} C^{(k)}\right)_{l,n} = c_{i,j}^l c_{k,n}^j.$$

17. Calculate the trace of this product:

$$K_{i,k} = \kappa\left(x_i, x_k\right) = \text{trace}\left([x_i, \cdot] \cdot [x_k, \cdot]\right) = \text{trace}\left(C^{(i)} C^{(k)}\right) = c_{i,j}^l c_{k,l}^j.$$

18. Make sure that this is indeed symmetric. Hint: interchanging $i \leftrightarrow k$ has no effect.
19. Write an algorithm to determine whether L is semisimple or not. Hint: use the above formula to calculate K. Then, calculate $\det(K)$. Finally, use Figure 21.1.

21.5 Simple Ideal Decomposition

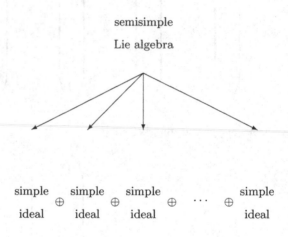

semisimple

Lie algebra

simple   simple   simple   simple
    $\oplus$ $\oplus$ $\oplus$ $\cdots$ $\oplus$
ideal   ideal   ideal   ideal

Fig. 21.2. A semisimple (finite-dimensional) Lie algebra could be written as a direct sum of simple ideals.

1. Let L be an m-dimensional semisimple Lie algebra. Show that $m \geq 2$. Hint: if $m = 1$, then L is Abelian, nilpotent, and solvable, and could never be semisimple any more (Figure 17.1),
2. Conclude that κ is nondegenerate, and K is nonsingular. Hint: see Figure 21.1.
3. Let $I \subset L$ be a nonzero ideal. Could I be solvable? Hint: if it were, then L could never be semisimple.
4. Could I be one-dimensional? Hint: if it were, then it would be Abelian, nilpotent, and solvable, and L could never be semisimple any more.

5. Let J contain those vectors "orthogonal" to I:

$$J \equiv \{l \in L \mid \kappa(I, l) = 0\} \equiv \{l \in L \mid \kappa(i, l) = 0, \ i \in I\}.$$

6. Is J a legitimate subspace of L? Hint: κ is bilinear.
7. Is J a legitimate ideal of L? Hint: pick three vectors: $i \in I$, $j \in J$, and $l \in L$. Is $[l, j] \in J$ as well? In other words, is $[l, j]$ "orthogonal" to i? Well,

$$\kappa(i, [l, j]) = \kappa([i, l], j) = 0$$

(because $[i, l]$ is sitll in the ideal I).
8. Look at the intersection $I \cap J$. Is it an ideal as well? Hint: since I is an ideal, $[I \cap J, L] \subset I$. Furthermore, since J is an ideal, $[I \cap J, L] \subset J$ as well. In summary, $[I \cap J, L] \subset I \cap J$, as required.
9. What is the Killing form of $I \cap J$ alone, disregarding all other vectors? Hint:

$$\kappa_{I \cap J}(I \cap J, I \cap J) = \kappa(I \cap J, I \cap J) = 0$$

(Section 21.2).
10. In particular,

$$\kappa_{I \cap J}([I \cap J, I \cap J], I \cap J) = 0$$

(because $[I \cap J, I \cap J] \subset I \cap J$).
11. Conclude that $I \cap J$ is solvable. Hint: use Cartan's criterion, as in Section 21.2.
12. What could $I \cap J$ be? Hint: it must vanish: L is semisimple, and includes no nonzero solvable ideal.
13. What is $[I, J]$? Hint: because both I and J are ideals, $[I, J] \subset I$ and $[I, J] \subset J$, so $[I, J] \subset I \cap J = \{\mathbf{0}\}$.
14. Assume that I has dimension $r \geq 2$. What is the dimension of

$$\bar{K}\bar{J} \equiv \{\bar{K}\bar{j} \mid \bar{j} \text{ is the complex conjugate of some } j \in J, \text{ expanded in our basis}\}?$$

Hint: $\bar{K}\bar{J}$ is the orthogonal complement of I. As such, it has dimension $m - r$.
15. What is the dimension of J? Hint: $m - r$ too, because K (and indeed $\bar{K}$) is an invertible matrix of nonzero determinant.
16. Conclude that

$$L = I + J.$$

17. Recall that L is semisimple. Must J be semisimple as well? Hint: let $T \subset J$ be a nonzero ideal of J. Is T an ideal of L too? To check on this, pick two vectors: $t \in T$, and $l \in L$. Write l as $l = i + j$, for some $i \in I$ and $j \in J$. This way,

$$[t, l] = [t, i + j] = [t, i] + [t, j] = \mathbf{0} + [t, j] = [t, j] \in T$$

(because T is an ideal of J). Thus, T is indeed an ideal (not only of J but also) of L. Since L is semisimple, T could never be solvable. This proves that J is semisimple as well.
18. Show that L could be written uniquely as a sum of the form

$$L = I_1 + I_2 + I_3 + \cdots + I_n,$$

where

$$I_i \cap I_j = [I_i, I_j] = \{\mathbf{0}\}, \quad 1 \le i < j \le n.$$

Hint: by mathematical induction: in the above, let I be a minimal nonzero ideal of L. Clearly, to be minimal, I must be simple. Define $I_1 \equiv I$. Finally, look at the above J: if it is neither zero nor simple, then apply the induction hypothesis to it.

19. Is above decomposition unique? Hint: see below.
20. Consider a nonzero ideal $H \subset L$. Pick some $1 \le j \le n$. Look at the intersection $H \cap I_j$. If it is nonzero, how could it look like? Hint: on one hand,

$$H \cap I_j \subset I_j.$$

On the other hand, $H \cap I_j$ is an ideal of L. As such, it must cover the simple ideal I_j completely:

$$H \cap I_j = I_j.$$

21. How does H look like? Hint: it is a sum of a few I_j's.
22. If H is also simple, how must it look like? Hint: in this case, H is a very small "sum:" just one I_j. Indeed, to be simple, H must have no inner ideal. For this purpose, H must be the same as one of the I_j's.
23. Conclude that the above decomposition is indeed unique.
24. The above decomposition is also called a direct sum, and denoted by

$$L = I_1 \oplus I_2 \oplus I_3 \oplus \cdots \oplus I_n$$

(Figure 21.2). Here, the '$\oplus$' symbol tells us that the I_j's never interact: they always commute, and never intersect, as discussed above.

25. How to study semisimple Lie algebras? Hint: it is sufficient to study simple Lie algebras (like A_l, B_l, C_l, and D_l in Chapter 15).
26. Pick some I_j. What is its derived algebra? Hint: clearly, $[I_j, I_j] \subset I_j$. Now, if $[I_j, I_j] = \{\mathbf{0}\}$, then I_j would be Abelian, and could never be simple any more (Figure 17.1). If, on the other hand, $[I_j, I_j] \ne I_j$, then it would make an inner ideal inside I_j, which is also forbidden. So, the only option is $[I_j, I_j] = I_j$.
27. What is the derived algebra of L? Hint: because the I_j's never interact,

$$
\begin{aligned}
[L, L] &= [I_1 \oplus I_2 \oplus I_3 \oplus \cdots \oplus I_n, I_1 \oplus I_2 \oplus I_3 \oplus \cdots \oplus I_n] \\
&= [I_1, I_1] \oplus [I_2, I_2] \oplus [I_3, I_3] \oplus \cdots \oplus [I_n, I_n] \\
&= I_1 \oplus I_2 \oplus I_3 \oplus \cdots \oplus I_n \\
&= L.
\end{aligned}
$$

28. Conclude that any ideal of L (like $H \subset L$ above) satisfies

$$[H, H] = H.$$

29. Conclude again that any ideal of L (like $H \subset L$ above) could never be solvable. Hint: its derived series remains constant.
30. Conclude again that any ideal of L (like $H \subset L$ above) must be semisimple as well. Hint: H may have no solvable ideal. Indeed, every ideal of H must have the same structure – a sum of a few I_j's. As such, it has a constant derived series, and can never be solvable.

Hamiltonian Mechanics:
Energy
and Angular Momentum

We are now ready to introduce a new Lie algebra. It will contain smooth functions on a diferentiable manifold, including a few special functions with an important physical meaning: energy, momentum, angular momentum, and more. In this Lie algebra, there are new Lie brackets: the Poisson brackets. Thanks to them, we can easily write the Hamilton equation, with its new geometrical interpretation [6, 35, 41, 54]. This will give us a few conservation laws for a few fundamental quantities: energy, momentum, angular momentum, and more.

Like every Lie brackets, the Poisson brackets are antisymmetric. Thanks to this algebraic property, we obtain Noether's theorem for free. Indeed, just look at things the other way around: the Hamiltonian not only makes new conserved quantities but also benefits from them, and takes new symmetry lines, on which it is invariant.

For this reason, in its orbit around the sun, the Earth keeps the same rate, energy, and angular momentum. This, however, is not the only conservation law. Our algebraic model leads to a useful algorithm to uncover new conserved quantities, and their symmetry lines.

22.1 Lie Algebra of Smooth Functions

22.1.1 Simplectic Manifold: Darboux Coordinates

We start with some geometrical preliminaries. Once the geometrical structure is ready, we give it an algebraic face. On top of this, physics will fit as easily as ever.

Let n be the dimension in our problem. (Say, $n = 3$ in the Cartesian space.) Consider a $2n$-dimensional manifold M, spanned by the variables $p_1, p_2, \ldots, p_n$ and $q^1, q^2, \ldots, q^n$. (These are just indices, not powers.) For $n = 3$, for example, these could stand for the (unknown) position and momentum in 3-D. Later on, we'll define the energy as a function of these variables. This way, the level set of the energy will give us the solution curve: what position and momentum in M are used during the motion. To find this curve, we'll need to differentiate functions defined on M.

Let $C^\infty(M)$ contain those real functions that are differentiable as many times as you like in M. Let $f \in C^\infty(M)$ be such a function. Its gradient (vector of partial derivatives) could be decomposed into two subvectors:

$$\nabla f \equiv \begin{pmatrix} \nabla_q f \\ \nabla_p f \end{pmatrix},$$

where ∇_q contains the partial derivatives with respect to the q^i's, and ∇_p with respect to the p_i's.

Our coordinates (the q^i's and p_i's) may be not global but only local: around each individual point $u \in M$, there is a spot $U \subset M$ (a little $2n$-dimensional domain), in which we've already picked a suitable coordinate system, in which our Hamilton equation will be written. These are called Darboux coordinates – local, not global: it is often impossible to "tie up" different spots and fit the local coordinate systems into one global coordinate system. To get around this problem, let's focus on just one spot $U \subset M$.

22.1.2 Hamiltonian Vector

Thanks to our Darboux coordinates, we can now use the $2n \times 2n$ antisymmetric matrix

$$S \equiv \begin{pmatrix} & I \\ -I & \end{pmatrix},$$

where I is the $n \times n$ identity matrix. This is the matrix used to introduce the simplectic Lie algebra in Chapter 15, Section 15.7. The only change is in the notation: n here plays the role of l there.

What is the Hamiltonian vector of a function? Take the gradient: ∇_q, followed by ∇_p. To this, apply S. This flips the subvectors, and also picks a minus sign: ∇_p, followed by $-\nabla_q$:

$$S\nabla = S \begin{pmatrix} \nabla_q \\ \nabla_p \end{pmatrix} = \begin{pmatrix} \nabla_p \\ -\nabla_q \end{pmatrix}.$$

Later on, we'll also discuss the geometrical meaning of this: this is orthogonal to the original gradient:

$$\nabla^t f S \nabla f = \begin{pmatrix} \nabla_q f \\ \nabla_p f \end{pmatrix}^t \begin{pmatrix} \nabla_p f \\ -\nabla_q f \end{pmatrix} = \nabla_q^t f \nabla_p f - \nabla_p^t f \nabla_q f = 0.$$

This could raise a new idea: why not use this as new Lie brackets? After all, the above already gives us one attractive property: alternativity.

22.1.3 The Poisson Brackets

Thanks to the Hamiltonian vector, we can now define a special kind of Lie brackets: the Poisson barackets. Traditionally, however, they are marked by braces rather than square brackets: for any two functions $f, u \in C^\infty(M)$,

$$\{f, u\} \equiv \nabla^t u S \nabla f = (S\nabla f)^t \nabla u.$$

This could also be written as an operator. For this purpose, just drop the 'u,' and obtain the adjoint:

$$\{f, \cdot\} \equiv (S\nabla f)^t \nabla.$$

What does this operator do? It differentiates at a specific direction: the direction pointed at by the vector $S\nabla f$.

As a differentiation, this satisfies the Leibniz rule. As a result, it also satisfies the Jacobi identity (see exercises below). Thus, with these new brackets, we have a new Lie algebra.

So, we have a complete algebraic structure, with all its attractive properties. Still, is it useful? What is its geometrical meaning?

22.2 Geometrical Point of View

22.2.1 The Phase Plane

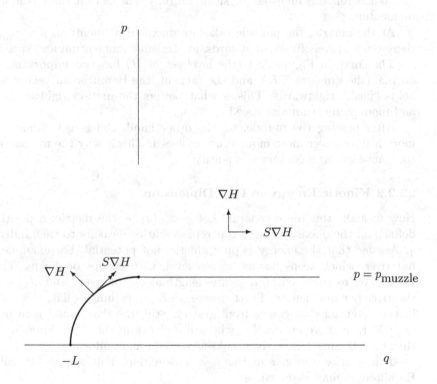

Fig. 22.1. The (q, p)-phase plane: position vs. momentum. At the initial state $(-L, 0)$, the particle is at $-L$ at rest. Upon advancing rightwards, it gains more and more momentum and speed, until reaching the muzzle at $(0, p_{\text{muzzle}})$. This makes a curve of constant energy in the phase plane: a level set of the Hamiltonian H. On this curve, the Hamiltonian vector $S\nabla H$ "pushes" rightwards. (This is just a virtual or nonphysical curve: physically, the particle remains on the horizontal q-axis.)

To see the geometrical meaning of our new Lie algebra, consider a one-dimensiona case ($n = 1$): a particle fired from a slingshot. Until it arrives at the muzzle, it behaves like a harmonic oscillator. This is illustrated in Figure 22.1.

In Newtonian physics, the position of the particle is often denoted by x. In Hamiltonian mechanics, on the other hand, x is often rpelaced by q: the horizontal q-axis. The momentum, on the other hand, is often denoted by p: the vertical p-axis. These are our dynamic variables that span the phase plane.

On the phase plane, we define the Hamiltonian: a real function $H \equiv H(p, q)$, which stands for the total energy: kinetic and potential alike. Its level set is a curve on which it is constant: for instance, the curve in Figure 22.1.

The curve starts at a point $(-L, 0)$, to the left of the origin. This is where the particle starts its journey: at rest, deep in the slingshot, with no momentum at all.

While travelling rightwards, the particle gains more and more momentum and speed. This is why, in Figure 22.1, the curve curves rightwards and upwards at the same time. Physically, however, the particle only moves rightwards: q increases. The p-axis remains invisible physically: it only tells us that there is more and more momentum.

At the muzzle, the particle takes its maximal momentum: $p = p_{\text{muzzle}}$. From there on, it travels freely rightwards, at the same momentum and speed.

The curve in Figure 22.1 (the level set of H) has two important vectors: the normal (the gradient ∇H), and the tangent: the Hamiltonian vector $S\nabla H$, which helps "push" rightwards. This is what powers the motion rightwards, with more and more momentum and speed.

After passing the muzzle, on the other hand, things get calmer: there is no more feul to power more momentum or speed. This is why the momentum remains constant forever: a conserved quantity.

22.2.2 Kinetic Energy in One Dimension

How to make this more general? Let $g \equiv g(q)$ be the metric: a positive function defined on the q-axis, giving a concrete weight or measure to each individual point q. Assume that the energy is only kinetic, not potential. For example, imagine a flat river, which keeps flowing at sea level, with no ups or downs. This way, the river serves as the q-axis: q measures length along the river., and $g(q)$ is the mass of the river per unit length. Furthermore, since everything is flat, there is no change in the potential that comes from gravity. Still, the river could gain in its metric: get wider, or more curvy. We, who sail and live in the river, know nothing about this: from the inside, the river looks as straight and uniform as ever. This is why we need the metirc g to give us this new information. This way, as we sail, the entire Hamiltonian may also change:

$$H \equiv \frac{p^2}{2g(q)} = \frac{1}{2}g(q)v^2,$$

where v is the speed of the flow.

22.2.3 Metric Increases Energy Indirectly

Here one may ask: how could the metric be in the denominator? After all, a bigger metric should make the energy bigger, not smaller! The answer is that the metric has mixed contributions. In the denominator, it indeed decreases H. Still, in the numerator, it has a much greater effect: it contributes to increasing p. This way, in total, it increases H indirectly.

22.2.4 Kinetic Energy Acting on Momentum

So far, we applied the Hamiltonian vector to a general function: u (Section 22.1.3). Now, instead of u, let's plug a more concrete function: the momentum $P \equiv p$. Why use a capital letter? Because P is a new thing: a physical quantity, not just a variable. As such, it deserves a new name. This way, it can also be differentiated. Fortunately, differentiation is easy: $\partial P/\partial q = 0$, and $\partial P/\partial p = 1$. Here is how the kinetic energy acts on the momentum:

$$\{H, P\} = \nabla^t P S \nabla H = (S \nabla H)^t \nabla P = \nabla_p^t H \nabla_q P - \nabla_q^t H \nabla_p P = -\nabla_q^t H = \frac{p^2 g'(q)}{2g^2(q)}$$

(because we are in one dimension: $n = 1$). In particular, when g is constant (river with constant curvature and width), this vanishes. This tells us that the momentum isn't "pushed" or raised any more. On the contrary: it remains constant: a conserved quantity. Indeed, in the phase plane, it shares the same level set with H: a horizontal line, on which p is constant. This way, its normal ∇P points upwards (just like ∇H): it only slides rightwards, but remains vertical.

22.3 Momentum and Its Conservation

22.3.1 The Hamilton Equation

What happens when g is not necessarily constant, and may change with q? In this case, the Hamilton equation tellls us how P may change in time:

$$\dot{P} = \dot{p} \equiv p'(t) = \{H, P\}.$$

Here, we are still in one dimension, and the dot stands for time derivative. (Later on, in higher dimension, p will stand for the vector of p_i's.)

For example, suppose we live inside a highly nonuniform curve (or river) of nonconstant curvature and density. From the inside, however, the curve seems as straight and uniform as ever. Fortunately, g tells us the missing information: say, g could increase with q: $g'(q) > 0$. This way, the Hamilton equation is a quadratic differential equation for $p(t)$, telling us how the momentum increases dynamically in time:

$$p'(t) = \{H, P\} = \frac{p^2 g'(q)}{2g^2(q)} > 0.$$

Still, to solve for p, we must also solve for q at the same time:

$$q'(t) = \{H, q\} = \nabla_p^t H = \frac{p}{g(q)}.$$

Together, these two equations could be solved for $q(t)$ and $p(t)$, and uncover the entire dynamics in one dimension.

22.3.2 Higher Dimension

Let's move on to the more interesting case of $n > 1$ (say, $n = 3$). In this case, it makes sense to introduce two new vectors:

$$q \equiv \begin{pmatrix} q^1 \\ q^2 \\ q^3 \\ \vdots \\ q^n \end{pmatrix} \quad \text{and} \quad p \equiv \begin{pmatrix} p_1 \\ p_2 \\ p_3 \\ \vdots \\ p_n \end{pmatrix}.$$

Furthermore, in higher dimension, the metric g (and its inverse g^{-1}) is no longer a scalar, but an $n \times n$ symmetric matrix, depending on the new vector q.

22.3.3 Hamilton System for The Momentum

Assume that our Hamiltonian is still kinetic:

$$H \equiv \frac{1}{2} p^t (g(q))^{-1} p.$$

Let's focus on the momentum in the kth coordinate:

$$P_k \equiv p_k,$$

for some fixed k. This way, its partial derivatives vanish, unless $j = k$:

$$\frac{\partial P_k}{\partial p_j} = \delta_k^j = \begin{cases} 1 & \text{if} \quad j = k \\ 0 & \text{if} \quad j \neq k. \end{cases}$$

We can now write the Hamilton equation for P_k:

$$\begin{aligned} \dot{P}_k &= \{H, P_k\} \\ &= -\nabla_q^t H \nabla_p P_k \\ &= -\frac{1}{2} p^t \left(\nabla_q^t \left(g^{-1} \right) \nabla_p P_k \right) p \\ &= \frac{1}{2} p^t \left(g^{-1} \nabla_q^t g \nabla_p P_k g^{-1} \right) p. \end{aligned}$$

(Here, the vector p is pulled out from H and thrown to the far right, because it participates neither in ∇_q nor in the inner product.) Better yet, do this for all $1 \leq k \leq n$ in one go. This will place columns like $\nabla_p P_k$ column by column, to form the identity matrix, and disappear. This will give a new vector equation, for the unknown vector p:

$$\dot{p} = \frac{1}{2} p^t \left(g^{-1} \left(\nabla_q g \right) g^{-1} \right) p.$$

This is a system of n coupled (quadratic) differential equations for the unknown momenta $p(t)$. Again, this can't be solved on its own, but only together with yet another system for $q(t)$:

$$\dot{q} = \{H, q\} = \nabla_p H = (g(q))^{-1} p.$$

Once these systems are solved together, the solutions $q(t)$ and $p(t)$ give us the entire dynamics, in higher dimension as well.

22.3.4 Conservation of Momentum: The Geodesic Equations

Still, there is one exception, in which there is only one system, not two. Indeed, what if we wanted the momentum to remain preserved, and never change? For this purpose, $\dot{p}$ must vanish. In other words, the Hamilton equation must be homogeneous:

$$\frac{1}{2}p^t \left(g^{-1}\left(\nabla_q g\right)g^{-1}\right)p = \dot{p} = 0.$$

Since p is assumed to be constant, it could be obtained from the initial state. Alternatively, one could use the other system to substitute $\dot{q}$ for $g^{-1}p$, and obtain

$$\dot{q}^t \left(\nabla_q g\right)\dot{q} = 0.$$

These are the geodesic equations: a system of n equations for the (unknown) "shortest" path $q \equiv q(t)$, along which the kinetic energy remains at its minimum.

22.4 Energy: Symmetry and Dynamics

22.4.1 Noether's Theorem: Translation Invariance

If the momentum is to remain constant, then our Hamiltonian system should be homoheneous:

$$-\nabla_q H = \{H, p\} = \dot{p} = 0.$$

In other words, $H \equiv H(p)$ depends on p only, not on q. Geometrically, this means that p shares the same level set with H. This way, it also shares the same normal, or direction of gradient. For this reason, the Hamiltonian vector $S\nabla H$ (tangent to the level set) is orthogonal to ∇p, and can never "push" it. As a result, the momentum remains conserved, as required.

Still, why not look at things the other way around? After all, thanks to anti-symmetry,

$$-\{p, H\} = \{H, p\} = 0.$$

Thus, $S\nabla p$ is also powerless, and can never "push" ∇H. What does this mean geometrically? Look at the right part of Figure 22.1: both H and p share the same (horizontal) level set, with the same tangent (pointing rightwards) and normal (pointing upwards). Thus, $S\nabla p$ points rightwards, along the q-axis, and has no effect on ∇H, which keeps pointing upwards. This way, H is translation invariant: it remains unchanged under any spatial shift in q.

This is indeed Noether's theorem. It applies not only to p but also to any other conserved quantity, giving us a new symmetry in the ($2n$-dimensional) q-p phase space. In fact, each new conserved quantity may introduce a new symmetry line, along which you could shift, leaving the Hamiltonian invariant. Later on, we'll come up with more examples.

22.4.2 Dynamic Evolution: Liouville Operator

Still, a general physical quantity u may be conserved or not. Indeed, in its general form, the Hamilton equation is not necessarily homogeneous:

$$\{H, u\} = \dot{u} \equiv \frac{du}{dt}(q(t), p(t)).$$

This is indeed a dynamic process: the dynamic variables $q \equiv q(t)$ and $p \equiv p(t)$ may change in time, affecting u indirectly. Thus, as time goes by, we move to a new state (q, p) in the phase space, with a new u defined there. This is indeed the dynamic change that can be observed experimentally.

How to solve for the unknown $u(q(t), p(t))$? For this, better drop u altogether, and write the adjoint:

$$\{H, \cdot\} = \frac{d}{dt}.$$

To integrate, write the Liouville operator:

$$L_H \equiv \exp(t\{H, \cdot\}).$$

Now, to the initial state $u_0(q, p)$ defined in the phase space at the initial time $t = 0$, apply the Liouville operator. This gives the evolution:

$$u(q(t), p(t)) = L_H(u_0)(q(0), p(0)) = \exp(t\{H, \cdot\})(u_0)(q(0), p(0)).$$

This will be the solution at a later time t. This will give the entire physical dynamics in terms of the observable u only, with no need to solve for $q(t)$ or $p(t)$ explicitly.

22.5 Angular Momentum and Its Conservation

22.5.1 Potential Energy

So far, the Hamiltonian contained a kinetic term only, as in a flat river, with no slope at all. But what about a river with a slight slope? In this case, thanks togravity, the potential is no longer constant. On the contrary: it may change, and produce more and more momentum, and enhance the flow downstream.

To help formulate this, let's use our position vector q, and its norm: $\|q\|$ (square root of sum of squares). This way, the Hamiltonian may contain not only kinetic energy (as in Section 22.3.2) but also potential:

$$H \equiv \frac{1}{2}p^t g^{-1}(q)p + V(q).$$

Let's focus on a practical application.

22.5.2 Kepler-Newton Gravity

Let's focus on three dimensions ($n = 3$): motion under Kepler-Newton's gravity laws. For this purpose, assume that our manifold is also Riemannian: the metric g is symmetric and positive definite. This way, it has n orthogonal eigenvectors, with positive eigenvalues. Define its square root: a new matrix $g^{1/2}$, with the same eigenvectors, but with new eigenvalues: the square root of those of g. We can now update our coordinate system, and substitute

$$p \leftarrow g^{-1/2}p.$$

This way, g is gone, and H simplifies to read

$$H \equiv \frac{\|p\|^2}{2} - \frac{K}{\|q\|},$$

where K is Kepler's constant, and q plays the same role as r in Chapter 2: the difference from the center of mass. We are now ready to prove an important law of nature: conservation of angular momentum.

22.5.3 Conservation of Angular Momentum

We can now study three physical quantities in one go: the three angular momenta, stored in one vector. In Chapter 2, this is denoted by $r \times p$. Here, on the other hand, r is replaced by q, so the angular momentum is $q \times p$.

How does the Hamiltonian act on it? To find out, let H act first on the three position coordinates in q:

$$\{H, q\} = \nabla_p H = p,$$

and also on the three momenta in p:

$$\{H, p\} = -\nabla_q H = -\frac{K}{\|q\|^3}q.$$

Thanks to Leibniz rule, H can now act on three new physical quantities – the angular momenta:

$$\{H, q \times p\} = \{H, q\} \times p + q \times \{H, p\} = p \times p - q \times \frac{K}{\|q\|^3}q = 0 - 0 = 0.$$

22.5.4 Conservation Law: Homogeneous Hamilton System

Thanks to this, the Hamilton equations become homogeneous:

$$\frac{d}{dt}(q(t) \times p(t)) = \{H, q \times p\} = \mathbf{0}.$$

This is what we wanted: $q \times p$ must remain constant in time. This is indeed conservation of angular momentum.

Is this enough to solve for both $q(t)$ and $p(t)$? Unfortunately not. We need three more equations, to have a big system of six equations in six dynamic variables.

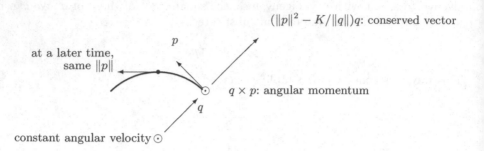

Fig. 22.2. The Earth orbits the sun at constant angular momentum and angular velocity vectors, pointing upwards, towards your eyes, as indicated by the '⊙.' On the orbit, the constant Hamiltonian H tells us the total energy. Thanks to Noether's theorem, the conserved quantities introduce symmetry lines, where H is invariant. This makes H angle invariant.

22.5.5 Hamiltonian and Its Symmetry

This is why the Earth orbits the sun at the same rate, or angular velocity. Thanks to Noether's theorem, in this orbit, the Hamiltonian also remains the same: it is angle invariant (or angle-symmetric). This is why you could shift angle, and observe no change. As a matter of fact, the symmetry is even stronger – spherical: you could shift angle in just any hypothetical orbit, and H wouldn't change. $q \times p$, on the other hand, is more sensitive: it would change direction.

In the right part of Figure 22.1, we've already seen a symmetry: on the horizontal line, both p and H are constant. They are translation invariant: you could shift in q, and notice no change.

In higher dimension, on the other hand, things are more subtle: each conserved quantity may have a slightly different level set. In Figure 22.2, for example, the level set of H is a sphere around the sun: both $\|q\|$, $\|p\|$, and H are constant there, as required. The angular momentum, on the other hand, has a different level set: the orbit only. Indeed, on any other orbit on the sphere, it would have a different direction. Next, to uncover more symmetries, we need a new conserved vector.

22.6 Conserved Quantities

22.6.1 Runge-Lenz Conserved Vector

Let's introduce three new physical quantities, stored in a new vector – the Runge-Lenz vector:

$$p \times (q \times p) - \frac{K}{\|q\|} q.$$

How does H act on it? First, let H act on the former term. Thanks to Leibniz rule,

$$\{H, p \times (q \times p)\} = \{H, p\} \times (q \times p) + p \times \{H, q \times p\}$$

$$= -\frac{K}{\|q\|^3} q \times (q \times p) + p \times \mathbf{0}$$

$$= -\frac{K}{\|q\|^3} q \times (q \times p).$$

Next, let H act on the latter term (with no minus sign):

$$\left\{H, \frac{K}{\|q\|} q\right\} = \frac{K}{\|q\|} \{H, q\} + Kq \left\{H, \frac{1}{\|q\|}\right\}$$

$$= \frac{K}{\|q\|} \nabla_p H + Kq \nabla_p^t H \nabla_q \left(\frac{1}{\|q\|}\right)$$

$$= \frac{K}{\|q\|} p + Kq \cdot p^t \nabla_q \left(\frac{1}{\|q\|}\right)$$

$$= \frac{K}{\|q\|^3} (q, q) p - Kq \cdot p^t \left(\frac{q}{\|q\|^3}\right)$$

$$= \frac{K}{\|q\|^3} ((q, q) p - (q, p) q).$$

Finally, subtract the latter from the former:

$$\left\{H, p \times (q \times p) - \frac{K}{\|q\|} q\right\} = -\frac{K}{\|q\|^3} q \times (q \times p) - \frac{K}{\|q\|^3} ((q, q) p - (q, p) q) = 0$$

(thanks to the exercises about triple vector product at the end of Chapter 2).

22.6.2 Complete System of Conservation Laws

As a result, the Hamilton equation becomes homogeneous, and makes a new conservation law for the Runge-Lenz vector:

$$\frac{d}{dt} \left(p \times (q \times p) - \frac{K}{\|q\|} q\right) = \left\{H, p \times (q \times p) - \frac{K}{\|q\|} q\right\} = \mathbf{0}.$$

Together with the three conservation laws for $q \times p$ in Section 22.5.4, we get a complete system of six equations for our six unknown dynamic variables: $q(t)$ and $p(t)$.

22.6.3 Conserved Quantity: Power Series in p

Let's go back to a more general case, with a more general potential V:

$$H = \frac{1}{2} p^t g^{-1} p + V(q).$$

Let's design V from scratch. For this purpose, let O be some conserved quantity, to be specified later. As such, it must obey the homogeneous Hamilton equation:

$$-\{O, H\} = \{H, O\} = \dot{O} = 0.$$

Now, how could O look like? Well, let's write it in powers of p, with coefficients that may depend on q:

$$O(q,p) = c(q) + \mathbf{c}^t(q)p + \frac{1}{2}p^t C(q)p + \cdots.$$

These (unknown) coefficients are of different types: c is a scalar, $\mathbf{c}$ (bold) is a vector, C is an $n \times n$ symmetric matrix, and so on. Let's design them cleverly, to obey the homogeneous Hamilton equation for any momentum p that could be introduced in practice later. On the way, we'll also make sure that the above power series is not really infinite, but terminates after a few terms, say three.

What conditions should the above coefficients satisfy? To find out, plug O in the homogeneous Hamilton equation, and make sure that all terms vanish, as required. This way, the equation will indeed be satisfied for any possible momentum. For a start, let's look at the free term that contains no p at all, and make sure that it indeed vanishes:

$$\begin{aligned}
0 &= \{O, H\} \\
&= \nabla_p^t O \cdot \nabla_q H - \nabla_q^t O \cdot \nabla_p H \\
&= \nabla_p^t O \cdot \nabla_q H - \nabla_p^t H \cdot \nabla_q O \\
&= (\mathbf{c} + Cp)^t \, \nabla_q V - p^t g^{-1} \left(\nabla_q c + \nabla_q \mathbf{c}^t p + \frac{1}{2}p^t \nabla_q Cp \right).
\end{aligned}$$

On the right-hand side, only the first term is free, and contains no p at all. Thus, it must vanish:

$$\mathbf{c}^t \nabla_q V = 0.$$

What is the physical meaning of this?

22.6.4 Cylindrical Angular Momentum

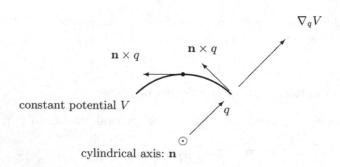

Fig. 22.3. Before the motion started, and before any momentum was introduced, the potential is already cylindrical: constant along a circle around $\mathbf{n}$, the cylindrical axis of symmetry.

What could $\mathbf{c}$ be in practice? Let's look at a familiar example: angular momentum, with g being the identity matrix. Unlike in Kepler-Newton gravity, where the angular momentum was three-dimensional, containing three momenta conserved on the orbit, here we assume a cylindrical angular momentum, conserved in one kind of orbit only: around a given unit vector $\mathbf{n}$. In other words, $q \times p$ is no longer conserved as a complete vector, but only its projection in the $\mathbf{n}$-direction:

$$(\mathbf{n}, q \times p) = \text{const.}$$

What is the left-hand side? This is a parallel prism, whose base is spanned by q and p. This way, the area of the base is $\|q \times p\|$, or $\|q\|$ times $\|p\|$ times the sine of the angle in between:

$$\triangle = \|q\| \cdot \|p\| \cdot \sin(q, p) = \|q \times p\|.$$

On top of this, the prism is also spanned by $\mathbf{n}$. To calculate its volume, we need to multiply by the height. Since $\|\mathbf{n}\| = 1$, the height of the prism is just the sine of the angle between $\mathbf{n}$ and the base:

$$\triangle \cdot \|\mathbf{n}\| \cos(\mathbf{n}, q \times p) = \triangle \cdot \cos(\mathbf{n}, q \times p)$$
$$= \|q \times p\| \cos(\mathbf{n}, q \times p)$$
$$= (\mathbf{n}, q \times p),$$

as asserted.

Still, we are not forced to pick the above base. Why not work the other way around: use $\mathbf{n}$ and q to span a new base, calculate its area, and use it to calculate the entire volume:

$$\|\mathbf{n} \times q\| \cdot \|p\| \cos(p, \mathbf{n} \times q) = (p, \mathbf{n} \times q).$$

What's so good about this approach? It gives the same volume as a linear combination of the p_i's. This could now be used to define the missing coefficient vector $\mathbf{c}$:

$$\mathbf{c} \equiv \mathbf{c}(q) \equiv \mathbf{n} \times q.$$

This gives us the missing physical information:

$$(\mathbf{n} \times q)^t \nabla_q V = \mathbf{c}^t \nabla_q V = 0.$$

22.6.5 Cylindrical Potential

Unlike in Kepler-Newton's gravity, our potential is no longer spherical, but cylindrical: it has a cylindrical symmetry, and mustn't change along a circle around $\mathbf{n}$ (Figure 22.3). This way, $\mathbf{n}$ makes a new cylindrical axis of symmetry. Around $\mathbf{n}$, neither O nor V can change.

By now, O already takes its final form: in the above expansion, only the second term survives – the linear term in p. We got all this information even before the motion started. In fact, the momentum will now have to adapt to our design. Indeed, in a motion around $\mathbf{n}$, p will have to keep tangent, and have a constant norm. This way, H will keep constant as well, as required. This way, the entire trajectory will be in the same level set not only of O and V but also of H, as predicted in Noether's theorem.

So far, we only focused on the free term in our homogeneous Hamilton equation, and made sure it vanished. What about the other terms? Fortunately, they could be forced to vanish as well. This will give us more physical information.

22.6.6 Killing Vector and Tensor

In the homogeneous Hamilton equation for the general O, let's collect the linear terms in p. To make sure that they vanish, we must force the following condition:

$$0 = (Cp)^t \nabla_q V - p^t g^{-1} \nabla_q c = p^t \left(C \nabla_q V - g^{-1} \nabla_q c \right).$$

But this should vanish for any p. To guarantee this, we must have

$$gC\nabla_q V = \nabla_q c.$$

Next, in the homogeneous Hamilton equation for O, let's look at the quadratic term in p. Like the previous terms, it should vanish for any p. To guarantee this, we must have

$$g^{-1} \nabla_q \mathbf{c}^t = (0).$$

To simplify this, multiply by g:

$$\nabla_q \mathbf{c}^t = (0).$$

A vector $\mathbf{c}$ that satisfies this is called a Killing vector.

Next, in the homogeneous Hamilton equation for O, let's look at the cubic term in p:

$$p^t g^{-1} \nabla_q \left(p^t C p \right) = 0.$$

Again, this should vanish for any p. Thus, you could drop the p, and obtain a tensor that should vanish. To simplify further, multiply by g:

$$\nabla_q C = (0).$$

A matrix C that satisfies this is called a Killing matrix.

Here, we stop after three terms only: O is quadratic in p. Still, we could continue yet more, and make O cubic in p. For this purpose, we could design a new (symmetric) tensor of three indices, to store the new coefficients (which may depend on q, but not on p). As before, it must satisfy a Killing condition: it must vanish under ∇_q. It is then called a Killing tensor. (Under ∇_p, on the other hand, this may produce a new quadratic term in p, which may change the above process a little.) This is Van-Holten's algorithm.

22.7 Exercises: The Poisson Brackets as Lie Brackets

22.7.1 The Poisson Brackets and The Jacobi identity

1. Recall that M is a $2n$-dimensional differentiable manifold, and that S is a $2n \times 2n$ antisymmetric matrix. Hint: see Sections 22.1.1–22.1.2.

2. Recall that, around each individual point $u \in M$, there is a spot $U \subset M$, in which Darboux coordinates guarantee that S is constant.
3. Place the partial derivatives (with respect to the q^i's and the p_i's) in a long $2n$-dimensional column vector:

$$\nabla \equiv \begin{pmatrix} \nabla_q \\ \nabla_p \end{pmatrix}.$$

4. Pick a real smooth function $f \in C^\infty(M)$.
5. What is ∇f? Hint: the gradient of f: a $2n$-dimensional column vector, containing the partial derivatives of f with respect to the q^i's and the p_i's:

$$\nabla f \equiv \begin{pmatrix} \nabla_q f \\ \nabla_p f \end{pmatrix}.$$

6. What is $\nabla^t f$? Hint: the transpose gradient of f: a $2n$-dimensional row vector, containing the partial derivatives of f with respect to the q^i's and the p_i's:

$$\nabla^t f \equiv (\nabla_q^t f, \nabla_p^t f).$$

7. What is $\nabla\nabla^t f$? Hint: the Hessian of f: a $2n \times 2n$ matrix, containing the second partial derivatives of f. It could be decomposed into four blocks:

$$\nabla\nabla^t f = \begin{pmatrix} \nabla_q \\ \nabla_p \end{pmatrix} (\nabla_q^t f, \nabla_p^t f) = \begin{pmatrix} \nabla_q \nabla_q^t f & \nabla_q \nabla_p^t f \\ \nabla_p \nabla_q^t f & \nabla_p \nabla_p^t f \end{pmatrix}.$$

8. In $\nabla\nabla^t f$, what are the off-diagonal elements? Hint: mixed partial derivatives.
9. Is $\nabla\nabla^t f$ symmetric? Hint: a mixed partial derivative is insensitive to the order in which the partial derivatives are taken.
10. Conclude that $S\nabla\nabla^t f S$ is symmetric as well.
11. Pick three smooth functions: $f, u, w \in C^\infty(M)$.
12. Write the Poisson brackets explicitly:

$$\{u, f\} \equiv \nabla^t f S \nabla u.$$

Hint: see Section 22.1.3.
13. Are the Poisson brackets antisymmetric? Hint: since S is antisymmetric,

$$\{u, f\} \equiv \nabla^t f S \nabla u = -\nabla^t u S \nabla f = -\{f, u\}.$$

14. Conclude that they are also alternative. Hint: $\{f, f\} = -\{f, f\} = 0$.
15. Write the triple Poisson brackets explicitly:

$$\{\{w, u\}, f\} = \nabla^t f S \nabla (\nabla^t u S \nabla w).$$

16. Use the antisymmetry of S and the Leibniz rule to write this yet more explicitly:

$$\begin{aligned} \{\{w, u\}, f\} &= \nabla^t f S \nabla (\nabla^t u S \nabla w) \\ &= \nabla^t f S \nabla \nabla^t u S \nabla w - \nabla^t f S \nabla \nabla^t w S \nabla u. \end{aligned}$$

17. Use the above to prove the Jacobi identity:

$$\{\{w,u\},f\} + \{\{f,w\},u\} + \{\{u,f\},w\}$$
$$= \nabla^t f S \nabla \nabla^t u S \nabla w - \nabla^t f S \nabla \nabla^t w S \nabla u$$
$$+ \nabla^t u S \nabla \nabla^t w S \nabla f - \nabla^t u S \nabla \nabla^t f S \nabla w$$
$$+ \nabla^t w S \nabla \nabla^t f S \nabla u - \nabla^t w S \nabla \nabla^t u S \nabla f$$
$$= 0.$$

Hint: in the right-hand side, the first term cancels with the last, the second with the third, and the fourth with the fifth.

Lie Algebras in Quantum Mechanics and Special Relativity

Thanks to Lie algebras, we can now look at quantum mechanics and special relativity from a new angle. In quantum mechanics, for example, we often use the Pauli matrices as observables. Here, however, we use them for yet another purpose: to span a new Lie algebra: su_2. For this purpose, however, they must first be multiplied by $i \equiv \sqrt{-1}$. This converts them from Hermitian to anti-Hermitian matrices. In this new form, they are ready to be exponentiated and *generate* a new Lie group: SU_2. This is useful to model two tricky observables: the spin of an electron, and the polarization of an electromagnetic wave. Furthermore, in special relativity, this helps model a Lorentz transformation from one system to another.

Thanks to the adjoint representation, su_2 can also take a new face: so_3. This is a new (isomorphic) Lie algebra, easier to understand geometrically. Like su_2, it contains anti-Hermitian matrices, ready for exponentiation.

Again, once exponentiated, so_3 *generates* a new Lie group: SO_3, which rotates the entire Cartesian space at a new angle, twice as big. This helps model the spin of a boson as well. In summary, thanks to Lie algebras, we have a transparent geometrical model to explain both spin and polarization at the same time.

To help introduce these new concepts, this chapter is arranged in exercises, followed by hints or complete solutions. This way, you get to see quite vividly how the theory develops from scratch, and how the new applications form before your very eyes.

23.1 Anti-Hermitian Matrix and Its Exponent

1. Let A be an anti-Hermitian matrix:

$$A^h = -A.$$

2. Conclude that A^h is anti-Hermitian as well. Hint:

$$\left(A^h\right)^h = A = -A^h.$$

3. Show that iA is Hermitian, where $i \equiv \sqrt{-1}$. Hint:

$$(iA)^h = \bar{i}A^h = (-i)(-A) = iA.$$

4. Conclude that iA has a diagonal form:

$$iA = ODO^h,$$

where O is unitary:

$$O^{-1} = O^h,$$

and D is real and diagonal:

$$D \equiv \operatorname{diag}(D_{1,1}, D_{2,2}, \ldots, D_{n,n}).$$

5. In this decomposition, what are the columns of O? Hint: the eigenvectors of iA.
6. What are their eigenvalues? Hint: the corresponding elements on the main diagonal of D.
7. Show that A has a diagonal form as well. Hint:

$$A = -i(iA) = -i\left(ODO^h\right) = O\left(-iD\right)O^h.$$

8. Likewise, show that A^h has a diagonal form as well. Hint:

$$A^h = -A = i(iA) = i\left(ODO^h\right) = O\left(iD\right)O^h.$$

9. Show that A has an exponent, with a diagonal form as well. Hint:

$$
\begin{aligned}
\exp(A) &= \sum_{j=0}^{\infty} \frac{A^j}{j!} \\
&= \sum_{j=0}^{\infty} \frac{\left(O(-iD)O^h\right)^j}{j!} \\
&= \sum_{j=0}^{\infty} \frac{O\left(-iD\right)^j O^h}{j!} \\
&= O\left(\sum_{j=0}^{\infty} \frac{(-iD)^j}{j!}\right) O^h \\
&= O\exp(-iD)O^h.
\end{aligned}
$$

10. In this triple product, write the middle matrix explicitly. Hint:

$$
\begin{aligned}
\exp(-iD) &= \sum_{j=0}^{\infty} \frac{(-iD)^j}{j!} \\
&= \sum_{j=0}^{\infty} \frac{\left(-i\operatorname{diag}(D_{1,1}, D_{2,2}, \ldots, D_{n,n})\right)^j}{j!} \\
&= \sum_{j=0}^{\infty} \frac{\operatorname{diag}\left((-iD_{1,1})^j, (-iD_{2,2})^j, \ldots, (-iD_{n,n})^j\right)}{j!} \\
&= \operatorname{diag}\left(\exp(-iD_{1,1}), \exp(-iD_{2,2}), \ldots, \exp(-iD_{n,n})\right).
\end{aligned}
$$

11. Conclude that this matrix is diagonal and unitary. Hint: the elements on the main diagonal have absolute value 1.

12. Conclude that $\exp(A)$ is unitary as well. Hint:

$$
\begin{aligned}
(\exp(A))^{-1} &= O(\exp(-iD))^{-1}O^h \\
&= O(\exp(-iD))^h O^h \\
&= (\exp(A))^h.
\end{aligned}
$$

13. Prove this in yet another way, using no diagonal form at all. Hint:

$$
\begin{aligned}
\exp(A)(\exp(A))^h &= \exp(A)\left(\sum_{j=0}^{\infty}\frac{A^j}{j!}\right)^h \\
&= \exp(A)\sum_{j=0}^{\infty}\frac{(A^h)^j}{j!} \\
&= \exp(A)\sum_{j=0}^{\infty}\frac{(-A)^j}{j!} \\
&= \exp(A)\exp(-A) \\
&= \exp(A-A) \\
&= \exp((0)) \\
&= I.
\end{aligned}
$$

23.2 Pauli Matrices

1. Recall the 2×2 Pauli matrices:

$$
\sigma_z \equiv \begin{pmatrix} 1 & 0 \\ 0 & -1 \end{pmatrix}
$$

$$
\sigma_x \equiv \begin{pmatrix} 0 & 1 \\ 1 & 0 \end{pmatrix}
$$

$$
\sigma_y \equiv \begin{pmatrix} 0 & -i \\ i & 0 \end{pmatrix}.
$$

2. Are they Hermitian?
3. Are they trace-free?
4. What is their determinant?
5. Multiply them by a new parameter θ, and calculate the exponent:

$$
\begin{aligned}
\exp(\theta\sigma_z) &= \exp\left(\begin{pmatrix} \theta & \\ & -\theta \end{pmatrix}\right) \\
&= \sum_{n=0}^{\infty}\frac{1}{n!}\begin{pmatrix} \theta & \\ & -\theta \end{pmatrix}^n \\
&= \sum_{n=0}^{\infty}\frac{1}{n!}\begin{pmatrix} \theta^n & \\ & (-\theta)^n \end{pmatrix}
\end{aligned}
$$

$$= \begin{pmatrix} \exp(\theta) & \\ & \exp(-\theta) \end{pmatrix}.$$

6. What is the determinant of this exponent?
7. Is this exponent unitary? Hint: are the columns orthonormal?
8. To make sure that it is, substitute $\theta \leftarrow i\eta$, where $i \equiv \sqrt{-1}$, and η is a new parameter.
9. If η is real, is the above exponent unitary?
10. Conclude that this exponent is a special unitary matrix: a unitary matrix of determinant 1.
11. In view of the above, should we have taken the exponent of $i\eta\sigma_z$ or $\eta\sigma_z$? Hint: $i\eta\sigma_z$ is better, because it is anti-Hermitian.

23.3 Lorentz Matrix in Special Relativity

1. Next, look at σ_x. Is it Hermitian?
2. Is it also anti-Hermitian? In what sense? Hint: see below.
3. To answer this, design an orthogonal Lie algebra: D_1. For this purpose, in Chapter 15, Section 15.6, set

$$S \equiv \begin{pmatrix} -1 & \\ & 1 \end{pmatrix}.$$

4. Show that σ_x is anti-Hermitian with respect to S:

$$S\sigma_x = -\sigma_x S = -\sigma_x^t S.$$

5. What is the physical meaning of S? Hint: in special relativity, the first coordinate t picks a minus sign (Figures 10.6 and 11.10).
6. Define the hyperbolic cosine:

$$\cosh(\theta) \equiv \frac{\exp(\theta) + \exp(-\theta)}{2}.$$

7. Write it in terms of even powers of θ:

$$\cosh(\theta) \equiv \frac{\exp(\theta) + \exp(-\theta)}{2}$$
$$= \frac{1}{2} \left(\sum_{n=0}^{\infty} \frac{\theta^n}{n!} + \sum_{n=0}^{\infty} (-1)^n \frac{\theta^n}{n!} \right)$$
$$= \sum_{n=0}^{\infty} \frac{\theta^{2n}}{(2n)!}.$$

Hint: see exercises at the end of Chapter 24.

8. How is the hyperbolic cosine related to the cosine? Hint: substitute $\theta \leftarrow i\eta$:

$$\cosh(i\eta) \equiv \frac{\exp(i\eta) + \exp(-i\eta)}{2} = \cos(\eta).$$

9. Define also the hyperbolic sine:

$$\sinh(\theta) \equiv \frac{\exp(\theta) - \exp(-\theta)}{2}.$$

10. Write it in terms of odd powers of θ:

$$\sinh(\theta) \equiv \frac{\exp(\theta) - \exp(-\theta)}{2}$$

$$= \frac{1}{2} \left(\sum_{n=0}^{\infty} \frac{\theta^n}{n!} - \sum_{n=0}^{\infty} (-1)^n \frac{\theta^n}{n!} \right)$$

$$= \sum_{n=0}^{\infty} \frac{\theta^{2n+1}}{(2n+1)!}.$$

11. How is the hyperbolic sine related to the sine? Hint: substitute $\theta \leftarrow i\eta$:

$$\sinh(i\eta) \equiv \frac{\exp(i\eta) - \exp(-i\eta)}{2} = i \sin(\eta).$$

12. Multiply σ_x by θ. What is the exponent? Hint:

$$\exp(\theta \sigma_x) = \exp \left(\begin{pmatrix} & \theta \\ \theta & \end{pmatrix} \right)$$

$$= \sum_{n=0}^{\infty} \frac{1}{n!} \begin{pmatrix} & \theta \\ \theta & \end{pmatrix}^n$$

$$= \sum_{n=0}^{\infty} \frac{1}{(2n)!} \begin{pmatrix} & \theta \\ \theta & \end{pmatrix}^{2n} + \sum_{n=0}^{\infty} \frac{1}{(2n+1)!} \begin{pmatrix} & \theta \\ \theta & \end{pmatrix}^{2n+1}$$

$$= \sum_{n=0}^{\infty} \frac{1}{(2n)!} \begin{pmatrix} \theta^{2n} & \\ & \theta^{2n} \end{pmatrix} + \sum_{n=0}^{\infty} \frac{1}{(2n+1)!} \begin{pmatrix} & \theta^{2n+1} \\ \theta^{2n+1} & \end{pmatrix}$$

$$= \begin{pmatrix} \cosh(\theta) & \\ & \cosh(\theta) \end{pmatrix} + \begin{pmatrix} & \sinh(\theta) \\ \sinh(\theta) & \end{pmatrix}$$

$$= \begin{pmatrix} \cosh(\theta) & \sinh(\theta) \\ \sinh(\theta) & \cosh(\theta) \end{pmatrix}.$$

13. Is this familiar? Hint: this is a Lorentz matrix (Chapter 10, Sections 10.4.2–10.4.3).
14. Verify that it indeed has determinant 1.
15. Here, θ may be real: there is no need to use η. Why? Hint: here, σ_x is already anti-Hermitian with respect to S, so there is no need to insert the coefficient i before taking the exponent.

23.4 Rotation Matrix

1. Next, look at σ_y. As before, multiply it by θ. What is the exponent? Hint:

$$\exp\left(\theta\sigma_y\right) = \exp\left(\begin{pmatrix} & -i\theta \\ i\theta & \end{pmatrix}\right)$$

$$= \sum_{n=0}^{\infty} \frac{1}{n!} \begin{pmatrix} & -i\theta \\ i\theta & \end{pmatrix}^n$$

$$= \sum_{n=0}^{\infty} \frac{1}{(2n)!} \begin{pmatrix} & -i\theta \\ i\theta & \end{pmatrix}^{2n} + \sum_{n=0}^{\infty} \frac{1}{(2n+1)!} \begin{pmatrix} & -i\theta \\ i\theta & \end{pmatrix}^{2n+1}$$

$$= \sum_{n=0}^{\infty} \frac{1}{(2n)!} \begin{pmatrix} \theta^{2n} & \\ & \theta^{2n} \end{pmatrix} + \sum_{n=0}^{\infty} \frac{1}{(2n+1)!} \begin{pmatrix} & -i\theta^{2n+1} \\ i\theta^{2n+1} & \end{pmatrix}$$

$$= \begin{pmatrix} \cosh(\theta) & \\ & \cosh(\theta) \end{pmatrix} + \begin{pmatrix} & -i\sinh(\theta) \\ i\sinh(\theta) & \end{pmatrix}$$

$$= \begin{pmatrix} \cosh(\theta) & -i\sinh(\theta) \\ i\sinh(\theta) & \cosh(\theta) \end{pmatrix}.$$

2. How to make this more familiar? Hint: substitute $\theta = i\eta$:

$$\exp\left(i\eta\sigma_y\right) = \exp\left(\theta\sigma_y\right)$$

$$= \begin{pmatrix} \cosh(\theta) & -i\sinh(\theta) \\ i\sinh(\theta) & \cosh(\theta) \end{pmatrix}$$

$$= \begin{pmatrix} \cosh(i\eta) & -i\sinh(i\eta) \\ i\sinh(i\eta) & \cosh(i\eta) \end{pmatrix}$$

$$= \begin{pmatrix} \cos(\eta) & \sin(\eta) \\ -\sin(\eta) & \cos(\eta) \end{pmatrix}.$$

3. Is this familiar? Hint: for a real η, this is a rotation matrix: it rotates the entire x-y plane by angle η clockwise (Figure 2.12).

4. Verify that it indeed has determinant 1.

5. In view of the above, should we have taken the exponent of $i\eta\sigma_y$ or $\eta\sigma_y$? Hint: $i\eta\sigma_y$ is better, because it is anti-Hermitian.

23.5 The Special Unitary Lie Group

1. Look at the matrices $i\sigma_x$, $i\sigma_y$, and $i\sigma_z$. Are they anti-Hermitian?

2. What is their commutator? Hint: it is cyclic:

$$[i\sigma_x, i\sigma_y] = -2i\sigma_z$$
$$[i\sigma_y, i\sigma_z] = -2i\sigma_x$$
$$[i\sigma_z, i\sigma_x] = -2i\sigma_y.$$

3. The anti-Hermitian matrices $i\sigma_x$, $i\sigma_y$, and $i\sigma_z$ span a new Lie algebra, denoted by su_2 (with real coefficients only).

4. Show that su_2 is simple. Hint: use the above cyclic commutator, and follow the proof for sl_2 in Chapter 17, Section 17.6.

5. Moreover, show that su_2 is the same as its derived algebra:

$$su_2 = [su_2, su_2].$$

6. Conclude that su_2 must be trace-free. Hint: a commutator must be trace-free (Chapter 15, Section 15.3).

7. Check on this. Hint: verify that $i\sigma_x$, $i\sigma_y$, and $i\sigma_z$ are indeed trace-free. Extend this linearly.

8. Conclude that its adjoint is trace-free as well. Hint: see Chapter 17, Section 17.4.

9. Can you verify this more explicitly? Hint: see below.

10. Look at $i\sigma_x$, $i\sigma_y$, and $i\sigma_z$ once again. Upon taking their exponent, these matrices *generate* the new group SU_2, containing the 2×2 *special unitary* matrices: unitary matrices of determinant 1.

11. For this purpose, they use three real parameters (angles). This is why they will be so useful to model the spin of an electron.

12. Mirror these matrices by standard unit vectors in the three-dimensional Cartesian space:

$$i\sigma_x \sim \begin{pmatrix} 1 \\ 0 \\ 0 \end{pmatrix}$$

$$i\sigma_y \sim \begin{pmatrix} 0 \\ 1 \\ 0 \end{pmatrix}$$

$$i\sigma_z \sim \begin{pmatrix} 0 \\ 0 \\ 1 \end{pmatrix}.$$

13. Thanks to this basis, the adjoint representation could be interpreted in terms of 3×3 anti-Hermitian matrices, or new vector-product operators:

$$[i\sigma_x, \cdot] \sim \begin{pmatrix} 0 & 0 & 0 \\ 0 & 0 & 2 \\ 0 & -2 & 0 \end{pmatrix} = -2 \begin{pmatrix} 1 \\ 0 \\ 0 \end{pmatrix} \times$$

$$[i\sigma_y, \cdot] \sim \begin{pmatrix} 0 & 0 & -2 \\ 0 & 0 & 0 \\ 2 & 0 & 0 \end{pmatrix} = -2 \begin{pmatrix} 0 \\ 1 \\ 0 \end{pmatrix} \times$$

$$[i\sigma_z, \cdot] \sim \begin{pmatrix} 0 & 2 & 0 \\ -2 & 0 & 0 \\ 0 & 0 & 0 \end{pmatrix} = -2 \begin{pmatrix} 0 \\ 0 \\ 1 \end{pmatrix} \times.$$

14. What is the physical meaning of these vector-product operators? Hint: rotation around the x-, y-, or z-axis (Figures 22.2–22.3).

15. Are these matrices familiar?

16. How are they related to the 3×3 matrices S_x, S_y, and S_z, defined right after Figures 12.10–12.11 to model the spin of a boson? Hint:

$$\hbar\,[i\sigma_x, \cdot] = 2iS_x$$
$$\hbar\,[i\sigma_y, \cdot] = 2iS_y$$
$$\hbar\,[i\sigma_z, \cdot] = 2iS_z.$$

17. From the above representation, it follows that iS_x, iS_y, and iS_z are trace-free and anti-Hermitian. Is this indeed true?

23.6 Polarization of a Photon

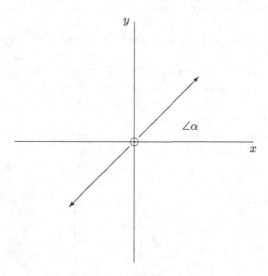

Fig. 23.1. A linearly-polarized electromagnetic wave: while propagating upwards, toward your eyes, it also oscillates in the x-y plane, along the oblique line that makes angle α with the positive part of the x-axis.

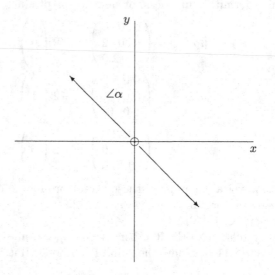

Fig. 23.2. Another linearly-polarized electromagnetic wave, orthogonal to the previous one: it makes angle α with the positive part of the y-axis.

1. How is the spin of the electron related to the spin of the boson? Hint: one represents the other:

$$\bar{h}\,[i\sigma_x, \cdot] = 2iS_x$$
$$\bar{h}\,[i\sigma_y, \cdot] = 2iS_y$$
$$\bar{h}\,[i\sigma_z, \cdot] = 2iS_z.$$

2. On the right-hand side, look at the coefficient 2. What is its physical meaning? Hint: upon taking the exponent, the angle doubles (see below).
3. In $\exp(i\eta\sigma_x)$, for example, the angle is η. In $\exp(2i\eta S_x/\bar{h})$, on the other hand, the angle is twice as big: 2η.
4. Use the double angle to model (nondeterministic) polarization in quantum mechanics.
5. For this purpose, consider an electromagnetic wave (or a photon), flying upwards (in the positive z-direction) at the speed of light.
6. At the same time, the wave also moves (or oscillates) forward and backward along an oblique line in the x-y plane. We then say that the wave is polarized linearly (Figure 23.1).
7. In this case, the x-oscillation is aligned with the y-oscillation: there is no phase-shift.
8. Now, design a polarizer: a machine that takes a given light ray (polarized at some unknown angle), and tells us how likely it is that this angle was α (where α is known: the angle of the polarizer). This is done by looking at the light that comes out, and measuring how weak it became. In the process, the light also changes forever: it becomes polarized at the new angle α rather than its original angle.
9. How does the polarizer work mathematically? Hint: see below.
10. The polarizer takes the form of a new observable. To design it, use the double angle:

$$\cos(2\alpha)\sigma_z + \sin(2\alpha)\sigma_x = \begin{pmatrix} \cos(2\alpha) & \sin(2\alpha) \\ \sin(2\alpha) & -\cos(2\alpha) \end{pmatrix}.$$

11. Why do we use the double angle 2α? Hint: the above addition actually takes place in the three-dimensional Cartesian space. Indeed, as discussed above, σ_z, σ_x, and σ_y are mirrored by

$$\sigma_z \sim \begin{pmatrix} 0 \\ 0 \\ 1 \end{pmatrix}$$

$$\sigma_x \sim \begin{pmatrix} 1 \\ 0 \\ 0 \end{pmatrix}$$

$$\sigma_y \sim \begin{pmatrix} 0 \\ 1 \\ 0 \end{pmatrix}.$$

Now, in three dimensions, the angle should be twice as big.
12. Is this a legitimate observable? Hint: is it Hermitian?

13. Does it have an eigenvector of eigenvalue 1? Hint: use a column vector that models a polarization at angle α (Figure 23.1):

$$\begin{pmatrix} \cos(2\alpha) & \sin(2\alpha) \\ \sin(2\alpha) & -\cos(2\alpha) \end{pmatrix} \begin{pmatrix} \cos(\alpha) \\ \sin(\alpha) \end{pmatrix}$$

$$= \begin{pmatrix} \cos^2(\alpha) - \sin^2(\alpha) & 2\sin(\alpha)\cos(\alpha) \\ 2\sin(\alpha)\cos(\alpha) & \sin^2(\alpha) - \cos^2(\alpha) \end{pmatrix} \begin{pmatrix} \cos(\alpha) \\ \sin(\alpha) \end{pmatrix}$$

$$= \begin{pmatrix} \cos(\alpha) \\ \sin(\alpha) \end{pmatrix}.$$

14. As a Hermitian matrix, this observable must also have yet another eigenvector, orthogonal to the previous one (Figure 23.2). What is its eigenvalue? Hint: -1:

$$\begin{pmatrix} \cos(2\alpha) & \sin(2\alpha) \\ \sin(2\alpha) & -\cos(2\alpha) \end{pmatrix} \begin{pmatrix} -\sin(\alpha) \\ \cos(\alpha) \end{pmatrix}$$

$$= \begin{pmatrix} \cos^2(\alpha) - \sin^2(\alpha) & 2\sin(\alpha)\cos(\alpha) \\ 2\sin(\alpha)\cos(\alpha) & \sin^2(\alpha) - \cos^2(\alpha) \end{pmatrix} \begin{pmatrix} -\sin(\alpha) \\ \cos(\alpha) \end{pmatrix}$$

$$= \begin{pmatrix} \sin(\alpha) \\ -\cos(\alpha) \end{pmatrix}$$

$$= -\begin{pmatrix} -\sin(\alpha) \\ \cos(\alpha) \end{pmatrix}.$$

15. What is the physical meaning of this? Hint: this random variable helps us guess how a given photon could look like: it could either be polarized at angle α (eigenvalue 1) or $\alpha + \pi/2$ (eigenvalue -1).

16. Before any experiment has been carried out, how likely is the photon to be polarized at angle α? Hint: this depends on its (normalized) state: a two-dimensional complex vector v. Take the inner product of v with the former eigenvector. Then, calculate the absolute value, and square it up:

$$\text{probability to be polarized at angle } \alpha: \quad \left| \frac{v^t}{\|v\|} \begin{pmatrix} \cos(\alpha) \\ \sin(\alpha) \end{pmatrix} \right|^2.$$

17. On the other hand, how likely is the photon to be polarized at angle $\alpha + \pi/2$? Hint: in the above calculation, take the inner product of the (normalized) state v with the latter eigenvector, not the former:

$$\text{probability to be polarized at angle } \alpha + \pi/2: \quad \left| \frac{v^t}{\|v\|} \begin{pmatrix} -\sin(\alpha) \\ \cos(\alpha) \end{pmatrix} \right|^2.$$

18. For example, assume that the photon was polarized at angle η. In this case, what is its state? Hint: in this case, v makes angle η with the positive part of the x-axis:

$$v = \begin{pmatrix} \cos(\eta) \\ \sin(\eta) \end{pmatrix}.$$

19. In this case, how likely is the photon to pass the polarizer? Hint: in this case,

$$\left| \frac{v^t}{\|v\|} \begin{pmatrix} \cos(\alpha) \\ \sin(\alpha) \end{pmatrix} \right|^2 = \left| (\cos(\eta), \sin(\eta)) \begin{pmatrix} \cos(\alpha) \\ \sin(\alpha) \end{pmatrix} \right|^2$$

$$= |\cos(\eta)\cos(\alpha) + \sin(\eta)\sin(\alpha)|^2$$

$$= \cos^2(\eta - \alpha)$$

$$= \cos^2(\alpha - \eta).$$

20. Design an experiment to observe this probability visually. Hint: design a beam of photons, all polarized at the same (unknown) angle η. Let the entire beam pass through the polarizer. The probability that η was the same as α is still calculated as before:

probability that η was the same as α: $\left| (\cos(\eta), \sin(\eta)) \begin{pmatrix} \cos(\alpha) \\ \sin(\alpha) \end{pmatrix} \right|^2$

$$= \cos^2(\alpha - \eta).$$

Thus, upon passing through the polarizer, the intensity of the light should fall by this probability. Only if $\eta = \alpha$ (the probability was 1) should the light pass in its entirety (Figure 23.1). On the other hand, only if $\eta = \alpha \pm \pi/2$ (the probability was 0) should no light pass at all (Figure 23.2). In all other (intermediate) cases, the intensity should fall somewhat, and be proportional to the above probability.

21. In summary, what is the only relevant angle? Hint: $\alpha - \eta$: the angle between the direction of the polarizer and the direction in which the photon was polarized originally.

22. In what case must the photon pass? Hint: if $\eta = \alpha$ (probability 1).

23. In what case can't the photon pass? Hint: if $\eta = \alpha \pm \pi/2$ (probability 0).

24. Consider now two polarizers: the first one is of angle α, as in Figure 23.1. Behind it, place yet another polarizer, of angle $\alpha + \pi/2$, as in Figure 23.2. Now, let a light ray pass through both of them, one by one. What comes out? Hint: no light comes out.

25. Now, in between, place a new polarizer, of angle $\alpha + \pi/4$. What comes out? Hint: after passing through the first polarizer, the light gets polarized at angle α, and its intensity falls somewhat. After passing through the second one, the light gets polarized at angle $\alpha + \pi/4$, and its intensity falls by $\cos^2(\pi/4) = 1/2$. Finally, after passing through the third one, the light gets polarized at angle $\alpha + \pi/2$, and its intensity falls by $\cos^2(\pi/4) = 1/2$ once again.

23.7 Phaseshift

1. For simplicity, define

$$\theta \equiv 2\alpha.$$

2. Modify your polarizer to detect a slightly different polarization: not quite linear, but also making a narrow ellipse in the horizontal x-y plane. Hint: introduce a small angle ϕ to design a new observable:

$$\cos(\theta)\sigma_z + \sin(\theta) \left(\cos(\phi)\sigma_x + \sin(\phi)\sigma_y \right).$$

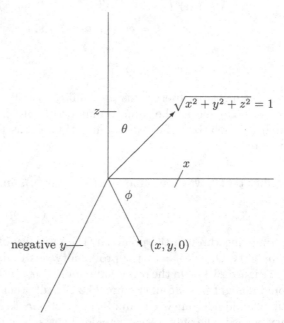

Fig. 23.3. A unit vector (x, y, z). It makes angle θ with the vertical z-axis. It also makes a "shadow" $(x, y, 0)$ on the horizontal x-y plane. This shadow makes angle ϕ with the positive part of the x-axis.

3. In this sum, why do we use the double angle $\theta = 2\alpha$? Hint: the above addition actually takes place in the three-dimensional Cartesian space, with σ_z, σ_x, and σ_y being mirrored by the z-, x-, and y-axes, respectively (Figure 23.3). As discussed above, in three dimensions, the angle is twice as big. Later on, we'll see this more explicitly in Poincare sphere.

4. Is this a legitimate observable? Hint: is it Hermitian?

5. What is the geometrical meaning of this? Hint: in the new polarizer, there is now a new phaseshift: the y-polarization is now at angle ϕ ahead of the x-polarization.

6. Show this in the eigenvector. Hint: the eigenvector is

$$\begin{pmatrix} \cos(\alpha) \\ \sin(\alpha)\exp(i\phi) \end{pmatrix}.$$

Indeed, for $\phi = 0$ (no phaseshift at all), we've already seen this before. Thus, for a nonzero ϕ too,

$$(\cos(\theta)\sigma_z + \sin(\theta)(\cos(\phi)\sigma_x + \sin(\phi)\sigma_y)) \begin{pmatrix} \cos(\alpha) \\ \sin(\alpha)\exp(i\phi) \end{pmatrix}$$

$$= \begin{pmatrix} \cos(\theta) & \sin(\theta)\exp(-i\phi) \\ \sin(\theta)\exp(i\phi) & -\cos(\theta) \end{pmatrix} \begin{pmatrix} \cos(\alpha) \\ \sin(\alpha)\exp(i\phi) \end{pmatrix}$$

$$= \begin{pmatrix} \cos(\alpha) \\ \exp(i\phi)\sin(\alpha) \end{pmatrix}.$$

7. What is the physical meaning of this? Hint: in our polarizer, $\cos(\alpha)$ is the x-polarization, whereas $\sin(\alpha)$ is the y-polarization, at a phaseshift of angle ϕ ahead.

8. While propagating upwards (in the positive z-direction), how does such a wave rotate? Hint: counterclockwise, along the narrow ellipse, in the horizontal x-y plane.

9. What is the orthogonal eigenvector? Hint:

$$\begin{pmatrix} -\sin(\alpha) \\ \cos(\alpha)\exp(i\phi) \end{pmatrix}.$$

Indeed, for $\phi = 0$ (no phaseshift at all), we've already seen this.

10. What is its eigenvalue? Hint: -1. Indeed,

$$(\cos(\theta)\sigma_z + \sin(\theta)(\cos(\phi)\sigma_x + \sin(\phi)\sigma_y)) \begin{pmatrix} -\sin(\alpha) \\ \cos(\alpha)\exp(i\phi) \end{pmatrix}$$

$$= \begin{pmatrix} \cos(\theta) & \sin(\theta)\exp(-i\phi) \\ \sin(\theta)\exp(i\phi) & -\cos(\theta) \end{pmatrix} \begin{pmatrix} -\sin(\alpha) \\ \cos(\alpha)\exp(i\phi) \end{pmatrix}$$

$$= -\begin{pmatrix} -\sin(\alpha) \\ \exp(i\phi)\cos(\alpha) \end{pmatrix}.$$

11. What is the physical meaning of this? Hint: in our polarizer, $-\sin(\alpha)$ is the x-polarization, whereas $\cos(\alpha)$ is the y-polarization, at a phaseshift of angle ϕ ahead.

12. While propagating upwards (in the positive z-direction), how does such a wave rotate? Hint: counterclockwise, along the narrow ellipse, in the horizontal x-y plane.

13. How to make the rotation clockwise rather than counterclockwise? Hint: replace $\phi \leftarrow -\phi$.

14. Why do we use the double angle $\theta = 2\alpha$? Hint: to design the above observable, the addition actually takes place in the three-dimensional Cartesian space, with σ_z, σ_x, and σ_y being mirrored by the z-, x-, and y-axes, respectively. As discussed above, in three dimensions, the angle is twice as big.

15. Let's see this not only algebraically but also geometrically.

23.8 Poincare Sphere

1. Indeed, this is illustrated in Poincare sphere. This is a sphere of radius 1.

2. Consider a three-dimensional vector of norm 1. Write it in terms of its spherical coordinates. Hint: in Figure 23.3, the point (x, y, z) could be written as

$$(x, y, z) = (\sin(\theta)\cos(\phi), \sin(\theta)\sin(\phi), \cos(\theta)).$$

3. What is the angle between (x, y, z) and the positive part of the z-axis? Hint: θ.

4. What is the range of θ? Hint: $0 \leq \theta \leq \pi$.

5. What "shadow" does (x, y, z) make on the horizontal x-y plane? Hint: $(x, y, 0)$.

6. What is the length of the shadow? Hint: $\sin(\theta)$.

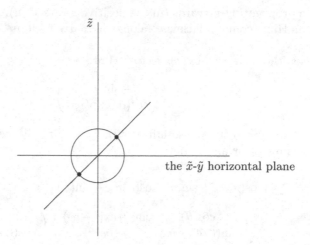

Fig. 23.4. The Poincare sphere: a view from the side. It uses new coordinates: $\tilde{x} \equiv z$, $\tilde{y} \equiv x$, and $\tilde{z} \equiv y$. On the sphere, we mark one point, at the top-right. It has an antipodal point, at the bottom-left.

7. What angle does the shadow make with the positive part of the x-axis? Hint: ϕ.

8. What is the range of ϕ? Hint: $-\pi < \phi \le \pi$.

9. Now, go back to SU_2. What happens to the angles? Hint: instead of θ, use just half of it: $\alpha = \theta/2$. The other angle ϕ, on the other hand, remains the same.

10. What is the associated eigenvector? Hint: we've already seen it in Section 23.7:

$$\begin{pmatrix} \cos(\alpha) \\ \sin(\alpha)\exp(i\phi) \end{pmatrix}.$$

11. What is its physical meaning? Hint: the polarization is no longer linear: there is a phaseshift of ϕ. As a result, as the wave propagates, it makes ellipses counterclockwise in the perpendicular plane.

12. The above point has an antipodal. Where is it located? Hint: symmetrically, on the other side of the sphere (Figure 23.4).

13. How to write the antipodal? Hint: just insert a minus sign.

14. What are its spherical coordinates (angles)? Hint: $\pi - \theta$ and $\phi \pm \pi$:

$$-(\sin(\theta)\cos(\phi), \sin(\theta)\sin(\phi), \cos(\theta))$$
$$= (-\sin(\theta)\cos(\phi), -\sin(\theta)\sin(\phi), -\cos(\theta))$$
$$= (\sin(\pi - \theta)\cos(\phi \pm \pi), \sin(\pi - \theta)\sin(\phi \pm \pi), \cos(\pi - \theta)).$$

15. Now, go back to SU_2. What happens to the angles? Hint: instead of $\pi - \theta$, use just half of it:

$$\frac{\pi - \theta}{2} = \frac{\pi}{2} - \frac{\theta}{2} = \frac{\pi}{2} - \alpha.$$

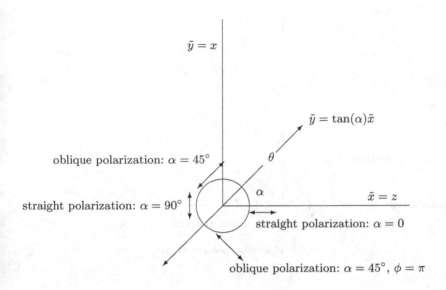

Fig. 23.5. The equator of the Poincare sphere: a view from above. At the equator, $\tilde{z} = 0$, so $\tilde{x}^2 + \tilde{y}^2 = 1$. In terms of the old x-y-z coordinates, on the other hand, this is a longitude: $y = 0$, and $z^2 + x^2 = 1$. Thus, there is no phaseshift at all: $\phi = 0$ or π. The polarization is therefore linear: forward and backward, along the oblique line $\tilde{y} = \pm \tan(\alpha)\tilde{x}$, where $\alpha = \theta/2$, and θ is the angle from the z-axis (or the $\tilde{x}$-axis).

The other angle $\phi \pm \pi$, on the other hand, remains the same.

16. What is the associated eigenvector? Hint: we've already seen it in Section 23.7:
$$\begin{pmatrix} \cos\left(\frac{\pi}{2} - \alpha\right) \\ \sin\left(\frac{\pi}{2} - \alpha\right) \exp(i(\phi \pm \pi)) \end{pmatrix} = \begin{pmatrix} \sin(\alpha) \\ -\cos(\alpha) \exp(i\phi) \end{pmatrix}.$$

17. What is its physical meaning? Hint: the polarization is no longer linear: there is a phaseshift of $\phi \pm \pi$. As a result, as the wave propagates, it makes ellipses in the perpendicular plane.

18. How are these eigenvectors related to each other? Hint: they are orthogonal to each other.

19. What is the physical meaning of this? Hint: they never interact or interfere with each other.

20. On the Poincare sphere (Figure 23.4), what is the relative angle between the original point and its antipodal? Hint: $\theta + (\pi - \theta) = \pi$.

21. In SU_2, on the other hand, what happens to this angle? Hint: it is halved to $\pi/2$. Indeed,
$$\alpha + \left(\frac{\pi}{2} - \alpha\right) = \frac{\pi}{2}.$$

22. What does this mean physically? Hint: thanks to orthogonality, these two kinds of polarization never interact or interfere with each other, but just pass through one another with no interruption whatsoever.

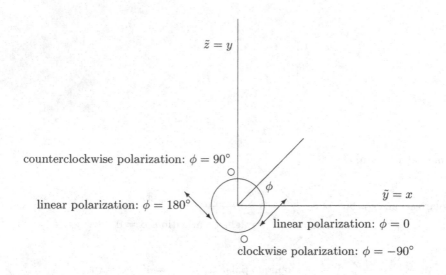

Fig. 23.6. A longitude in the Poincare sphere: a view from the right. On it, $\tilde{x} = 0$, so $\tilde{y}^2 + \tilde{z}^2 = 1$. In terms of the old x-y-z coordinates, on the other hand, this is the equator: $z = 0$, and $x^2 + y^2 = 1$. On it, the phaseshift changes from $\phi = 0$ on the right, to $\pi/2$ at the top, to π on the left, and to $-\pi/2$ at the bottom.

23. For instance, in Poincare sphere, look at the equator. Hint: note that the Poincare sphere uses new (shifted) coordinates: $\tilde{x} \equiv z$, $\tilde{y} \equiv x$, and $\tilde{z} \equiv y$ (Figure 23.4).

24. In the Poincare sphere, do the new $\tilde{x}$-$\tilde{y}$-$\tilde{z}$-axes make a right-hand system?

25. Indeed, could they be obtained by rotating the original x-y-z-axes?

26. Indeed, could this be done by just two planar (cylindrical) rotations? Hint: in the exercises at the end of Chapter 2, use just two (right) Euler angles. (The third angle is just zero.)

27. Now, on the equator of the Poincare sphere, look at the original point. What is its physical meaning? Hint: on the equator, $\tilde{z} = y = 0$, so $\phi = 0$ or π. This means no phaseshift: linear polarization, as in Figure 23.5.

28. Where is the antipodal? Hint: on the equator too, on the other side, across from the original point.

29. What does this mean physically? Hint: the original point stands for linear polarization at angle α (Figure 23.1). Its antipodal, on the other hand, stands for the orthogonal wave: linear polarization at angle $\alpha + \pi/2$ (Figure 23.2).

30. If two such waves meet, what happens? Hint: nothing: thanks to orthogonality, they pass through one another smoothly, and continue.

31. In particular, look at the rightmost point in Poincare sphere: $\tilde{x} = 1$. What happens there? Hint: here, $z = \tilde{x} = 1$, so $\alpha = \theta = 0$. This means linear polarization: rightward and leftward in Figure 23.5, time and again.

32. What happens at the antipodal, the leftmost point in Poincare sphere? Hint: here, $z = \tilde{x} = -1$, so $\theta = \pi$, and $\alpha = \pi/2$. This means linear polarization: upward and downward in Figure 23.5, time and again.

33. If two such waves meet each other, what happens? Hint: nothing: thanks to orthogonality, they never interact, but just pass through one another smoothly, and continue.

34. For yet another example, in Poincare sphere, look at the north pole. Hint: set $\tilde{z} = y = 1$, and $z = x = 0$. This way, $\theta = \pi/2$, so $\alpha = \pi/4$. Furthermore, the phaseshift is $\phi = \pi/2$.

35. What does this mean physically? Hint: circular polarization: as the wave propagates, it makes circles counterclockwise in the perpendicular plane.

36. Where is the antipodal? Hint: at the south pole of the Poincare sphere: at $\tilde{z} = y = -1$, and $z = x = 0$. Here too, $\theta = \pi/2$, so $\alpha = \pi/4$. On the other hand, the phaseshift is now $\phi = -\pi/2$.

37. What does this mean physically? Hint: circular polarization, the other way around: as the wave propagates, it makes circles clockwise rather than counterclockwise.

38. What is the relative angle between the north and south pole? Hint: π.

39. In terms of physical polarization, what is the angle in between? Hint: just half: $\pi/2$.

40. What does this mean physically? Hint: these two polarizations are orthogonal to one another: never interact or interfere with each other.

41. If two such waves propagate in perpendicular directions and meet, what happens? Hint: nothing: they pass through one another smoothly, and continue.

42. In Poincare sphere, look now at the longitude $\tilde{x} = 0$ (Figure 23.6). What happens there? Hint: in terms of the old x-y-z coordinates, this is now the equator: $z = 0$, and $x^2 + y^2 = 1$. This way, $\theta = \pi/2$, and $\alpha = \pi/4$. The phaseshift, on the other hand, could vary. At $x = 1$, for a start, there is no phaseshift at all: linear polarization, as in Figure 23.1. Now, let's move along the equator $x^2 + y^2 = 1$. As y increases, the phaseshift increases as well. As a result, the polarization is not quite linear any more, but makes a narrow ellipse in the plane perpendicular to the wave propagation. Once we arrive at $y = 1$ and $x = 0$, the ellipse gets as "fat" as a circle. Furthermore, once we arrive at $x = -1$ and $y = 0$, it gets as narrow as an oblique line again (as in Figure 23.2). Once we arrive at $x = 0$ and $y = -1$, it gets as "fat" as a circle again, and the polarization gets circular again, although clockwise rather than counterclockwise.

23.9 Spin of an Electron

1. The same framework could also be used to model the spin of an electron (or any other fermion).

2. For this purpose, we must first work in three dimensions. Assume that we have a detector, pointing at a fixed (normalized) direction u:

$$u \equiv \begin{pmatrix} u_1 \\ u_2 \\ u_3 \end{pmatrix} \in \mathbb{R}^3, \quad \|u\| = 1.$$

3. How likely is a given electron to spin-up in the u-direction? In other words, how likely is it to spin (counterclockwise) around u? Hint: this depends on the angle θ between u and w (the direction around which the electron "really" spins in the three-dimensional Cartesian space).

4. Unfortunately, θ is often unknown.

5. To calculate the above probability, we better go back to SU_2, where the Pauli matrices originally belong. (After all, we have an electron, not a boson.)

6. In the process, θ halves, to yield $\theta/2$.

7. In summary, the probability to spin (counterclockwise) around u is $\cos^2(\theta/2)$.

8. Still, this is rather theoretical. After all, θ is often unknown. How to calculate this probability in practice? Hint: work in su_2: look at the observable

$$u_1\sigma_x + u_2\sigma_y + u_3\sigma_z.$$

Calculate its eigenvector of eigenvalue 1. Normalize it to have norm 1 in $\mathbb{C}^2$. Take its inner product with the (normalized) state of the electron in $\mathbb{C}^2$. Take the absolute value of this inner product, and square it up.

9. What is the geometrical meaning of this? Give an example. Hint: look at things in $\mathbb{R}^3$ again, as is done below.

10. Assume that, in $\mathbb{R}^3$, u and w are perpendicular to each other:

$$\theta = \frac{\pi}{2}.$$

What does this mean physically? Hint: the electron has a neutral state: it is equally likely to spin clockwise or counterclockwise around u.

11. Confirm this not only geometrically but also algebraically. Hint: indeed,

$$\cos^2\left(\frac{\theta}{2}\right) = \cos^2\left(\frac{\pi}{4}\right) = \frac{1}{2},$$

as required. Thus, the electron is indeed equally likely to spin clockwise or counterclockwise around u.

23.10 Electron in a Magnetic Field

1. Next, let's work the other way around: go back to three dimensions. In the process, the angle $\theta/2$ should double to θ again. This will give us the full geometrical picture: the spin of the electron in three dimensions. Still, this can never be deterministic, but only stochastic: a random variable, and its expectation.

2. Consider a constant magnetic field, pointing upwards, in the positive z-direction (Figure 23.7):

$$\mathbf{B} \equiv \|\mathbf{B}\| \begin{pmatrix} 0 \\ 0 \\ 1 \end{pmatrix}.$$

3. In this magnetic field, let's guess a dynamic (time-dependent) state for our electron:

$$v \equiv v(t) \equiv \begin{pmatrix} \cos\left(\frac{\theta}{2}\right)\exp(-i\phi t) \\ \sin\left(\frac{\theta}{2}\right)\exp(i\phi t) \end{pmatrix}.$$

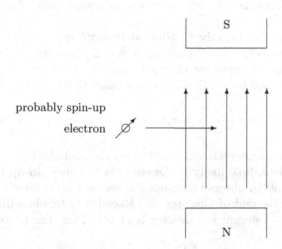

Fig. 23.7. The uniform magnetic field $\mathbf{B} \equiv \|\mathbf{B}\|(0, 0, 1)^t$, pointing upwards, in the positive z-direction.

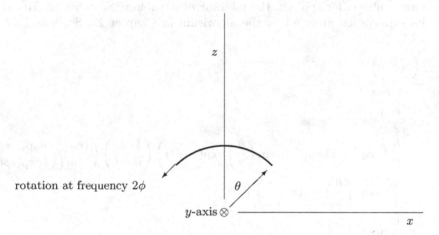

Fig. 23.8. Larmor precession: a view from the side. The expectations of $\tilde{S}_x$, $\tilde{S}_y$, and $\tilde{S}_z$ make a complete three-dimensional vector, rotating around the positive part of the z-axis at frequency 2ϕ.

4. Later on, we'll see that this is much more than just a guess: it actually solves the Schrodinger equation.

5. Here, ϕ is the Larmor frequency. It is proportional to $\|\mathbf{B}\|$.

6. What is the norm of v? Hint: 1, as required.

7. What is the phaseshift between the first and second components of v? Hint: $2\phi t$.

8. At what time is the phaseshift exactly π? Hint: at time $\pi/(2\phi)$.

9. During this time, what happened to v? Hint: up to a scalar multiple, v changed only a little: only its second component changed sign.

10. Algebraically, what happened to v? Hint: σ_z was applied to it:

$$v\left(\frac{\pi}{2\phi}\right) = \sigma_z v(0)$$

(up to an outer phase change, which has no effect on the probabilities).

11. In the above magnetic field, how likely is the electron to have spin-up (in the positive z-direction)? Hint: to observe spin-up, the relevant observable is $\tilde{S}_z = \hbar\sigma_z/2$. (See exercises at the end of Chapter 12.) Recall that this is a diagonal 2×2 matrix. As such, its relevant eigenvector is $(1,0)^t$. Thus, the probability for this is

$$\left|v^t \begin{pmatrix} 1 \\ 0 \end{pmatrix}\right|^2 = \left|\cos\left(\frac{\theta}{2}\right)\right|^2 = \cos^2\left(\frac{\theta}{2}\right).$$

12. What is the expectation of the spin-up random variable at v? Hint: as discussed above, for spin-up, the relevant observable is $\tilde{S}_z = \hbar\sigma_z/2$. To calculate its expectation at v, follow the algorithm in Chapter 12, Sections 12.7.4 and 12.11.4:

$$\left(v, \tilde{S}_z v\right)$$
$$= \bar{v}^t \tilde{S}_z v$$
$$= \frac{\hbar}{2}\bar{v}^t \sigma_z v$$
$$= \frac{\hbar}{2}\left(\cos\left(\frac{\theta}{2}\right)\exp(i\phi t), \sin\left(\frac{\theta}{2}\right)\exp(-i\phi t)\right)\begin{pmatrix} 1 & 0 \\ 0 & -1 \end{pmatrix}\begin{pmatrix} \cos\left(\frac{\theta}{2}\right)\exp(-i\phi t) \\ \sin\left(\frac{\theta}{2}\right)\exp(i\phi t) \end{pmatrix}$$
$$= \frac{\hbar}{2}\left(\cos^2\left(\frac{\theta}{2}\right) - \sin^2\left(\frac{\theta}{2}\right)\right)$$
$$= \frac{\hbar}{2}\cos(\theta).$$

13. During this calculation, what happened to the angle? Hint: it has doubled from $\theta/2$ back to θ, as required in three dimensions.

14. Next, what is the expectation of the spin-right random variable (in the positive x-direction)? Hint: here, the relevant observable is $\tilde{S}_x = \hbar\sigma_x/2$:

$$\left(v, \tilde{S}_x v\right)$$
$$= \bar{v}^t \tilde{S}_x v$$
$$= \frac{\hbar}{2}\bar{v}^t \sigma_x v$$

$$= \frac{\bar{h}}{2} \left(\cos\left(\frac{\theta}{2}\right) \exp(i\phi t), \sin\left(\frac{\theta}{2}\right) \exp(-i\phi t) \right) \begin{pmatrix} 0 & 1 \\ 1 & 0 \end{pmatrix} \begin{pmatrix} \cos\left(\frac{\theta}{2}\right) \exp(-i\phi t) \\ \sin\left(\frac{\theta}{2}\right) \exp(i\phi t) \end{pmatrix}$$

$$= \frac{\bar{h}}{2} \left(\cos\left(\frac{\theta}{2}\right) \sin\left(\frac{\theta}{2}\right) \exp(2i\phi t) + \sin\left(\frac{\theta}{2}\right) \cos\left(\frac{\theta}{2}\right) \exp(-2i\phi t) \right)$$

$$= \frac{\bar{h}}{2} 2\sin\left(\frac{\theta}{2}\right) \cos\left(\frac{\theta}{2}\right) \frac{\exp(2i\phi t) + \exp(-2i\phi t)}{2}$$

$$= \frac{\bar{h}}{2} \sin(\theta) \cos(2\phi t).$$

15. During this calculation, what happened to the angles? Hint: they have doubled: $\theta/2$ has doubled to θ, and ϕ has doubled to 2ϕ, as required in three dimensions.
16. Next, what is the expectation of the spin-in random variable (deep into the page, in the positive y-direction)? Hint: here, the relevant observable is $\tilde{S}_y = \bar{h}\sigma_y/2$:

$$\left(v, \tilde{S}_y v \right)$$

$$= \bar{v}^t \tilde{S}_y v$$

$$= \frac{\bar{h}}{2} \bar{v}^t \sigma_y v$$

$$= \frac{\bar{h}}{2} \left(\cos\left(\frac{\theta}{2}\right) \exp(i\phi t), \sin\left(\frac{\theta}{2}\right) \exp(-i\phi t) \right) \begin{pmatrix} 0 & -i \\ i & 0 \end{pmatrix} \begin{pmatrix} \cos\left(\frac{\theta}{2}\right) \exp(-i\phi t) \\ \sin\left(\frac{\theta}{2}\right) \exp(i\phi t) \end{pmatrix}$$

$$= \frac{\bar{h}}{2} \left(\cos\left(\frac{\theta}{2}\right) \sin\left(\frac{\theta}{2}\right) (-i) \exp(2i\phi t) + \sin\left(\frac{\theta}{2}\right) \cos\left(\frac{\theta}{2}\right) i \exp(-2i\phi t) \right)$$

$$= \frac{\bar{h}}{2} 2\sin\left(\frac{\theta}{2}\right) \cos\left(\frac{\theta}{2}\right) \frac{\exp(2i\phi t) - \exp(-2i\phi t)}{2i}$$

$$= \frac{\bar{h}}{2} \sin(\theta) \sin(2\phi t).$$

17. During this calculation, what happened to the angles? Hint: they have doubled: $\theta/2$ has doubled to θ, and ϕ has doubled to 2ϕ, as required in three dimensions.
18. What is the physical meaning of $2\phi t$? Hint: this is the phaseshift.
19. By now, we have three expectations. Place them in a complete three-dimensional vector (Figure 23.8):

$$\frac{\bar{h}}{2} \left(\sin(\theta) \cos(2\phi t), \sin(\theta) \sin(2\phi t), \cos(\theta) \right).$$

20. How does this look like? Hint: this is a rotation around the z-axis. θ is the angle from the positive z-direction. 2ϕ is the frequency of the rotation around the positive part of the z-axis.
21. This is called Larmor precession.
22. In classical electromagnetics, a spinning charged ball would also behave in the same way.
23. This is a special case of a more general theorem: the expectations in quantum mechanics obey the same laws as in classical physics.

23.11 The Special Orthogonal Lie Group

1. Algebraically, the spin of the electron is modeled best by the Pauli matrices.
2. Geometrically, on the other hand, it is visualized better in three dimensions.
3. How to have a good algebraic framework in three dimensions as well? Hint: for this purpose, recall the 3×3 Hermitian matrices S_x, S_y, and S_z, defined right after Figures 12.10–12.11 to model the spin of a boson.
4. Multiply them by i, to obtain new anti-Hermitian matrices: iS_x, iS_y, and iS_z.
5. These anti-Hermitian matrices span a new Lie algebra, denoted by so_3 (with real coefficients only).
6. How is so_3 related to su_2? Hint: they are isomorphic to each other. Indeed, so_3 is the adjoint of su_2, with the same dimension: 3.
7. Being anti-Hermitian, these new matrices are now ready for exponentiation. This way, they *generate* the new group SO_3, containing the 3×3 rotation matrices: orthogonal matrices of determinant 1, as in the exercises at the end of Chapter 2.
8. For this purpose, they use three real parameters (angles). This is why they are so useful to model the spin of a boson.
9. For this purpose, look again at the 3×3 Hermitian matrices S_x, S_y, and S_z. Are they legitimate observables?
10. Use them to model the spin of a boson. Hint: see exercises at the end of Chapter 12.

23.12 Dynamics: Schrodinger's Equation and Conservation

1. In a quantum-mechanical system, consider a state v (wave function). How does v look like? Hint: a complex vector of norm 1.
2. Assume now that $v \equiv v(t)$ is time-dependent. How does it evolve in time? Hint: it must solve the Schrodinger equation:

$$v'(t) \equiv \frac{dv}{dt}(t) = -\frac{i}{\hbar} H v(t),$$

where H is the Hamiltonian of the system.
3. How does $v(t)$ look like? Hint: in terms of the initial state, it takes the form

$$v(t) = \exp\left(-\frac{i}{\hbar} Ht\right) v(0).$$

4. What is the evolution operator that produces this kind of dynamics? Hint: $\exp\left(-iHt/\hbar\right)$.
5. Is this familiar? Hint: compare this to the Liouville operator in Hamiltonian mechanics (Chapter 22, Section 22.4.2).
6. Write H in its diagonal form:

$$H = ODO^h,$$

where O is unitary, and D is real and diagonal.

7. Look at the elements on the main diagonal of D. What are they? Hint: the eigenvalues of H.
8. What are they physically? Hint: the energy levels.
9. What are the columns of O? Hint: the eigenvectors of H. For instance, let $u \equiv (1, 0, 0, \ldots, 0)^t$ be the first standard unit vector. Then, Ou is the first column in O. This is indeed an eigenvector of eigenvalue $D_{1,1}$:

$$H(Ou) = ODO^hOu = ODu = D_{1,1}Ou.$$

10. What is this physically? Hint: Ou is a state of constant energy level: $D_{1,1}$.
11. Use the above to write the evolution operator more explicitly. Hint: thanks to the diagonal form of H, the evolution operator has its own diagonal form as well:

$$
\begin{aligned}
\exp\left(-\frac{i}{\hbar}Ht\right) &= \sum_{j=0}^{\infty} \frac{1}{j!}\left(-\frac{i}{\hbar}Ht\right)^j \\
&= \sum_{j=0}^{\infty} \frac{1}{j!}\left(-\frac{i}{\hbar}ODO^ht\right)^j \\
&= \sum_{j=0}^{\infty} \frac{1}{j!}O\left(-\frac{i}{\hbar}Dt\right)^j O^h \\
&= O\left(\sum_{j=0}^{\infty} \frac{1}{j!}\left(-\frac{i}{\hbar}Dt\right)^j\right) O^h \\
&= O\exp\left(-\frac{i}{\hbar}Dt\right) O^h.
\end{aligned}
$$

12. Conclude that the evolution operator is unitary. Hint: $\exp(-iDt/\hbar)$ is unitary, because it is diagonal, and its main-diagonal elements have absolute value 1.
13. Is this as expected? Hint: the exponent of an anti-Hermitian matrix is indeed unitary (Section 23.1).
14. Is this good? Hint: thanks to this, $v(t)$ keeps norm 1, as required.
15. Apply the evolution operator to the initial state $v(0)$. Hint: this produces the state at time t:

$$v(t) = \exp\left(-\frac{i}{\hbar}Ht\right)v(0) = O\exp\left(-\frac{i}{\hbar}Dt\right)O^hv(0).$$

16. On the right-hand side, look at the vector $O^hv(0)$. In this vector, look at the first component, for example. What is this? Hint: this is the inner product of $v(0)$ with Ou (the first column of O, or the first state of constant energy).
17. How does this individual component evolve in time? Hint: it only precesses clockwise at frequency $D_{1,1}/\hbar$.
18. Answer the same question for the other components as well.
19. Use this new insight to interpret the above formula better. Hint: in the above formula, $v(0)$ has been decomposed in terms of columns of O: states of constant energy level. In this decomposition, the coefficients are stored in $O^hv(t)$ ($t \geq 0$). This way, their sum of squares is 1, as required. As time goes by, each coefficient

precesses at its own frequency, changing phase, but not magnitude. This way, the sum of squares is still 1, as required.

20. Consider now a new observable. Could it have a conserved state? How could it look like? Hint: its conserved state must also be a column of O. Indeed, such a column remains the same (apart from a phaseshift). A sum of two columns, on the other hand, wouldn't do. After all, they precess at different frequencies, thus interfere with each other, producing a new state, with a new observation.

21. How should a conserved observable be related to H? Hint: it should have the same eigenvectors: the columns of O. In other words, it should have the same diagonal form (with the same O, but not necessarily the same D). In other words, it should commute with H.

22. Is this familiar? Hint: in the Hamilton equation (Chapter 22, Section 22.3.1), commutativity means that the right-hand side vanishes. Therefore, the left-hand side vanishes as well. Thus, the new physical quantity is time-independent and conserved.

23. For example, look at the harmonic oscillator in Chapter 12, Section 12.12.1. Could momentum be conserved? Hint: only if there is no potential at all. In this case, there is no force, so the momentum remains constant. Indeed, the energy is only kinetic, and the Hamiltonian is proportional to the momentum squared, which of course commutes with the momentum, as required. If, on the other hand, the Hamiltonian also contains a potential term (such as position that doesn't commute with momentum), then momentum is not conserved any more.

24. Is this familiar? Hint: this is as in Hamiltonian mechanics.

25. For yet another example, consider the physical system in Figure 23.7: an electron in a uniform magnetic field. What is the Hamiltonian? Hint: since the magnetic field points in the (positive) z-direction, the magnetic energy is proportional to spin-up. Furthermore, it is also proportional to the strength of the magnetic field:

$$H = \|\mathbf{B}\|\tilde{S}_z.$$

26. What are its eigenvectors? Hint: the standard unit vectors: $(1,0)^t$ and $(0,1)^t$.

27. What are their eigenvalues? Hint: $\pm\|\mathbf{B}\|\bar{h}/2$.

28. What is the physical meaning of this? Hint: a spin-up electron aligns with the magnetic field, thus has a positive magnetic energy. A spin-down electron, on the other hand, has a negative magnetic energy. (Compare with Figure 12.2.)

29. In this example, what is O? Hint: the 2×2 identity matrix.

30. Thanks to the Schrodinger equation, how should the initial state evolve in time? Hint: as discussed in Section 23.10,

$$v \equiv v(t) \equiv \begin{pmatrix} \cos\left(\frac{\theta}{2}\right)\exp(-i\phi t) \\ \sin\left(\frac{\theta}{2}\right)\exp(i\phi t) \end{pmatrix}.$$

31. In this solution, what is the phaseshift? Hint: $2\phi t = \|\mathbf{B}\|t$.

32. At what frequency does the phaseshift precess? Hint: $2\phi = \|\mathbf{B}\|$.

33. In this system, what spin is conserved? Hint: only spin-up is conserved, because only $\tilde{S}_z$ commutes with the Hamiltonian (which is also proportional to $\tilde{S}_z$). Spin-right and spin-in, on the other hand, can't be conserved, because $\tilde{S}_x$ and $\tilde{S}_y$ don't commute with the Hamiltonian.

Part VII

Appendix:
Background
in Calculus

Appendix: Background in Calculus

To make the book self-contained, here is the required background in linear algebra and elementary calculus. (The background in discrete math was already introduced above.) For this purpose, we start with functions and their derivatives.

In discrete math, we've focused on finite structures, defined recursively, level by level. This was used throughout the book: in chaos theory, general relativity, and more. As a matter of fact, discrete math was even used in calculus: to prove the Bolzano–Weierstrass theorem (Chapter 4).

In discrete math, we have a powerful tool: mathematical induction. This was used to develop the binomial formula, used in statistical mechanics, and Lie algebras, used in quantum and Hamiltonian mechanics.

In this part, on the other hand, we introduce elementary calculus more systematically: differentiation, integration, partial derivatives, and more. This material is vital in Newtonian and modern physics alike. Finally, we also discuss matrices and their eigenvalues, useful in quantum mechanics.

Functions
and Their Derivatives

In this chapter, we introduce elementary calculus from scratch. For this purpose, we start with a new mathematical object: a function.

In set theory, a function has just a narrow face: a mapping. In calculus, on the other hand, a function is much more: an analytical and geometrical tool. Indeed, to the original function, we can now apply new analytical operations: differentiation and integration [10, 34]. In particular, we differentiate a product of two functions, a composite function, and the inverse function. Furthermore, we use the binomial formula to study a few important functions: the exponent function, the logarithm function, and the power function.

This material is vital in Newtonian and modern physics alike. For example, thanks to the exponent function, the original Lie algebra *generates* a new Lie group, useful to model spin and polarization alike.

24.1 Functions in Physics

24.1.1 Functions

What is a function? It is just a mapping: it maps a given set S to some other set. This way, each individual element $x \in S$ is mapped to some $f(x)$:

$$x \to f(x), \quad x \in S.$$

For example, assume that

$$S = \mathbb{R}$$

(the real axis). Assume also that f is a real function:

$$f : \mathbb{R} \to \mathbb{R}.$$

In other words, for each real number $x \in \mathbb{R}$, $f(x)$ is a real number as well. Let's study this kind of function geometrically and analytically. This is indeed calculus.

Unfortunately, only a few functions can be analyzed (differentiated). Still, the differentiable functions are the most interesting ones: they are good enough to model physical phenomena like motion, acceleration, and force.

A differentiable function $f(x)$ has a derivative: a new function $f'(x)$ that tells us how fast f grows in a short interval around x. This rate of growth may depend on x: at some x's, f may grow rapidly. At some other x's, on the other hand, f may remain nearly flat, or even decrease. At such x's, the derivative $f'(x)$ may be nearly zero, or even negative.

24.1.2 Time, Position, and Speed

For example, x could stand for time. This way, at time x, $f(x)$ stands for position (or location, or distance): how far the particle got from the starting point (the origin). In this case, $f'(x)$ tells us how fast the distance grows: the speed (or velocity) at time x.

The speed is a function of x in its own right. After all, it depends on x as well, and changes with time.

24.1.3 Acceleration

Furthermore, the velocity $f'(x)$ is a differentiable function in its own right. In fact, its derivative $f''(x)$ tells us how fast the speed grows. This is a familiar physical quantity: acceleration.

Moreover, the acceleration is a function of x in its own right. After all, it may depend on x as well, and change with time.

Thus, thanks to calculus, we have the analytical tools required to model motion and dynamics. Moreover, we can go ahead and define fundamental physical concepts: force, work, and energy. Thanks to calculus, physical phenomena become easier to study, calculate, and comprehend, not only analytically but also geometrically.

24.1.4 Real Function in an Open Domain

Consider the real axis $\mathbb{R}$. Let $\Omega \subset \mathbb{R}$ be an open domain.

What does this mean? It means that Ω has no endpoint: each individual number $x \in \Omega$ is surrounded by other numbers in Ω. In fact, x is surrounded by a complete neighborhood: an open interval, with x in the middle:

$$x \in (x - \eta, x + \eta) \subset \Omega \subset \mathbb{R},$$

for some positive number $\eta > 0$, which may depend on x.

For example, Ω could be the entire real axis:

$$\Omega \equiv \mathbb{R}.$$

In this case, η is not necessarily small: it could be as large as you like.

For yet another example, Ω could be the open semiaxis:

$$\Omega \equiv (-\infty, 0) \quad \text{or} \quad \Omega \equiv (0, \infty).$$

In this case, η can no longer be picked arbitrarily. In fact, it now depends on x: it must be as small as $|x|$, say

$$\eta \equiv \eta(x) \equiv \frac{|x|}{2}.$$

For yet another example, Ω could be the union of just any number of open intervals.

What can't Ω be? Well, Ω mustn't be a closed interval. After all, a closed interval has an endpoint. Now, let x be this endpoint. For this x, no η is good enough to produce the required neighborhood. For this reason, a closed interval is *not* an open domain.

Assume now that Ω is a proper open domain. For example, Ω could be the union of a few open intervals, none of them closed (or even half-closed). Now, on Ω, define a new real function $f(x)$:

$$f : \Omega \to \mathbb{R}.$$

For each individual $x \in \Omega$, f is defined not only at x but also in a small neighborhood of x. In such a neighborhood, we can now estimate how fast f changes: increases or decreases.

24.2 Derivative

24.2.1 Rate of Change: A Geometrical Approach

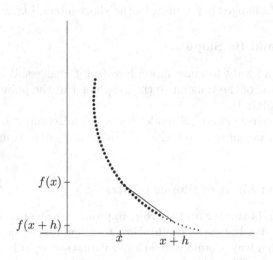

Fig. 24.1. On the graph of the function f, draw a string from $(x, f(x))$ to $(x+h, f(x+h))$.

Consider some point $x \in \Omega$. How fast does f change at x? To estimate this, pick a small parameter h. Look at two points on the real axis: x and $x + h$ (Figure 24.1).

For a sufficiently small $|h|$, both are in Ω:

$$x, x + h \in \Omega.$$

This way, we can now calculate both $f(x)$ and $f(x+h)$, and subtract them from one another:

$$f(x+h) - f(x).$$

To see how fast f changes at x, we need to estimate this difference. If it is big, then f changes sharply at x. If, on the other hand, it is small (in magnitude), then f is nearly flat at x.

24.2.2 Divided Difference

Unfortunately, this difference is still meaningless. In fact, it is not well defined. After all, h was never specified. How to pick h? How small should h be? After all, h has no scale at all!

Thus, we must estimate not how f changes in absolute terms, but how f changes per h-unit, in some h-scale. For this purpose, divide the above difference by h, to obtain the divided difference

$$\frac{f(x+h) - f(x)}{h}.$$

24.2.3 String and Its Slope

What does this mean geometrically? Look at the string leading from $(x, f(x))$ to $(x+h, f(x+h))$, on the graph of f (Figure 24.1). Now, look at the slope of this string. This is the same as the above divided difference! This is indeed what we wanted: how fast f changes per h-unit, in the short interval $[x, x+h]$, on average.

24.2.4 Tangent and Its Slope

Or is it? After all, we want to know more: how fast f changes at x itself! Geometrically, this is the slope of the tangent to the graph of f at the point $(x, f(x))$ (Figure 24.2). How to calculate this?

Unfortunately, our geometrical model helps visualize and comprehend, but not calculate. For this, we must define the derivative not only geometrically but also analytically.

24.2.5 Derivative: Limit of Slopes of Strings

So far, our string gets shorter and shorter, approaching the tangent geometrically. This is nice and visual, but no so useful. How to calculate the slope of the tangent? Better work the other way around: for each individual string, calculate its own slope: the divided difference. Then, let h approach zero. This way, the string approaches the tangent, and its slope approaches the slope of the tangent, as required. This is the derivative of f at x:

$$f'(x) \equiv \lim_{h \to 0} \frac{f(x+h) - f(x)}{h}.$$

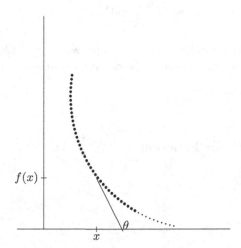

Fig. 24.2. At the point $(x, f(x))$, draw the tangent to the graph of f. Its slope is the derivative: $f'(x) \equiv \tan(\theta)$. In the present picture, f decreases at x, so $f'(x) \equiv \tan(\theta) < 0$.

24.2.6 Limit: Small Error

If this limit exists, then it deserves a new name: $f'(x)$. In this case, we say that f is differentiable at x, and that its derivative at x is $f'(x)$.

How to make sure that $f'(x)$ is indeed a legitimate limit? This is a game between you and me. You pick a positive number $\varepsilon > 0$, as small as you like. Then, you ask me to make sure that the error is as small as ε. In return, I must supply yet another positive number $\delta > 0$, so small that

$$|h| < \delta \;\Rightarrow\; \left| f'(x) - \frac{f(x+h) - f(x)}{h} \right| < \varepsilon.$$

(My δ may depend on your ε: $\delta \equiv \delta(\varepsilon)$.) If we can go on play this game without fail, then the limit indeed exists, and is denoted by $f'(x)$.

24.2.7 The Second Derivative

If $f'(x)$ exists at every $x \in \Omega$, then the derivative is a real function in its own right:

$$f' : \Omega \to \mathbb{R}.$$

At each individual x, we may now ask whether f' is differentiable or not. If it is, then it has its own derivative at x:

$$f'' \equiv (f')'.$$

This is the second derivative of f: the derivative of the derivative. More generally, the nth derivative of f (if exists) is defined by mathematical induction:

$$f^{(0)} \equiv f$$
$$f^{(n+1)} \equiv \left(f^{(n)}\right)' \qquad n \geq 0.$$

Here, n is called the order of the derivative. This way, $f^{(n)}$ is also called the derivative of f of order n. Below, we compute the derivatives of some useful functions.

24.3 Examples

24.3.1 The Hyperbolic Function

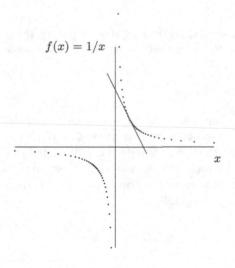

$$f(x) = 1/x$$

Fig. 24.3. The hyperbolic function $f(x) = 1/x$ is defined at each $x \neq 0$. The graph of this function has a steep tangent: its slope is as sharp as $f'(x) = -1/x^2$.

Consider the following open domain:

$$\Omega \equiv \mathbb{R} \setminus \{0\}$$

(the real axis, excluding the origin). In it, define the hyperbolic function:

$$f(x) \equiv \frac{1}{x} \equiv x^{-1}$$

(Figure 24.3). What is its derivative? Well, at each individual $x \neq 0$, look at the slope of a string:

$$\frac{1}{h}\left(\frac{1}{x+h} - \frac{1}{x}\right) = \frac{x-(x+h)}{h(x+h)x} = \frac{-1}{(x+h)x} \xrightarrow{h \to 0} \frac{-1}{x^2}.$$

This limit is the slope of the tangent:

$$(x^{-1})' = -x^{-2} \quad (x \neq 0).$$

Next, let's differentiate a few trigonometric functions. For this purpose, we need complex numbers.

24.4 Trigonometric Functions

24.4.1 Sine, Cosine, and Polar Decomposition

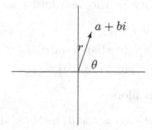

Fig. 24.4. The complex number $a + bi$ (where $i \equiv \sqrt{-1}$ is the imaginary number). In its polar decomposition, it is written as $r(\cos(\theta) + \sin(\theta)i)$, where $r \equiv |a + bi|$ is its absolute value, and θ is the angle it makes with the positive part of the real axis.

To differentiate the sine and cosine functions, we need complex numbers. Fortunately, we've already seen complex numbers, and their arithmetic operations (Figure 3.17). Let's introduce them in their polar decomposition (Figure 24.4). This will give us some useful formulas about sines and cosines.

In what follows, i stands for the imaginary number:

$$i^2 = -1.$$

In the complex plane, the unit circle contains those complex numbers with absolute value 1 (or radius 1, or distance 1 from the origin). On the unit circle, consider two complex numbers:

- $\cos(\theta) + \sin(\theta)i$, which makes angle θ with the positive part of the real axis,

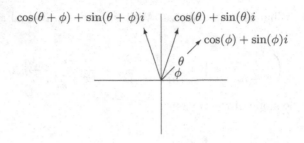

Fig. 24.5. How to multiply $\cos(\theta) + \sin(\theta)i$ by $\cos(\phi) + \sin(\phi)i$? Just add θ and ϕ, to obtain $\cos(\theta + \phi) + \sin(\theta + \phi)i$.

- and $\cos(\phi) + \sin(\phi)i$, which makes angle ϕ with the positive part of the real axis (Figure 24.5). What is their product? Well, this is easy to see geometrically: their product is on the unit circle as well, making angle $\theta + \phi$ with the positive part of the real axis.

Fortunately, this product can be interpreted not only geometrically but also algebraically, using linearity, or the distributive and commutative laws:

$$\cos(\theta + \phi) + \sin(\theta + \phi)i$$
$$= (\cos(\theta) + \sin(\theta)i)(\cos(\phi) + \sin(\phi)i)$$
$$= (\cos(\theta)\cos(\phi) - \sin(\theta)\sin(\phi)) + (\sin(\theta)\cos(\phi) + \cos(\theta)\sin(\phi))i.$$

Now, look at the real part alone:

$$\cos(\theta + \phi) = \cos(\theta)\cos(\phi) - \sin(\theta)\sin(\phi).$$

Likewise, look at the imaginary part alone:

$$\sin(\theta + \phi) = \sin(\theta)\cos(\phi) + \cos(\theta)\sin(\phi).$$

In the real part, substitute $-\phi$ for ϕ:

$$\cos(\theta - \phi) = \cos(\theta)\cos(-\phi) - \sin(\theta)\sin(-\phi) = \cos(\theta)\cos(\phi) + \sin(\theta)\sin(\phi),$$

and add them to one another:

$$\cos(\theta + \phi) + \cos(\theta - \phi) = 2\cos(\theta)\cos(\phi).$$

Likewise, in the imaginary part above, substitute $-\phi$ for ϕ:

$$\sin(\theta - \phi) = \sin(\theta)\cos(-\phi) + \cos(\theta)\sin(-\phi) = \sin(\theta)\cos(\phi) - \cos(\theta)\sin(\phi),$$

and add to the original formula:

$$\sin(\theta + \phi) + \sin(\theta - \phi) = 2\sin(\theta)\cos(\phi).$$

These formulas will be useful later.

24.4.2 Trigonometric Functions and Their Derivative

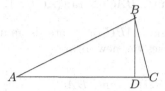

Fig. 24.6. Look at the isosceles triangle $\triangle ABC$, in which $AB = AC$. In it, draw the new right-angled triangle $\triangle ABD$, in which $\angle ADB = 90°$. Note that the angle $\angle CBD$ is twice as small as the angle $\angle BAC$. Therefore, their tangents satisfy the inequality $\tan(\angle CBD) < \tan(\angle BAC)$.

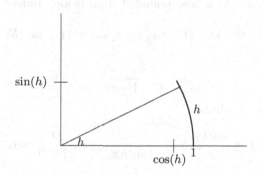

Fig. 24.7. For a small angle (or arc) h, we see that $1 - \cos(h)$ is much smaller than $\sin(h)$. Therefore, as h approaches 0, $\sin(h)/h$ approaches 1.

Thanks to these formulas, we can now differentiate the sine and cosine functions. For this purpose, note that, as h approaches 0, $\cos(h)$ approaches 1:

$$\cos(h) \to_{h \to 0} \cos(0) = 1.$$

How fast is this limit process? Very fast: as fast as superlinear: faster than the rate by which h approaches 0. Indeed, look at the isosceles triangle $\triangle ABC$ in Figure 24.6:

$$AB = AC \quad \text{and} \quad \angle ABC = \angle ACB.$$

In this triangle, the head angle (vertexed at A) can be written as

$$\angle BAC = 180° - \angle ABC - \angle ACB = 2\left(90° - \angle ACB\right) = 2\angle CBD > \angle CBD.$$

In short, we have the inequality

$$\angle BAC > \angle CBD.$$

Because the tangent function is monotonically increasing, we also have

$$\tan\left(\angle BAC\right) > \tan\left(\angle CBD\right).$$

Now, copy the original triangle $\triangle ABC$ right into the unit circle in Figure 24.7. This way, the above inequality takes the new form

$$\frac{1-\cos(h)}{\sin(h)} = \tan(\angle CBD) < \tan(\angle BAC) = \frac{\sin(h)}{\cos(h)} = \tan(h) \to_{h\to 0} 0.$$

Let's use this to prove that $\cos(h)$ approaches 1 superlinearly: faster than $h \to 0$. For this purpose, we only need to replace $\sin(h)$ by h, or show that both approach zero at the same rate.

Indeed, thanks to the above estimate, we can also compare $\sin(h)$ to h (the length of the arc in Figure 24.7). On one hand, $\sin(h)$ is clearly shorter than this arc:

$$\sin(h) \le h.$$

On the other hand, $\sin(h)$ is also bounded from below. Indeed, as can be seen in Figure 24.7,

$$h \le \sin(h) + (1 - \cos(h)) \le \sin(h)\,(1 + \tan(h)),$$

or

$$\sin(h) \ge \frac{h}{1 + \tan(h)}.$$

Upon dividing by h, we obtain

$$1 \ge \frac{\sin(h)}{h} \ge \frac{1}{1 + \tan(h)} \to_{h\to 0} \frac{1}{1+0} = 1,$$

so

$$\frac{\sin(h)}{h} \to_{h\to 0} 1.$$

This is what we wanted:

$$\frac{1-\cos(h)}{h} = \frac{1-\cos(h)}{\sin(h)} \cdot \frac{\sin(h)}{h} \to_{h\to 0} 0 \cdot 1 = 0,$$

meaning that $\cos(h)$ indeed approaches 1 superlinearly: faster than the rate by which h approaches 0.

We are now ready to use our formulas to help differentiate the sine function:

$$\frac{\sin(x+h) - \sin(x)}{h}$$

$$= \frac{\sin(x)\cos(h) + \cos(x)\sin(h) - \sin(x)}{h}$$

$$= \sin(x)\frac{\cos(h) - 1}{h} + \cos(x)\frac{\sin(h)}{h}$$

$$\to_{h\to 0} \cos(x).$$

This gives us the derivative
$$\sin'(x) = \cos(x).$$

Likewise, we can also use our formulas to differentiate the cosine function:

$$\frac{\cos(x+h) - \cos(x)}{h}$$

$$= \frac{\cos(x)\cos(h) - \sin(x)\sin(h) - \cos(x)}{h}$$

$$= \cos(x)\frac{\cos(h) - 1}{h} - \sin(x)\frac{\sin(h)}{h}$$

$$\rightarrow_{h \to 0} -\sin(x).$$

This gives us the derivative
$$\cos'(x) = -\sin(x).$$

24.5 Powers and Their Derivative

24.5.1 Using The Binomial Formula

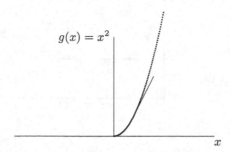

Fig. 24.8. The quadratic function $g(x) = x^2$ has a familiar graph: a parabola. Its tangent has a linear slope: $g'(x) = 2x$.

Assume now that f is a power of x:

$$f(x) \equiv x^n,$$

where $n > 0$ is a constant natural number. (For $n = 2$, this is a parabola, as in Figure 24.8.) To differentiate f at x, let's use Newton's binomial formula (Chapter 8, Section 8.5.1):

$$\frac{f(x+h) - f(x)}{h} = \frac{(x+h)^n - x^n}{h}$$

$$= \frac{\sum_{i=0}^{n} \binom{n}{i} x^i h^{n-i} - x^n}{h}$$

$$= \frac{\sum_{i=0}^{n-1} \binom{n}{i} x^i h^{n-i}}{h}$$

$$= \sum_{i=0}^{n-1} \binom{n}{i} x^i h^{n-i-1}$$

$$\to_{h \to 0} \binom{n}{n-1} x^{n-1}$$

$$= nx^{n-1}.$$

In summary, we have the derivative

$$(x^n)' = nx^{n-1}.$$

Not convinced? Later on, we'll prove this in yet another way.

24.5.2 The Constant Function

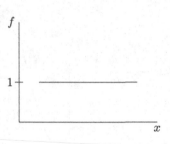

Fig. 24.9. The constant function $f \equiv 1$, and its flat graph. Its tangent has a zero slope: $f' \equiv 0$.

The above formula works for every natural number $n > 0$. Does it work even for $n = 0$? To check on this, look at the constant function

$$f(x) \equiv x^0 \equiv 1.$$

How does it look like geometrically? Well, its graph is completely flat (Figure 24.9). Thus, its tangent has a zero slope. In other words, it has the zero derivative:

$$f'(x) \equiv 0.$$

This is easy to see, not only geometrically but also algebraically:

$$\frac{f(x+h) - f(x)}{h} = \frac{1-1}{h} = \frac{0}{h} = 0.$$

This agrees with our formula (with $n = 0$, and $x \neq 0$):

$$\left(x^0\right)' = 0 \cdot x^{0-1} = 0.$$

24.5.3 The Linear Identity Function

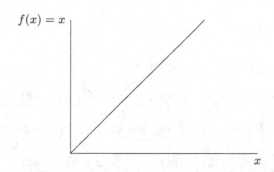

Fig. 24.10. The identity function $f(x) = x$, and its linear graph. Its tangent has a constant slope: $f' \equiv 1$.

Let's check our formula in yet another simple case: $n = 1$. This is the identity function:

$$f(x) = x^1 = x.$$

How does it look like geometrically? Well, its graph is linear (Figure 24.10). This way, its tangent has a constant slope:

$$f' = 1.$$

This is easy to see, not only geometrically but also algebraically:

$$\frac{f(x + h) - f(x)}{h} = \frac{x + h - x}{h} = \frac{h}{h} = 1.$$

This also agrees with our formula (with $n = 1$):

$$\left(x^1\right)' = 1 \cdot x^{1-1} = 1 \cdot x^0 = 1.$$

Let's prove our formula in yet another way, using a new principle: Leibniz rule.

24.6 Leibniz Rule – The Product Rule

24.6.1 Product of Two Functions

Let f and g be two given functions. Consider an x for which both $f(x)$ and $g(x)$ are well defined. The product function fg is defined by

$$(fg)(x) \equiv f(x)g(x).$$

Furthermore, if both f and g are differentiable at x, then fg is differentiable as well, and its derivative can be written in terms of f, f', g, and g':

$$(fg)' = f'g + fg'.$$

Indeed,

$$\frac{(fg)(x+h) - (fg)(x)}{h}$$

$$= \frac{f(x+h)g(x+h) - f(x)g(x)}{h}$$

$$= \frac{f(x+h)g(x+h) - f(x)g(x+h) + f(x)g(x+h) - -f(x)g(x)}{h}$$

$$= \frac{f(x+h)g(x+h) - f(x)g(x+h)}{h} + \frac{f(x)g(x+h) - -f(x)g(x)}{h}$$

$$= g(x+h)\frac{f(x+h) - f(x)}{h} + f(x)\frac{g(x+h) - -g(x)}{h}$$

$$\to_{h \to 0} f'(x)g(x) + f(x)g'(x).$$

This way, if both f and g are available with their derivatives at x, then fg and its derivative are available as well.

24.6.2 Leibniz Rule and Mathematical Induction

Thanks to Leibniz rule, we have a new way to prove our formula:

$$(x^n)' = nx^{n-1}, \quad n \geq 0.$$

(For $n = 0$, this is valid only for $x \neq 0$.) Indeed, by mathematical induction: for $n = 0$, differentiate as in Section 24.5.2:

$$(x^0)' = 1' = \lim_{h \to 0} \frac{1-1}{h} = \lim_{h \to 0} \frac{0}{h} = 0.$$

This agrees with our formula. Likewise, for $n = 1$, differentiate as in Section 24.5.3

$$x' = \lim_{h \to 0} \frac{x+h-x}{h} = \lim_{h \to 0} \frac{h}{h} = \lim_{h \to 0} 1 = 1.$$

This agrees with our formula as well. Next, let's carry out the induction step. For this purpose, assume that the formula holds for some $n \geq 1$. (This is the induction hypothesis.) To prove the formula for $n + 1$ as well. just differentiate x^{n+1} as the product $x^{n+1} = x^n x$:

$$\left(x^{n+1}\right)' = (x^n x)' = (x^n)'x + x^n = (nx^{n-1})x + x^n = nx^n + x^n = (n+1)x^n,$$

as required. This completes the induction step, and indeed the entire proof.

24.6.3 Differentiation: A Linear Operator

Differentiation (or derivation) is an additive operator: you can open parentheses:

$$(f+g)' = f' + g'.$$

This is easy to see:

$$\begin{aligned}(f+g)' &= \lim_{h \to 0} \frac{f(x+h) + g(x+h) - f(x) - g(x)}{h}\\ &= \lim_{h \to 0} \left(\frac{f(x+h) - f(x)}{h} + \frac{g(x+h) - g(x)}{h} \right)\\ &= \lim_{h \to 0} \frac{f(x+h) - f(x)}{h} + \lim_{h \to 0} \frac{g(x+h) - g(x)}{h}\\ &= f' + g'.\end{aligned}$$

Furthermore, differentiation is also multiplicative: for any constant number a, you can pull a out of parentheses:

$$(ag)' = ag'.$$

Indeed, in Leibniz rule, let f be the constant function $f \equiv a$, which has a zero derivative:

$$(ag)' = a'g + ag' = ag'.$$

In summary, differentiation is a linear operator: additive and multiplicative at the same time. Let's use this important property in the binomial formula.

24.6.4 Using The Binomial Coefficients

By now, we've seen that differentiation is linear: additive and multiplicative. Thanks to this property, we can now go ahead and calculate the second derivative of x^n:

$$(x^n)'' = (nx^{n-1})' = n(x^{n-1})' = n(n-1)x^{n-2}.$$

This is valid for $n \geq 2$. For $n < 2$, on the other hand, this still makes sense, provided that $x \neq 0$. In summary,

$$(x^n)'' = \begin{cases} n(n-1)x^{n-2} & \text{if } n \geq 2 \\ 0 & \text{if } 0 \leq n < 2. \end{cases}$$

Next, let's go ahead and differentiate many more times. This way, we can calculate the kth derivative as well:

$$(x^n)^{(k)} = \begin{cases} \frac{n!}{(n-k)!}x^{n-k} & \text{if } n \geq k \geq 0 \\ 0 & \text{if } 0 \leq n < k. \end{cases}$$

Thanks to the notation in Chapter 8, Section 8.6.1, this can also be written as

$$(x^n)^{(k)} = C_{n,k}x^{n-k}.$$

This formula will be used later.

24.6.5 Mirroring The Binomial Formula

By now, we've analyzed the power function x^n, and calculated its high-order derivatives. For this purpose, we've used the binomial coefficients, studied in discrete math.

Still, this could be done not only for the power function but also for a general product of two functions. In fact, if both f and g are available with k derivatives at some x (for some integer $k \geq 0$), then their product fg is available with k derivatives as well. This mirrors Newton's binomial formula:

$$(fg)^{(k)} = \sum_{i=0}^{k} \binom{k}{i} f^{(i)} g^{(k-i)}.$$

After all, to differentiate fg k times, one must differentiate f i times and g $k - i$ times, multiply, and sum up over $0 \leq i \leq k$. For each particular i, it doesn't matter whether f is differentiated before or after g: the same term is produced in either case. Thus, as in Chapter 8, Section 8.5.1, the same term appears $\binom{k}{i}$ times.

24.6.6 Mirroring The Trinomial Formula

Assume now that there is yet another function u, available with k derivatives at the same point x. Consider now the triple product fgu. What is its kth derivative at x? For this purpose, mirror the trinomial formula in Chapter 8, Section 8.6.3:

$$(fgu)^{(k)} = \sum_{0 \leq i,j,l \leq k,\ i+j+l=k} \frac{k!}{i!j!l!} f^{(i)} g^{(j)} u^{(l)}.$$

In this sum, how many terms are there? Thanks to Chapter 8, Section 8.3.5, we already know the answer:

$$\binom{k+3-1}{3-1} = \binom{k+2}{2} = \frac{(k+1)(k+2)}{2}.$$

24.7 Composite Function and Its Derivative

24.7.1 Composite Function

So far, both f and g had the same status: they took part in the product fg. Now, let's go ahead and compose them on top of each other.

For this purpose, let g be a given function, well-defined (and differentiable) at x. Likewise, assume that f is well-defined (and differentiable) at $g(x)$. This way, we can now define the composite function $f \circ g$:

$$(f \circ g)(x) \equiv f(g(x)).$$

What is its derivative at x? To calculate it, define a new limit process:

$$\triangle g \equiv g(x + h) - g(x) \to_{h \to 0} 0.$$

24.7.2 Derivative of The Composite Function

Thanks to this, we can now write

$$\frac{(f \circ g)(x + h) - (f \circ g)(x)}{h}$$

$$= \frac{f(g(x + h)) - f(g(x))}{h}$$

$$= \frac{f(g(x + h)) - f(g(x))}{g(x + h) - g(x)} \cdot \frac{g(x + h) - g(x)}{h}$$

$$= \frac{f(g(x) + \Delta g) - f(g(x))}{\Delta g} \cdot \frac{g(x + h) - g(x)}{h}$$

$$\to_{h \to 0} f'(g(x))g'(x),$$

Here, g is differentiated at x. f, on the other hand, is differentiated at $g(x)$, not x. For this reason, g' is evaluated at x, but f' is evaluated at $g(x)$, not x. In summary,

$$(f \circ g)'(x) = f'(g(x))g'(x).$$

Let's rewrite this formula in a more informative way, to remind us of the reasoning behind it.

24.7.3 The Differentiation Operator

The differentiation operator maps the original function f to its derivative f':

$$f \to f'.$$

In Section 24.6.3, we've already seen that this is a linear operator. Let's give it a more informative name. d/dx:

$$\frac{df}{dx} \equiv \frac{d}{dx}(f) \equiv f'.$$

Here, d/dx is neither a fraction nor a ratio, but just a new notation. Why is it better than the prime used so far? Because it reminds us of two important things:

- The differentiation is with respect to x.
- The derivative has a geometrical meaning: the limit of slopes of strings (divided differences).

By applying the operator time and again, we can now produce high-order derivatives as well. For this purpose, let's go ahead and differentiate f n times (for some $n \geq 0$). This will produce the nth derivative:

$$\frac{d^n f}{dx^n} \equiv \left(\frac{d}{dx}\right)^n (f) \equiv f^{(n)}.$$

24.7.4 Differentiation of The Composite Function

Could our new notation be used in a composite function as well? Well, in this case, things are a bit more complicated: f depends on x only indirectly, through g: f depends on g, which depends on x. Thus, to differentiate f with respect to x, one should actually differentiate $f \circ g$:

$$\frac{df}{dx} \equiv \frac{d(f \circ g)}{dx} = \frac{df}{dg} \cdot \frac{dg}{dx},$$

where dg/dx is evaluated at x, and df/dg is evaluated at $g(x)$.

Again, this writing style has nothing to do with fraction or ratio: it stands for differentiation only. Still, it reminds us of the original proof in Section 24.7.2, which uses divided differences. Furthermore, this formula is easier to remember: the dg in the numerator "cancels out" with the dg in the denominator, just like $\triangle g$ in the original proof.

24.8 The Inverse Function and Its Derivative

24.8.1 The Inverse Function

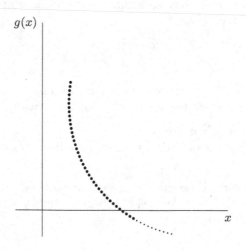

Fig. 24.11. The monotone function g, whose derivative g' never vanishes.

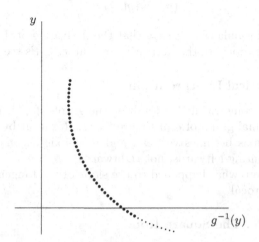

Fig. 24.12. How to obtain the inverse function g^{-1}? Well, in the original graph of g, just interchange the roles of the dependent and independent variables.

How to use the above formula in practice? Let's look at an interesting application. This is indeed deduction (Figures 7.1–7.2).

Assume that g is a monotone function, so g' never vanishes. In Figure 24.11, for example, g' is always negative.

In the composite function introduced above, consider now a special case:

$$f = g^{-1}.$$

(Don't confuse this with the reciprocal $1/g$: it has nothing to do with it.) This is the inverse mapping: it undoes g, and maps back to x:

$$(f \circ g)(x) = (g^{-1} \circ g)(x) = g^{-1}(g(x)) = x.$$

This way, the composition $g^{-1} \circ g$ is just the identity function that does nothing.

How to analyze g^{-1} without calculating it explicitly? For this purpose, look at Figure 24.12. It mirrors Figure 24.11, and looks at things the other way around. Indeed, the axes have now different roles: the vertical axis stands now for the independent variable y, whereas the horizontal axis stands now for the dependent variable $g^{-1}(y)$. Later on, we'll see the geometrical meaning of this: the reciprocal slope. Here, however, we work algebraically.

24.8.2 Differentiating The Inverse Function

In the above equation. differentiate both sides. The right-hand side is easy enough to differentiate: the identity function has derivative 1 (Figure 24.10). To differentiate the left-hand side, on the other hand, we need to work a little more, and use the formula in Section 24.7.2:

$$\left(g^{-1}\right)'(g(x))g'(x) = 1,$$

or

$$\left(g^{-1}\right)'(g(x)) = \frac{1}{g'(x)}.$$

What does this formula say? It says that the derivative of the inverse function g^{-1} at $g(x)$ is the reciprocal of the derivative of g at x. Let's see this geometrically.

24.8.3 Geometrical Interpretation

How to see this geometrically? Let's draw the graph of g^{-1}. How to do this easily? Look at the original graph of g in Figure 24.11. Rotate it by 90° counterclockwise. This way, its x-axis becomes vertical, and its y-axis becomes horizontal, with its positive part pointing leftwards, not rightwards.

In this rotation, what happened to the slope of the tangent? It was transformed to the own reciprocal.

24.8.4 Example: The Square Root

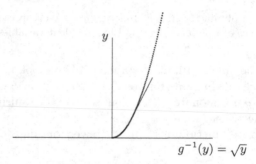

$$g^{-1}(y) = \sqrt{y}$$

Fig. 24.13. To obtain the graph of the square-root function, take the original graph of the quadratic function $y = g(x) = x^2$, and interchange the roles of the dependent and independent variables. The slope of the string is now $h/\triangle g$: the reciprocal of $\triangle g/h$, used in the original derivative.

For example, consider the quadratic function:

$$g(x) = x^2, \quad x > 0$$

(Figure 24.8). In this case, the inverse function is the square root:

$$g^{-1}(y) = \sqrt{y}, \quad y > 0.$$

There is no need to draw its graph explicitly. In Figure 24.13, we obtain it for free, by looking at the axes the other way around: y serves now as the independent variable, whereas $x = \sqrt{y}$ serves now as the dependent variable. The slope is now the cotangent: the reciprocal of the original slope in Figure 24.8. Indeed, in each string, the original slope is $\triangle g/h$, whereas the new slope is $h/\triangle g$.

24.8.5 The Tangent Point

In Figure 24.8, the tangent point is at $(x, g(x))$. In Figure 24.13, on the other hand, the tangent point is still the same, but has a new name:

$$(g^{-1}(y), y) = (x, g(x)).$$

In terms of the new independent variable y, the derivative should therefore be evaluated at $y = g(x)$, as is indeed done above.

24.8.6 Differentiating The Square Root

Let's use this in practice. In our example, g is the quadratic function. In this case, the original derivative is

$$(x^2)' = 2x.$$

The inverse function, on the other hand, is the square root. What is its derivative? Well, at $y = x^2$, it is the reciprocal:

$$(\sqrt{y})' = \frac{1}{2x} = \frac{1}{2\sqrt{y}} = \frac{1}{2}y^{-1/2}.$$

This agrees with the formula in Section 24.5.1, with n being even a fraction: $n = 1/2$. Later on, we'll extend this even more.

24.9 Application: Power Functions

24.9.1 Inverse Powers

In the composite function $f \circ g$, we picked $f \equiv g^{-1}$, to help analyze the inverse function. Next, let's pick

$$g(x) = x^n \quad \text{and} \quad f(x) = \frac{1}{x}$$

(for some natural number n, and a nonzero $x \neq 0$). This way,

$$(f \circ g)(x) = f(g(x)) = \frac{1}{x^n} = x^{-n}.$$

From Sections 24.3.1 and 24.7.2, we then have

$$(x^{-n})' = f'(g(x))g'(x) = \frac{-1}{g^2(x)}g'(x) = \frac{-1}{x^{2n}}nx^{n-1} = -nx^{-n-1}.$$

To simplify this, substitute n for $-n$:

$$(x^n)' = nx^{n-1}$$

(where n is now negative). In summary, the same formula holds not only for a positive but also for a negative n.

Next, let's see that n is not necessarily integer: it could be a fraction as well.

24.9.2 Roots

So far, we used the inverse function to differentiate the square-root function. Let's extend this to any root.

For a natural number n (and a positive $x > 0$), set now

$$g(x) = x^n.$$

This is a monotone function. What is its inverse function? This is the nth root:

$$g^{-1}(y) = +y^{1/n}.$$

(For $n = 2$, for example, we've already seen this: the square root in Figure 24.13.) Fortunately, we already know how to differentiate the inverse function:

$$\left(g^{-1}\right)'(g(x)) = \frac{1}{g'(x)}.$$

In our special case, in which $y = g(x) = x^n$,

$$\left(y^{1/n}\right)' = \frac{1}{nx^{n-1}} = \frac{1}{n}x^{1-n} = \frac{1}{n}y^{(1-n)/n} = \frac{1}{n}y^{1/n-1}.$$

To simplify this, substitute x for y, and α for $1/n$:

$$(x^\alpha)' = \alpha x^{\alpha-1}$$

(where α is now yet more general than before: either an integer number, or the reciprocal of a natural number). Let's extend this to any rational number α.

24.9.3 Rational Powers

In the above formula, α is a special kind of rational number: its numerator is 1. How to extend this to any rational number α, with just any numerator?

For this purpose, let n and m be two natural numbers. For $x > 0$, consider now the composite function $f \circ g$, where

$$g(x) = x^{1/m} \quad \text{and} \quad f(x) = x^n.$$

What is the composition? It is the rational power

$$(f \circ g)(x) = f(g(x)) = \left(x^{1/m}\right)^n = x^{n/m}.$$

Fortunately, we already know how to differentiate a composite function:

$$\begin{aligned}
(f \circ g)'(x) &= f'(g(x))g'(x) \\
&= n(x^{1/m})^{n-1} \cdot \frac{1}{m}x^{1/m-1} \\
&= \frac{n}{m}x^{n/m-1/m+1/m-1} \\
&= \frac{n}{m}x^{n/m-1}.
\end{aligned}$$

To simplify this, substitute α for n/m:

$$(x^\alpha)' = \alpha x^{\alpha-1}$$

(for every (positive) rational number α). To extend this to a negative α, use the same trick as in Section 24.9.1.

Could α be irrational as well? Yes, it could! To see this, we need a special function: the exponent function.

24.10 The Exponent Function

24.10.1 Infinite Series

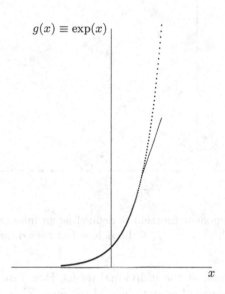

Fig. 24.14. The graph of the exponent function $g(x) \equiv \exp(x) \equiv e^x$, and its tangent, whose slope is $g'(x) = \exp(x)$ as well.

A rational number can be written as a ratio of two natural numbers: n/m. An irrational number, on the other hand, cannot. Two famous examples are π and e: the natural exponent.

e could be defined in two equivalent ways, each with its own advantages and applications. To see this, we better study e in a wider context: the exponent function (Figure 24.14).

The exponent function $\exp(x)$ (or e^x) can be defined in two limit processes. The first definition uses an infinite series:

$$\exp(x) \equiv \sum_{i=0}^{\infty} \frac{x^i}{i!}.$$

24.10.2 The Individual Term: Fast Decay

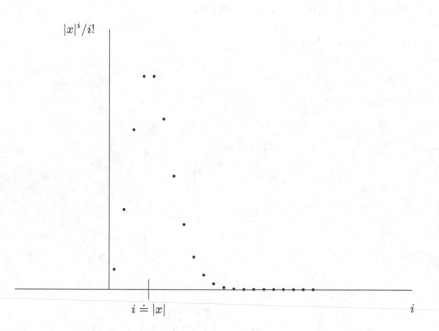

Fig. 24.15. The exponent function is defined as an infinite series. Here are 21 leading terms: $1, |x|, |x|^2/2, \ldots, |x|^{20}/20!$. Look how fast they decay with i.

In this series, look at the individual terms. How fast do they decay? Very fast: as fast as any geometrical power series of the form $\sum q^i$, with $q > 0$ as small as you like (Figure 24.15).

Indeed, the ith term could be written in terms of the previous term:

$$\frac{x^i}{i!} = \frac{x}{i} \cdot \frac{x^{i-1}}{(i-1)!}, \quad i \geq 1.$$

In other words, the ith term is x/i times the previous term. Now, even for a very large $|x|$, one could still pick a k as large as

$$k \geq \frac{|x|}{q}.$$

This way, for every $i \geq k$,

$$\frac{|x|}{i} \leq \frac{|x|}{k} \leq q.$$

Thus, from the kth term onward, each term is as small as q times the previous term. Thus, the terms decay rapidly, and approach zero quite fast:

$$\left|\frac{x^i}{i!}\right| = \frac{|x|^i}{i!} \to_{i \to \infty} 0.$$

24.10.3 The Tail: Fast Decay

Furthermore, the tail (that starts from the kth term onward) is nearly as small as the kth term alone:

$$\left|\sum_{i=k}^{\infty} \frac{x^i}{i!}\right| \leq \sum_{i=k}^{\infty} \frac{|x|^i}{i!}$$

$$\leq \frac{|x|^k}{k!} \sum_{i=0}^{\infty} \left(\frac{|x|}{k}\right)^i$$

$$\leq \frac{|x|^k}{k!} \sum_{i=0}^{\infty} q^i$$

$$= \frac{|x|^k}{k!} \cdot \frac{1}{1-q}$$

(Chapter 7, Section 7.5.3). To make things more concrete, pick

$$q \equiv \frac{1}{2}.$$

Then, for each given x, pick k as large as

$$k \geq 2|x|.$$

This way, the entire tail is as small as twice the kth term alone (in absolute value). Thus, the infinite series indeed converges, and our definition is legitimate.

24.10.4 A New Tail: How Small Is It?

Assume now that we're given an arbitrarily small number $\varepsilon > 0$. Now, let's pick a new k, bigger than before: so big that the individual term is as small as

$$\frac{|x|^k}{k!} \leq \frac{\varepsilon}{6}.$$

Since the terms decay, this is indeed possible. Still, k must now depend not only on $|x|$ but also on ε:

$$k \equiv k(|x|, \varepsilon).$$

With this new k, we can still carry out the above estimate, and obtain that the new tail is now as small as

$$\left|\sum_{i=k}^{\infty}\frac{x^i}{i!}\right| \le 2\frac{|x|^k}{k!} \le \frac{\varepsilon}{3}.$$

Why do we want the tail to be as small as $\varepsilon/3$? Because, later on, we'll split the error into three parts, each as small as $\varepsilon/3$. For this purpose, let's pick a yet bigger k: so big that

$$2k^{-2}\exp(|x|) \le \frac{\varepsilon}{3}.$$

As before, this new k may depend on both $|x|$ and ε:

$$k \equiv k(|x|,\varepsilon).$$

Fortunately, picking a bigger k doesn't spoil what we did so far. This will be useful later.

24.11   The Exponent: New Definition

24.11.1   Sequence and Its Limit

So far, we defined the exponent as an infinite sum. What is this? This is a limit process: a sequence of finite sums, getting longer and longer:

$$\exp(x) \equiv \sum_{i=0}^{\infty}\frac{x^i}{i!} \equiv \lim_{n\to\infty}\sum_{i=0}^{n}\frac{x^i}{i!}.$$

But this is not the only sequence. Here is another one:

$$\exp(x) \equiv \lim_{n\to\infty}\left(1+\frac{x}{n}\right)^n.$$

24.11.2   Bank Interest

What is the meaning of this limit? Suppose that you have some spare money, and want to invest it. Your bank offers you an annual interest of

$$x = 0.01 = 1\%.$$

Is this a good deal? Unfortunately, the interest is paid monthly: at the end of each month. So, in one year, your deposit would multiply by

$$\left(1+\frac{x}{12}\right)^{12}.$$

Is this a good deal? Well, across the road, there is another bank, which offers the same annual interest, paid daily: at the end of each day. With this bank, after one year, your deposit would multiply by

$$\left(1+\frac{x}{365}\right)^{365}.$$

This is more attractive, but still not enough. After all, this is your money, and it should work for you! After each second, the interest should be paid on spot, with no delay! So, if the bank were honest, then they'd offer a yet better deal: after one year, your original deposit should multiply by

$$\exp(x) \equiv \lim_{n\to\infty}\left(1+\frac{x}{n}\right)^n.$$

24.11.3 Using The Binomial Formula

So far, the exponent function has been defined in two different limit processes. Are these limits the same? Yes, they are! Indeed, thanks to the binomial formula, the above sequence could also be written in terms of longer and longer finite sums:

$$\left(1 + \frac{x}{n}\right)^n = \sum_{i=0}^{n} \binom{n}{i} \frac{x^i}{n^i}$$

$$= \sum_{i=0}^{n} \frac{n!}{(n-i)!i!} \cdot \frac{x^i}{n^i}$$

$$= \sum_{i=0}^{n} \frac{n!}{(n-i)!n^i} \cdot \frac{x^i}{i!}.$$

Isn't this familiar? This is quite similar to what we had before. Here, however, the ith term contains a new coefficient:

$$\frac{n!}{(n-i)!n^i} = \frac{n(n-1)(n-2)\cdots(n-i+1)}{n^i}$$

$$= \left(1 - \frac{1}{n}\right)\left(1 - \frac{2}{n}\right)\cdots\left(1 - \frac{i-1}{n}\right).$$

24.11.4 The Coefficient: Smaller Than 1

Clearly, this coefficient is smaller than 1. Therefore, if $n \geq k$, then the tail is as small as the original tail:

$$\left|\sum_{i=k}^{n} \binom{n}{i} \frac{x^i}{n^i}\right| \leq \sum_{i=k}^{n} \binom{n}{i} \frac{|x|^i}{n^i} \leq \sum_{i=k}^{\infty} \frac{|x|^i}{i!} \leq \leq \frac{\varepsilon}{3}.$$

24.11.5 The Leading Terms

So, we can focus on the k leading terms, for which i is as small as

$$i < k.$$

For these, let's show that the coefficient is very close to 1.

By now, k is already picked and fixed. It is now time to pick n as well. Let's pick n as large as

$$n \geq k^4.$$

24.11.6 The Coefficient: Is It Nearly 1?

This way, the coefficient is nearly 1. Indeed, upon opening parenthesis, from each factor we must pick either 1 or a fraction of the form $-1/n$, or $-2/n$, or $-3/n$, etc. This way, the coefficient is now written as a new sum of products:

$$\left(1 - \frac{1}{n}\right)\left(1 - \frac{2}{n}\right)\cdots\left(1 - \frac{i-1}{n}\right)$$

$$= 1 - \left(\sum_{j=1}^{i-1} \frac{j}{n}\right) + \left(\sum_{j<l=1}^{i-1} \frac{jl}{n^2}\right) - \left(\sum_{j<l<m=1}^{i-1} \frac{jlm}{n^3}\right) + \cdots.$$

What do we have here? Well, first of all, we have 1, as required. Next, we have a sum that contains $i - 1$ terms, each as small as

$$\frac{i}{n} < \frac{k}{n} \le k^{-3}.$$

So, in total, this sum is as small as

$$\left|\sum_{j=1}^{i-1} \frac{j}{n}\right| < i \cdot k^{-3} < k \cdot k^{-3} = k^{-2}.$$

Next, we have a bigger sum, containing as many as $(i-1)(i-2)/2$ terms, each as small as

$$\frac{i^2}{n^2} < \frac{k^2}{n^2} \le k^{-6}.$$

So, in total, this second sum is as small as

$$\left|\sum_{j<l=1}^{i-1} \frac{jl}{n^2}\right| < i^2 k^{-6} < k^2 k^{-6} = k^{-4},$$

and so on. Together, these sums are as small as a new power series:

$$\left|\left(\sum_{j=1}^{i-1} \frac{j}{n}\right) - \left(\sum_{j<l=1}^{i-1} \frac{jl}{n^2}\right) + \left(\sum_{j<l<m=1}^{i-1} \frac{jlm}{n^3}\right) - \cdots\right|$$

$$\le \left(\sum_{j=1}^{i-1} \frac{j}{n}\right) + \left(\sum_{j<l=1}^{i-1} \frac{jl}{n^2}\right) + \left(\sum_{j<l<m=1}^{i-1} \frac{jlm}{n^3}\right) + \cdots$$

$$\le \sum_{i=1}^{\infty} \left(k^{-2}\right)^i$$

$$= k^{-2} \frac{1}{1 - k^{-2}}$$

$$\le 2k^{-2}.$$

24.11.7 Leading Terms: Small Error

In summary, in our new definition, in each of the k leading terms (for which $i < k$) the error is as small as

$$\left|\frac{n!}{(n-i)!n^i} \cdot \frac{x^i}{i!} - \frac{x^i}{i!}\right| \le 2k^{-2}\frac{|x|^i}{i!}.$$

Together, these errors are as small as

$$2k^{-2}\exp(|x|) \le \frac{\varepsilon}{3}.$$

24.11.8   Equivalent Definitions

To see that both definitions are the same, we need to show that the error is as small as you like. Well, we just did! In fact, we have just split the error into three parts, each as small as $\varepsilon/3$:

- the tail of the new definition,
- the total error in the k leading terms,
- and the tail of the original definition.

Thus, the total error is as small as

$$
\left| \left(1 + \frac{x}{n}\right)^n - \sum_{i=0}^{\infty} \frac{x^i}{i!} \right|
$$

$$
= \left| \sum_{i=k}^{n} \binom{n}{i} \frac{x^i}{n^i} + \sum_{i=0}^{k-1} \binom{n}{i} \frac{x^i}{n^i} - \sum_{i=0}^{k-1} \frac{x^i}{i!} - \sum_{i=k}^{\infty} \frac{x^i}{i!} \right|
$$

$$
= \left| \sum_{i=k}^{n} \binom{n}{i} \frac{x^i}{n^i} + \sum_{i=0}^{k-1} \left(\binom{n}{i} \frac{x^i}{n^i} - \frac{x^i}{i!} \right) - \sum_{i=k}^{\infty} \frac{x^i}{i!} \right|
$$

$$
\leq \left| \sum_{i=k}^{n} \binom{n}{i} \frac{x^i}{n^i} \right| + \sum_{i=0}^{k-1} \left| \binom{n}{i} \frac{x^i}{n^i} - \frac{x^i}{i!} \right| + \left| \sum_{i=k}^{\infty} \frac{x^i}{i!} \right|
$$

$$
\leq \frac{\varepsilon}{3} + \frac{\varepsilon}{3} + \frac{\varepsilon}{3}
$$

$$
= \varepsilon.
$$

This is true for every n as large as

$$
n \geq k^4,
$$

where k is defined as in Section 24.10.4, and depends on both $|x|$ and ε.

This could be done for every $\varepsilon > 0$. Since ε is arbitrarily small, both definitions must agree.

24.11.9   The Natural Exponent

As a special case, set $x = 1$. This is the natural exponent, denoted by $e = e^1 = \exp(1)$:

$$
e \equiv \lim_{n \to \infty} \left(1 + \frac{1}{n}\right)^n = \sum_{i=0}^{\infty} \frac{1}{i!}.
$$

24.12   The Exponent and Its Derivative

24.12.1   Uniform Convergence

Consider an $\varepsilon > 0$, as small as you like. Consider also a given x. Assume that we already worked hard to find a suitable

$$k \equiv k(|x|, \varepsilon),$$

as in Section 24.10.4. Now, what about a smaller number y, for which

$$|y| \leq |x|?$$

Can we use the same k? Fortunately, we can. After all, k satisfies the same requirements not only for the original x but also for every $y \in [-|x|, |x|]$:

•

$$k \geq 2|x| \geq 2|y|,$$

•

$$\frac{|y|^k}{k!} \leq \frac{|x|^k}{k!} \leq \frac{\varepsilon}{6},$$

• and

$$2k^{-2} \exp(|y|) \leq 2k^{-2} \exp(|x|) \leq \frac{\varepsilon}{3}.$$

Thus, the same k could be picked for the entire interval $[-|x|, |x|]$. We then say that the convergence is uniform in $[-|x|, |x|]$: it has the same rate or "speed" for all y's in this interval.

Better yet, pick a bigger k, suitable for the bigger interval $[-2|x| - 1, 2|x| + 1]$:

$$k \equiv k(2|x| + 1, \varepsilon).$$

This new k is suitable not only for our original x but also for a complete neighborhood around x. This will allow differentiation term by term.

24.12.2 Differentiation Term by Term

In Section 24.6.3, we've already seen that differentiation is additive:

$$(f + g)' = f' + g',$$

where f and g are two differentiable functions. What about three functions? What about four, or even more? Easy: use mathematical induction to prove additivity for just any finite sum, containing as many functions as you like. Still, what about an infinite sum? Could it be differentiated term by term?

For example, look at the exponent function (in its original definition). This is an infinite sum of powers of x. Could it be differentiated term by term? If it could, then

$$\exp'(x) = \sum_{i=0}^{\infty} \left(\frac{x^i}{i!}\right)'$$

$$= \sum_{i=1}^{\infty} \frac{i x^{i-1}}{i!}$$

$$= \sum_{i=1}^{\infty} \frac{x^{i-1}}{(i-1)!}$$

$$= \sum_{i=0}^{\infty} \frac{x^i}{i!}$$

$$= \exp(x).$$

This way, the exponent function is quite unique: it remains unchanged under differentiation:

$$\exp'(x) = \exp(x).$$

This is quite useful in physics. Let's go ahead and prove it.

24.12.3 The Error in The Derivative

What is the derivative of $\exp(x)$? If differentiation term by term is legitimate, then the derivative is $\exp(x)$ as well. To prove this, let's make sure that it indeed behaves like a proper derivative: limit of slopes of strings.

For this purpose, let's use the same trick as before: let $\varepsilon > 0$ be arbitrarily small. We need to pick an h so small that the error in our derivative (slope of string minus slope of tangent at x) is as small as ε:

$$\left| \frac{\exp(x+h) - \exp(x)}{h} - \sum_{i=0}^{\infty} \frac{x^i}{i!} \right|$$

$$= \left| \sum_{i=1}^{\infty} \frac{(x+h)^i - x^i}{i!h} - \sum_{i=0}^{\infty} \frac{x^i}{i!} \right|$$

$$\leq \varepsilon.$$

What do we have here? This is the difference between two infinite sums. The latter is already familiar: the exponent function (in its original definition). As a matter of fact, we've already picked a k for which its tail is as small as $\varepsilon/3$.

24.12.4 Divided Difference: Slope of String

The former, on the other hand, is new. This is the divided difference of the exponent function, calculated term by term. Let's see that its tail is also as small as $\varepsilon/3$.

Fortunately, k was already picked cleverly, to be good enough not only for x but also for the entire interval $[-2|x| - 1, 2|x| + 1]$. As a matter of fact, this big k is good enough even for many different numbers

$$y_i \in [-2|x| - 1, 2|x| + 1].$$

Indeed, even with these new y_i's, the tail is still small:

$$\left| \sum_{i=k}^{\infty} \frac{y_i^i}{i!} \right| \leq \sum_{i=k}^{\infty} \frac{|y_i|^i}{i!} \leq \sum_{i=k}^{\infty} \frac{(2|x| + 1)^i}{i!} \leq \frac{\varepsilon}{3}.$$

This is uniform convergence in its strong form. In particular, these new y_i's could be in a neighborhood around x. Let's use this to estimate the divided difference: slope of string.

$$x \qquad x + h$$

Fig. 24.16. What is the slope of the string leading from $(x, f(x))$ to $(x + h, f(x + h))$?
Well, it is in between two numbers: the minimal derivative, and the maximal derivative in
$[x, x + h]$.

24.12.5 String and Its Maximal Slope

For this purpose, look at the ith term. This is just a power function. For this kind
of function, what is the slope of a string? Well, as can be seen in Figure 24.16, it is
less than the slope of the tangent at $|x| + h$:

$$\left| \frac{(x + h)^i - x^i}{i!h} \right| \leq \max_{y \in [x-h, x+h]} \left| \frac{iy^{i-1}}{i!} \right|$$

$$= \max_{y \in [x-h, x+h]} \left| \frac{y^{i-1}}{(i-1)!} \right|$$

$$\leq \frac{(|x| + h)^{i-1}}{(i-1)!}.$$

24.12.6 Slope: Small Tail

Let's sum the above estimates over $i = k + 1, k + 2, \ldots$. This will show that the
slope has a small tail:

$$\left| \sum_{i=k+1}^{\infty} \frac{(x + h)^i - x^i}{i!h} \right| \leq \sum_{i=k+1}^{\infty} \left| \frac{(x + h)^i - x^i}{i!h} \right|$$

$$\leq \sum_{i=k+1}^{\infty} \left| \frac{(|x|+h)^{i-1}}{(i-1)!} \right|$$

$$= \sum_{i=k}^{\infty} \left| \frac{(|x|+h)^{i}}{i!} \right|$$

$$\leq \frac{\varepsilon}{3}.$$

24.12.7 Slope: Leading Terms

So, we can now focus on the k leading terms, for which $1 \leq i \leq k$. Each is just a power function, which we already know how to differentiate: just let the string approach the tangent at x. In other words, in the divided difference, pick h_i's so small that

$$\left| \frac{(x+h_i)^i - x^i}{i! h_i} - \frac{x^{i-1}}{(i-1)!} \right| \leq \frac{\varepsilon}{3k}, \qquad 1 \leq i \leq k.$$

After all, thanks to uniform convergence, k was already picked and fixed. Finally, define h as the minimal h_i:

$$h \equiv \min_{1 \leq i \leq k} h_i.$$

24.12.8 The Total Error in The Derivative

This way, the total error in our derivative splits into three parts, each as small as $\varepsilon/3$:

$$\left| \frac{\exp(x+h) - \exp(x)}{h} - \sum_{i=0}^{\infty} \frac{x^i}{i!} \right|$$

$$= \left| \sum_{i=1}^{\infty} \frac{(x+h)^i - x^i}{i! h} - \sum_{i=0}^{\infty} \frac{x^i}{i!} \right|$$

$$= \left| \sum_{i=k+1}^{\infty} \frac{(x+h)^i - x^i}{i! h} + \sum_{i=1}^{k} \frac{(x+h)^i - x^i}{i! h} - \sum_{i=0}^{k-1} \frac{x^i}{i!} - \sum_{i=k}^{\infty} \frac{x^i}{i!} \right|$$

$$= \left| \sum_{i=k+1}^{\infty} \frac{(x+h)^i - x^i}{i! h} + \sum_{i=1}^{k} \left(\frac{(x+h)^i - x^i}{i! h} - \frac{x^{i-1}}{(i-1)!} \right) - \sum_{i=k}^{\infty} \frac{x^i}{i!} \right|$$

$$\leq \left| \sum_{i=k+1}^{\infty} \frac{(x+h)^i - x^i}{i! h} \right| + \sum_{i=1}^{k} \left| \frac{(x+h)^i - x^i}{i! h} - \frac{x^{i-1}}{(i-1)!} \right| + \left| \sum_{i=k}^{\infty} \frac{x^i}{i!} \right|$$

$$\leq \frac{\varepsilon}{3} + k \frac{\varepsilon}{3k} + \frac{\varepsilon}{3}$$

$$= \varepsilon.$$

This could be done for every $\varepsilon > 0$, as small as you like. For each such ε, a new k has to be picked and fixed, good enough for the entire interval $[-2|x| - 1, 2|x| + 1]$. Thanks to uniform convergence, this is indeed possible.

In summary, we can now play our ε-h game, without fail. Thus, the latter infinite sum

$$\exp(x) = \sum_{i=0}^{\infty} \frac{x^i}{i!}$$

is indeed the correct derivative. Thus, it was indeed a good idea to differentiate term by term in the first place.

24.12.9 The Natural Logarithm

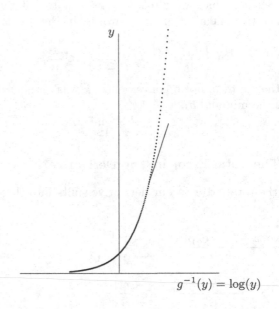

Fig. 24.17. How to obtain the graph of the natural logarithm for free? In the original graph of the exponent function, interchange the roles of the dependent and independent variables. This way, the divided difference is now $h/\triangle g$: the reciprocal of the original divided difference $\triangle g/h$.

In the exercises below, we'll see that the exponent function is positive, and monotonically increasing. As such, it has an inverse function: the natural logarithm $\log(y)$ (for $y > 0$), defined implicitly by

$$\log(\exp(x)) = x$$

(Figure 24.17). Fortunately, we already know how to differentiate an inverse function (Section 24.8.2):

$$\log'(\exp(x)) = \frac{1}{\exp'(x)} = \frac{1}{\exp(x)}.$$

To simplify this, substitute x for $\exp(x)$:

$$\log'(x) = \frac{1}{x}$$

for every positive number $x > 0$.

By now, we're quite experienced in differentiation. Let's use this in a new subject: integration.

24.13 Integration

24.13.1 How to Calculate Area?

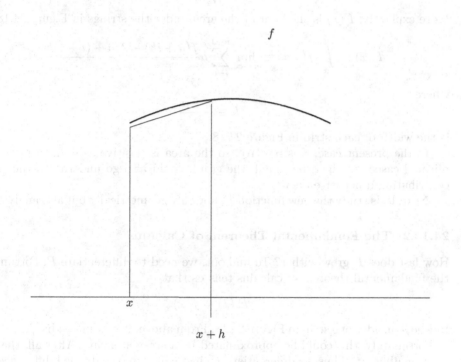

Fig. 24.18. The fundamental theorem of calculus: $f(x)$ is the derivative of $F(x) \equiv \int_a^x f(s)ds$.

So far, we've discussed differentiation. Let's go ahead and use it in integration: calculating an area. For this purpose, look at the graph of the function f. For simplicity, assume that the entire graph is positive, as in Figure 24.18. What is the area under the graph?

To be more precise, recall that f is defined in the open domain

$$\Omega \subset \mathbb{R}.$$

Consider a closed interval

$$[a, x] \subset \Omega,$$

where f is well-defined. Now, from the x-axis, issue two parallel verticals upwards: one issuing from $(a, 0)$, to hit the graph at $(a, f(a))$, and the other issuing from $(x, 0)$, to hit the graph at $(x, f(x))$.

Now, consider the area under the graph. This area is also bounded by three straight lines: a vertical from the left, another vertical from the right, and the horizontal x-axis from below.

What is the right edge? It is the vertical leading from $(x, 0)$ to $(x, f(x))$. Thus, as x grows, the area gets bigger and bigger.

How fast does the area grow? To calculate this, let's write the area as a new function of x:

$$F(x) \equiv \int_a^x f(s)ds.$$

More explicitly, $F(x)$ is the limit of the area under the strings in Figure 24.18:

$$F(x) \equiv \int_a^x f(s)ds \equiv \lim_{n \to \infty} \sum_{i=0}^{n-1} h \frac{f(a + ih) + f(a + (i+1)h)}{2},$$

where

$$h \equiv \frac{x - a}{n}$$

is the width of each strip in Figure 24.18.

In the present case, f is positive, so the area is positive as well. In more complicated cases, on the other hand, the graph could also go underneath the x-axis, contributing a negative area.

Next, let's study the new function F, not only geometrically but also analytically.

24.13.2   The Fundamental Theorem of Calculus

How fast does F grow with x? To find out, we need to differentiate F. Fortunately, the fundamental theorem of calculus tells us that

$$F'(x) = f(x).$$

Indeed, consider one strip in Figure 24.18. From above, it is bounded by the graph of f. Fortunately, this could be approximated by a straight string. After all, the error is as small as h^2. Thus, to differentiate F, just calculate the divided difference, and let h approach zero:

$$F'(x) \equiv \lim_{h \to 0} \frac{F(x + h) - F(x)}{h}$$
$$= \lim_{h \to 0} \frac{\int_x^{x+h} f(s)ds}{h}$$
$$= \lim_{h \to 0} \frac{f(x) + f(x + h)}{2h} h$$
$$= \lim_{h \to 0} \frac{f(x) + f(x + h)}{2}$$
$$= f(x),$$

as asserted. This completes the proof.

In practice, however, F is often unknown. How to uncover it?

24.13.3 Primitive Function (Antiderivative)

This is an inverse problem. Indeed, in differentiation, we are given a function, and seek its derivative. Here, on the other hand, we already have the derivative, and seek the original function, which is still unknown. This is why the original function is called the primitive function (or antiderivative): a new function F, whose derivative is f. There may be many good candidates: we only need one.

This will help solve our practical problem: for a given function f, how to calculate the area under its graph? Easy: find one primitive function G, satisfying

$$G' = f.$$

How different is G from the desired F? Not much:

$$(G - F)' = G' - F' = f - f = 0,$$

so

$$G - F = K,$$

for some constant K (Figure 24.9).

How to find K? Easy: just set $x = a$:

$$G(a) = K + F(a) = K + \int_a^a f(s)ds = K + 0 = K.$$

This uncovers F, as required:

$$F(x) = G(x) - K = G(x) - G(a).$$

24.13.4 Definite vs. Indefinite Integral

A primitive function like G is also called an *indefinite* integral of f. Such a function is often denoted by

$$\int^x f(x)dx.$$

This is just a new notation: it stands for no area as yet.

To have an area, we must also specify a. This way, the area has a different notation:

$$\int_a^x f(s)ds$$

(Section 24.13.1). This also has a different name: the *definite* integral of f over the interval $[a, x]$.

Thus, to calculate the *definite* integral (or the area), just find some *indefinite* integral G. The answer follows immediately:

$$\int_a^x f(s)ds = G(x) - G(a).$$

This answer is often denoted by

$$\int_a^x f(s)ds = G \mid_a^x .$$

This is most useful in physics.

24.14 Exercises: Taylor Series

24.14.1 Lipschitz Continuity

1. Consider two converging sequences:

$$a_n \to_{n\to\infty} a$$
$$b_n \to_{n\to\infty} b.$$

Show that the limit of products is the product of limits:

$$a_n b_n \to_{n\to\infty} ab.$$

Hint: let $\varepsilon > 0$ be arbitrarily small. Pick n so large that both errors are very small:

$$|a_n - a| \le \frac{\varepsilon}{2|b| + 1}$$
$$|b_n - b| \le \frac{\varepsilon}{2|a| + 2}$$
$$|a_n| \le |a| + 1.$$

This way, the total error splits into two parts, each as small as $\varepsilon/2$:

$$
\begin{aligned}
|a_n b_n - ab| &= |a_n b_n - a_n b + a_n b - ab| \\
&\le |a_n b_n - a_n b| + |a_n b - ab| \\
&= |a_n| \cdot |b_n - b| + |b| \cdot |a_n - a| \\
&\le |a_n| \frac{\varepsilon}{2|a| + 2} + |b| \frac{\varepsilon}{2|b| + 1} \\
&\le \frac{\varepsilon}{2} + \frac{\varepsilon}{2} \\
&= \varepsilon.
\end{aligned}
$$

2. Let f be a given function, defined in an open domain $\Omega \subset \mathbb{R}$. Assume also that f is differentiable in the closed interval

$$[a, b] \subset \Omega.$$

Show (geometrically) that, for every two points

$$x, x + h \in [a, b],$$

the divided difference is as small as the maximal derivative:

$$\left| \frac{f(x + h) - f(x)}{h} \right| \le \max_{y \in [a,b]} |f'(y)|.$$

Hint: in Figure 24.16, estimate the slope of the string.

3. Multiply this by $|h|$, to obtain

$$|f(x+h) - f(x)| \leq \max_{y \in [a,b]} |f'(y)| \cdot |h|.$$

4. We then say that f is Lipschitz-continuous in $[a,b]$. This property will be useful below.

5. Prove this once again, using integrals. Hint:

$$|f(x+h) - f(x)| = \left| \int_x^{x+h} f'(s)ds \right|$$

$$\leq \left| \int_x^{x+h} |f'(s)|ds \right|$$

$$\leq \max_{y \in [a,b]} |f'(y)| \cdot \left| \int_x^{x+h} ds \right|$$

$$= \max_{y \in [a,b]} |f'(y)| \cdot |h|.$$

24.14.2 The Exponent and Its Properties

1. Look at the exponent function, in its second definition:

$$\exp(x) \equiv \lim_{n \to \infty} \left(1 + \frac{x}{n} \right)^n$$

(Section 24.11.1). What is it good for?

2. Use it to show that

$$\exp(x) \geq 0, \quad -\infty < x < \infty.$$

Hint: even for a negative x, one could still pick n so large that

$$1 + \frac{x}{n} > 0.$$

3. Use it also to show that

$$\exp(0) = 1.$$

4. Use it also to show that, for any number x,

$$0 \leq \exp(-|x|) \leq 1 \leq \exp(|x|).$$

5. Conclude that $\exp(x)$ is monotonically increasing with x (at least weakly). Hint: its derivative is either positive or zero:

$$\exp'(x) = \exp(x) \geq 0, \quad -\infty < x < \infty.$$

(Later on, we'll prove that this is positive.)

6. Prove that

$$\exp(x+y) = \exp(x)\exp(y),$$

where x and y are any two real (or even complex) numbers. Hint: follow the exercises below, one by one.

7. Let x be a fixed number. For $n = 1, 2, 3, \ldots$, show (in two different ways) that the sequence

$$\left(1 + \frac{|x|}{n}\right)^n$$

increase monotonically with n (even strictly). In other words,

$$1 + |x| < \left(1 + \frac{|x|}{2}\right)^2 < \left(1 + \frac{|x|}{3}\right)^3 < \cdots \leq \exp(|x|).$$

Hint: use either the bank interest in Section 24.11.2, or the representation in Section 24.11.3, obtained from the binomial formula.

8. Let n be a fixed natural number. Let x be a given number. Use the binomial formula to show that

$$\left(1 + \frac{x}{n}\right)^{n-1} = \sum_{i=0}^{n-1} \binom{n-1}{i} \frac{x^i}{n^i}.$$

9. Conclude that

$$\left|\left(1 + \frac{x}{n}\right)^{n-1}\right| \leq \left(1 + \frac{|x|}{n}\right)^{n-1}.$$

10. This is the derivative of what? In other words, what is its primitive function? Hint: see next exercise.

11. Differentiate the composite function

$$\left(1 + \frac{x}{n}\right)^n.$$

Hint: the inner fraction has derivative

$$\left(\frac{x}{n}\right)' = \frac{1}{n}.$$

Therefore, the composite function has derivative

$$\left(\left(1 + \frac{x}{n}\right)^n\right)' = \frac{1}{n} n \left(1 + \frac{x}{n}\right)^{n-1} = \left(1 + \frac{x}{n}\right)^{n-1}.$$

12. Conclude that this function is Lipschitz-continuous: for every h satisfying $|h| < 1$,

$$\left|\left(1 + \frac{x+h}{n}\right)^n - \left(1 + \frac{x}{n}\right)^n\right| \leq \max_{y \in [x-1, x+1]} \left|\left(1 + \frac{y}{n}\right)^{n-1}\right| \cdot |h|$$

$$\leq \max_{y \in [x-1, x+1]} \left(1 + \frac{|y|}{n}\right)^{n-1} |h|$$

$$\leq \max_{y \in [x-1, x+1]} \left(1 + \frac{|y|}{n}\right)^n |h|$$

$$\leq \max_{y \in [x-1, x+1]} \exp(|y|)|h|$$

$$\leq \exp(|x| + 1)|h|.$$

13. Now, let x and y be two given numbers. In the above inequality, substitute $x+y$ for x, to read

$$\left| \left(1 + \frac{x+y+h}{n} \right)^n - \left(1 + \frac{x+y}{n} \right)^n \right| \leq \exp(|x+y|+1)|h|.$$

14. Now, in the above inequality, set

$$h \equiv \frac{xy}{n}.$$

This way, for a sufficiently large n, $|h| < 1$, as required. Conclude that

$$\left| \left(1 + \frac{x+y}{n} + \frac{xy}{n^2} \right)^n - \left(1 + \frac{x+y}{n} \right)^n \right| \leq \exp(|x+y|+1) \left| \frac{xy}{n} \right| \to_{n \to \infty} 0.$$

15. Conclude that

$$\exp(x)\exp(y) = \left(\lim_{n\to\infty} \left(1 + \frac{x}{n} \right)^n \right) \left(\lim_{n\to\infty} \left(1 + \frac{y}{n} \right)^n \right)$$

$$= \lim_{n\to\infty} \left(1 + \frac{x}{n} \right)^n \left(1 + \frac{y}{n} \right)^n$$

$$= \lim_{n\to\infty} \left(\left(1 + \frac{x}{n} \right) \left(1 + \frac{y}{n} \right) \right)^n$$

$$= \lim_{n\to\infty} \left(1 + \frac{x}{n} + \frac{y}{n} + \frac{xy}{n^2} \right)^n$$

$$= \lim_{n\to\infty} \left(1 + \frac{x+y}{n} \right)^n$$

$$= \exp(x+y).$$

16. Conclude that

$$\exp(x)\exp(-x) = \exp(x-x) = \exp(0) = 1.$$

17. Conclude that the reciprocal always exists:

$$\exp(-x) = \frac{1}{\exp(x)}.$$

18. Conclude that

$$\exp(x) > 0.$$

19. Conclude that $\exp(x)$ is monotonically increasing (even strictly). Hint: it has a positive derivative:

$$\exp'(x) = \exp(x) > 0.$$

20. Conclude that the exponent function indeed has an inverse function: the natural logarithm.

24.14.3 Ratio and Its Derivative

1. Let $g(x)$ be a nonzero differentiable function. This way, its reciprocal $1/g$ is well-defined. Differentiate it:

$$\left(\frac{1}{g(x)}\right)' = \frac{-g'(x)}{g^2(x)}.$$

Hint: recall that, for $x \neq 0$,

$$\left(\frac{1}{x}\right)' = -\frac{1}{x^2}$$

(Section 24.3.1). Now, differentiate the composite function $f \circ g$ (Section 24.7.2) with $f(x) = 1/x$:

$$(f \circ g)' = f'(g(x))g'(x) = \frac{-g'(x)}{g^2(x)}.$$

2. Let $\alpha > 0$ be a positive rational number:

$$\alpha \equiv \frac{n}{m} > 0,$$

for some natural numbers n and m. In Section 24.9.3, we've already seen that

$$(x^\alpha)' = \alpha x^{\alpha-1}, \quad x > 0.$$

Show that

$$(x^{-\alpha})' = -\alpha x^{-\alpha-1}, \quad x > 0.$$

Hint: in the previous exercise, set $g(x) = x^\alpha$.

3. In the above formula, substitute $-\alpha$ for α, and obtain the extended formula

$$(x^\alpha)' = \alpha x^{\alpha-1},$$

where α can now be just any rational number, positive or negative:

$$\alpha \equiv \pm\frac{n}{m}.$$

4. Let $f(x)$ and $g(x) \neq 0$ be two differentiable functions. Differentiate their ratio

$$\left(\frac{f(x)}{g(x)}\right)' = \frac{f'(x)g(x) - f(x)g'(x)}{g^2(x)}.$$

Hint: in Leibniz rule (Section 24.6.1), substitute $1/g$ for g:

$$\left(\frac{f(x)}{g(x)}\right)' = \left(f(x)\frac{1}{g(x)}\right)'$$

$$= f'(x)\frac{1}{g(x)} + f(x)\left(\frac{1}{g(x)}\right)'$$

$$= \frac{f'(x)}{g(x)} + f(x)\frac{-g'(x)}{g^2(x)}$$

$$= \frac{f'(x)g(x) - f(x)g'(x)}{g^2(x)}.$$

5. Write this formula more concisely:

$$\left(\frac{f}{g}\right)' = \frac{f'g - fg'}{g^2}.$$

6. Show that, for $x \neq -1$, the function

$$\frac{1 + 2x}{1 + x}$$

is monotonically increasing (even strictly). Hint: thanks to the above formula,

$$\left(\frac{1 + 2x}{1 + x}\right)' = \frac{(1 + 2x)'(1 + x) - (1 + 2x)(1 + x)'}{(1 + x)^2}$$

$$= \frac{2(1 + x) - (1 + 2x)}{(1 + x)^2}$$

$$= \frac{2 + 2x - 1 - 2x}{(1 + x)^2}$$

$$= \frac{1}{(1 + x)^2}$$

$$> 0.$$

7. Conclude that the nonlinear process in Chapter 7, Section 7.9.1, is unstable: in each step, the relative error multiplies by a monotonically increasing factor, much larger than 1.

24.14.4 Power Series and Its Derivative

1. Let x be a real (or even complex) number. Show that

$$(1 - x) \sum_{i=0}^{n} x^i = 1 - x^{n+1}.$$

Hint: open parentheses.

2. For $x \neq 1$, divide this by $1 - x$, and obtain a formula for the finite power series:

$$\sum_{i=0}^{n} x^i = \frac{1 - x^{n+1}}{1 - x}.$$

3. For $|x| < 1$, show that

$$|x|^n \to_{n \to \infty} 0.$$

4. Conclude that, for $|x| < 1$, the infinite geometrical power series converges:

$$\sum_{i=0}^{\infty} x^i = \frac{1}{1 - x}.$$

5. For any natural number k, multiply the above formula by x^k, to obtain the tail:

$$\sum_{i=k}^{\infty} x^i = x^k \sum_{i=0}^{\infty} x^i = \frac{x^k}{1 - x}.$$

6. For $x \neq 1$, show that

$$\left(\frac{1}{1-x}\right)' = \frac{1}{(1-x)^2}.$$

Hint: differentiate the composite function $f \circ g$, where $g(x) = 1 - x$, and $f(x) = 1/x$.

7. Prove the above in yet another way. Hint: differentiate the ratio f/g, where $g(x) = 1 - x$, and $f \equiv 1$.

8. Assume again that $|x| < 1$. Show that

$$\sum_{i=0}^{\infty} (-1)^i x^i = \frac{1}{1+x}.$$

Hint: in the above infinite series, substitute $-x$ for x.

9. In the formula

$$\left(\frac{1}{1-x}\right)' = \frac{1}{(1-x)^2}$$

proved above, is it legitimate to substitute $-x$ for x, and obtain

$$\left(\frac{1}{1+x}\right)' = \frac{1}{(1+x)^2}?$$

Hint: no! The differentiation is with respect to x, not $-x$.

10. Differentiate more carefully:

$$\left(\frac{1}{1+x}\right)' = \frac{-1}{(1+x)^2}.$$

24.14.5 Uniform Convergence

1. For $n = 1, 2, 3, \ldots$, consider the infinite sequence

$$\frac{n+1}{n} = 1 + \frac{1}{n}.$$

Show that it has the following properties:
- It is monotonically decreasing with n.
- 1 is a lower bound:

$$\frac{n+1}{n} = 1 + \frac{1}{n} \geq 1.$$

- 1 is the maximal lower bound.
- 1 is also the limit:

$$\frac{n+1}{n} = 1 + \frac{1}{n} \to_{n \to \infty} 1.$$

Hint: let $\varepsilon > 0$ be a given (arbitrarily small) number. Pick n so large that

$$\frac{1}{n} \leq \varepsilon.$$

2. Let p be a fixed natural number. For $n = 1, 2, 3, \ldots$, consider the new infinite sequence

$$\left(\frac{n+1}{n}\right)^p = \frac{(n+1)^p}{n^p}.$$

3. Use the binomial formula to estimate it:

$$\frac{(n+1)^p}{n^p} = \frac{\sum_{i=0}^{p} \binom{p}{i} n^{p-i}}{n^p}$$

$$= \sum_{i=0}^{p} \binom{p}{i} n^{-i}$$

$$= 1 + \sum_{i=1}^{p} \binom{p}{i} n^{-i}$$

$$\leq 1 + n^{-1} \sum_{i=1}^{p} \binom{p}{i}$$

$$= 1 + \frac{2^p - 1}{n}.$$

4. Use the binomial formula once again, and show that it has the same properties:
   - It is monotonically decreasing with n.
   - 1 is a lower bound:
   $$\frac{(n+1)^p}{n^p} \geq 1.$$
   - 1 is the maximal lower bound.
   - 1 is also the limit:
   $$\frac{(n+1)^p}{n^p} \to_{n \to \infty} 1.$$

5. Let
$$|x| \leq L < M < 1.$$

Use the above properties to show that, for a sufficiently large k, the (polynomial) growth thanks to k^p is overridden by the (geometrical) decay due to L^k:

$$\left| k^p x^k \right| = k^p |x|^k \leq k^p L^k < M^k.$$

Hint: as in Section 24.10.1, estimate the ratio between $(k+1)^p L^{k+1}$ and $k^p L^k$:

$$\frac{(k+1)^p L^{k+1}}{k^p L^k} = \frac{(k+1)^p}{k^p} L \to_{k \to \infty} L < M < 1.$$

6. Consider the polynomial-geometrical series

$$\sum_{i=0}^{\infty} i^p x^i.$$

Conclude that, for a sufficiently large k, it has a small tail:

$$\left| \sum_{i=k}^{\infty} i^p x^i \right| \leq \sum_{i=k}^{\infty} i^p |x|^i$$

$$\leq \sum_{i=k}^{\infty} i^p L^i$$

$$< \sum_{i=k}^{\infty} M^i$$

$$= \frac{M^k}{1-M}.$$

7. Here is yet another polynomial-geometrical series:

$$\sum_{i=0}^{\infty} (i+1)(i+2) \cdots (i+p) x^i.$$

Show that, for a sufficiently large k, it has a small tail:

$$\left| \sum_{i=k}^{\infty} (i+1)(i+2) \cdots (i+p) x^i \right| \leq \sum_{i=k}^{\infty} (i+1)(i+2) \cdots (i+p)|x|^i$$

$$\leq \sum_{i=k}^{\infty} (i+1)(i+2) \cdots (i+p) L^i$$

$$< \sum_{i=k}^{\infty} M^i$$

$$= \frac{M^k}{1-M}.$$

Hint: here too, the ratio is now as small as

$$\frac{(i+2)(i+3) \cdots (i+p+1) L^{i+1}}{(i+1)(i+2) \cdots (i+p) L^i} = \frac{i+p+1}{i+1} L \rightarrow_{i \to \infty} L < M < 1.$$

8. Conclude that th convergence is uniform in the entire interval $[-L, L]$: the same large k could be picked to have an arbitrarily small tail, no matter which x is used in each individual term.

24.14.6 Differentiation Term by Term

1. Look again at the power series

$$\frac{1}{1-x} = \sum_{i=0}^{\infty} x^i, \quad |x| < 1.$$

Use the same technique as in Sections 24.12.2–24.12.8 to differentiate it term by term, not only once but also p times:

$$\frac{p!}{(1-x)^{p+1}} = \left(\frac{1}{1-x}\right)^{(p)}$$

$$= \left(\sum_{i=0}^{\infty} x^i\right)^{(p)}$$

$$= \sum_{i=0}^{\infty} \left(x^i\right)^{(p)}$$

$$= \sum_{i=p}^{\infty} \left(x^i\right)^{(p)}$$

$$= \sum_{i=0}^{\infty} \left(x^{i+p}\right)^{(p)}$$

$$= \sum_{i=0}^{\infty} (i+1)(i+2)\cdots(i+p)x^i$$

$$= \sum_{i=0}^{\infty} C_{i+p,p} x^i,$$

where $|x| < 1$, and $C_{i+p,p}$ is defined as in Chapter 8, Section 8.6.1.

2. Divide this by $p!$, to obtain

$$(1-x)^{-p-1} = \sum_{i=0}^{\infty} \frac{C_{i+p,p}}{p!} x^i$$

$$= \sum_{i=0}^{\infty} \frac{(i+p)!}{i!p!} x^i$$

$$= \sum_{i=0}^{\infty} \frac{C_{i+p,i}}{i!} x^i.$$

3. Here, the coefficient has three equivalent forms:

$$\frac{C_{i+p,i}}{i!} = \frac{C_{i+p,p}}{p!} = \binom{i+p}{p}.$$

Why are these the same?

4. What is the geometrical meaning of this number? Hint: this is the total number of points in the discrete p-dimensional simplex of size i (Chapter 8, Section 8.3.4).

5. Estimate this number. Hint: this discrete simplex is smaller than the discrete p-dimensional hypercube of size i:

$$\binom{i+p}{p} \le (i+1)^p.$$

24.14.7 The Extended Binomial Formula

1. Extend the definition of $C_{a,i}$ to apply to a negative a as well. Hint: see Chapter 8, Section 8.6.1.

2. Use this to obtain

$$
\begin{aligned}
C_{-p-1,i} &\equiv (-p-1)(-p-2)\cdots(-p-i)\\
&= (-1)^i (p+1)(p+2)\cdots(p+i)\\
&= (-1)^i (p+i)(p+i-1)\cdots(p+1)\\
&= (-1)^i C_{p+i,i}.
\end{aligned}
$$

3. Substitute this in the above series:

$$
\begin{aligned}
(1-x)^{-p-1} &= \sum_{i=0}^{\infty} \frac{C_{i+p,i}}{i!} x^i\\
&= \sum_{i=0}^{\infty} (-1)^i \frac{C_{-p-1,i}}{i!} x^i.
\end{aligned}
$$

4. Extend also the definition of the binomial coefficient:

$$
\binom{n}{i} \equiv \frac{C_{n,i}}{i!},
$$

   where n can now be not only positive but also negative.
5. With this new definition, what happens when $i > n \geq 0$? Hint: the binomial coefficient still vanishes, as before.
6. Substitute this new definition in the above series:

$$
(1-x)^{-p-1} = \sum_{i=0}^{\infty} (-1)^i \binom{-p-1}{i} x^i.
$$

7. Substitute n for $-p-1$, and x for $-x$. This gives the extended binomial formula

$$
(1+x)^n = \sum_{i=0}^{\infty} \binom{n}{i} x^i,
$$

   where n can now be not only positive but also negative.
8. Show that, for $n \geq 0$, most of these terms vanish, yielding the original binomial formula once again.

24.14.8 Taylor Expansion

1. Let f be a given function, defined in an open domain Ω, containing the origin

$$
0 \in \Omega \subset \mathbb{R}.
$$

2. Assume that, at 0, f is differentiable as many times as you like. In other words, at 0, f has infinitely many derivatives:

$$
f(0),\ f'(0),\ f''(0),\ f'''(0),\ f''''(0),\ \ldots.
$$

3. Divide them by suitable factorials:

$$f(0), \; f'(0), \; \frac{f''(0)}{2}, \; \frac{f'''(0)}{3!}, \; \frac{f''''(0)}{4!}, \; \ldots$$

4. Use these as coefficients in a new power series, defined in a neighborhood of 0. For this purpose, assume that there is an $\eta > 0$, for which

$$f(x) = \sum_{i=0}^{\infty} \frac{f^{(i)}(0)}{i!} x^i, \qquad |x| < \eta.$$

5. This is the Taylor expansion of f around 0. It tells us how f looks like at

$$x \in (-\eta, \eta),$$

but not elsewhere.

6. Show that the (extended) binomial formula is actually a Taylor expansion. Hint: to see this, set

$$f(x) \equiv (1+x)^n, \qquad |x| < 1,$$

where n is a fixed integer number, positive or negative. In this case, the coefficients are

$$\frac{f^{(i)}(0)}{i!} = \frac{C_{n,i}}{i!}(1+0)^{n-i} = \binom{n}{i},$$

as in the (extended) binomial formula.

7. Let

$$\alpha \in \mathbb{Q}$$

be some rational number. Define the new function

$$f(x) \equiv (1+x)^\alpha, \qquad x > 0.$$

Hint: see Section 24.9.3.

8. What is its Taylor expansion around 0? Hint: in this case, the coefficients are

$$\frac{f^{(i)}(0)}{i!} = \frac{C_{\alpha,i}}{i!}(1+0)^{\alpha-i} = \binom{\alpha}{i}$$

(the extended binomial coefficient, defined in Chapter 8, Section 8.6.2). This way, the Taylor expansion is actually an extended binomial formula:

$$f(x) \equiv (1+x)^\alpha = \sum_{i=0}^{\infty} \binom{\alpha}{i} x^i, \quad |x| < 1.$$

24.14.9 Power Function: General Definition

1. In the above, the power function was defined only for a rational α. What about an irrational α? How to define

$$(1+x)^\alpha?$$

2. Define this new function as follows:

$$(1+x)^\alpha \equiv \sum_{i=0}^\infty \binom{\alpha}{i} x^i, \quad |x| < 1.$$

In this definition, α is any real (or even complex) number.

3. How to differentiate this new function? Hint: term by term.

4. Why is this legitimate? Hint: use the same technique as in Sections 24.12.2–24.12.8.

5. Conclude that the derivative is

$$
\begin{aligned}
((1+x)^\alpha)' &= \sum_{i=0}^\infty \binom{\alpha}{i} (x^i)' \\
&= \sum_{i=1}^\infty \binom{\alpha}{i} i x^{i-1} \\
&= \sum_{i=1}^\infty \frac{C_{\alpha,i}}{i!} i x^{i-1} \\
&= \sum_{i=1}^\infty \frac{C_{\alpha,i}}{(i-1)!} x^{i-1} \\
&= \sum_{i=1}^\infty \frac{\alpha C_{\alpha-1,i-1}}{(i-1)!} x^{i-1} \\
&= \alpha \sum_{i=1}^\infty \frac{C_{\alpha-1,i-1}}{(i-1)!} x^{i-1} \\
&= \alpha \sum_{i=0}^\infty \frac{C_{\alpha-1,i}}{i!} x^{i} \\
&= \alpha \sum_{i=0}^\infty \binom{\alpha-1}{i} x^{i} \\
&= \alpha (1+x)^{\alpha-1}.
\end{aligned}
$$

6. This extends the formula

$$((1+x)^\alpha)' = \alpha(1+x)^{\alpha-1}$$

to any real (or even complex) α (for $|x| < 1$).

7. Repeat this time and again: differentiate $(1+x)^\alpha$ not only once but also i times with i as large as you like:

$$((1+x)^\alpha)^{(i)} = C_{\alpha,i}(1+x)^{\alpha-i}, \quad |x| < 1, \quad 0 \le i < \infty.$$

8. Evaluate this derivative at $x = 0$, and divide by $i!$, to obtain the new coefficient

$$\frac{C_{\alpha,i}}{i!} = \binom{\alpha}{i}, \quad 0 \le i < \infty.$$

9. Use these coefficients to produce the Taylor expansion of $(1+x)^\alpha$ around 0 (with $\eta \equiv 1$):

$$(1+x)^\alpha = \sum_{i=0}^{\infty} \binom{\alpha}{i} x^i, \quad |x| < 1.$$

10. Is this familiar? Hint: this is a yet more general binomial formula (which is the same as the original definition).

24.14.10   Complex Numbers: Polar Decomposition

1. Show that the original definition

$$\exp(x) \equiv \sum_{i=0}^{\infty} \frac{x^i}{i!}, \quad |x| < \infty,$$

coincides with the Taylor expansion of $\exp(x)$ around 0 (with $\eta \equiv \infty$). Hint: in this case, the coefficients are

$$\frac{\exp^{(i)}(0)}{i!} = \frac{\exp 0)}{i!} = \frac{1}{i!}.$$

2. What is the Taylor expansion of $\sin(x)$ around 0?
3. Show that it contains odd powers only:

$$\sin(x) = \sum_{i=0}^{\infty} \frac{(-1)^i x^{2i+1}}{(2i+1)!}, \quad |x| < \infty.$$

Hint: use the formula

$$\sin''(x) = \cos'(x) = -\sin(x)$$

(Section 24.4.2).
4. What is the Taylor expansion of $\cos(x)$ around 0?
5. Show that it contains even powers only:

$$\cos(x) = \sum_{i=0}^{\infty} \frac{(-1)^i x^{2i}}{(2i)!}, \quad |x| < \infty.$$

Hint: use the formula

$$\cos''(x) = -\sin'(x) = -\cos(x)$$

(Section 24.4.2).
6. Differentiate term by term.
7. Why is this legitimate? Hint: as in the exponent function.
8. After differentiating term by term, look at the result. Is it the correct derivative?
9. Multiply the expansion of $\sin(x)$ by $i \equiv \sqrt{-1}$. Add this to the expansion of $\cos(x)$.
10. What is this?

11. Show that this is
$$\cos(x) + i\sin(x) = \exp(ix).$$

Hint: check this, term by term.

12. What is the absolute value of this? Hint:

$$|\exp(ix)| = \cos^2(x) + \sin^2(x) = 1.$$

13. In the complex plane, where is $\exp(ix)$? Hint: on the unit circle (Figures 24.5 and 3.17).

14. Conclude that, for $0 \le \theta < 2\pi$, the arrow leading from the origin to $\exp(i\theta)$ makes angle θ with the positive part of the real axis. Hint: this is the polar decomposition (Section 24.4.1).

Polynomials
and Partial Derivatives

So far, we focused on a function of just one independent variable: x. Let's move on to a more complicated case: a function of two (or even three) independent variables: x and y (and even z). To simplify things, we focus on a a simple function: a polynomial. Still, the discussion could be easily extended to a more general function as well.

Geometrically, we are now in a higher dimension: not only the one-dimensional line $\mathbb{R}$, but also the two-dimensional plane $\mathbb{R}^2$ (and even the three-dimensional space $\mathbb{R}^3$). This gives us new features: there are now not only one but also infinitely many directions.

Thus, one could differentiate not only in one but also in many different directions. This produces new partial (and even directional) derivatives, useful in classical and modern physics.

25.1 Polynomials and Their Arithmetic Operations

25.1.1 Polynomial of One Variable

What is a polynomial? It is a real function

$$p : \mathbb{R} \to \mathbb{R},$$

defined by

$$p(x) \equiv a_0 + a_1 x + a_2 x^2 + \cdots + a_n x^n \equiv \sum_{i=0}^{n} a_i x^i,$$

where n is a nonnegative integer number (the degree of the polynomial), and $a_0, a_1, a_2, \ldots, a_n$ are fixed real numbers (the coefficients of the polynomial). Usually, it is assumed that $a_n \neq 0$. Otherwise, it could drop.

How to define a concrete polynomial of degree n? Just specify its coefficients: $a_0, a_1, a_2, \ldots, a_n$. Thus, the polynomial is mirrored by the $(n+1)$-dimensional vector

$$(a_0, a_1, a_2, \ldots, a_n) \in \mathbb{R}^{n+1}.$$

25.1.2 Real vs. Complex Polynomial

This is a real polynomial. What is a complex polynomial? Well, it is different in one aspect only: the coefficients $a_0, a_1, a_2, \ldots, a_n$ (and the independent variable x) could now be not only real but also complex numbers. This way, the polynomial is a complex function

$$p : \mathbb{C} \to \mathbb{C},$$

rather than a mere real function $p : \mathbb{R} \to \mathbb{R}$.

In what follows, we focus on real polynomials. Still, Complex polynomials could be treated in the same way.

Polynomials are algebraic objects: they support arithmetic operations like addition, multiplication, and composition. Later on, we'll see that they are also analytical objects: they can be differentiated and integrated.

We start with polynomials of just one independent variable: x. Later on, we'll extend this to more complicated polynomials of two (and even three) independent variables: x and y (and even z). These too will be not only algebraic but also analytical, with addition, multiplication, differentiation, and integration.

25.1.3 Addition

In the above, the original polynomial $p(x)$ was mirrored by an $(n+1)$-dimensional vector. What is this good for? It is useful in arithmetic operations like addition. For this purpose, let $q(x)$ be yet another polynomial (of degree m):

$$q(x) \equiv \sum_{i=0}^{m} b_i x^i.$$

Without loss of generality, assume that $m \leq n$. (Otherwise, just interchange the roles of p and q.) This way, there are still two possibilities: if $m = n$, then we are ready to add p and q. If, on the other hand, $m < n$, then we must first define $n - m$ fictitious zero coefficients:

$$b_{m+1} \equiv b_{m+2} \equiv \cdots \equiv b_n \equiv 0,$$

to let q have $n + 1$ coefficients as well.

We're now ready to add, term by term. This defines the new polynomial $p + q$

$$(p + q)(x) \equiv p(x) + q(x) = \sum_{i=0}^{n} (a_i + b_i) x^i.$$

This is also mirrored in the underlying vectors of coefficients: they are added to one another, coefficient by coefficient, to produce a new vector of new coefficients. This works in subtraction as well:

$$(p - q)(x) \equiv p(x) - q(x) = \sum_{i=0}^{n} (a_i - b_i) x^i.$$

Again, this is carried out term by term. This defines a new polynomial: $p - q$, with a new vector of new coefficients.

25.1.4 Scalar Multiplication

How to multiply the original polynomial p by a given scalar c? Term by term, of course! This produces the new polynomial cp:

$$(cp)(x) \equiv c \cdot p(x) = c \sum_{i=0}^{n} a_i x^i = \sum_{i=0}^{n} (ca_i) x^i.$$

This is also mirrored in the vector of coefficients: it has been multiplied by c, coefficient by coefficient, to produce the new vector of the new coefficients. Thus, polynomials are well-mirrored by vectors, not only in addition but also in scalar multiplication.

25.1.5 Multiplying Polynomials: Convolution

By now, we've seen that a polynomial of degree n is mirrored by the $(n + 1)$-dimensional vector that contains its coefficients. Indeed, arithmetic operations like addition, subtraction, and scalar multiplication are linear, as in vectors. Still, the polynomial is more than that: it also supports a new algebraic operation: multiplication.

To see this, consider two given polynomials: p and q. Let's multiply them by one another. In other words, define the new function pq:

$$(pq)(x) \equiv p(x)q(x).$$

After all, polynomials are just functions, which can be multiplied (as in Chapter 24, Section 24.6.1). The result is a new function: the product pq. Still, is pq a *polynomial* in its own right? To see this, we must produce its new vector of new coefficients.

By now, pq is only a new function. This means that, if you give me an x, then I could return the value of pq at x. How? Easy:

- calculate $p(x)$,
- then calculate $q(x)$,
- then multiply.

Still, is pq more than that? Does pq make a new polynomial, with a new vector of new coefficients?

Fortunately, the product of

$$p(x) \equiv \sum_{i=0}^{n} a_i x^i \quad \text{and} \quad q(x) \equiv \sum_{j=0}^{m} b_j x^j$$

could be written as

$$(pq)(x) \equiv p(x)q(x) = \sum_{i=0}^{n} a_i x^i \sum_{j=0}^{m} b_j x^j = \sum_{i=0}^{n} \sum_{j=0}^{m} a_i b_j x^{i+j}.$$

This is not good enough. After all, this double sum scans the $(n+1) \times (m+1)$ grid

$$\{(i, j) \mid 0 \leq i \leq n, \ 0 \leq j \leq m\}$$

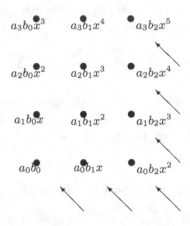

Fig. 25.1. Multiply $p(x) = a_0 + a_1x + a_2x^2 + a_3x^3$ by $q(x) = b_0 + b_1x + b_2x^2$: sum the terms diagonal by diagonal, where the kth diagonal ($0 \le k \le 5$) contains terms with x^k only.

(Figure 25.1). In this rectangular grid, each point of the form (i, j) contributes a term of the form $a_ib_jx^{i+j}$. This is not quite what we wanted. A legitimate polynomial must have a yet simpler form: not a double sum that uses two indices, but a standard sum that uses just one index: k.

For this purpose, scan the grid diagonal by diagonal, as in Figure 25.1. After all, on the kth diagonal, each point of the form (i, j) satisfies $i + j = k$, and contributes $a_ib_jx^{i+j} = a_ib_{k-i}x^k$.

Thus, the grid is scanned diagonal by diagonal, not row by row. The diagonals are indexed by the new index $k = i + j = 0, 1, 2, \ldots, n + m$. This designs a new (outer) sum over k.

In the inner sum, on the other hand, we scan one individual diagonal: the kth diagonal. For this purpose, we need an inner index. Fortunately, we already have a good index: the original row-index i. It mustn't exceed the original grid:

$$(pq)(x) = \sum_{i=0}^{n}\sum_{j=0}^{m} a_ib_jx^{i+j} = \sum_{k=0}^{n+m}\sum_{i=\max(0,k-m)}^{\min(k,n)} a_ib_{k-i}x^k.$$

This is what we wanted. In this form, pq is a legitimate polynomial. Indeed, it has a new vector of new coefficients:

$$(c_0, c_1, c_2, c_3, \ldots, c_{n+m}) \equiv (c_k)_{k=0}^{n+m},$$

where the new coefficient c_k is defined in the inner sum:

$$c_k \equiv \sum_{i=\max(0,k-m)}^{\min(k,n)} a_i b_{k-i}.$$

This is also called convolution (see exercises below).

25.1.6 Example: Scalar Multiplication

For example, consider a special case, in which q is a very short polynomial, of degree $m = 0$:

$$q(x) \equiv b_0.$$

In this (degenerate) case, c_k is just

$$c_k = \sum_{i=k}^{k} a_i b_{k-i} = a_k b_0$$

$(0 \le k \le n)$. Thus, in this case,

$$(pq)(x) = \sum_{k=0}^{n} c_k x^k = \sum_{k=0}^{n} b_0 a_k x^k = b_0 \sum_{i=0}^{n} a_i x^i = b_0 \cdot p(x).$$

This is just a scalar multiplication (Section 25.1.4).

25.2 Polynomial and Its Value

25.2.1 Value at a Given Point

What is a polynomial p? It is just a special kind of a function: it maps x to $p(x)$. In other words, for each given argument x, it returns a new value: $p(x)$. How to calculate $p(x)$ efficiently?

25.2.2 The Naive Method

The naive method uses three stages.

- Calculate the individual powers of x:

$$x^2, x^3, x^4, \ldots, x^n.$$

Fortunately, this could be done recursively:

$$x^i = x \cdot x^{i-1} \quad (i = 2, 3, 4, \ldots, n).$$

Is this expensive? Not too much: just $n - 1$ arithmetic operations: scalar-times-scalar multiplications.
- Next, multiply each power of x by the corresponding coefficient, to obtain the terms:

$$a_1 x, a_2 x^2, a_3 x^3, \ldots, a_n x^n.$$

This costs n more arithmetic operations: scalar-times-scalar multiplications.

- Finally, sum up:

$$a_0 + a_1 x + a_2 x^2 + \cdots + a_n x^n = p(x).$$

This costs n more arithmetic operations: scalar-plus-scalar additions.

Thus, the total cost (or the complexity) is $2n - 1$ multiplications and n additions.

Could this be improved? Fortunately, it could. For this purpose, introduce parentheses, and take a common factor out of them.

25.2.3 Using the Distributive Law

Consider a simple problem: let a, b, and c be some given numbers. Calculate

$$ab + ac.$$

How to do this? The naive method uses three arithmetic operations:

- Calculate ab.
- Calculate ac.
- Add up: $ab + ac$.

Could this be improved? Yes! Just use the distributive law: introduce parentheses and take the common factor a out of them:

$$ab + ac = a(b + c).$$

This is easier to calculate: it requires just two arithmetic operations:

- Add $b + c$.
- Multiply $a(b + c)$.

Let's use this idea in our original polynomial as well.

25.2.4 Recursion: Horner's Algorithm

Let's use the above to help calculate

$$p(x) = \sum_{i=0}^{n} a_i x^i$$

(where x is a given argument). How to take x out of parentheses? For this purpose, write p in a new form:

$$p(x) = a_0 + x p_1(x),$$

where $p_1(x)$ is a new polynomial:

$$p_1(x) = a_1 + a_2 x + a_3 x^2 + \cdots + a_n x^{n-1} = \sum_{i=0}^{n-1} a_{i+1} x^i.$$

Fortunately, p_1 has a lower degree: $n - 1$. Therefore, the value of $p_1(x)$ could be calculated recursively by the same method itself. This is indeed Horner's algorithm.

25.2.5 Complexity: Mathematical Induction

Is this more efficient? Yes, it is! In fact, Horner's algorithm requires less arithmetic operations: just n scalar-times-scalar multiplications, and n scalar-plus-scalar additions.

To prove this, use mathematical induction on the degree n. For $n = 0$, this is clearly true: $p(x)$ is just the constant function

$$p(x) \equiv a_0.$$

To "calculate" this, one needs to do nothing: just $n = 0$ multiplications and $n = 0$ additions, as asserted. Now, for $n = 1, 2, 3, \ldots$, assume that the induction hypothesis holds: to calculate $p_1(x)$, one needs $n-1$ multiplications and $n-1$ additions. Thanks to this, to calculate the original polynomial

$$p(x) = a_0 + x p_1(x),$$

one needs just two more arithmetic operations: one more multiplication to calculate $x p_1(x)$, and one more addition to calculate $a_0 + x p_1(x)$. Together, this makes a total of n multiplications and n additions, as asserted. This completes the induction step, and indeed the entire proof.

25.3 Composition

25.3.1 Mathematical Induction

So far, we've used Horner's algorithm to calculate the value of a polynomial. Let's use it for yet another purpose: to compose two polynomials.

The composition of the polynomials p and q is defined by

$$(p \circ q)(x) \equiv p(q(x)).$$

This is indeed how functions are composed (Chapter 24, Section 24.7.1).

This formula is useful to calculate the value of $p \circ q$ at any given argument x:

- calculate $q(x)$,
- and use it as an argument in p, to calculate $p(q(x))$.

This defines $p \circ q$ as a new *function*. Still, is it a legitimate *polynomial* as well? In other words, does it have its own vector of coefficients?

To design this, let's use mathematical induction on the degree of p: n. For $n = 0$, p is just the constant function $p(x) \equiv a_0$, so

$$(p \circ q)(x) = p(q(x)) = a_0$$

as well. This has just one coefficient, which could be placed in a new "vector:"

$$(a_0).$$

This is a rather short "vector." Still, it is as legitimate as ever.

25.3.2 The Induction Step

Now, for $n = 1, 2, 3, \ldots$, assume that the induction hypothesis holds: for every polynomial p_1 of degree $n-1$ (including the specific polynomial p_1 defined in Section 25.2.4), we already have the entire vector of coefficients of $p_1 \circ q$. Fortunately, we already know how to multiply polynomials (Section 25.1.5). So, we can go ahead and multiply q times $p_1 \circ q$, to obtain the entire vector of coefficients of the product

$$q \cdot (p_1 \circ q).$$

Finally, add a_0 to the first coefficient, to obtain the entire vector of coefficients of the required polynomial:

$$p \circ q = a_0 + q \cdot (p_1 \circ q).$$

25.3.3 Recursion: A New Horner Algorithm

This completes the induction step. This completes the inductive (or recursive) Horner algorithm for composing two polynomials.

25.4 Natural Number as a Polynomial

25.4.1 Decimal Polynomial

Why are polynomials so useful? Because every natural number k is actually a polynomial. After all, k must lie somewhere in between

$$10^n \leq k < 10^{n+1},$$

for some nonnegative integer number n. Therefore, k could be written as a decimal number, containing $n + 1$ digits:

$$a_n a_{n-1} a_{n-2} \cdots a_1 a_0.$$

What is this? This is actually a polynomial:

$$k = a_0 + a_1 \cdot 10 + a_2 \cdot 10^2 + \cdots + a_n \cdot 10^n = \sum_{i=0}^{n} a_i \cdot 10^i = p(10).$$

This is a special kind of polynomial: a decimal polynomial, with coefficients that are digits between 0 and 9. To obtain k, just evaluate this polynomial at the argument 10.

25.4.2 Binary Polynomial

In the above, we've used base 10. Instead, we could also use base 2. After all, the original natural number k must lie somewhere in between

$$2^m \leq k < 2^{m+1},$$

for some nonnegative integer number m. Therefore, k could also be written in terms of $m + 1$ binary digits (0 or 1):

$$b_m b_{m-1} b_{m-2} \cdots b_1 b_0.$$

What is this? This is actually a binary polynomial:

$$k = b_0 + b_1 \cdot 2 + b_2 \cdot 2^2 + \cdots + b_m \cdot 2^m = \sum_{j=0}^{m} b_j 2^j = q(2).$$

To obtain k, just evaluate this polynomial at the argument 2.

This formula is quite practical: it may help store an (arbitrarily long) natural number efficiently. (This is useful in cryptography.) Let's see an interesting application.

25.5 Monomial and Its Value

25.5.1 Monomial

By now, we've represented the natural number k in its binary form:

$$k = q(2),$$

where q is a binary polynomial, with coefficients that are binary digits (bits): either 0 or 1. Let's see an interesting application: calculating an individual monomial.

Let x be a given number. For a given polynomial, we already know how to calculate its value at x: by Horner's algorithm. Still, what about a polynomial that contains one term only? Is there a better method that exploits its special form?

25.5.2 A Naive Method

This is actually a monomial: x^k. How to calculate its value? The naive approach is sequential: for $i = 2, 3, 4, \ldots, k$, calculate

$$x^i = x \cdot x^{i-1}.$$

This is a bit expensive: it requires $k - 1$ multiplications. Still, it also gives a bonus: not only one but also many monomials: x^2, x^3, x^4, $\ldots$, x^k. But what if all we need is x^k alone? Is there a more efficient way?

25.5.3 Horner Algorithm: Implicit Form

Fortunately, there is: Horner's algorithm, applied to the binary polynomial that represents k. Indeed, thanks to Horner's algorithm, k could be written as

$$k = q(2) = b_0 + 2q_1(2),$$

where q is a binary polynomial of degree m, and q_1 is yet another binary polynomial, of degree $m - 1$. Fortunately, there is no need to specify q or q_1 explicitly: it is sufficient to know that they do exist.

25.5.4 Mathematical Induction

Thanks to this (implicit) representation of k, we can now calculate x^k cheaply: in just $2m$ multiplications! Indeed, by mathematical induction on m: for $m = 0$, no work is needed at all. After all, in this case, $k \equiv b_0$ could take only two possible values: either 0 or 1. In either case, there is no work at all: $x^k = x^0 = 1$ or $x^k = x^1 = x$.

25.5.5 The Induction Step

Now, assume that the induction hypothesis holds: we already have a good algorithm for polynomials of degree $m - 1$ (such as q_1 above). This is relevant to just any argument: not only x but also x^2. Thus, if x^2 was available, then we could calculate

$$\left(x^2\right)^{q_1(2)}$$

in just $2(m - 1)$ multiplications. (This is the induction hypothesis.) Let's use this to calculate our original monomial:

$$x^k = x^{q(2)} = x^{b_0 + 2q_1(2)} = x^{b_0} x^{2q_1(2)} = x^{b_0} \left(x^2\right)^{q_1(2)} = \begin{cases} x \cdot \left(x^2\right)^{q_1(2)} & \text{if } b_0 = 1 \\ \left(x^2\right)^{q_1(2)} & \text{if } b_0 = 0. \end{cases}$$

25.5.6 Complexity: Total Cost

To calculate this right-hand side, we may need two more multiplications:

- Calculate $x^2 = x \cdot x$. Later on, this will help calculate $(x^2)^{q_1(2)}$ recursively.
- Finally, multiply $(x^2)^{q_1(2)}$ by x (if $b_0 = 1$).

Thus, the total cost is as low as

$$2 + 2(m - 1) = 2m$$

multiplications. This completes the induction step.

25.5.7 Recursion Formula

To simplify this recursion, let's get rid of q and q_1 altogether. For this purpose, note that b_0 is just the unit binary digit in

$$k = q(2) = b_0 + 2q_1(2).$$

Thus, if k is even, then

$$b_0 = 0 \quad \text{and} \quad q_1(2) = \frac{k}{2}.$$

If, on the other hand, k is odd, then

$$b_0 = 1 \quad \text{and} \quad q_1(2) = \frac{k - 1}{2}.$$

Thus, the above recursion could be written more explicitly as

$$x^k = \begin{cases} x \cdot (x^2)^{(k-1)/2} & \text{if } k \text{ is odd} \\ (x^2)^{k/2} & \text{if } k \text{ is even.} \end{cases}$$

This is useful not only to calculate x^k. Later on, we'll mirror this to obtain q^k where q is a complete polynomial.

25.6 Differentiation

25.6.1 Derivative of a Polynomial

The polynomial is an algebraic object, with arithmetic operations: addition and multiplication. It is also a mapping: at each individual argument x, it has a value, which can be calculated efficiently. Still, this isn't the whole story: the polynomial is also an analytic object, which can be differentiated and integrated.

For this purpose, consider again a polynomial of degree n:

$$p(x) = \sum_{i=0}^{n} a_i x^i.$$

What is its derivative? Well, it depends: if $n = 0$, then the derivative is zero. If, on the other hand, $n > 0$, then the derivative is a polynomial of degree $n - 1$:

$$p'(x) \equiv \begin{cases} 0 & \text{if } n = 0 \\ \sum_{i=1}^{n} a_i i x^{i-1} = \sum_{i=0}^{n-1} (i+1) a_{i+1} x^i & \text{if } n > 0. \end{cases}$$

25.6.2 Second Derivative

In either case, this is a new polynomial of degree $\max(0, n - 1)$. As such, it could be differentiated as well, to produce the second derivative of p (Chapter 24, Section 24.2.7):

$$p''(x) = \frac{d^2 p}{dx^2}(x) = (p'(x))'.$$

25.6.3 Higher Derivatives

This is a new polynomial of degree $\max(0, n - 2)$. As such, it could be differentiated as well, to produce the third derivative of p, and so on. In general, for $i = 0, 1, 2, \ldots$, the ith derivative of p is defined recursively by

$$p^{(i)} \equiv \begin{cases} p & \text{if } i = 0 \\ \left(p^{(i-1)}\right)' & \text{if } i > 0. \end{cases}$$

In this notation, the zeroth derivative is just the function itself:

$$p^{(0)} \equiv p,$$

and its derivative is

$$p^{(1)} \equiv p'.$$

Furthermore, the nth derivative of p is just a constant, or a polynomial of degree 0:

$$p^{(n)} = a_n n!.$$

For this reason, every higher derivative must vanish:

$$p^{(i)} = 0, \quad i > n.$$

25.7 Integration

25.7.1 Indefinite Integral

Let's move on to yet another analytic operation: integration. For this purpose consider again our original polynomial

$$p(x) = \sum_{i=0}^{n} a_i x^i.$$

As discussed in Chapter 24, Section 24.13.2, it has a primitive function (or antiderivative, or indefinite integral): a new polynomial of degree $n + 1$:

$$P(x) \equiv \sum_{i=0}^{n} \frac{a_i}{i+1} x^{i+1} = \sum_{i=1}^{n+1} \frac{a_{i-1}}{i} x^i.$$

Indeed, the derivative of P is just the original polynomial p:

$$P'(x) = p(x),$$

as required.

25.7.2 Definite Integral Over an Interval

The indefinite integral can now be used in a new (geometrical) task: to calculate the area underneath the graph of our original polynomial p. More precisely, let's bound the area from all four sides. For simplicity, assume that the graph of p is above the x-axis. This way, the graph bounds the area from above, and the x-axis bounds the area from below.

How to bound the area from the left and right? For this, issue two parallel verticals from the x-axis upwards: at $x = a$ on the left, and at $x = b$ on the right, where $a < b$ are some fixed real numbers.

How to calculate this area? It is the definite integral:

$$\int_a^b p(x)dx = P(b) - P(a)$$

(Chapter 24, Sections 24.13.2–24.13.4). If p is mostly negative, so its graph is mostly underneath the x-axis, then this area may be negative.

25.7.3 Examples

Here are some elementary examples. If $p(x) = x$ (a linear polynomial), then $P(x) = x^2/2$, so

$$\int_a^b x\,dx = \frac{b^2 - a^2}{2} = (b - a)\frac{a + b}{2}.$$

What is this? Just a product: the length of the interval, times the value of p at its midpoint. This is the trapezoidal (or the trapezoid, or the trapezium) rule.

If, on the other hand, $p(x) = x^2$ (a quadratic polynomial), then $P(x) = x^3/3$, so

$$\int_a^b x^2 dx = \frac{b^3 - a^3}{3}.$$

Finally, if $p(x) \equiv 1$ is just the constant function, then $P(x) = x$, so

$$\int_a^b dx = b - a.$$

What is this? Just the length of the original interval $[a, b]$.

25.7.4 Definite Integral Over The Unit Interval

Consider again our general polynomial p. Now, set $a \equiv 0$, and $b \equiv 1$. This is the unit interval $[0, 1]$. What is the definite integral of p over it? Thanks to the above formula, the result is

$$\int_0^1 p(x)dx = P(1) - P(0) = P(1) = \sum_{i=1}^{n+1} \frac{a_{i-1}}{i}.$$

25.8 Sparse Polynomials

25.8.1 Sparse Polynomials

So far, we've considered a dense polynomial, with many nonzero terms. To calculate such a polynomial, it makes sense to use Horner's algorithm (Section 25.2.4). But what about a sparse polynomial, with only a few nonzero terms? For such a polynomial, Horner's algorithm should better be modified.

How could a sparse polynomial look like? Well, it could contain just one nonzero term. This is just a monomial:

$$p(x) = x^k.$$

For a given x, how to calculate this? Well, Horner's algorithm is no good: it reduces to the naive algorithm in Section 25.5.2, which requires as many as $k - 1$ multiplications. By now, we already have a better way:

$$x^k = \begin{cases} 1 & \text{if } k = 0 \\ x & \text{if } k = 1 \\ x \cdot (x^2)^{(k-1)/2} & \text{if } k > 1 \text{ and } k \text{ is odd} \\ (x^2)^{k/2} & \text{if } k > 1 \text{ and } k \text{ is even.} \end{cases}$$

Could Horner's algorithm be rewritten in a new form, which will reduce to this method, rather than the naive method?

25.8.2 How to Write The Sparse Polynomial?

For this purpose, write the sparse polynomial as follows:

$$p(x) = a_0 + a_k x^k + a_l x^l + \cdots + a_n x^n.$$

In this writing style, we specify three terms only:

- The first term a_0 is either zero or nonzero.
- The next nonzero term is $a_k x^k$, where $a_k \neq 0$ is the next nonzero coefficient $(k \geq 1)$.
- The next nonzero term is $a_l x^l$, where $a_l \neq 0$ is the next nonzero coefficient $(l > k)$, and so on.

25.8.3 The Improved Horner Algorithm

In the original Horner algorithm, we pulled x out of parentheses. Here, on the other hand, we do better: pull not only x but also x^k:

$$p(x) = a_0 + x^k p_1(x),$$

where the new polynomial

$$p_1(x) \equiv a_k + a_l x^{l-k} + \cdots + a_n x^{n-k}$$

has degree $n - k$ only. This leads to the improved Horner algorithm:

$$p(x) = \begin{cases} a_0 + x^k p_1(x) & \text{if } a_0 \neq 0 \\ x^k p_1(x) & \text{if } a_0 = 0, \end{cases}$$

where $p_1(x)$ is calculated recursively by the same algorithm itself, and x^k is calculated efficiently as before:

$$x^k = \begin{cases} x & \text{if } k = 1 \\ x \cdot (x^2)^{(k-1)/2} & \text{if } k > 1 \text{ and } k \text{ is odd} \\ (x^2)^{k/2} & \text{if } k > 1 \text{ and } k \text{ is even.} \end{cases}$$

Clearly, if p is dense, then this is just the good old version. Let's mirror this in yet another task: composition.

25.8.4 Power of a Polynomial

So far, our task was easy: to calculate a polynomial. Next, let's move on to a more complicated task: constructing the entire vector of coefficients.

Let q be some polynomial, dense or sparse. Assume that we already have its complete vector of coefficients. How to have the vector of coefficients of q^k as well? Easy: just mirror x^k:

$$q^k = \begin{cases} 1 & \text{if } k = 0 \\ q & \text{if } k = 1 \\ q \cdot (q^2)^{(k-1)/2} & \text{if } k > 1 \text{ and } k \text{ is odd} \\ (q^2)^{k/2} & \text{if } k > 1 \text{ and } k \text{ is even.} \end{cases}$$

In this recursion, we use q^2. This is the vector of coefficients of q^2, calculated as in Section 25.1.5.

25.8.5 Composition

Assume that p is still sparse, and still has the same form as before:

$$p(x) = a_0 + x^k p_1(x).$$

How to compose $p \circ q$, and have all coefficients at the same time? Just mirror Horner's algorithm, in its improved version:

$$p \circ q = a_0 + q^k \cdot (p_1 \circ q).$$

Fortunately, p_1 is of degree $n - k < n$. Therefore, $p_1 \circ q$ could be obtained recursively (including all coefficients). This should be multiplied by q^k, to produce $q^k \cdot (p_1 \circ q)$ (including all coefficients). Finally, if $a_0 \neq 0$, then add a_0 to the first coefficient, as required.

25.9 Two Independent Variables

25.9.1 Polynomial of Two Variables

So far, we've discussed polynomials of one independent variable: x. Let's move on to polynomials of two independent variables: x and y.

What is a polynomial of two variables? It is a real function of the form

$$p : \mathbb{R}^2 \to \mathbb{R},$$

defined by

$$p(x, y) = \sum_{i=0}^{n} a_i(x) y^i,$$

where $a_i(x)$ ($0 \leq i \leq n$) is a polynomial in one variable: x only.

This way, $p(x, y)$ is a real polynomial. A complex polynomial, on the other hand, is slightly different: a complex function of the form

$$p : \mathbb{C}^2 \to \mathbb{C}$$

with the same structure as above, except that x, y, and $a_i(x)$ can now be not only real but also complex.

In what follows, we'll mostly discuss real polynomials. Complex polynomials could be treated in the same way.

25.9.2 Arithmetic Operations

How to carry out arithmetic operations between polynomials of two variables? Just as before (Sections 25.1.3–25.1.5). The only difference is that the coefficients are no longer constants, but polynomials in x. Fortunately, we already know how to add such polynomials. Let's go ahead and use this .

25.9.3 Addition

For this purpose, let

$$p(x,y) = \sum_{i=0}^{n} a_i(x)y^i \quad \text{and} \quad q(x,y) = \sum_{j=0}^{m} b_j(x)y^j$$

be two given polynomials of two variables (for some natural numbers $m \leq n$). To add them to each other, define dummy zero b_j's:

$$b_{m+1} \equiv b_{m+2} \equiv \cdots \equiv b_n \equiv 0.$$

We're now ready to add, term by term:

$$(p+q)(x,y) = p(x,y) + q(x,y) = \sum_{i=0}^{n} (a_i + b_i)(x)y^i.$$

After all, we already know how to add polynomials of one variable:

$$(a_i + b_i)(x), \quad 0 \leq i \leq n.$$

These are complete vectors of coefficients, as required.

25.9.4 Multiplication

Likewise, we already know how to multiply polynomials of one variable, to obtain a product like $a_i b_j$ (a new vector of coefficients). Let's use this to have the product of p and q as well:

$$(pq)(x,y) = p(x,y)q(x,y)$$

$$= \left(\sum_{i=0}^{n} a_i(x)y^i \right) \left(\sum_{j=0}^{m} b_j(x)y^j \right)$$

$$= \sum_{i=0}^{n} \left(\sum_{j=0}^{m} (a_i b_j)(x)y^{i+j} \right),$$

which is just the sum of $n+1$ polynomials of two variables.

25.10 Partial Differentiation

25.10.1 Partial Derivatives

So far, the polynomial

$$p(x,y) = \sum_{i=0}^{n} a_i(x)y^i$$

was viewed as an algebraic object, with arithmetic operations: addition and multiplication. Next, let's view it as an analytic object, with a new analytic operation: partial differentiation.

For this purpose, consider a particular point

$$(x, y) \in \mathbb{R}^2.$$

At this point, how fast does p grow in the x-direction? To answer this, view y as a fixed parameter, and differentiate p as a function of x alone. The result is called the partial derivative of p with respect to x:

$$p_x(x, y) \equiv \sum_{i=0}^{n} a_i'(x) y^i,$$

where $a_i'(x)$ is the derivative of $a_i(x)$.

This tells us how fast p grows in the x-direction, with y kept unchanged. Next, let's work the other way around: interchange the roles of x and y. This time, we ask: how fast does p grow in the y-direction? To find out, view x as a fixed parameter, and differentiate p as a function of y alone. The result is called the partial derivative of p with respect to y:

$$p_y(x, y) \equiv \begin{cases} 0 & \text{if } n = 0 \\ \sum_{i=1}^{n} a_i(x) i y^{i-1} = \sum_{i=0}^{n-1} (i+1) a_{i+1}(x) y^i & \text{if } n > 0. \end{cases}$$

Thee partial derivatives are polynomials of two variables in their own right. As such, they depend on the particular point (x, y) under consideration. After all, the rate of growth may change from point to point.

Thus, the partial derivatives could be differentiated as well, with respect to x or y. This will produce the second partial derivatives of p. Later on, we'll continue this process on and on, producing higher and higher partial derivatives.

25.10.2 Second Partial Derivatives

By now, we've differentiated p with respect to x and y, to produce the partial derivatives p_x and p_y. Actually, we should write $p_x(x, y)$ and $p_y(x, y)$, but we often drop the "(x, y)" for short.

Fortunately, both p_x and p_y are polynomials of two variables in their own right, which could be differentiated as well. For example, look at p_x, and differentiate it with respect to y. This produces the mixed partial derivative of p:

$$p_{xy}(x, y) \equiv (p_x(x, y))_y.$$

Here, we differentiated p with respect to x, then with respect to y. Still, order doesn't matter. In fact, we could also work the other way around: differentiate p with respect to y, then with respect to x:

$$p_{yx}(x, y) = \sum_{i=0}^{n-1} (i+1) a_{i+1}'(x) y^i = p_{xy}(x, y).$$

This is also called the $(1,1)$st partial derivative of p. After all, $x^1 = x$, and $y^1 = y$, so we could also write

$$p_{xy}(x,y) = p_{x^1y^1}(x,y).$$

With this notation, the $(0,0)$th partial derivative is nothing but p itself:

$$p_{x^0y^0}(x,y) = p(x,y).$$

This will be useful below.

25.10.3 High-Order Partial Derivatives

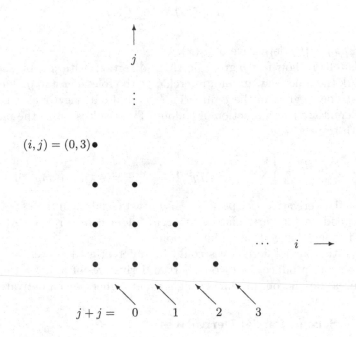

Fig. 25.2. To define partial derivatives of order $m = i + j = 0, 1, 2, 3, \ldots$, march in the two-dimensional grid: start from the origin, and march diagonal by diagonal. After all, each diagonal depends only on the previous diagonal below it.

Let's go ahead and continue this process. For example, the $(2,1)$st partial derivative of p is

$$p_{x^2y^1}(x,y) \equiv p_{xxy}(x,y) \equiv (p_{xx}(x,y))_y.$$

How to extend this further? Recursively, of course! For this purpose, look at the two-dimensional grid in Figure 25.2. Split it into parallel diagonals: the mth diagonal contains those partial derivatives of order m. What are these? These are the (i,j)th partial derivatives for which

$$i + j = m.$$

To define them, just start from the origin, and march diagonal by diagonal. After all, the mth diagonal depends on he $(m-1)$st diagonal only. This is nothing but mathematical induction on the order $m = 1, 2, 3, \ldots$:

$$p_{x^i y^j} \equiv \begin{cases} p & \text{if } i = j = 0 \\ \left(p_{x^{i-1}y^j}\right)_x & \text{if } i > 0 \\ \left(p_{x^i y^{j-1}}\right)_y & \text{if } j > 0. \end{cases}$$

Fortunately, there is no ambiguity: if both $i > 0$ and $j > 0$, then both definitions agree. Indeed, prove this in the same inductive process:

$$\left(p_{x^i y^{j-1}}\right)_y = \left(p_{x^{i-1}y^{j-1}}\right)_{xy} = \left(p_{x^{i-1}y^{j-1}}\right)_{yx} = \left(p_{x^{i-1}y^j}\right)_x.$$

This completes the induction step, and indeed the entire mathematical induction.

25.10.4 Using a Discrete Triangle

How many partial derivatives of order m or less are there? Let's mirror them by a discrete triangle of size m (Figure 8.5)). Thanks to Chapter 8, Section 8.3.4, the total number of points in it is

$$\binom{m+2}{2} = \frac{(m+2)!}{2! \cdot m!} = \frac{(m+1)(m+2)}{2}.$$

How many partial derivatives of order m are there? Thanks to Chapter 8, Section 8.3.5, the answer is

$$\binom{m+2-1}{2-1} = \binom{m+1}{1} = m + 1.$$

Indeed, here they are:

$$p_{x^0 y^m}, \ p_{x^1 y^{m-1}}, \ p_{x^2 y^{m-2}}, \ \ldots, \ p_{x^m y^0}.$$

25.10.5 Degree

In a polynomial of one variable, the degree is n: the highest power. In a polynomial of two variables like

$$p(x, y) = \sum_{i=0}^{n} a_i(x) y^i,$$

on the other hand, the degree is not necessarily n. To find the degree, some more work is needed.

For this purpose, write each polynomial $a_i(x)$ more explicitly:

$$a_i(x) = \sum_j a_{i,j} x^j,$$

where the $a_{i,j}$'s are some numbers. This way,

$$p(x, y) = \sum_i a_i(x) y^i = \sum_i \sum_j a_{i,j} x^j y^i.$$

The degree of p is the maximal sum $i + j$ for which $a_{i,j} \neq 0$. Geometrically, this is the maximal m in Figure 25.2 (the index of the last diagonal). This is why we often denote the degree by m. It is often greater than n.

25.10.6 Degree and Discrete Triangle

Look at a polynomial $p(x, y)$ of degree m. How many terms could it contain? Let's mirror it by a discrete triangle of size m (Figure 8.5)). Thanks to Chapter 8, Section 8.3.4, the total number of points in it is

$$\binom{m+2}{2} = \frac{(m+2)!}{m! \cdot 2!} = \frac{(m+1)(m+2)}{2}.$$

25.11 Three Independent Variables

25.11.1 Polynomial of Three Variables

So far, we've considered a polynomial of two variables: x and y. Let's go ahead and introduce one more variable: z.

A polynomial of three independent variables is defined by

$$p(x, y, z) \equiv \sum_{i=0}^{n} a_i(x, y) z^i,$$

where the coefficients a_i are now polynomials of two independent variables: x and y.

How to add, subtract, or multiply polynomials of three variables? Just as before. There is just one change: the a_i's are now polynomials of two variables in their own right. Fortunately, we already know how to "play" with them algebraically.

25.12 Partial Differentiation

25.12.1 Partial Derivatives

So, the algebraic bit is the same as before. What about the analytic bit? It is the same as before too. Indeed, to have the x-partial derivative, view both y and z as fixed parameters, and differentiate $p(x, y, z)$ as a function of x alone:

$$p_x(x, y, z) \equiv \sum_{i=0}^{n} (a_i)_x(x, y) z^i.$$

In this sum, the a_i's are differentiated with respect to x, as required. Fortunately, we already know how to do this.

Likewise, to have the y-partial derivative, view both x and z as fixed parameters, and differentiate $p(x, y, z)$ as a function of y alone:

$$p_y(x, y, z) \equiv \sum_{i=0}^{n} (a_i)_y(x, y) z^i.$$

Finally, to have the z-partial derivative, view both x and y as fixed parameters, and differentiate $p(x, y, z)$ as a function of z alone:

$$p_z(x, y, z) \equiv \begin{cases} 0 & \text{if } n = 0 \\ \sum_{i=1}^{n} a_i(x, y) i z^{i-1} = \sum_{i=0}^{n-1} (i+1) a_{i+1}(x, y) z^i & \text{if } n > 0. \end{cases}$$

25.12.2 Directional Derivative

So far, we've differentiated in a Cartesian direction: x or y or z. How about an oblique direction? How fast does p grow in such a direction? To find out, we must carry out a new analytic operation: directional differentiation.

For this purpose, let $\mathbf{n}$ be a fixed three-dimensional vector:

$$\mathbf{n} \equiv \begin{pmatrix} n_1 \\ n_2 \\ n_3 \end{pmatrix} \in \mathbb{R}^3.$$

Assume that $\mathbf{n}$ is a unit vector:

$$\|n\|^2 \equiv n_1^2 + n_2^2 + n_3^2 = 1.$$

How to differentiate p in direction $\mathbf{n}$? This is indeed the directional derivative of p, in direction n:

$$p_n(x,y,z) \equiv n_1 p_x(x,y,z) + n_2 p_y(x,y,z) + n_3 p_z(x,y,z).$$

This depends on the particular point (x, y, z) under consideration. After all, the partial derivatives may change from point to point. Still, this dependence goes without saying, and is often dropped:

$$p_n \equiv n_1 p_x + n_2 p_y + n_3 p_z.$$

Let's write this in a slightly different style.

25.12.3 Differential Operators

Let's define three new differential operators that map p to its partial derivative:

$$\frac{\partial}{\partial x} : p \to p_x$$

$$\frac{\partial}{\partial y} : p \to p_y$$

$$\frac{\partial}{\partial z} : p \to p_z.$$

This way, the directional-differentiation operator could now be written as a linear combination of these three:

$$\frac{\partial}{\partial n} \equiv n_1 \frac{\partial}{\partial x} + n_2 \frac{\partial}{\partial y} + n_3 \frac{\partial}{\partial z}.$$

How to have the directional derivative? Just apply this to p. This could be done not only once but also many times. For example, how to have the ith directional derivative? Just apply this i times. To open parentheses, use the trinomial formula.

25.12.4 Using The Binomial Formula

To make this more concrete, let's look at a simple example:

$$\mathbf{n} \equiv \frac{1}{\sqrt{2}}(1,1,0).$$

In this case,

$$\frac{\partial}{\partial n} = \frac{1}{\sqrt{2}}\left(\frac{\partial}{\partial x} + \frac{\partial}{\partial y}\right).$$

What is the ith directional derivative of a monomial of the form $x^a y^b$? Thanks to the binomial formula,

$$\frac{\partial^i}{\partial n^i}(x^a y^b) = \left(\frac{1}{\sqrt{2}}\left(\frac{\partial}{\partial x} + \frac{\partial}{\partial y}\right)\right)^i (x^a y^b)$$

$$= \frac{1}{2^{i/2}} \sum_{k=0}^{i} \binom{i}{k}\left(\frac{\partial}{\partial x}\right)^k \left(\frac{\partial}{\partial y}\right)^{i-k}(x^a y^b)$$

$$= \frac{1}{2^{i/2}} \sum_{k=0}^{i} \binom{i}{k} C_{a,k} C_{b,i-k} x^{a-k} y^{b-(i-k)},$$

where $C_{a,k}$ is as in Chapter 8, Section 8.6.1. For example, look at a special point at which $x = y$:

$$(x,y,z) = (x,x,0).$$

At this point, the above result simplifies to read

$$\frac{x^{a+b-i} C_{a+b,i}}{2^{i/2}}$$

(Chapter 8, Section 8.6.1).

25.12.5 Using The Trinomial Formula

Consider yet another simple example:

$$\mathbf{n} \equiv \frac{1}{\sqrt{3}}(1,1,1),$$

In this case,

$$\frac{\partial}{\partial n} = \frac{1}{\sqrt{3}}\left(\frac{\partial}{\partial x} + \frac{\partial}{\partial y} + \frac{\partial}{\partial z}\right).$$

To have the ith directional derivative, we need the ith power of this. To open parentheses, use the trinomial formula (Chapter 8, Section 8.6.3). Let's go ahead and apply this to a monomial of the form $x^a y^b z^c$:

$$\frac{\partial^i}{\partial n^i}(x^a y^b z^c) = \left(\frac{1}{\sqrt{3}}\left(\frac{\partial}{\partial x} + \frac{\partial}{\partial y} + \frac{\partial}{\partial z}\right)\right)^i (x^a y^b z^c)$$

$$= \frac{1}{3^{i/2}} \sum_{0 \le l,j,m \le i,\ l+j+m=i} \frac{i!}{l!j!m!}\left(\frac{\partial}{\partial x}\right)^l \left(\frac{\partial}{\partial y}\right)^j \left(\frac{\partial}{\partial z}\right)^m (x^a y^b z^c)$$

$$= \frac{1}{3^{i/2}} \sum_{0 \le l,j,m \le i,\ l+j+m=i} \frac{i!}{l!j!m!} C_{a,l} C_{b,j} C_{c,m} x^{a-l} y^{b-j} z^{c-m}.$$

Let's evaluate this at a special point, at which $x = y = z$:

$$(x, y, z) = (x, x, x).$$

At this point, the above result simplifies to read

$$\frac{x^{a+b+c-i} C_{a+b+c,i}}{3^{i/2}}$$

(Chapter 8, end of Section 8.6.3).

25.12.6   High-Order Partial Derivatives

We've already seen high-order partial derivatives in two places: in two variables, we've defined them recursively (Section 25.10.3). I three variables, on the other hand, we've used them implicitly in differential operators (Sections 25.12.4–25.12.5). Let's define them more explicitly.

For example, here is a mixed partial derivative: look at p_x, and differentiate it with respect to z:

$$p_{xz}(x, y, z) \equiv (p_x(x, y, z))_z.$$

From Section 25.12.1, order doesn't matter. In fact, you could equally well work the other way around: differentiate p with respect to z, and then with respect to x:

$$p_{zx}(x, y, z) = \sum_{i=0}^{n-1} (i+1)\,(a_{i+1})_x\,(x, y)z^i = p_{xz}(x, y, z).$$

In general, the (i, j, k)th partial derivative is obtained by a differential operator:

$$p_{x^i y^j z^k}(x, y, z) = \left(\left(\frac{\partial}{\partial x} \right)^i \left(\frac{\partial}{\partial y} \right)^j \left(\frac{\partial}{\partial z} \right)^k p \right)(x, y, z).$$

For example, the $(2, 1, 0)$th partial derivative of p is just

$$p_{x^2 y^1 z^0}(x, y, z) = p_{xxy}(x, y, z).$$

In particular, the $(0, 0, 0)$th partial derivative of p is just p itself:

$$p_{x^0 y^0 z^0}(x, y, z) \equiv p(x, y, z).$$

25.12.7   Mathematical Induction Triangle by Triangle

How to define partial derivatives of a yet higher order? This is illustrated geometrically in Figure 25.3. In the three-dimensional grid, the point (i, j, k) mirrors the (i, j, k)th partial derivative. What is the order of this partial derivative? It is just the sum: $m = i + j + k$. For a fixed order m, we have an oblique triangle, mirroring those partial derivatives of order m. To define them, use mathematical induction on $m = 1, 2, 3, \ldots$. After all, the mth triangle depends on the $(m-1)$st triangle only:

$$p_{x^i y^j z^k} \equiv \begin{cases} p & \text{if } i = j = k = 0 \\ \left(p_{x^{i-1} y^j z^k} \right)_x & \text{if } \quad i > 0 \\ \left(p_{x^i y^{j-1} z^k} \right)_y & \text{if } \quad j > 0 \\ \left(p_{x^i y^j z^{k-1}} \right)_z & \text{if } \quad k > 0. \end{cases}$$

As discussed in Section 25.10.3, there is no ambiguity here. This could be proved in the same inductive process.

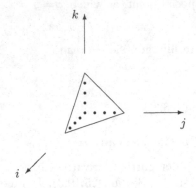

Fig. 25.3. To define partial derivatives of order $m = i + j + k = 1, 2, 3, \ldots$, march in the three-dimensional grid: start from the origin, and march triangle by triangle. After all, each triangle depends only on the previous triangle below it.

25.12.8 Using a Discrete Tetrahedron

How many partial derivatives of order m or less are there? Well, they are already mirrored by a discrete tetrahedron of size m (Figures 25.3 and 8.6. The total number of points in it is

$$\binom{m+3}{3} = \frac{(m+3)!}{3! \cdot m!} = \frac{(m+1)(m+2)(m+3)}{6}.$$

How many partial derivatives of order m are there? Thanks to Chapter 8, Section 8.3.5, the answer is

$$\binom{m+3-1}{3-1} = \binom{m+2}{2} = \frac{(m+2)!}{2! \cdot m!} = \frac{(m+1)(m+2)}{2}.$$

25.12.9 Degree and Discrete Tetrahedron

As discussed in Section 25.10.5, the inner polynomials $a_i(x, y)$ could be written more explicitly as

$$a_i(x, y) = \sum_j \sum_k a_{i,j,k} x^k y^j,$$

where the $a_{i,j,k}$'s are some numbers. Using this formulation, the original polynomial could be written as

$$p(x, y, z) = \sum_{i=0}^{n} \sum_j \sum_k a_{i,j,k} x^k y^j z^i.$$

Thus, the degree of p could be much larger than n: it is the maximal sum $i + j + $ for which $a_{i,j,k} \neq 0$. We often denote the degree by m.

Look at a polynomial $p(x, y, z)$ of degree m. How many terms could it contain Fortunately, they are mirrored by a discrete tetrahedron of size m (Figures 25.3 and 8.6), The total number of points in it is

$$\binom{m+3}{3} = \frac{(m+3)!}{m! \cdot 3!} = \frac{(m+1)(m+2)(m+3)}{6}.$$

25.13 Exercises: Convolution

25.13.1 Convolution and Polynomials

1. Let
$$u \equiv (u_0, u_1, u_2, \ldots, u_n) \equiv (u_i)_{0 \le i \le n}$$
be an $(n + 1)$-dimensional vector. Likewise, let
$$v \equiv (v_0, v_1, v_2, \ldots, v_m) \equiv (v_i)_{0 \le i \le m}$$
be an $(m + 1)$-dimensional vector. Complete both u and v into $(n + m + 1)$-dimensional vectors by adding dummy zero components:
$$u_{n+1} \equiv u_{n+2} \equiv \cdots \equiv u_{n+m} \equiv 0,$$
and
$$v_{m+1} \equiv v_{m+2} \equiv \cdots \equiv v_{n+m} \equiv 0.$$

2. The convolution of u and v (denoted by $u * v$) is a new $(n+m+1)$-dimensional vector, with the following components:
$$(u * v)_k \equiv \sum_{i=0}^{k} u_i v_{k-i}, \qquad 0 \le k \le n + m.$$

3. Show that convolution is commutative:
$$u * v = v * u.$$

4. Show that convolution is also associative.
5. Show that convolution is also distributive.
6. Use u as a vector of coefficients to help define a new polynomial p of degree n:
$$p(x) \equiv \sum_{i=0}^{n} u_i x^i = \sum_{i=0}^{n+m} u_i x^i.$$

7. Likewise, use v as a vector of coefficients to help define a new polynomial q of degree m:
$$q(x) \equiv \sum_{i=0}^{m} v_i x^i = \sum_{i=0}^{n+m} v_i x^i.$$

8. Consider the product pq. Is it a legitimate polynomial as well?
9. If so, what is its vector of coefficients? Hint: the convolution vector $u * v$:
$$(pq)(x) = \sum_{k=0}^{n+m} (u * v)_k x^k.$$

10. These are isomorphic algebras: polynomials (with their product) mirror vectors (with their convolution).
11. Thus, both algebras have the same algebraic properties.
12. In particular, both are commutative, associative, and distributive.
13. For polynomials, this is easy to prove. Use this to prove once again that convolution of vectors is indeed commutative, associative, and distributive as well.

25.13.2   Polar Decomposition

1. The infinite Taylor expansion of the exponent function around zero is

$$\exp(x) = 1 + x + \frac{x^2}{2} + \frac{x^3}{3!} + \frac{x^4}{4!} + \cdots = \sum_{n=0}^{\infty} \frac{x^n}{n!}.$$

   (See exercises at the end of Chapter 24.)
2. Truncate this series after $k + 1$ terms. This approximates $\exp(x)$ by a new polynomial of degree k:

$$\exp(x) \doteq \sum_{n=0}^{k} \frac{x^n}{n!}.$$

3. For a moderate $|x|$, this is indeed a good approximation.
4. For a big $|x|$, on the other hand, increase k until the tail is small enough.
5. To calculate this, design a new version of Horner's algorithm. Hint: avoid dividing by $n!$. The solution can be found in Chapter 1 in [44].
6. Likewise, the infinite Taylor expansion of the sine function around zero is

$$\sin(x) = x - \frac{x^3}{3!} + \frac{x^5}{5!} - \frac{x^7}{7!} + \cdots = \sum_{n=0}^{\infty} (-1)^n \frac{x^{2n+1}}{(2n+1)!}.$$

7. Do the same to it too.
8. The infinite Taylor expansion of the cosine function around zero is

$$\cos(x) = 1 - \frac{x^2}{2} + \frac{x^4}{4!} - \frac{x^6}{6!} + \cdots = \sum_{n=0}^{\infty} (-1)^n \frac{x^{2n}}{(2n)!}.$$

9. Do the same to it too.
10. Conclude that

$$\exp(i\theta) = \cos(\theta) + i\sin(\theta),$$

   where $i \equiv \sqrt{-1}$ is the imaginary number. This is the polar decomposition of the complex number $\exp(i\theta)$, where $0 \le \theta < 2\pi$ is the angle that it makes with the positive part of the real axis (Figures 24.4–24.5). Hint: add the above expansions, term by term. (See exercises at the end of Chapter 24.)

Matrices
and Their Eigenvalues

Throughout the book, we use matrices in many places. In special relativity, we often use a small 2×2 matrix: the Lorentz matrix (Chapter 10, Section 10.4.3). In general relativity, on the other hand, we use a 4×4 matrix: the metric (Chapter 11, Section 11.12.3). In quantum mechanics, on the other hand, we often use a yet bigger (and even infinite) matrix to model an observable like position, momentum, energy, angular momentum, spin, or polarization (Chapters 12–13 and 23). This is indeed how linear algebra helps understand physics better. For example, the eigenvalues of the infinite energy matrix have a new physical meaning: energy levels.

This is quite practical: it can be used to model the distribution of electrons in the atom. For this purpose, we need yet another algebraic tool: the determinant of a matrix (Chapter 13, Section 13.6.2).

In this chapter, we introduce eigenvalues and eigenvectors from scratch, with their algebraic properties. In particular, we focus on a few important kinds of matrices: Hermitian, orthogonal, unitary, or positive definite. These are particularly useful in both classical and quantum mechanics.

26.1 Complex Matrix and Its Hermitian Adjoint

26.1.1 The Hermitian Adjoint

We start with a rectangular matrix (not necessarily square). We assume that we already know how to multiply matrix-times-matrix and matrix-times-vector [45].

Often, one only needs a real matrix, with real elements (entries) in $\mathbb{R}$. Here, on the other hand, we consider a more general case: an $m \times n$ complex matrix, with complex elements in $\mathbb{C}$.

Fortunately, complex matrices still have the same algebraic properties, including the distributive and associative laws. Instead of the transpose, we now have a better notion: the Hermitian adjoint.

Let $i \equiv \sqrt{-1}$ be the imaginary number. Recall that the complex number

$$c \equiv a + bi$$

has the complex conjugate

$$\bar{c} \equiv a - bi$$

(Figure 3.18). This helps define the Hermitian adjoint of a given complex matrix. This could be done in two stages:

- First, take the transpose.
- Then, for each individual element, take the complex conjugate.

This way, the original $m \times n$ matrix

$$A \equiv (a_{i,j})_{1 \leq i \leq m, \ 1 \leq j \leq n}$$

makes a new $n \times m$ matrix, denoted by A^h (or A^*, or $A^\dagger$):

$$A^h_{j,i} \equiv \bar{a}_{i,j}, \quad 1 \leq i \leq m, \ 1 \leq j \leq n.$$

In particular, if A happens to be real, then the complex conjugate is the same as the original number. In this case, the Hermitian adjoint is just the good old transpose:

$$A^h = A^t.$$

Thus, the Hermitian adjoint actually extends the transpose to the wider class of complex matrices. This is why it also inherits a few nice properties:

$$(A^h)^h = A,$$

and

$$(BA)^h = A^h B^h,$$

where the number of columns in B is the same as the number of rows in A.

26.1.2 Hermitian (Self-Adjoint) Matrix

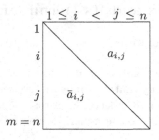

Fig. 26.1. In a Hermitian matrix, the lower triangular part mirrors the upper triangular part, with a bar on top.

Consider now a square matrix A, of order $m = n$. We say that A is Hermitian (or self-adjoint) if it is the same as its Hermitian adjoint:

$$A = A^h.$$

In this case, the main diagonal acts like a mirror (Figure 26.1). In fact, the lower triangular part mirrors the upper triangular part, with a bar on top:

$$a_{j,i} = \left(A^h\right)_{j,i} = \bar{a}_{i,j}, \quad 1 \le i,j \le n.$$

This is why the main-diagonal must be real:

$$a_{i,i} = \bar{a}_{i,i}, \quad 1 \le i \le n.$$

26.2 Inner Product and Norm

26.2.1 Inner Product

Let

$$u \equiv \begin{pmatrix} u_1 \\ u_2 \\ \vdots \\ u_n \end{pmatrix} \quad \text{and} \quad v \equiv \begin{pmatrix} v_1 \\ v_2 \\ \vdots \\ v_n \end{pmatrix}$$

be two column vectors in $\mathbb{C}^n$. They can also be viewed as narrow $n \times 1$ "matrices," with just one column. This way, they also have their own Hermitian adjoint:

$$u^h = (\bar{u}_1, \bar{u}_2, \ldots, \bar{u}_n) \quad \text{and} \quad v^h = (\bar{v}_1, \bar{v}_2, \ldots, \bar{v}_n).$$

These are $1 \times n$ "matrices," or n-dimensional row vectors.

Thus, u^h has n "columns," and v has n "rows." Fortunately, this is the same number. So, we can now go ahead and multiply u^h times v. The result is a new 1×1 "matrix," or just a new (complex) scalar:

$$(u, v) \equiv u^h v = \sum_{j=1}^{n} \bar{u}_j v_j.$$

This is called the scalar (or inner) product of u and v.

26.2.2 Bilinearity

What are the algebraic properties of the inner product? It is bilinear! To see this, let $c \in \mathbb{C}$ be some complex number. Now, in our inner product, we could "pull" c out (possibly with a bar on top):

$$(cu, v) = \bar{c}(u, v), \quad \text{and} \quad (u, cv) = c(u, v).$$

Furthermore, let w be yet another vector. Then, we also have a distributive law:

$$(u + w, v) = (u, v) + (w, v) \quad \text{and} \quad (u, v + w) = (u, v) + (u, w).$$

Thus, the inner product is bilinear.

26.2.3 Skew-Symmetry

Moreover, the inner product is skew-symmetric: interchanging u and v just places a new bar on top. Indeed,

$$(v, u) = \sum_{j=1}^{n} \bar{v}_j u_j$$

is just the complex conjugate of (u, v).

26.2.4 Norm

What is the inner product of v with itself? This is a real nonnegative number:

$$(v, v) = \sum_{j=1}^{n} \bar{v}_j v_j = \sum_{j=1}^{n} |v_j|^2 \geq 0.$$

Could this vanish? Only if v was the zero vector:

$$(v, v) = 0 \quad \Leftrightarrow \quad v = \mathbf{0}.$$

Thus, (v, v) has a square root. Let's use it to define the norm (or length, or magnitude) of v:

$$\|v\| \equiv +\sqrt{(v, v)}.$$

This way,

$$\|v\| \geq 0,$$

and

$$\|v\| = 0 \quad \Leftrightarrow \quad v = \mathbf{0},$$

as expected from a magnitude.

26.2.5 Normalization

Furthermore, every complex number $c \in \mathbb{C}$ could be "pulled" out of the $\| \cdot \|$ sign:

$$\|cv\| = \sqrt{(cv, cv)} = \sqrt{\bar{c}c(v, v)} = \sqrt{|c|^2(v, v)} = |c|\sqrt{(v, v)} = |c| \cdot \|v\|.$$

This way, if v is a nonzero vector, then it could be normalized. Indeed, since $\|v\| > 0$ one could pick $c = 1/\|v\|$, to produce the normalized unit vector $v/\|v\|$: the unique vector of norm 1 that is proportional to v.

26.2.6 Other Norms

The norm $\|v\|$ defined above is also called the l_2-norm, and is also denoted by $\|v\|_2$ to distinguish it from other norms: the l_1-norm

$$\|v\|_1 \equiv \sum_{i=1}^{n} |v_i|,$$

and the l_∞- norm (or the maximum norm)

$$\|v\|_\infty \equiv \max_{1 \leq i \leq n} |v_i|.$$

Like the l_2-norm defined above, these norms also make sense: they vanish if and only if v is the zero vector. Furthermore, every complex number $c \in \mathbb{C}$ could be pulled out:

$$\|cv\|_1 = |c| \cdot \|v\|_1, \quad \text{and} \quad \|cv\|_\infty = |c| \cdot \|v\|_\infty.$$

In what follows, however, we mostly use the l_2-norm. This is why we denote it simply by $\|v\|$ for short, rather than $\|v\|_2$.

26.2.7 Inner Product and The Hermitian Adjoint

Assume again that A is an $m \times n$ rectangular complex matrix. Let u and v be complex vectors of different dimensions: u is m-dimensional, and v is n-dimensional. This way, Av is m-dimensional as well. We can now take the inner product of two m-dimensional vectors:

$$(u, Av) \equiv u^h Av.$$

This is a well-defined complex number (Section 26.2.1).

Furthermore, A^h is an $n \times m$ rectangular complex matrix. This way, A^h can be applied to u: the result $A^h u$ is n-dimensional. For this reason, the inner product $(A^h u, v)$ is a well-defined complex number as well. Is it the same as (u, Av)?

To find out, recall that both u and v could also be viewed as narrow matrices. Thanks to associativity, we now have

$$(u, Av) = u^h(Av) = (u^h A)v = (A^h u)^h v = (A^h u, v).$$

Let's use this result in a special case: a square matrix.

26.2.8 Inner Product and Hermitian Matrix

Let's look at an interesting case: $m = n$, so A is now a square matrix. Assume that A is also Hermitian:

$$A = A^h$$

(Section 26.1.2). In this case, the above formula reduces to

$$(u, Av) = (Au, v),$$

for every two n-dimensional complex vectors u and v.

Next, let's work the other way around. Assume that we didn't know that A was Herminian. Instead we only knew that A was a square complex matrix of order $m = n$, satisfying

$$(u, Av) = (Au, v)$$

for every two n-dimensional complex vectors u and v. Could we then deduce that A was Hermitian?

To find out, let's pick u and v cleverly. For each pair of natural numbers $1 \leq i, j \leq n$, let u and v be standard unit vectors, with one nonzero component only:

$$u_k = \begin{cases} 1 & \text{if } k = i \\ 0 & \text{if } k \neq i \end{cases} \qquad (1 \leq k \leq n)$$

$$v_k = \begin{cases} 1 & \text{if } k = j \\ 0 & \text{if } k \neq j \end{cases} \qquad (1 \leq k \leq n).$$

With this choice,

$$a_{i,j} = (u, Av) = (Au, v) = \bar{a}_{j,i},$$

implying that A is indeed Hermitian (Figure 26.1).

26.3 Orthogonal and Unitary Matrix

26.3.1 Inner Product of Column Vectors

Assume again that A is an $m \times n$ rectangular complex matrix. For $1 \leq j \leq n$ let $v^{(j)}$ denote the jth column in A: an m-dimensional column vector. Now, for $1 \leq i, j \leq n$, what is the inner product of the ith and jth columns? Well, it is the same as the (i, j)th element in $A^h A$. Indeed,

$$\begin{aligned}
(A^h A)_{i,j} &= \sum_{k=1}^{m} \left(A^h\right)_{i,k} a_{k,j} \\
&= \sum_{k=1}^{m} \bar{a}_{k,i} a_{k,j} \\
&= \sum_{k=1}^{m} \bar{v}_k^{(i)} v_k^{(j)} \\
&= \left(v^{(i)}, v^{(j)}\right).
\end{aligned}$$

This formula will be useful below.

26.3.2 Orthogonal and Orthonormal Column Vectors

Consider two complex vectors u and v, of the same dimension. We say that they are *orthogonal* to each other if their inner product vanishes:

$$(u, v) = 0.$$

Furthermore, we also say that u and v are *orthonormal* if they are not only orthogonal but also have norm 1:

$$(u, v) = 0, \quad \text{and} \quad \|u\| = \|v\| = 1.$$

Consider again our $m \times n$ rectangular complex matrix, written column by column

$$A \equiv \left(v^{(1)} \mid v^{(2)} \mid \cdots \mid v^{(n)}\right),$$

for some $m \geq n$. Assume that its columns are orthonormal:

$$\left(v^{(i)}, v^{(j)}\right) = 0, \quad 1 \le i, j \le n, \; i \ne j,$$

and

$$\left\|v^{(i)}\right\| = 1, \quad 1 \le i \le n.$$

In this case, what is the (i, j)th element in $A^h A$? Well, it is either zero or one:

$$(A^h A)_{i,j} = \left(v^{(i)}, v^{(j)}\right) = \begin{cases} 1 & \text{if } i = j \\ 0 & \text{if } i \ne j \end{cases}$$

(Section 26.3.1). In other words, $A^h A$ is just the identity matrix of order n:

$$A^h A = I.$$

As a result, A preserves the inner product of any two n-dimensional vectors u and v:

$$(Au, Av) = (A^h Au, v = (Iu, v)) = (u, v)$$

(Section 26.2.7). In particular, by picking $u = v$, A preserves norm as well:

$$\|Av\|^2 = (Av, Av) = (v, v) = \|v\|^2.$$

Next, let's work the other way around. Assume that we didn't know that A had orthogonal columns. Instead, we only knew that $A^h A$ was the identity matrix:

$$A^h A = I.$$

Could we then deduce that A must also have orthonormal columns? Fortunately, yes. Indeed, from Section 26.3.1,

$$\left(v^{(i)}, v^{(j)}\right) = (A^h A)_{i,j} = I_{i,j} = \begin{cases} 1 & \text{if } i = j \\ 0 & \text{if } i \ne j, \end{cases}$$

as asserted. In summary, A has orthonormal columns if and only if

$$A^h A = I$$

is the identity matrix of order $n \le m$.

26.3.3 Projection Matrix and Its Null Space

So far, we've studied the product $A^h A$. Next, let's multiply the other way around: AA^h. This is a different matrix: after all, the commutative law doesn't hold. Still, is AA^h special in any way?

Well, thanks to the above assumptions ($m \ge n$, and A has orthonormal columns), it sure is. To see how, multiply the above equation by A from the left:

$$A\left(A^h A\right) = AI = A.$$

Thanks to associativity,

$$\left(AA^h\right) A = A\left(A^h A\right) = AI = A.$$

Thus, AA^h has no effect on the columns of A: it leaves them unchanged. In this respect, it behaves quite like the identity matrix.

Still, this is not the whole story. Since $m \geq n$, there are also vectors that are orthogonal to all columns of A. Let v be such a vector:

$$A^h v = \mathbf{0}.$$

What is the effect of AA^h on v? To see this, multiply the above equation by A from the left:

$$A\left(A^h v\right) = A\mathbf{0} = \mathbf{0}.$$

Thanks to associativity,

$$\left(AA^h\right) v = A\left(A^h v\right) = A\mathbf{0} = \mathbf{0}.$$

Thus, v is in the null space of AA^h. In summary, AA^h is nearly the identity matrix: on the columns of A, it behaves like the identity matrix. On vectors that are orthogonal to the columns of A, on the other hand, it acts like the zero matrix.

As a matter of fact, we could characterize AA^h more concisely. To do this, look again at our original equation

$$AA^h A = A.$$

Next, multiply this by A^h from the right. Thanks to associativity, we obtain

$$\left(AA^h\right)^2 = \left(AA^h\right)\left(AA^h\right) = AA^h.$$

Thus, AA^h is a projection matrix: it is the same as its own square.

26.3.4 Unitary and Orthogonal Matrix

Assume now that A is a square matrix of order $m = n$. If A has orthonormal columns, then A is called a unitary matrix. In this case, the above formulas tell us that

$$A^h A = AA^h = I$$

(the identity matrix of order n). This could also be written in terms of the inverse matrix:

$$A^h = A^{-1}, \quad \text{and} \quad A = \left(A^h\right)^{-1}.$$

Now, the inverse of the transpose is just the transpose of the inverse. Therefore, one oculd drop these parentheses, and simply write

$$A = A^{-h}.$$

If A is also real, then it is also called an orthogonal matrix. In this case,

$$A^t A = AA^t = I,$$

so

$$A^t = A^{-1}, \quad \text{and} \quad A = (A^t)^{-1} = A^{-t}.$$

26.4 Eigenvalues and Eigenvectors

26.4.1 Eigenvectors and Their Eigenvalues

Let A be a square complex matrix of order n. An eigenvector of A is a nonzero vector $v \in \mathbb{C}^n$ satisfying

$$Av = \lambda v,$$

for some number $\lambda \in \mathbb{C}$, called the eigenvalue. In other words, applying A to v has the same effect as multiplying v by λ. Thus, λ is the eigenvalue of A associated with the eigenvector v.

Note that v is not unique: it could be scalaed. In fact, v could be multiplied by just any nonzero number $c \in \mathbb{C}$, producing a "new" eigenvector: cv. Indeed, cv satisfies the above equation as well:

$$A(cv) = cAv = c\lambda v = \lambda(cv).$$

Thus, the eigenvector associated with λ is not defined uniquely, but only up to a (nonzero) scalar multiple.

What is the best way to pick c? Well, since $v \neq \mathbf{0}$, $\|v\| > 0$ (Section 26.2.4). Thus, best pick $c \equiv 1/\|v\|$. This would "normalize" v, and produce a "new" unit eigenvector:

$$A\left(\frac{v}{\|v\|}\right) = \lambda\left(\frac{v}{\|v\|}\right), \qquad \left\|\frac{v}{\|v\|}\right\| = 1.$$

This is the unique unit eigenvector proportional to v.

26.4.2 Singular Matrix and Its Null Space

What are the algebraic properties of the eigenvector v? First of all, it is nonzero:

$$v \neq \mathbf{0},$$

where $\mathbf{0}$ is the n-dimensional zero vector. Furthermore, v also satisfies

$$(A - \lambda I)\, v = \lambda v - \lambda v = \mathbf{0},$$

where I is the identity matrix of order n. Thus, the matrix $A - \lambda I$ maps v to $\mathbf{0}$. In other words, v is in the null space of $A - \lambda I$.

For this reason, $A - \lambda I$ must be singular (noninvertible). Indeed, by contradiction: if there were an inverse matrix $(A - \lambda I)^{-1}$, then we could apply it to the zero vector, to map it back to v. On the other hand, from the very definition of matrix-times-vector, this must be the zero vector:

$$v = (A - \lambda I)^{-1}\mathbf{0} = \mathbf{0},$$

in violation of the very definition of v as a nonzero eigenvector. Thus, $A - \lambda I$ must indeed be singular, as asserted. Let's use this to design an eigenvalue for A^h as well.

26.4.3 Eigenvalues of The Hermitian Adjoint

So far, we discussed an eigenvalue of A: λ (associated with the eigenvector v). What about A^h? It also has an eigenvalue: $\bar{\lambda}$.

Indeed, A, λ, and v have a joint algebraic property: $A - \lambda I$ is singular. What about its Hermitian adjoint

$$(A - \lambda I)^h = A^h - \bar{\lambda} I?$$

Is it singular as well? It sure is. (After all, it has the same determinant: 0. Still, let's prove this without using determinants.)

Indeed, by contradiction: if it were nonsingular (invertible), then there would be some nonzero vector u, mapped to v:

$$\left(A^h - \bar{\lambda} I\right) u = v.$$

This would lead to a contradiction:

$$0 = (\mathbf{0}, u) = ((A - \lambda I) v, u) = \left(v, \left(A^h - \bar{\lambda} I\right) u\right) = (v, v) > 0.$$

So, $A^h - \bar{\lambda} I$ must be singular as well. As such, it must map some nonzero vector $w \neq \mathbf{0}$ to the zero vector:

$$\left(A^h - \bar{\lambda} I\right) w = \mathbf{0},$$

or

$$A^h w = \bar{\lambda} w.$$

Thus, $\bar{\lambda}$ must be an eigenvalue of A^h. In summary, the complex conjugate of any eigenvalue of A is an eigenvalue of A^h.

26.4.4 Eigenvalues of a Hermitian Matrix

Assume now that A is also Hermitian:

$$A = A^h.$$

Thanks to Section 26.2.8,

$$\lambda(v, v) = (v, \lambda v) = (v, Av) = (Av, v) = (\lambda v, v) = \bar{\lambda}(v, v).$$

Now, since v is a nonzero vector, $(v, v) > 0$ (Section 26.2.4). Thus, we can divide by (v, v):

$$\lambda = \bar{\lambda},$$

so λ is real:

$$\lambda \in \mathbb{R}.$$

In summary, a Hermitian matrix has real eigenvalues only. As a matter of fact, in the "world" of matrices, a Hermitian matrix plays the role of a real number.

26.4.5 Eigenvectors of a Hermitian Matrix

So far, we've discussed the eigenvalues of a Hermitian matrix, and saw that they must be real. What about the eigenvectors? What are their algebraic properties?

To answer this, let u and v be two eigenvectors of the Hermitian matrix A:

$$Av = \lambda v \quad \text{and} \quad Au = \mu u,$$

where $\lambda \neq \mu$ are two distinct eigenvalues. What is the relation between u and v? Well, it turns out that they are orthogonal to each other. Indeed, thanks to Sections 26.2.8 and 26.4.4,

$$\mu(u, v) = \bar{\mu}(u, v) = (\mu u, v) = (Au, v) = (u, Av) = (u, \lambda v) = \lambda(u, v).$$

Now, since $\lambda \neq \mu$, we must have

$$(u, v) = 0,$$

as asserted. Furthermore, we can normalize both u and v, to obtain the orthonormal eigenvectors $u/\|u\|$ and $v/\|v\|$. This will be useful later.

26.5 Positive (Semi)definite Matrix

26.5.1 Positive Semidefinite Matrix

Consider a Hermitian matrix A. Assume that, for every complex vector v,

$$(v, Av) \equiv v^h Av \geq 0.$$

We then say that A is positive semidefinite.

In particular, we could pick v to be an eigenvector of A, with the eigenvalue λ. This would give

$$\|v\|^2 \lambda = (v, v)\lambda = (v, \lambda v) = (v, Av) \geq 0.$$

As an eigenvector, v is a nonzero vector, with a nonzero norm. Therefore, we could divide by $\|v\|^2$:

$$\lambda \geq 0.$$

Thus, all eigenvalues of A must be nonnegative.

Next, let's work the other way around: if all eigenvalues of a Hermitian matrix are nonnegative, then it must be positive semidefinite. Indeed, given a complex vector v, just decompose it as a linear combination of (orthogonal) eigenvectors.

26.5.2 Positive Definite Matrix

Consider again a Hermitian matrix A. Assume that, for every nonzero vector $v \neq \mathbf{0}$, we even have a strict inequality:

$$(v, Av) \equiv v^h Av > 0.$$

We then say that A is not only positive semidefinite but also positive definite.

In particular, we could pick v to be an eigenvector of A, with the eigenvalue λ This would give

$$\|v\|^2\lambda = (v, v)\lambda = (v, \lambda v) = (v, Av) > 0.$$

Next, divide by $\|v\|^2 > 0$:

$$\lambda > 0.$$

Therefore, all eigenvalues of A are positive. Thus, A must be nonsingular: 0 is no an eigenvalue.

Next, let's work the other way around: if all eigenvalues of a Hermitian matri are positive (so it is positive semidefinite and nonsingular at the same time), ther it must be positive definite. Indeed, given a nonzero vector $v \neq \mathbf{0}$, just decompos it as a linear combination of (orthogonal) eigenvectors.

26.6 Exercises: Generalized Eigenvalues

26.6.1 The Cauchy–Schwarz Inequality

1. Let u and v be n-dimensional (complex) vectors:

$$u \equiv (u_1, u_2, \ldots, u_n)^t \text{ and } v \equiv (v_1, v_2, \ldots, v_n)^t.$$

Prove the Cauchy–Schwarz inequality:

$$|(u, v)| \leq \|u\| \cdot \|u\|,$$

or

$$\left|\sum_{i=1}^n \bar{u}_i v_i\right|^2 = |(u, v)|^2 \leq (u, u)(v, v) = \left(\sum_{i=1}^n |u_i|^2\right)\left(\sum_{i=1}^n |v_i|^2\right).$$

Hint: pick two different indices between 1 and n: $0 \leq i \neq j \leq n$. Note that

$$
\begin{aligned}
0 &\leq |u_i v_j - u_j v_i|^2 \\
&= (\bar{u}_i \bar{v}_j - \bar{u}_j \bar{v}_i)(u_i v_j - u_j v_i) \\
&= \bar{u}_i \bar{v}_j u_i v_j - \bar{u}_i \bar{v}_j u_j v_i - \bar{u}_j \bar{v}_i u_i v_j + \bar{u}_j \bar{v}_i u_j v_i \\
&= |u_i|^2 |v_j|^2 - (\bar{u}_i v_i)(u_j \bar{v}_j)) - (\bar{u}_j v_j)(u_i \bar{v}_i) + |u_j|^2 |v_i|^2,
\end{aligned}
$$

so

$$(\bar{u}_i v_i)(u_j \bar{v}_j)) + (\bar{u}_j v_j)(u_i \bar{v}_i) \leq |u_i|^2 |v_j|^2 + |u_j|^2 |v_i|^2.$$

Do the same with the plus sign, to obtain

$$|(\bar{u}_i v_i)(u_j \bar{v}_j)) + (\bar{u}_j v_j)(u_i \bar{v}_i)| \leq |u_i|^2 |v_j|^2 + |u_j|^2 |v_i|^2.$$

Finally, sum this over $1 \leq i < j \leq n$.

2. Could the Cauchy–Schwarz inequality be an exact equality as well? Hint: onl if u is proportional to v (or is a scalar multiple of v).

26.6.2 The Triangle Inequality

1. Conclude the triangle inequality:

$$\|u + v\| \leq \|u\| + \|v\|.$$

Hint:

$$
\begin{aligned}
\|u + v\|^2 &= (u + v, u + v) \\
&= (u, u) + (u, v) + (v, u) + (v, v) \\
&= \|u\|^2 + (u, v) + (v, u) + \|v\|^2 \\
&\leq \|u\|^2 + 2|(u, v)| + \|v\|^2 \\
&\leq \|u\|^2 + 2\|u\| \cdot \|v\| + \|v\|^2 \\
&= (\|u\| + \|v\|)^2 .
\end{aligned}
$$

26.6.3 Generalized Eigenvalues

1. Let A be a square complex matrix of order n. Let B be a Hermitian matrix of order n. Let v be an n-dimensional vector satisfying

$$Av = \lambda Bv \quad \text{and} \quad (v, Bv) \neq 0.$$

Then the (complex) scalar λ is called a generalized eigenvalue of A.
2. Could v be the zero vector?
3. Conclude that $A - \lambda B$ maps v to the zero vector.
4. Conclude that v is in the null space of $A - \lambda B$.
5. Is $A - \lambda B$ singular? Hint: otherwise,

$$v = (A - \lambda B)^{-1} \mathbf{0} = \mathbf{0},$$

in violation of $(v, Bv) \neq 0$.
6. Is $A^h - \bar{\lambda} B$ singular as well? Hint: otherwise, there would be a nonzero vector u mapped to v:

$$\left(A^h - \bar{\lambda} B \right) u = v,$$

leading to a contradiction:

$$0 = (\mathbf{0}, u) = ((A - \lambda B) v, u) = \left(v, \left(A^h - \bar{\lambda} B \right) u \right) = (v, v) > 0.$$

7. Prove this in yet another way. Hint:

$$\det \left(A^h - \bar{\lambda} B \right) = \overline{\det (A - \lambda B)} = 0.$$

8. Conclude that $A^h - \bar{\lambda} B$ has a nontrivial null space.
9. Conclude that $\bar{\lambda}$ is a generalized eigenvalue of A^h.
10. Assume now that A is Hermitian as well. Must the generalized eigenvalue λ be real? Hint:

$$\lambda(v, Bv) = (v, \lambda Bv) = (v, Av) = (Av, v) = (\lambda Bv, v) = \bar{\lambda}(Bv, v) = \bar{\lambda}(v, Bv).$$

Finally, divide this by $(v, Bv) \neq 0$.

11. Let u and v be two such vectors:

$$Av = \lambda Bv \quad \text{and} \quad Au = \mu Bu,$$

where $\lambda \neq \mu$ are two distinct generalized eigenvalues. Show that

$$(u, Bv) = 0.$$

Hint:

$$\begin{aligned}
\mu(u, Bv) &= \mu(Bu, v) \\
&= \bar{\mu}(Bu, v) \\
&= (\mu Bu, v) \\
&= (Au, v) \\
&= (u, Av) \\
&= (u, \lambda Bv) \\
&= \lambda(u, Bv).
\end{aligned}$$

12. Assume now that B is not only Hermitian but also positive definite. Define the new "inner product:"
$$(\cdot, \cdot)_B \equiv (\cdot, B\cdot).$$

Show that this is indeed a legitimate inner product. Hint: show that it is indeed bilinear, skew-symmetric, and positive, and vanishes only for the zero vector.

13. Look at the new matrix $B^{-1}A$. Show that it is "Hermitian" with respect to the new "inner product:"

$$\left(B^{-1}A\cdot, \cdot\right)_B = (A\cdot, \cdot) = (\cdot, A\cdot) = \left(\cdot, B^{-1}A\cdot\right)_B.$$

14. Show that the above u and v are eigenvectors of $B^{-1}A$.
15. Show that they are indeed orthogonal with respect to the new "inner product."
16. Normalize them with respect to the new "inner product."
17. For example, set $B \equiv I$. Is this familiar?

References

1. Adachi, H., Mukoyama, T. and Kawai, J.: *Hartree-Fock-Slater Method for Materials Science: the DV-X Alpha Method for Design and Characterization of Materials.* Springer, 2006.
2. Alcubierre, M.: Introduction to 3+1 Numerical Relativity. Oxford Univ. Press (2012).
3. Anthony, M., and Harvey, M.: *Linear Algebra: Concepts and Methods.* Cambridge University Press, Cambridge, 2012.
4. Baumgarte, T.W., and Shapiro, S.L.: Numerical Relativity: Solving Einstein Equations on the Computer. Cambridge Univ. Press (2010).
5. Bransden, B.H., and Joachain, C.J.: *Quantum Mechanics.* Prentice-Hall, 2000.
6. Calkin, M.G.: *Lagrangian and Hamiltonian Mechanics.* WSPC, 1996.
7. Carlson, B.: *Lecture Notes on Quantum Mechanics,* 2019.
8. Carroll, S.: *Something Deeply Hidden: Quantum Worlds and the Emergence of Spacetime.* Oneworld Publications (2019).
9. Chirvasa, M.: Finite Difference Methods in Numerical Relativity. VDM Verlag (2010).
10. Courant, R. and John, F.: *Introduction to Calculus and Analysis* (vol. 1-2). Springer, N.Y., 1998–1999.
11. Einstein, A.: *Relativity: The Special and the General Theory.* Martino Fine Books, 2010.
12. Fischer, C.F.: *The Hartree-Fock Method for Atoms: a Numerical Approach.* Wiley, 1977.
13. Fowles, G.R.: *Introduction to Modern Optics.* Dover Publications, N.Y., 1989.
14. French, A.P.: *Newtonian Mechanics.* Norton, 1971.
15. Gibson, C.C.: *Elementary Euclidean Geometry: An Introduction.* Cambridge University Press, 2003.
16. Gilmore, R.: *Lie Groups, Lie Algebras, and Some of Their Applications.* Dover, UK, 2005.
17. Glendinning, P.: 1994 *Stability, Instability and Chaos: An Introduction to the Theory of Nonlinear Differential Equations.* Cambridge University press (1994).
18. Goeke, K., and Reinhard, P.G.: *Time-Dependent Hartree-Fock and Beyond: Lecture Notes in Physics: 171.* Springer-Verlag, 1983.
19. Griffiths, D.J., and Schroeter, D.F.: *Introduction to Quantum Mechanics* (3rd edition). Cambridge University Press, 2018.
20. Hall, B.: *Lie Groups, Lie Algebras, and Representations: an Elementary Introduction (2nd edition).* Springer, NY, 2015.
21. Hecht, E.: *Optics, 4th Edition.* Addison-Wesley, Boston, MA, 2001.
22. Henrici, P.: *Applied and Computational Complex Analysis,* Vol. 1-2. Wiley-Interscience, N.Y., 1977.

23. Henrici, P. and Kenan, W.R.: *Applied and Computational Complex Analysis: Power Series-Integration-Conformal Mapping-Location of Zeros*. Wiley-IEEE, Piscataway NJ, 1988.

24. Hrabovsky, C.: em Classical Mechanics: The Theoretical Minimum. 2014.

25. Humphreys, J.E.: *Introduction to Lie Algebras and Representation Theory*. Springer N.Y., 1972.

26. d'Inverno, R.: Approaches to Numerical Relativity. Cambridge Univ. Press (2008).

27. Jones, H.F.: *Groups, Representations and Physics* (Second Edition). CRC Press N.Y., 1998.

28. Kac, V.G., Raina, A.K.:, and Rozhkovskaya, N.: *Bombay Lectures on Highest Weight Representations of Infinite Dimensional Lie Algebras (Second Edition)*. World Scientific, Singapore, (2013).

29. Kadanoff, L.P.: *Quantum Statistical Mechanics*. Taylor and Francis, 2019.

30. Kannan, A.: *Lecture Notes on Lie Algebras,* 2019.

31. Karatzas, I. and Shreve, S.E.: *Brownian Motion and Stochastic Calculus* (second edition). Springer, New York, 1991.

32. Kibble, T.: *Classical Mechanics (5th Edition)*. Imperial College Press, London, 2004.

33. Knudsen, J.M., and Hjorth, P.G.: *Elements of Newtonian Mechanics: Including Non linear Dynamics*. Springer, NY, 2013.

34. Kuratowski, K., and Musielak, J.: *Introduction to Calculus*. Pergamon Press, 1961.

35. Meloney, A.: Lecture Notes on Classical Mechanics, 2010.

36. Meloney, A.: Lecture Notes on General Relativity, 2019.

37. Robinett, R.: *Quantum Mechanics: Classical Results, Modern Systems, and Visualized Examples (Second Edition)*. Oxford Univ. Press, 2006.

38. Romano, A., and Furnari, M.M.: *The Physical and Mathematical Foundations of th Theory of Relativity: A Critical Analysis*. Birkhouser, NY, 2019.

39. Rosenlicht, M.: *Introduction to Analysis*. Dover Publications, N.Y., 1985.

40. Schieve, W.C., and Horwitz, L.P.: *Quantum Statistical Mechanics*. Cambridge Univ Press, 2009.

41. Schuller, M.F.: *Lectures on the Geometrical Anatomy of Theoretical Physics and Quantum Theory*. Lecture notes, independently published, 2019.

42. Schwerdtfeger, H.: *Geometry of Complex Numbers: Circle Geometry, Moebius Transformation, Non-Euclidean Geometry*. Dover Publications, N.Y., 1980.

43. Shapira, Y.: *Mathematical Objects in C++: Computational Tools in a Unified Object Oriented Approach*. CRC press, Boca Raton, FL, 2009.

44. Shapira, Y.: *Solving PDEs in C++: Numerical Methods in a Unified Object–Oriente Approach (Second Edition)*. SIAM, Philadelphia, PA, 2012.

45. Shapira, Y.: *Linear Algebra and Group Theory for Physicists and Engineers*. Birkhouser, Springer Nature, N.Y. (2019).

46. Shibata, M.: Numerical Relativity. World Scientific Publishing (2016).

47. Steinbauer, R.: *The Penrose and Hawking Singularity Theorems Revisited*. Math Faculty, Vienna Univ., 2006.

48. Strang, G.: *Introduction to Applied Mathematics*. Wellesley-Cambridge Press, 1986.

49. Strang, G.: *Introduction to Linear Algebra* (third edition). SIAM, Philadelphia, 200

50. Strocchi, F.: *An Introduction to the Mathematical Structure of Quantum Mechanics a Short Course for Mathematicians (Second Edition)*. World Scientific, Singapore (2008).

51. Strogatz, S.H.: *Nonlinear Dynamics and Chaos: With Applications to Physics, Biology, Chemistry, and Engineering (Second Edition)*. Westview Press (2014).

52. Tauvel, P., and Yu, R.W.T.: *Lie Algebras and Algebraic Groups*. Springer, NY, 2005.

53. Taylor, J.R.: *Classical Mechanics*. University Science Books, Herndon, VA, 2005.

54. Torres del Castillo, G.F.: *An Introduction to Hamiltonian Mechanics*. Birkhouser NY, 2018.

55. Vaisman, I.: *Analytical Geometry.* World Scientific, 1997.
56. Varchenko, A.: *Multidimensional Hypergeometric Functions and Representation Theory of Lie Algebras and Quantum Groups.* World Scientific, Singapore, (1995).
57. Vazirani, U.V.: *Lecture Notes on Quantum Mechanics and Quantum Computation.* UC, Berkeley.
58. Woodhouse, N.M.J.: *Special Relativity., (2nd edition.* Springer, London, 2007.

Index

Printed in the United States
by Baker & Taylor Publisher Services